Erratum

Publisher's Erratum to:

South American Primates: Comparative Perspectives in the Study of Behavior, Ecology, and Conservation

Paul A. Garber, Alejandro Estrada, Júlio César, Bicca-Marques, Eckhard W. Heymann and Karen B. Strier (Editors), 2009

Dr. Eckhard W. Heymann was incorrectly affiliated with the University of Göttingen in the front matter of Dr. Paul Garber's *South American Primates: Comparative Perspectives in the Study of Behavior, Ecology, and Conservation*. His correct affiliation and email address are as follows:

Abteilung Verhaltensökologie & Soziobiologie,
Deutsches Primatenzentrum, Göttingen,
Germany
eheyman@gwdg.de

South American Primates

DEVELOPMENTS IN PRIMATOLOGY: PROGRESS AND PROSPECTS

Series Editor: Russell H. Tuttle, University of Chicago, Chicago, Illinois

This peer-reviewed book series melds the facts of organic diversity with the continuity of the evolutionary process. The volumes in this series exemplify the diversity of theoretical perspectives and methodological approaches currently employed by primatologists and physical anthropologists. Specific coverage includes: primate behavior in natural habitats and captive settings; primate ecology and conservation; functional morphology and developmental biology of primates; primate systematics; genetic and phenotypic differences among living primates; and paleoprimatology.

RINGAILED LEMUR BIOLOGY: LEMUR CATTA IN MADAGASCAR
Edited by Alison Jolly, Robert W. Sussman, Naoki Koyama and Hantanirina Rasamimanana

PRIMATES OF WESTERN UGANDA
Edited by Nicholas E. Newton-Fisher, Hugh Notman, James D. Paterson and Vernon Reynolds

PRIMATE ORIGINS: ADAPTATIONS AND EVOLUTION
Edited by Matthew J. Ravosa and Marian Dagosto

LEMURS: ECOLOGY AND ADAPTATION
Edited by Lisa Gould and Michelle L. Sauther

PRIMATE ANTI-PREDATOR STRATEGIES
Edited by Sharon L. Gursky and K.A.I. Nekaris

CONSERVATION IN THE 21ST CENTURY: GORILLAS AS A CASE STUDY
Edited by T.S. Stoinski, H.D. Steklis and P.T. Mehlman

ELWYN SIMONS: A SEARCH FOR ORIGINS
Edited by John G. Fleagle and Christopher C. Gilbert

THE BONOBOS: BEHAVIOR, ECOLOGY, AND CONSERVATION
Edited by Takeshi Furuichi and Jo Thompson

PRIMATE CRANIOFACIAL FUNCTION AND BIOLOGY
Edited by Chris Vinyard, Matthew J. Ravosa, and Christine E. Wall

THE BABOON IN BIOMEDICAL RESEARCH
Edited by John L. VandeBerg, Sarah Williams-Blangero, and Suzette D. Tardif

NEW PERSPECTIVES IN THE STUDY OF MESOAMERICAN PRIMATES: DISTRIBUTION, ECOLOGY, BEHAVIOR, AND CONSERVATION
Edited by Alejandro Estrada, Paul A. Garber, Mary Pavelka, and LeAndra Luecke

SOUTH AMERICAN PRIMATES: COMPARATIVE PERSPECTIVES IN THE STUDY OF BEHAVIOR, ECOLOGY, AND CONSERVATION
Edited by Paul A. Garber, Alejandro Estrada, Júlio César Bicca-Marques, Eckhard W. Heymann, Karen B. Strier

Paul A. Garber · Alejandro Estrada · Júlio César
Bicca-Marques · Eckhard W. Heymann ·
Karen B. Strier

Editors

South American Primates

Comparative Perspectives in the Study
of Behavior, Ecology, and Conservation

 Springer

Editors

Paul A. Garber
University of Illinois at Urbana
IL, USA
p-garber@illinois.edu

Júlio César Bicca-Marques
Pontifícia Universidade
Católica do Rio Grande do
Sul, Porto Alegre, RS, Brazil
jcbicca@pucrs.br

Karen B. Strier
University of Wisconsin
Madison, WI, USA
kbstrier@wisc.edu

Alejandro Estrada
Universidad Nacional Autonoma de Mexico
Mexico
aestrada@primatesmx.com

Eckhard W. Heymann
Universität Göttingen
Germany
eheyman@gwdg.de

ISBN: 978-0-387-78704-6   e-ISBN: 978-0-387-78705-3
DOI 10.1007/978-0-387-78705-3

Library of Congress Control Number: 2008938925

Cover illustrations by Michelle Bezanson

Printed on acid-free paper

springer.com

Acknowledgment

We acknowledge and gratefully thank the following scholars for reviewing earlier drafts of chapters in this volume: Jorge A. Ahumada, K. Christopher Beard, Nancy G. Caine, Shannon Digweed, Julia Fischer, Mauro Galetti, Walter Hartwig, Michael Heistermann, Phyllis C. Lee, Araceli Lima, Jessica Lynch Alfaro, Ripan Mahli, Salvador Mandujano, Laura Marsh, Ricardo Mondragón Ceballos, Salvador Montiel, Fernando C. Passos, Ernesto Rodríguez-Luna, Alfred L. Rosenberger, Antônio Rossano Mendes Pontes, Juan Carlos Serio Silva, Wilson R. Spironello, Kathryn Stoner, Suzette D. Tardif, and Eleonore Zulnara Freire Setz

Contents

Contributors

Maria Aparecida de O. Azevedo-Lopes Secretaria de Estado de Meio Ambiente, Estado do Acre, Brazil, cida.lopes@ac.gov.br

Júlio César Bicca-Marques Laboratório de Primatologia, Departamento de Biodiversidade e Ecologia Faculdade de Biociências, Pontifícia Universidade Católica do Rio Grande do Sul, Porto Alegre, RS 90619-90 Brazil, jcbicca@pucrs.br

Gregory E. Blomquist Department of Anthropology, University of Missouri, Columbia, MO, USA, blomquistg@missouri.edu

Rafael Bueno Laboratório de Biologia da Conservação Universidade Estadual Paulista (UNESP), CP 199 Rio Claro, SP, Brazil 13506-900, rafabrc@yahoo.com.br, rafachyteles@yahoo.com.br

Richard W. Byrne School of Psychology, University of St. Andrews, St. Andrews, UK, rwb@st-andrews.ac.uk

Sarah Carnegie Department of Anthropology, University of Calgary, Calgary, Alberta, Canada, sdcarneg@ucalgary.ca

Nancy L. Conklin-Brittain Department of Anthropology, Harvard University, Cambridge, MA, USA, nconklin@fas.harvard.edu

Siobhán B. Cooke The Graduate Center, The City University of New York, and The New York Consortium in Evolutionary Primatology (NYCEP), New York, NY, USA, scooke@gc.cuny.edu

Anthony Di Fiore Center for the Study of Human Origins, Department of Anthropology, New York University, 25 Waverly Place, New York, NY 10003, USA, anthony.difiore@nyu.edu

Alejandro Estrada Estación de Biología Tropical Los Tuxtlas, Instituto de Biología, Universidad Nacional Autónoma de México, México, aestrada@primatesmx.com

Stephen F. Ferrari Department of Biology, Federal University of Sergipe, Sergipe, Brazil, ferrari@pq.cnpq.br, ferrari@pitheciineactiongroup.org

Mauro Galetti Laboratório de Biologia da Conservação, Universidade Estadual Paulista (UNESP), CP 199, 13506-900 Rio Claro, SP, Brazil, mgaletti@rc.unesp.br, galetti@mac.com

Paul A. Garber Department of Anthropology, University of Illinois at Urbana-Champaign, IL, USA, p-garber@illinois.edu

Thomas R. Gillespie Departments of Anthropology and Pathobiology, University of Illinois, Urbana, IL, USA; The Global Health Institute and Department of Environmental Studies, Emory University, Atlanta, GA, USA, thomas.gillespie@emory.edu

Rogério Grassetto Teixeira da Cunha Caixa Postal 17011, CEP 02340-970, São Paulo, Brazil, rogcunha@hotmail.com

Eckhard Heymann Abteilumg Verhaltensoekologie & Sociobiologie, Deutsches Primatenzentrum GmbH (DPZ), Leibniz-Institut fur Primatenforschung, Kellnerweg 4, D-37077 Goettingen, Germany, eheyman@gwdg.de

Patrícia Izar Department of Experimental Psychology, University of São Paulo, São Paulo Brazil, patrizar@usp.br

Timothy H. Keitt Section of Integrative Biology, University of Texas, Austin, TX, 78712, USA, tkeitt@mail.utexas.edu

Martin M. Kowalewski Estacion Biologica Corrientes-MACN, Corrientes, Argentina; Department of Anthropology, University of Illinois, Urbana-Champaign, IL, USA, mkowalew@illinois.edu

Joanna E. Lambert Department of Anthropology, University of Wisconsin, Madison, WI, USA, jelambert@wisc.edu

Jesse R. Lasky Section of Integrative Biology, University of Texas, Austin, TX 78712, USA, jesserlasky@mail.utexas.edu

Steven R. Leigh Department of Anthropology, University of Illinois, Urbana-Champaign, IL, USA, sleigh@illinois.edu

Gabriel Marroig Departmento de Genética e Biologia Evolutiva, Instituto de Biociências, Universidade de São Paulo, Rua do Matão, 277, 05508-900, São Paulo, Brazil, gmarroig@usp.br

Sérgio L. Mendes Departamento de Ciências Biológicas, Universidade Federal do Espírito Santo, Brazil, slmendes1@gmail.com

Russell A. Mittermeier Conservation International, 2011 Crystal Drive, Suite 500, Arlington, VA 22202, USA, r.mittermeier@conservation.org

Eder Cassola Molina Departamento de Geofísica, Instituto de Astronomia, Geofísica e Ciências Atmosféricas, Universidade de São Paulo, Rua do Matão, 1226, 05508-900, São Paulo, Brazil, eder@iag.usp.br

Érica S. Nakai Department of Experimental Psychology, University of São Paulo, São Paulo Brazil, ericanakai@usp.br

Marilyn A. Norconk Department of Anthropology and School of Biomedical Sciences, Kent State University, Kent, OH 44242, USA, mnorconk@kent.edu

Felipe Bandoni de Oliveira Departmento de Genética e Biologia Evolutiva, Instituto de Biociências, Universidade de Säo Paulo, Rua do Matäo, 277, 05508-900, Säo Paulo, Brazil, fbo@ib.usp.br

Stephen Pekar Queens College, The City University of New York, 65-30 Kissena Blvd., Flushing, NY, USA; and Lamont Doherty Earth Observatory of Columbia University, Palisades NY, USA, stephen.pekar@qc.cuny.edu

Carlos A. Peres School of Environmental Sciences, University of East Anglia, Norwich, NR4 7TJ, UK, C.Peres@uea.ac.uk

Naiara Pinto Jet propulsion Laboratary, 4800 Oak Grove MS 300-325, Pasadena, CA, 91105, sardinra@jpl.nasa.gov

Cécile Richard-Hansen National Game Agency, French Guiana, Cecile.Richard-Hansen@ecofog.pf

Alfred L. Rosenberger Brooklyn College, The City University of New York, Department of Anthropology and Archaeology; The Graduate Center, The City University of New York; The American Museum of Natural History, Department of Mammalogy; New York Consortium in Evolutionary Primatology (NYCEP), New York, NY, USA, alfredr@brooklyn.cuny.edu

Anthony B. Rylands Center for Applied Biodiversity Science, Conservation International, 2011 Crystal Drive, Suite 500, Arlington, VA 22202, USA, a.rylands@conservation.org

Anita Stone Department of Biology, Grand Valley State University, Allendale, MI, USA, stonean@gvsu.edu

Karen B. Strier Department of Anthropology, University of Wisconsin-Madison, Madison, WI 53706, USA, kbstrier@wisc.edu

Marcelo F. Tejedor Department of Mammalogy, American Museum of Natural History and The New York Consortium in Evolutionary Primatology (NYCEP), New York, NY. Current address: Facultad de Ciencias Naturales, Sede Esquel Universidad Nacional de la Patagonia "San Juan Bosco" Sarmiento 849, (9200) Esquel, Argentina, mtejedor@unpata.edu.ar

Benoit de Thoisy Kwata NGO, Association Kwata "Study and Conservation of French Guianan Wildlife," BP 672, F-97335 Cayenne cedex, French Guiana, thoisy@nplus.gf

Sarie Van Belle Department of Zoology, University of Wisconsin-Madison, Madison, WI 53706, USA, sarievanbelle@primatesmx.com

Christopher J. Vinyard Department of Anatomy and Neurobiology, NEOUCOM, Rootstown, OH, USA, cvinyard@neoucom.edu

Kevina Vulinec Department of Agriculture and Natural Resources, Delaware State University, Dover, DE 19901-2277 USA, kvulinec@desu.edu

Barth W. Wright Department of Anatomy, Kansas City University of Medicine and Biosciences, Kansas City, MO, USA, Bwright@kcumb.edu

Toni E. Ziegler Wisconsin National Primate Research Center, Department of Psychology, University of Wisconsin-Madison, WI 53715, USA, ziegler@primate.wisc.edu

Part I
Introduction

Chapter 1
Advancing the Study of South American Primates

Paul A. Garber and Alejandro Estrada

1.1 Introduction

Given the recent publication of several texts offering a comprehensive review of
the behavior and ecology of each genus or major taxonomic group of New World
primates (Campbell et al. 2007: Barnett et al. in press, Ford et al. in press), our goals
in developing this volume are (1) to test and evaluate recent theories of sexual selection, population genetics, socioecology, predation risk, ontogeny and life history,
reproductive endocrinology, foraging strategies, cognition and problem-solving, and
conservation biology based on data derived from studies of South American primates, (2) to produce a resource of important scholarly information and intellectual
encouragement for the expanding set of South American scientists with interests
in primatology, tropical ecology, evolutionary biology, and conservation (more than
half of the contributors to this volume are from Latin America), and (3) to encourage
researchers focusing on similar or related theoretical issues in other animal taxa
including avians, chiropterans, rodents, carnivores, and in particular, Old World primates to expand their use of the published literature on South American primates
to inform their studies. For example, based on a review of 60 randomly selected
research articles published between 2005 and 2007 in 15 issues of the *American
Journal of Primatology* (Table 1.1), only 8.9% of the citations in studies of prosimians, 7.5% of the citations in studies of Old World monkeys, and less than 4% of the
citations in studies of apes refer to the relevant literature on New World primates.
Although, it is possible that this could be explained by the fact that publications on
New World primates are under-represented in the literature, this is not the case. Of
the total number of taxonomically-oriented research articles published in these 15
journal issues, 34% were on New World monkeys, 19.3% on prosimians, 20.4% on
Old World monkeys, and 26.1% on apes. In addition, given that two forthcoming
volumes on South American primates focus exclusively on the callitrichids (Ford
et al.) and Pitheciines (Barnett et al.) a major challenge of this volume is to highlight

P.A. Garber (✉)
Department of Anthropology, University of Illinois at Urbana-Champaign, IL, USA
e-mail: p-garber@illinois.edu

P.A. Garber et al. (eds.), *South American Primates,* Developments in Primatology:
Progress and Prospects, DOI 10.1007/978-0-387-78705-3_1,
© Springer Science+Business Media, LLC 2009

Table 1.1 Citation Bias in a Select Sample of Recent Articles Published on Primate Behavior and Ecology

	Prosimian	NW monkey	OW monkey	Ape	Human	Other	Total references
Prosimian							
% references	**58.9**	6.7	7.6	1.2	0.4	25.0	670
% species references[1]	**78.6**	8.9	10.1	1.6	0.6		
NW Monkey							
% references	1.0	**42.5**	14.0	5.5	7.0	29.8	636
% species references[1]	1.4	**60.5**	20.0	7.9	10.0		
OW Monkey							
% references	1.3	5.5	**49.3**	6.3	10.8	26.6	680
% species references[1]	1.8	7.5	**67.2**	8.6	14.8		
Ape							
% references	0.8	2.5	7.5	**52.3**	11.8	25.0	630
% species references[1]	1.0	3.3	10.0	**69.7**	15.7		

[1] References listed under Other are omitted from the calculation. These include articles focused on nonprimate taxa and theoretical issues in which data from a broad range of taxa (primate and nonprimate) are included.

[2] Issues of the American Journal of Primatology used in this analysis are: 2005 volume 65 no. 1, 2, and 3, volume 66 no. 1 and 2, volume 67 no. 1 and 3; 2006 volume 68 no. 2, 7, 9, 10, and 12; 2007 volume 69 no. 3, 4, and 5.

[3] 34% of all research articles published in these issues of AJP were on New World monkeys, 19.3% were on prosimians, 20.4% were on Old World monkeys, and 26.1% were on apes.

recent theoretical advances in the study of South American primates and encourage primatologists, biologists, ecologists, and conservationists to use insights gained from studies of a broad range of platyrrhine species in their own research.

1.1.1 South American Primates

As an island continent separated from Africa, North America, Central America, and Asia for most of the past 100 million years, South America has witnessed the evolution of several distinct indigenous animal and plant communities, including the platyrrhini or New World monkeys. The earliest fossil evidence of platyrrhines on the continent (*Branisella-boliviana*) dates to the Deseadan (late Oligocene) of Bolivia, approximately 26 mya (Rosenberger et al. this volume; Fleagle and Tejedor, 2002). Based on biogeography, comparative anatomy, and molecular evidence, however, it is likely that primates first reached South America some 10 million years earlier (Poux et al. 2006), possibly the outcome of a rafting event across the South Atlantic from Africa or a rafting event through the Caribbean Sea from North America (de Oliveira et al. this volume).

Currently, there are 19 genera, 7 subfamilies, and 199 recognized species and subspecies of New World monkeys (Rylands et al. this volume), making platyrrhines one of the most taxonomically, behaviorally, and anatomically diverse primate radiations. Modern South American primates vary in adult body size by a factor of over 100, with the smallest species, the pygmy marmoset (*Cebuella pygmaea*) weighing 120 gm and the largest species, the muriqui (*Brachyteles arachnoides*) and the gray woolly monkey (*Lagothrix cana*) weighing 10–12 kg (Di Fiore and Campbell, 2007). South American primates are characterized by a number of different foraging strategies, patterns of habitat utilization, and anatomical adaptations of their dental, masticatory, digestive, sensory, and locomotor systems (*see* chapter by Norconk et al.) that enable them to efficiently exploit food types such as insects, small vertebrates, immature and mature leaves, hard unripe fruits and soft ripe fruits, nuts, seeds, exudates, fungi, and floral nectar. *Aotus*, the night monkey is the only taxon of higher primate that includes both nocturnal and cathemeral species (Fernandez Duque, 2007). Monkeys of the genus *Cebus* are reported to co-operatively hunt and share vertebrate prey (Rose, 1997) and to frequently break open difficult to obtain foods, by pounding them against hard substrates (Rose, 1997; Panger, 1998). A capuchin species, *Cebus libidinosus*, has been documented using large stones as tools in the wild to open hard palm fruits (Fragaszy et al. 2004a).

1.1.2 South American Primate Mating and Social Systems

South American primates are also characterized by extreme diversity in reproductive biology, mating strategies, and social systems. There are species of the genus *Aotus* and *Callicebus* that live in small pair-bonded social units (Fernandez Duque, 2007); species of the genera *Saguinus*, *Callithrix*, *Leontopithecus*, *Cebuella*, and *Mico* that produce twin infants, exhibit a polyandrous-polygynous mating system with cooperative care of infants and reproductive suppression of subordinate females (Garber, 1997; Digby et al. 2007; Heymann, 2000; Zeigler and Strier this volume); species of the genus *Saimiri* that live in sex-segregated social groups of 25–60 individuals in which males attain a "fatted stage" and increase body mass by 20% during a short breeding season (Boinski, 1999; Stone, 2006; Izar et al. this volume); species of the genus *Alouatta* that are found in either small one male-multifemale groups or small multimale-multifemale groups (Kowalewski, 2007; Di Fiore and Campbell, 2007); species of the genera *Lagothrix, Chiropotes, Cacajao, Brachyteles* that live in large multimale-male multifemale groups or communities from over 20–100 individuals (Ayres, 1989; Barnett et al. 2005; Defler, 2001; Jack, 2007; Strier, 1997) , and species of the genus *Ateles* that live in fission-fusion communities in which adult males patrol the borders of their range in an attempt to maintain access to a set of adult females (Di Fiore and Campbell, 2007). In addition, adult male to adult female sex ratios in established groups of many species of South American primates more closely approaches 1:1 than is generally found in Old World monkeys and apes, in which adult sex ratios are highly biased toward females (Jack and Fedigan, 2006). The presence of a roughly equal number of adult males and females in established social groups, coupled with the fact that in several species of New World primates the number of female breeding positions appears to be limited, is likely to have an important effect on sexual selection, sexual coercion, reproductive strategies, infanticide risk, and the opportunity for co-operative mate defense by resident males (Garber and Kowalewski, in press).

Compared with Old World monkeys, many species of South American primates exhibit bisexual dispersal and weak social bonds among adult females (Strier, 2000, 2004). An important result of bisexual dispersal is that social groups often are composed of several unrelated adults of both sexes. Although it is generally argued that kinship is the primary basis for dyadic affiliative and cooperative behaviors in primates (Silk, 2007), theories of reciprocity, biological markets, byproduct mutual, and partner competence all outline the mutual advantages that both related and unrelated individuals obtain through coordinated, tolerant, and cooperative interactions and as members of a functioning and cohesive social unit. (Dugatkin, 1997; Barrett and Henzi, 2006; Chapais, 2006; Sussman and Garber, 2007). This may help to explain observations of strong adult male-male tolerance and social bonds that characterize many platyrrhine lineages (Strier, 2000; Di Fiore and Campbell, 2007). In the case of muriquis, for example, data collected by Strier (2000: 73) highlight "the benefits that larger groups of male kin may gain in competition with other groups of related males over access to females. . ." (Strier, 2000: 73), whereas in the case of *Alouatta*, *Chiropotes*, *Saguinus*, and *Callithrix* both related and unrelated resident males may co-operate in ways that enable each to receive social, dietary, and reproductive benefits (Wang and Milton, 2003; Digby et al. 2007; Garber and Kowalewski, in press; Kowalewski, 2007; Kowalewski and Garber, submitted).

1.1.3 South American Primate Conservation

Finally, as in other parts of the world, in South America high rates of human population growth associated with anthropogenic disturbance resulting from deforestation for timber, agriculture, and cattle ranching and increased susceptibility to new vectors of disease in changing landscapes exert a strong negative impact on the sustainability of nonhuman primate populations (Estrada, this volume; Kowalewski and Gillespie, this volume). However, although humans have been a fundamental part of the primate ecological community in Africa and Asia for several million years, it has been only in the past 10,000–20,000 years that populations of humans entered South America and encountered nonhuman primates as an important source of dietary protein. In this regard, the impact of humans as predators of South American primates is relatively recent and its evolutionary effect on the behavior, ecology, and group size of platyrrhines remains unclear (*see* chapters by de Thoisy et al. and Ferrari, this volume). What is clear, is that overhunting by native and nonnative communities, deforestation, and rapid increases in human population growth have resulted in a serious decline in the biomass and survivorship of many primate populations leading to recent local extinction, especially in the case of several ateline species, as well as changes in patterns and processes of seed disperal, pollination, and forest regeneration (de Thoisy et al. this volume; Estrada, this volume; Vulinec and Lambert, this volume; Peres and Palacios, 2007). In this regard, South American primates offer important behavioral and ecological models for addressing contemporary theoretical issues in evolutionary biology, community ecology, and conservation.

1.2 Organization of the Volume

This volume on South American Primates is divided into three main areas of inquiry. Part I: Taxonomy, Distribution, Evolution, and Historical biogeography of South American primates, Part II: Recent Theoretical Advances in Primate Behavior, Ecology, and Biology, and Part III: Conservation and Management of South American Primates. We end the volume with a concluding chapter that focuses on research priorities and conservation imperatives.

1.2.1 Part I: Taxonomy, Distribution, Evolution, and Historical Biogeography of South American Primates

Following the Introduction to the volume, Chapter 2 by Rylands and Mittermeier focuses on issues of platyrrhine taxonomy, acknowledging the substantial efforts and contributions of the late Phillip Hershkovitz to this endeavor. The chapter includes a brief historical discussion of New World primate classification, and then outlines recent revisions to platyrrhine taxonomy adopting a Phylogenetic Species Concept (PSC). PSC integrates morphological, geographical, genetic, and chromosomal lines of evidence to assign species distinctions. These authors caution that although current taxonomic evaluations will undoubtedly change in the future, "a neatly explained taxonomy with a well-drawn map unfortunately tends to inspire complacency." Both species' definitions and geographic distributions are hypotheses that require continuous testing; evaluating the quality and quantity of information upon which they are based.

In Chapter 3, Oliveira et al. examine recent tectonic, biogeographical, geological and paleocurrent information to re-evaluate the question of how and when New World primates first arrived in South America. Given the current fossil evidence, there remain two plausible scenarios for the origin of South American primates. The presence of basal anthropoids in Asia (Eosimiids) that date to 45 mya (Beard, 2002), offer the possibility that an early population of Asian anthropoids migrated into North America and later via a water route dispersed into South America. It also is possible that such a population migrated first to Africa and then via a water dispersal route from Africa to South America. Alternatively, there is fossil evidence of pre-platyrrhine anthropoids in Africa (Parapithecids) dated at approximately 37 mya (Fleagle, 1999) offering the possibility that a population of endemic early African anthropoids may have migrated across the South Atlantic to South America.

Oliveria et al. explore three dispersal models for an African origin of platyrrhines. These are (1) a continuous land bridge connection between Africa and South

America, (2) island hopping, or (3) dispersal on floating islands which are large buoyant masses of soil, roots, shrubs, and entangled trees that have eroded from a terrace of land. Based on ocean paleocurrents, tectonic movements, and sea-floor subsidence movements, Oliveria et al. reconstruct a paleodistance of 1000 km between West Africa and eastern-Brazil at 40–50 mya, and argue that the most plausible scenario is that ancestral platyrrhines crossed the transatlantic on large floating islands at the time. Ultimately additional fossil evidence is required to test competing theories of an African or a North American origin for platyrrhines.

In Chapter 4, Rosenberger and colleagues integrate issues of phylogeny, geology, paleontology, paleoclimate and paleoecology in developing a new perspective on platyrrhine evolution. Although it has been argued that the radiation of extant platyrrhines is principally Amazonian in origin, with several modern lineages traced to ancestral forms that inhabited forested regions of Amazonia 12–16 mya (Hartwig, 2007), these authors identify four different regions in South America (the Amazonian, Atlantic, Patagonia, Caribbean) where individual primate taxa appear to have first evolved. For example, lineages such as *Alouatta, Cebus, Aotus,* and *Callicebus* may have originated outside of Amazonia in drier and more marginal forested habitats analogous to present day semi-arid savanna, cerrado and caatinga vegetation. These genera today are characterized by an extremely widespread geographic distribution ranging from southern Argentina (*Aotus, Cebus,* and *Alouatta*) and Paraguay (*Callicebus* and *Aotus*), through the Amazon Basin, Colombia, the Guianan shield, and into Panama (*Aotus, Cebus,* and *Alouatta,* but not *Callicebus*) (Fernandez Duque, 2007; Cortes-Ortiz et al. 2003; Fragaszy et al. 2004b). The distribution of *Cebus* extends west to Honduras and the distribution of *Alouatta* continues across Mesoamerica into Mexico (Rylands et al. 2006). These authors present evidence that during much of the past 15 million years, Amazonia was severely flooded and part of a giant riverbed or lake. They argue that in response to this ecological condition, several platyrrhine lineages evolved positional adaptations such as a prehensile tail, claw-like nails, and trunk-to-trunk leaping in order to exploit subcanopy resources.

1.2.2 Part II: Recent Theoretical Advances in Primate Behavior, Ecology, and Biology

This Part of the volume begins with Chapter 5, a paper by Blomquist et al. on primate life history evolution. Recent studies of platyrrhine life histories offer a critical perspective from which to examine questions of ontogeny, maternal investment, brain growth, and developmental trajectories (Garber and Leigh, 1997; Leigh and Blomquist, 2007). For example, among both New and Old World primates there are species characterized by extensive prenatal brain growth and delayed postnatal somatic growth, as well as species that reach adult body mass relatively early in development but fail to reach adult brain size until late in development (Garber and Leigh, 1997; Leigh and Blomquist, 2007). Perhaps more importantly, it has become clear that classifying species as having either a "fast" or "slow" life history fails to account for tradeoffs in the timing and duration of growth of energetically expensive tissues (Leigh and Blomquist, 2007) and disassociations in the ontogeny and development of individual traits. Relative to body mass, platyrrhine lineages such as *Cebus, Ateles, Brachyteles, Lagothrix,* and *Saimiri* exhibit a delay in certain life history traits such as late age at first reproduction, long interbirth interval, and extended period of gestation that are analogous to those reported for apes and humans. Some of these species, however are characterized by accelerated locomotor development and the attainment of adult-like foraging skills (Bezanson, 2006; Stone, 2006). As indicated by Blomquist et al. (pp. 124) "This dissociation of developing structures is a core concept for understanding how ontogeny can be molded into adaptive patterns, and contrasts remarkably with traditional 'fast vs. slow' models for mammalian life history evolution in which development is entirely absent or is the vacant space between neonatal and adult endpoints." In addition, these authors argue that demographic modeling of life history traits in Old and New World primates has the potential to identify those variables with the greatest effects on population growth rates, and thus, also has considerable relevance for determining effective policies of conservation and management.

In Chapter 6, Strier and Mendes outline the unique insights that long-term field studies make to our understanding of primate demography, group structure, and for testing theories of fitness and natural selection. The first extended field study of a primate in the wild was conducted on mantled howling monkeys (*Alouatta palliata*) in Panama and published by Clarence Raymond Carpenter in 1934. In the mid-to-late 1950s and early 1960s field studies of Old World monkeys and apes began in earnest, resulting in several species now studied continuously or nearly continuously for over 40 years. Although historically, long-term studies of New World primates have lagged behind their Old World counterparts, Strier and Mendes point out that using the criteria of "the number of primate generations" a study spans, we have reached a point at which detailed long-term data exist for species of the genera *Saguinus, Leontopithecus, Brachyteles, Alouatta,* and *Cebus.* These data enable researchers to address critical questions concerning long-term relationships between individual reproductive success, age at dispersal, dominance, social affiliations, and survivorship, as well as how individuals, groups, and species respond behaviorally to proximate changes in predator pressure, group size and composition, and food availability. In addition, long-term studies of primate groups offer an important framework for modeling the ability of local populations to recover from natural environmental perturbations such as drought, disease, and hurricanes, as well as anthropogenic change such as deforestation (Strier and Mendes this volume; Estrada this volume; Kowalewski and Gillespie, this volume; Pavelka and Chapman, 2006).

Chapter 7 by Izar et al. examines issues of sexual selection, mate choice, male coercion, female promiscuity, male reproductive tenure, and female avoidance strategies in understanding primate reproductive behavior. Using data on capuchins and squirrel monkeys, these authors test a series of hypotheses concerning the set of conditions under which females exercise mate choice associated with a preference for particular male qualities and the set of conditions under which female mating behavior reflects behavioral tactics designed to reduce infanticide risk. In examining patterns of sexual conflict, these authors argue that in species in which males provide females minimal direct benefits of infant care and protection, resource defense, or territorial defense (*Saimiri sciureus*), male-male breeding competition rather than female mate choice is a primary driver of mating behavior. In the case of two capuchin species (*Cebus nigritus* and *Cebus capucinus*), patterns of female mating were found not to closely reflect differences in the level of infanticide risk. Although female mating patterns in each species were found to include paternity concentration and paternity confusion, in *Cebus capucinus* resident males were tolerant of each other and females mated promiscuously whereas in *Cebus nigritus*, male social interactions were more despotic and females mated principally with the alpha male. Overall, this study supports the contention that intersexual conflict plays an important role in the primate mating strategies.

In Chapter 8, Ziegler, Strier, and Van Belle focus on recent advances using non-invasive techniques to measure endocrine profiles in wild and captive primate populations to examine the effect of ecological and social factors on male and female fertility. Unlike many mammals, mating behavior in anthropoid primates "is not restricted to the periovulatory period" (Zeigler et al. this volume pp. 204). Females in many primate species mate during all phases of their reproductive cycle (ovulation, pregnancy, and lactation). This creates a wider opportunity for female mate choice and a reduction in male mating competition if females (a) preferentially mate with a particular male or males when they are most fertile, (b) reinforce a sociosexual bond with individual male group members by mating throughout the year, or (c) mate promiscuously such that males collectively defend resident females from males in neighboring groups (Garber and Kowalewski, in press). Zeigler et al. present endocrine data on in several New World primate species outlining a set of conditions under which conception is most likely to occur, identify species differences in the occurrence and duration of nonconceptive ovulatory cycles, and discuss factors that may affect the cessation of ovulatory cycling in adult females. These authors (pp. 205) stress the importance of female nutrition, steroid hormone production, and the social environment in understanding "the factors that regulate the onset of ovarian cycling and conception, and environmental influences on reproductive patterns."

In Chapter 9, Di Fiore provides a comprehensive review of analytical techniques using molecular genetic data to address questions of kinship and within-group social behavior in primates. He identifies two critical factors, dispersal patterns (solitary, paired, or the common migration of individuals from one group or population into a second group or population; sex-biased dispersal, bisexual dispersal) and individual reproductive tactics (partner fidelity, promiscuity, sexual coercion, or rank related effects on reproductive success) as primary determinants of genetic relatedness among individuals in the same group and among individuals in neighboring groups. Di Fiore argues that (pp. 212) "Given the widespread acceptance of kinship as a key explanatory principle underlying and structuring much of primate social lives, it is imperative that future primate studies pay more attention to exploring the link between relatedness and individual behavior using molecular data." To this end, genetic information enables researchers to examine the success of individual male and female reproductive strategies by distinguishing between the mating group (i.e., the set of individuals that engage in sexual behavior during both fertile and non-fertile periods) and the breeding group (i.e., the set of individuals that successfully contribute genes to the next generation), as well as determine the degree to which kinship, familiarity, and/or partner competency offer more robust explanations of affiliative and agonistic social interactions among group members. Di Fiore also presents data based on an ongoing study of woolly monkey and spider monkey populations in Ecuador. Using genetic information extracted from tissue and fecal samples he found that for woolly monkeys, both males and females commonly dispersed from their natal groups, that adult males residing in the same group or local population are not more closely related to each other than are females, and that some males and some females were found to reside in groups with closely related same-sex kin. In the case of spider monkeys, however, the genetic evidence indicates that in one study population males are philopatric, dispersal is strongly female-biased, and groups were composed of closely-related males, whereas in a second population (this volume, pp. 240) "many adult females seemed to reside with likely close kin and the mean degree of relatedness among both adult males and females was close to zero." The presence of both related and unrelated same-sex individuals co-residing in the same groups offers an important opportunity to more directly examine the effects of kinship, partner competency, and familiarity on primate social interactions.

In Chapter 10, Ferrari reviews data on predation risk in South American primates in order to better understand the relationship between antipredator strategies and group structure. He presents information on primate body mass, predator type and behavior, primate pelage coloration, patterns of habitat utilization, group size, and antipredator behavior (mobbing, crypticity, alarm calls, branch shaking and object throwing, selection of sleeping sites, limited reuse of sleeping sites). "From both an ecological and an evolutionary viewpoint, predation events are rare and unpredictable." (pp. 267). However, studies of predator nests and feces indicate that primates are preyed upon by raptors and mammals at a considerably higher rate than observed by primate field researchers. Smaller bodied species appear to be more vulnerable than larger bodied species to raptors and possibly snakes, whereas felids may be the most frequent predator of larger bodied playtrrhines. Ferrari also suggests that predation rates on primates appear to be higher in fragmented and edge habitats, and more common near or on the ground. This has important implications for conservation and the size and design (relative area of the center to edge) of reserves. Finally, intragroup cooperative behavior associated with vigilance, mobbing, alarm calling, and chasing potential predators is argued to be an important factor in reducing predation risk in primates.

In Chapter 11, Norconk et al. examine the challenges that primate foragers face in obtaining, dentally processing, and digesting plant tissues that vary in their mechanical properties. These authors integrate three critical areas of investigation, namely dental and masticatory morphology, the toughness of food items, and the nutrient quality (metabolizable energy) of foods ingested. In a comparison of 16 New World primate genera, Norconk and colleagues present evidence of a size related decrease in the bite force produced at M1 and the incisors. This relationship was especially pronounced in callitrichines and may be related to gouging into tree trunks during exudate feeding (Vinyard and Ryan, 2006), and opening large, tough legume pods with their anterior dentition (Tornow et al. 2006). Hard-object feeders such as *Cebus*, *Pithecia*, *Chiropotes*, and *Cacajao* were found to produce the greatest mechanical advantage with their incisors, canines, and M1. This may be required to process difficult to open foods such as unripe fruits, nuts, and seeds. Leaf-eating platyrrhines (*Brachyteles* and *Alouatta*) produced considerably lower bite force with their anterior dentition. These authors also found that although platyrrhines tend to exhibit a generalized digestive tract, in many cases smaller-bodied taxa are characterized by longer gut retention times than closely related larger bodied forms.

Finally, an examination of the nutritional composition of 128 plant species indicates that (pp. 301) "seeds are more energy dense than are fruit, flowers or leaves....." "fungi were equivalent to fruit pulp in terms of crude protein....." "flowers are moderately high in fiber too (*as are whole fruit with pulp and seeds chewed together*), and their protein content is as high as leaves." In integrating data on metabolizable energy intake, masticatory anatomy, and dietary toughness in South American primates these authors identify several distinct evolutionary trajectories that vary across body mass and dietary pattern. For example, on average both *Callithrix* (chewing bark to obtain plant exudates) and *Alouatta* (chewing leaves) consume the toughest foods, but neither taxa exhibit the most robust jaws. However, *Cebus* consumes both less tough and extremely tough foods. In this regard, critical function or the use of fallback foods during times of resource scarcity may offer important insight into a species' ecomorphology and dietary adaptations.

In Chapter 12, Vulinec and Lambert examine and compare the predictive value of neutral models and niche models for understanding tree species assemblages, richness, species abundance, and community ecology, using the example of primates as seed dispersers and seed predators. Several authors have argued that primates have played a critical evolutionary role in shaping fruit and seed characteristics of many species of flowering plants (Janson, 1983; Gautier-Hion et al. 1985; Julliot, 1996). These authors have defined a suite of fruit and seed traits (size, shape, phenology, pulp weight, color, number of seeds) or "syndrome" that represent primate dispersed fruits. Vulinec and Lambert challenge this assumption and present data on primate densities, tree species distributions, and the fate of seeds secondarily dispersed by dung beetles in a forest community in the State of Amazonas, Brazil. They argue (pp. 323) high variance in factors such as the diversity and biomass of frugivores in an area, the manner in which primates treat seeds (swallow whole, drop under the parent tree, consume), the time of day or night seeds are voided, the conditions of the site at which the seed is deposited, the fruiting patterns of nearby trees, and the behavior of secondary dispersers and predators (fungal pathogens, rodents, ants, dung beetles) "can swamp directional selection pressure and effectively neutralize competitive interactions and resulting species assemblages." Thus, rather than the efforts of a single species or taxon, it is the combined effects of fruit and seed handling by primary dispersers, seed predators, secondary dispersers, and post-secondary dispersal seed and seedling fate that act to determine patterns of forest regeneration and tree species characteristics (Garber and Lambert, 1998). Vulinec and Lambert conclude that with increasing deforestation reducing mammalian and avian communities in the tropics, effective policies of rainforest conservation will need to determine the effects of stochasticity or neutrality on plant and animal species assemblages.

In Chapter 13, da Cunha and Byrne examine a critical set of research questions concerning primate cognition, group movement, spatial cohesion, intentionality, and the function of primate vocalizations (call and answer systems). These authors argue that theory of mind or the ability of a caller to know information possessed by other callers and to manipulate that information is unlikely to offer the most accurate explanation of primate call and answer systems. These authors examine two main hypotheses; the personal-status hypothesis in which "calls are not given with the intent of maintaining contact or informing the whereabouts of the group to the separated animal(s)" but rather reflect "the state of mind of the 'responder' itself," (pp. 346), and the reunion hypothesis which assumes that both caller and responder share the same goal of reuniting or rejoining. Both hypotheses are tested using the paradigm of intentionality. Whereas theory of mind requires 2nd order intentionality, these other hypotheses require either zero order or 1st order intentionally. Authors da Cunha and Byrne use the example of the moo call in black howler monkeys (*Alouatta caraya*), which appears to function for purposes of reuniting individuals, to explore these cognitive hypotheses and to outline a framework to describe the function of contact calls in primates. The scheme has as its first functional level, calls that serve to maintain the cohesiveness of the group. The second functional level involves targeted calls that serve to maintain or attain close proximity with a particular individual. Calls that serve to maintain group cohesion are further divided into those designed to keep contact, regaining lost contact, and coordinating group movement. The scheme proposed by da Cunha and Byrne can be used to examine the specific function of contact calls within and across primate species. These authors suggest that call-and-answer systems in primates are generally consistent with first-order intentionality and the reunion hypothesis.

Part II ends with Chapter 14 by Garber et al. presenting a series of controlled experimental field studies designed to examine the ability of two species of tamarin monkeys (*Saguinus imperator* and *Saguinus fuscicollis*) to integrate ecological information (spatial and temporal predictability of the location of baited feeding sites and expectations concerning the quantity of food available at feeding sites) and social information (identity, tolerance, and dominance status of co-feeders) in foraging decisions. Virtually all primates are social foragers, and therefore individuals are commonly faced with decisions whose effective solutions require the ability to integrate both social and ecological information. Under the conditions of these field experiments, individuals could act as searchers (use ecological information to encounter a feeding site), joiners (co-feed or usurp food patches found by others using social information), or opportunists (more evenly distribute their time and energy budgets to both searching for food and visiting food sites located by conspecifics) and integrate social information and ecological information. Experimental feeding sites varied systematically both in the quantity of the food reward and the monopolizability of the food reward (either 2 or 8 individual platforms contained concealed food rewards). In each species, individuals flexibly switched from searcher, joiner, and opportunist foraging patterns under changing conditions of food availability and distribution. However, species-specific differences in social tolerance influenced the degree to which dominant and subordinate individuals were successful when adopting a joiner strategy or opportunist strategy. At productive feeding sites, differences in individual feeding success were minimal and not influenced by the strategy adopted by individuals in either species. However at small and monopolizable feeding sites, searchers had higher feeding success than joiners or opportunists regardless of rank. The results indicate that primate foragers attend to the behavior of conspecifics and cognitively solve problems of food acquisition by integrating both social information and ecological information in their decision making.

1.2.3 Part III: Conservation and Management of South American Primates

This Part of the volume begins with Chapter 15 by de Thoisy et al. examining factors that impact and regulate subsistence hunting of Amazonian primates by indigenous Amazonian groups. These authors compiled a database of case studies from 41 sites in French Guiana and 70 sites in the lowland Amazon Basin of Brazil. Data on the degree to which primates were hunted (heavily, moderately, or minimally), information on primate harvest rate, group size, level of habitat disturbance, and the duration of hunts (single day or multi-day) are presented. These authors found (pp. 401) "that hunting effort allocated to multi-day expeditions in infrequently hunted areas primarily attempts to maximize yield of preferred (and locally depleted) prey species rather than the overall bag size (or biomass) of all potential prey species . . . thereby diluting their impact on a per area basis." Nevertheless, hunting was found to frequently result in the local extinction of larger bodied primate species such as *Ateles* and *Lagothrix*, significant population declines in *Pithecia*, *Chiropotes*, and *Cebus*, and, on occasion to overhunt but maintain sustainable populations of *Alouatta* and *Cebus*. Changes in hunting pressure due to an increase in the number of indigenous neighboring communities, incursion by nontribal peoples, and logging make urgent the need for the protection of indigenous land-rights and incentives to promote long-term resource sustainability and management.

In Chapter 16, Pinto et al. synthesize data on the differential impact of anthropogenic change on the density and distribution of individual primate species inhabiting the Atlantic forests of Brazil. This area is characterized by extremely high rates of endemism in both plant and animal communities. The Atlantic forest of Brazil contains a human population of over 130 million people, with less than 8% of the original habitat remaining and distributed as scattered and isolated fragments. Using geographical information systems (GIS) technology and a statistical method (Regression Trees) designed to control for non-linear interactions among environmental predictors of primate population size (forest type, rainfall, elevation, fragment size), these authors found that the five primate species examined differed in response to particular changes in the habitat availability including the characteristics of "hot spots" or areas of high density. This is critical for developing successful conservation and management programs. For example, the population density of *Callicebus* was positively affected by the amount of nearby land devoted to agriculture, whereas in *Alouatta* the impact was negative. The density of capuchin monkeys was found to increase in the vicinity of industrialized cities. The ability of some primate species to survive in landscapes that contain agricultural fields, archaeological sites, or sites of ecotourism (*see* Estrada et al. 2006a, b) offer hope for the development of sustainable practices of land use that maintain both human and nonhuman primates. The results of this study indicate that detailed information on "patterns of land use and social indicators from municipalities where fragments are located" including income and hunting, serve to identify locations that have the potential to support and maintain primate populations (pp. 422).

In Chapter 17, Kowalewski and Gillespie examine the role of parasites (gastrointestinal, blood, and ectoparasites) in community biodiversity, primate heath, and opportunities for cross-transmission in undisturbed versus fragmented habitats. Similar to baboons, macaques, and some colobines, several genera of South American primates (e.g., marmosets, howler monkeys, and capuchins) are able to survive in highly modified environments, including areas adjacent to cattle pasture, gardens, agricultural fields, parks set aside for ecotourism, and urban areas (Bicca-Marques, 2003; Frões, 2006; Sabbatini et al. 2006). Capuchins, for example, are known to raid crops such as bananas, coconuts, and maize (Garber, pers. obser), and in doing so come into contact with humans and domesticated animals. Similarly, howler monkeys represent an important model for the "dynamics of infectious disease transmission among wild primates, humans, and domesticated animals" because of their frequent proximity to human settlements and susceptibility to parasitic, bacterial, and viral diseases found in humans and their livestock (Kowalewski and Gillespie, this volume pp. 434). The threat of disease transmission across humans, domesticated animals, and nonhuman primates remains an extremely serious public health issue throughout the New and Old World tropics, as well as a serious environmental and conservation issue as human populations expand and encroach more and more on areas once only inhabited by nonhuman primates. Kowalweski and Gillespie conclude that (pp. 449) "almost 86% of gastrointestinal parasites, and 100% of blood-borne parasites found in howlers are found in humans. Our results provide a baseline for understanding causative factors for patterns of parasitic infections in wild primate populations and may alert us to iminent threats to primate conservation."

In Chapter 18, Estrada details the complex and specific set of human (population increase and poverty), economic, sociopolitical, historical, and ecological drivers that impact patterns of land use, biodiversity, and conservation pressures in South America. Integrating data bases obtained from several governmental, United Nations, and Development agencies for 10 South American countries located within and outside of the Amazon Basin, his analysis reveals that increases in human population growth (1.7% per year), poverty (25% of the population of South America live on less than US $2 per day), and conversion of forest to cattle pasture and agricultural fields in response to global markets for meat, biofuels, food production, and timber have resulted in the loss of 37 million ha of forest per year, and significant reductions in biodiversity. Countries with the largest amounts of deforestation are Brazil, Bolivia, Venezuela, and Ecuador. These are countries that harbor a diverse and increasingly threatened primate community. Using a model of exponential decay, Estrada projects the magnitude of forest loss in South America over the next 50–100 years, stresses the responsibility of national governments to make decisions in improving the standard of living of their people, to protect the environment and natural patrimony, to establish and maintain national parks, to restore native habitats, and to expand research on the (pp. 467) "ways indigenous populations manage their forests and primate wildlife, on their traditional ecological knowledge, and on ways to incorporate their interest in conservation plans." The chapter concludes with a list of high priority issues for primate conservation in South America.

The final Chapter 19 of the volume, by Estrada and Garber, recognizes the contributions of early biologists and mamalogists, as well as more recent primate specialists in advancing the development of Primatology in South America. This chapter provides a chronological overview of the richness of published scientific information on South American primates by country and major taxa in order to assess the current state of accumulated knowledge available to scholars and researchers. Finally, the authors identify and detail critical research priorities and conservation imperatives for the study and preservation of South American primates and their habitats, and emphasize the importance of working with indigenous scholars and the local human communities in order to achieve this goal.

Acknowledgments Paul A. Garber wishes to thank Chrissie, Sara, and Jenni for their love and for the many ways by which they inspired him.

References

Ayres, J.M. (1989). Comparative feeding ecology of the uakari and bearded saki, *Cacajao* and *Chiropotes*. *Journal of Human Evolution*, 18, 697–716.

Barnett, A.A., Volkmar de Castilho, C., Shapley, R.L., Anicacio, A. (2005). Diet, habitat selection and natural history of Cacajao melanocephalus ouakary in Jau National Park, Brazil. *International Journal of Primatology*, 26, 949–969.

Barnett, A.A., Veiga, L.M., Ferrari, S.F., Norconk, M.A. (in press). *Evolutionary Biology and Conservation of Titis, Sakis and Uacaris*. Cambridge: Cambridge University Press.

Barrett, L., & Henzi, S.P. (2006). Monkeys, markets, and minds: Biological markets and primate sociality. In P.M. Kappeler & C.P. van Schaik (Eds.), *Cooperation in Primates and Humans: Mechanisms and Evolution* (pp. 209–232). Berlin: Springer-Verlag.

Beard, K.C. (2002). Basal anthropoids. In W.C. Harwtig (Ed.), *The Primate Fossil Record* (pp. 133–150). Cambridge: Cambridge University Press.

Bezanson, M.F. (2006). Leap, bridge, or ride? Ontogenetic influences on positional behavior in *Cebus* and *Alouatta*. In A. Estrada, P.A. Garber, M.S.M. Pavelka, L. Luecke (Eds.), *New Perspectives in the Study of Mesoamerican Primates: Distribution, Ecology, Behavior, and Conservation* (pp. 333–348). New York: Springer.

Bicca-Marques, J.C. (2003). How do howler monkeys cope with habitat fragmentation? In L.K. Marsh (Ed.), *Primates in Fragments: Ecology and Conservation* (pp. 283–303). New York: Kluwer Academic/Plenum Publishers.

Boinski, S. (1999). The social organization of squirrel monkeys: Implications for ecological models of social evolution. *Evolutionary Anthropology*, 8, 101–112.

Campbell, C.J., Fuentes, A., MacKinnon, K.C., Panger, M., Bearder, S.K. (2007). *Primates in Perspective*. New York: Oxford University Press.

Chapais, B. (2006). Kinship, competence and cooperation in primates. In P.M. Kappeler & C.P. van Schaik (Eds.), *Cooperation in Primates and Humans: Mechanisms and Evolution* (pp. 47–64). Berlin: Springer-Verlag.

Cortes-Ortiz, L., Bermingham, E., Rico, C., Rodriguez-Luna, E., Sampaio, L., Ruiz-Garcia, M. (2003). Molecular systematics and biogeography of the Neotropical monkey genus, Alouatta. *Molecular Phylogenetics and Evolution*, 26, 64–81.

Defler, T.R. (2001). *Cacajao melanocephalus ouakary* densities on the lower Apaporis River, Colombian Amazon. *Neotropical Primates*, 61, 31–36.

Di Fiore, A., & Campbell, C.J. (2007). The Atelines: Variation in ecology, behavior, and social organization. In C.J. Campbell, A. Fuentes, K.C. MacKinnon, M. Panger, S.K. Bearder (Eds.), *Primates in Perspective* (pp. 155–185). New York: Oxford University Press.

Digby, L.J., Ferrari, S.F., Saltzman, W. (2007). Callitrichines: The role of competition in cooperatively breeding species. In C.J. Campbell, A. Fuentes, K.C. MacKinnon, M. Panger, S.K. Bearder (Eds.), *Primates in Perspective* (pp. 85–106). New York: Oxford University Press.

Dugatkin, L.A. (1997). *Cooperation among Animals: An Evolutionary Perspective*. New York: Oxford University Press.

Estrada, A., Saenz, J., Harvey, C., Naranjo, E., Munoz, D., Rosales-Meda, M. (2006a). Primates in agroecosystems: Conservation value of some agricultural practices in Mesoamerican landscapes. In A. Estrada, P.A. Garber, M.S.M. Pavelka, L. Luecke (Eds.), *New Perspectives in the Study of Mesoamerican Primates: Distribution, Ecology, Behavior, and Conservation* (pp. 437–470). New York: Springer.

Estrada, A., Van Belle, S., Luecke, L., Rosales-Meda, M. (2006b). Primate populations in the protected forests of Maya archaeological sites in southern Mexico and Guatemala. In A. Estrada, P.A. Garber, M.S.M. Pavelka, L. Luecke (Eds.), *New Perspectives in the Study of Mesoamerican Primates: Distribution, Ecology, Behavior, and Conservation* (pp 471–488). New York: Springer.

Fernandez-Duque, E. (2007). Aotinae Social monogamy in the only nocturnal haplorhine. In C.J. Campbell, A. Fuentes, K.C. MacKinnon, M. Panger, S.K. Bearder (Eds.), *Primates in Perspective* (pp. 139–154). New York: Oxford University Press.

Fleagle, J.G. (1999). *Primate Adaptation and Evolution*, 2nd ed. San Diego: Academia Press.

Fleagle, J., & Tejedor, M.F. (2002). Early platyrrhines of southern South America. In W.C. Hartwig (Ed.), *The Primate Fossil Record* (pp. 161–173). Cambridge: Cambridge University Press.

Ford, S., Porter, L.M., Davis, L. (in press). The Smallest Anthropoids: The Marmoset/Callimico Radiation. New York: Springer Science+Business Media, Inc.

Fragaszy, D., Izar, P., Visalberghi, E., Ottoni, E.B., Oliveira, M.G. (2004a). Wild capuchin monkeys (*Cebus libidinosus*) use anvils and stone pounding tools. *American Journal of Primatology*, 64, 359–366.

Fragaszy, D.M., Visalberghi, E., Fedigan, L.M. (2004b). *The Complete Capuchin*. Cambridge: Cambridge University Press.

Frões, A.P. (2006). Micos vivendo em ilhas verdes num mar de concreto: fatores que influenciam a distribuição de *Callithrix penicillata* (mico estrela) em parques urbanos de Belo Horizonte – MG. Dissertação de Graduação, Pontifícia Universidade Católica de Minas Gerais, Brazil. 62p.

Garber, P.A. (1997). One for all and breeding for one: cooperation and competition as a tamarin reproductive strategy. *Evolutionary Anthropology*, 5(6), 187–199.

Garber, P.A., & Kowalewski M.M. (in press). Male cooperation in pitheciines: the reproductive costs and benefits to individuals of forming large mulitmale and multifemale groups. In

L. Veiga, A. Barnett, M.A. Norconk (Eds.), *Evolutionary Biology and Conservation of Titis, Sakis and Uakaris.* Cambridge: Cambridge University Press.

Garber, P.A., & Lambert, J.E. (1998). Primates as seed dispersers: Ecological processes and directions for future research. *American Journal of Primatology,* 45, 3–8.

Garber, P.A., & Leigh, S.R. (1997). Ontogenetic variation in small-bodied New World primates: Implications for patterns of reproduction and infant care. *Folia Primatologica,* 68, 1–22.

Gautier-Hion, A., Duplantier, J.M., Quris, R., Feer, F., Sourd, C., Decoux, J.P., Dubost, G., Emmons, L., Erard, C., Hecketsweiler, P., Moungazi, A., Roussilhon, C., Thiollay, J.M. (1985). Fruit characters as a basis of fruit choice and seed dispersal in a tropical forest vertebrate community. *Oecologia,* 65, 324–337.

Heymann, E.W. (2000). The number of adult males in callitrichine groups and its implications for callitrichine social evolution. In P.M. Kappeler (Ed.), *Primate Males: Causes and Consequences of Variation in Group Composition* (pp. 64–71). Cambridge: Cambridge University Press.

Jack, K.M. (2007). The Cebines: Toward an explanation of variable social structure. In C.J. Campbell, A. Fuentes, K.C. MacKinnon, M. Panger, S.K. Bearder. *Primates in Perspective* (pp. 107–122). New York: Oxford Univesity Press.

Jack, K., & Fedigan, L.M. (2006). Why be alpha male? Dominance and reproductive success in wild, white-faced capuchins (*Cebus capucinus*). In A. Estrada, P.A. Garber, M.S.M. Pavelka, L. Luecke (Eds.), *New Perspectives in the Study of Mesoamerican Primates: Distribution, Ecology, Behavior, and Conservation* (pp. 367–386). New York: Springer.

Janson, C. H. (1983). Adaptation of fruit morphology to dispersal agents in a neotropical forest. *Science,* 219, 187–189.

Julliot, C. (1996). Fruit choice by red howler monkeys (*Alouatta seniculus*) in a tropical rain forest. *American Journal of Primatology,* 40, 261–282.

Kowalewski, M.M. (2007). Patterns of affiliation and co-operation in howler monkeys: an alternative model to explain social organization in non-human primates. Ph. D. thesis, Dept. of Anthropology, University of Illinois at Urbana-Champaign

Kowalewski, M.M., & Garber, P.A. (in prep). Mating promiscuity, energetics, and reproductive tactics in black and gold howler monkeys (*Alouatta caraya*).

Leigh, S.R., & Blomquist, G.E. (2007). Life history. In C.J. Campbell, A. Fuentes, K.C. MacKinnon, M. Panger, S.K. Bearder (Eds.), *Primates in Perspective* (pp. 396–407). New York: Oxford Univesity Press.

Miranda, G.H.B., & Faria, D.S. (2001). Ecological aspects of black-pincelled marmosets (*Callithrix penicillata*) in the cerradão and dense cerrado of the Brazilian Central Plateau. *Brazilian Journal of Biology,* 61, 397–404.

Panger M.A. (1998). Object-use in free-ranging white-faced capuchins (*Cebus capucinus*) in Costa Rica. *American Journal of Physical Anthropology,* 106, 311–321.

Pavelka, M.S.M., & Chapman, C.A. (2006). Population structure of black howlers (*Alouatta pigra*) in southern Belize and responses to Hurrican Iris. In A. Estrada, P.A. Garber, M.S.M. Pavelka, L. Luecke (Eds.), *New Perspectives in the Study of Mesoamerican Primates: Distribution, Ecology, Behavior, and Conservation* (pp. 143–164). New York: Springer.

Peres, C. A., & Palacios, E. (2007). Basin-wide effects of game harvest on vertebrate population densities in Amazonian forests: implications for animal-mediated seed dispersal. *Biotropica,* 39, 304–315.

Poux, C., Chevret, P., Huchon, D., de Jong, W.W., Douzery, E.J.P. (2006). Arrival and diversification of caviomorph rodents and platyrrhine primates in South America. *Syst. Biol.* 55, 228–244.

Rose, L.M. (1997). Vertebrate predation and food-sharing in *Cebus* and *Pan. International Journal of Primatology,* 18, 727–765.

Rylands, A.B., Groves, C.P., Mittermeier, R.A., Cortes-Ortiz, L., Hines, J.J.H. (2006). Taxonomy and distributions of Mesoamerican primates. In A. Estrada, PA Garber, M. Pavelka, L. Luecke (Eds.), *New Perspectives in the Study of Mesoamerican Primates: Distribution, Ecology, Behavior, and Conservation* (pp. 29–79). New York: Springer Science+Business Media, Inc.

Sabbatini, G., Stammati, M., Tavares, M.C.H., Giuliani, M.V., Visalberghi, E. (2006). Interactions between humans and capuchin monkeys (*Cebus libidinosus*) in the Parque Nacional de Brasília, Brazil. *Applied Animal Behaviour Science*, 97, 272–283.

Silk, J.B. (2007). The adaptive value of sociality in mammalian groups. *Philosophical Transactions of the Royal Society B*, 362, 539–559.

Stone, A. I. (2006). Foraging ontogeny is not linked to delayed maturation in squirrel monkeys. *Ethology*, 112, 105–115.

Strier, K.B. (1997). Mate preferences of wild muriqui monkeys (Brachyteles arachnoids): reproductive and social correlates. *Folia Primatologica*, 68, 120–133.

Strier, K.B. (2000). From binding brotherhoods to short-term sovereignty: the dilemma of male Cebidae. In P.M. Kappeler (Ed), *Primate Males: Causes and Consequence of Variation in Group Composition* (pp. 72–83). Cambridge: Cambridge University Press.

Strier, K.B. (2004). Patrilineal kinship and primate behavior. In B. Chapais & C.M. Berman (Eds.), *Kinship and Behavior in Primates* (pp. 177–199). Oxford: Oxford University Press.

Sussman, R.W., & Garber, P.A. (2007). Cooperation and competition in primate social interactions. In C.J. Campbell, A. Fuentes, K.C. MacKinnon, M. Panger, S.K. Bearder (Eds.), *Primates in Perspective* (pp. 636–651). New York: Oxford University Press.

Tornow, M.A., Ford, S.M., Garber, P.A., de sa Sauerbrunn, E. (2006). The dentition of moustached tamarins (Saguinus mystax mystax) from Padre Isla, Peru. Part 1: quantitative variation. *American Journal of Physical Anthropology*, 130, 352–363.

Veiga, L. M., Pinto, L. P., Ferrari, S. F. (2006). Fission-fusion sociality in bearded sakis (*Chiropotes albinasus* and *Chiropotes satanas*) in Brazilian Amazonia. *International Journal of Primatology*, 27(Suppl 1), Abst. 224.

Veiga, L.M., & Silva, S.S.B (2005). Relatives or just good friends? Affiliative relationships among male southern bearded sakis (Chiropotes satanas). Livro de Resumos, XI Congresso Brasileiro de Primatologia, Porto Alegre, 13 a 18 de fevereiro de 2005, p. 174.

Vilela, S.L., & Faria, D.S. (2004). Seasonality of the activity pattern of *Callithrix penicillata* (Primates, Callitrichidae) in the cerrado (scrub savanna vegetation). *Brazilian Journal of Biology*, 64, 363–370.

Vinyard, C.J., & Ryan, T.M. (2006). Cross-sectional bone distribution in the mandibles of gouging and non-gouging platyrrhines. *International Journal of Primatology*, 27, 1461–1490.

Wang, E., & Milton, K. (2003). Intragroup social relationships of male Alouatta palliata on Barro Colorado Island, Republic of Panana. *International Journal of Primatology*, 24, 1227–1244.

Part II
Taxonomy, Distribution, Evolution, and Historical Biogeography of South American Primates

Chapter 2
The Diversity of the New World Primates (Platyrrhini): An Annotated Taxonomy

Anthony B. Rylands and Russell A. Mittermeier

2.1 Introduction

The modern taxonomy of the Infraorder Platyrrhini is deeply influenced by the numerous publications of the late Philip Hershkovitz (1909–1997). This has meant that in many aspects platyrrhine taxonomy has been extraordinarily stable over the last two decades, while his work has at the same time provided the wherewithal for considerable refinement and adjustments. Hershkovitz laid the foundation for the modern taxonomy of the New World primates first in his monumental treatise on the Families Callitrichidae and Callimiconidae (1977) (supplemented with revisions of the emperor tamarins, *Saguinus imperator* [1979] and black-mantle tamarins, *S. nigricollis* [1982]), and subsequently with a number of papers, results of his revisions of the systematics of most of the remaining extant platyrrhines that he lumped in a third family, the Cebidae: the saki monkeys *Pithecia* (1987a); the night monkeys, *Aotus* (1983); the squirrel monkeys, *Saimiri* (1984); the bearded sakis, *Chiropotes* (1985); the uacaris, *Cacajao* (1987b); and the titi monkeys, *Callicebus* (1988, 1990). Hershkovitz was working on the remaining genera for the second volume of his treatise, but his findings were never published. The foundations for the modern taxonomies of the capuchin monkeys (*Cebus*), howling monkeys (*Alouatta*), spider monkeys (*Ateles*), woolly monkeys (*Lagothrix*) and muriquis (*Brachyteles*) have had to depend, therefore, on studies such as those Kellogg and Goldman (1944) for the spider monkeys, Hershkovitz (1949) for *Cebus* in particular, Cabrera (1957) and Hill (1960, 1962) who covered all the platyrrhines, and Fooden (1963) for the woolly monkeys.

It may well be that the legacy of Hershkovitz is the cause of there currently being more species and subspecies of primates in the New World than in Africa or Asia, providing as he did the capacity to compare findings with what is known, both in terms of the physiognomy of the primates under scrutiny and their supposed distributions. The latest taxonomies of the non-human primates indicate approximately

A.B. Rylands (✉)
Center for Applied Biodiversity Science, Conservation International, Arlington, VA 22202, USA
e-mail: a.rylands@conservation.org

P.A. Garber et al. (eds.), *South American Primates,* Developments in Primatology: Progress and Prospects, DOI 10.1007/978-0-387-78705-3_2,
© Springer Science+Business Media, LLC 2009

657 species and subspecies in 71 genera and 16 families. Of these, we list here five families, 19 genera and 199 species and subspecies in the Neotropics—31% of the primates. At present, Africa has 169 species and subspecies, Asia 186 and Madagascar 100 (Grubb et al. 2003, Brandon-Jones et al. 2004, Mittermeier et al. in press).

Two further tendencies deserve mention. The first is associated with the desire to conserve the full diversity of primates, an aspect which drags taxonomy from the realm of cataloguing and academic pursuit into the applied sciences. It is of paramount importance that the full diversity of primates be recognized and mapped. The second is related to our increased knowledge of the geography of the phenotypes we observe in situ that has made it increasingly difficult to accept single definitions or dichotomies of species and subspecies. This and the new insights resulting from molecular genetics and chromosome studies have promoted the adoption of the Phylogenetic Species Concept, and the gradual rejection of the often arbitrary interpretations of variation using the category of subspecies (see Groves 2001, 2004).

The basis for the taxonomy we present here can be found in two recent compilations. The first is that of Rylands et al. (2000); the result of the workshop "Primate Taxonomy for the New Millennium," organized by the IUCN/SSC Primate Specialist Group and held in Orlando, Florida, 25–29 February 2000. The second is the remarkable and timely revision of the taxonomy of the Order Primates by Groves (2001, 2005). Our concern in this chapter is not merely to present a taxonomic list, but to indicate the scientific sources and reasoning upon which it is based, and to indicate, if only summarily, changes or divergences from the hypotheses of other authors, most notably those of Groves (2001), due to new studies and information.

2.2 Families, Subfamilies and Genera

Previously considered to comprise just two families (the Callitrichidae, formerly Hapalidae [marmosets and tamarins], and Cebidae [the rest]), credit is due to Rosenberger (1980, 1981) for breaking with tradition and suggesting an arrangement based on morphological affinity and phylogenetic relationships. While the callitrichines, pitheciines (*Chiropotes, Pithecia, Cacajao*), and atelines (*Alouatta, Ateles, Lagothrix, Brachyteles*) continued as well-defined groupings, revolutionary was his demonstration of the affiliation of the titi monkeys (*Callicebus*) with the pitheciines and, likewise, the capuchin (*Cebus*) and squirrel monkeys (*Saimiri*) with the callitrichines (placed in a redefined, family Cebidae). Confirmed by the genetic evidence, Rosenberger's (1981) proposal provided the basis of platyrrhine systematics at the family and subfamily level as accepted today (Schneider and Rosenberger 1996, Rylands et al. 2000, Groves 2001, 2005).

Groves (2001) argued for the priority of the following family and subfamily names: Hapalinae Gray, 1821 for the marmosets and tamarins; Chrysotrichinae Cabrera, 1900 for the squirrel monkeys; Nyctipithecidae Gray, 1870 for the

Table 2.1 Families, subfamilies and genera of New World primates—the taxonomy of Groves (2001, 2005) and that proposed here

Groves (2001)	Rylands and Mittermeier (this paper)
Family Cebidae Gray, 1821	Family Callitrichidae
Subfamily Callitrichinae Gray, 1821	
Callithrix Erxleben, 1777	*Callithrix*
(Cebuella) Gray, 1866	*Cebuella*
(Callibella) Van Roosmalen & Van Roosmalen, 2003	*Callibella*
(Mico) Lesson, 1840	*Mico*
Saguinus Hoffmannsegg, 1807	*Saguinus*
Leontopithecus Lesson, 1840	*Leontopithecus*
Callimico Miranda-Ribeiro, 1911	*Callimico*
	Family Cebidae
Subfamily Cebinae Bonaparte, 1821	Subfamily Cebinae
Cebus Erxleben, 1777	*Cebus*
Subfamily Saimiriinae Miller, 1812	Subfamily Saimiriinae
Saimiri Voigt, 1831	*Saimiri*
Family Aotidae Elliott, 1913	Family Aotidae
Aotus Illiger, 1811	*Aotus*
Family Pitheciidae Mivart, 1865	Family Pitheciidae
Subfamily Pitheciinae Mivart, 1965	Subfamily Pitheciinae
Pithecia Desmarest, 1804	*Pithecia*
Chiropotes Lesson, 1840	*Chiropotes*
Cacajao Lesson, 1840	*Cacajao*
Subfamily Callicebinae Pocock, 1925	Subfamily Callicebinae
Callicebus Thomas, 1903	*Callicebus*
Family Atelidae Gray, 1825	Family Atelidae
Subfamily Alouattinae Trouessart, 1897	Subfamily Alouattinae
Alouatta Lacepede, 1799	*Alouatta*
Subfamily Atelinae Gray, 1825	Subfamily Atelinae
Ateles É. Geoffroy, 1806	*Ateles*
Brachyteles Spix, 1823	*Brachyteles*
Lagothrix É. Geoffroy, 1812	*Lagothrix*
Oreonax Thomas, 1927	*Oreonax*

night monkeys, and Mycetinae Gray, 1825 for the howling monkeys. These he later retracted (Brandon-Jones and Groves 2002), and Callitrichinae, Saimiriinae, Aotidae, and Alouattinae were the respective names used by Groves (2005). The taxonomy at the family level that we maintain here (Table 2.1) is that of Groves (2001, 2005), except that we place the marmosets and tamarins in their own family, Callitrichidae, rather than as a subfamily of the Cebidae.

Hershkovitz (1977) placed *Callimico* in its own family, Callimiconidae, and others have since placed it in its own subfamily within the Callitrichidae. Cronin and Sarich (1978), Seuánez et al. (1989), Schneider et al. (1993), Pastorini et al. (1998), Chaves et al. (1999) and Canavez et al. (1999) have demonstrated that *Callimico* is more closely related to *Callithrix* than it is to the tamarins, *Saguinus*. Placing *Callimico* in its own subfamily, therefore, is no longer correct, unless, *Leontopithecus* and *Saguinus* are independently separated from *Callithrix* by their own subfamily.

Regarding genera, there is also concordance between our vision and that of Groves (2001, 2005), except for his use of subgenera for the marmosets. *Cebuella, Callibella* and *Mico* he lists as subgenera of *Callithrix* Erxleben, 1777 (see Table 2.1). Silva Jr. (2001, 2002) argued that the tufted capuchins and the untufted capuchins (*sensu* Hershkovitz 1949, 1955) are distinct in their morphology and should be considered separate genera. *Cebus* Erxleben, 1777 is referable to the untufted group, and *Sapajus* Kerr, 1792 is the name available for the tufted capuchins.

2.3 Species and Subspecies

We emphasize that the taxonomic list we present here (Table 2.2) will change considerably, even in the near future, with further discoveries of new forms, genetic and phylogenetic analyses, and the revision of genera and species groups based on morphology and the study of museum specimens. We use the Phylogenetic Species Concept (see Groves 2001, 2004) as the basis for our determination of the taxonomic status of each of the forms we list, but have not changed subspecies to species automatically. We are doing so only when we can take recourse to a recent revision of the group combining a careful review of the taxonomic characters and clear explanations as to why the forms are to be considered "good species" rather than geographic variants or subspecies. Gregorin (2006) recently eliminated subspecies from the taxonomy of the Brazilian howler monkeys (*Alouatta*), fruit of some years of detailed morphological studies, geographical analyses, and investigation of the taxonomic history, types and type localities. Likewise, Silva Jr. (2001) proposed a taxonomy of *Cebus* without the use of subspecies. Genetic research on the tamarins (*Saguinus*) and some major revisions underway for the sakis (*Pithecia*) and uacaris (*Cacajao*), combined with genetic research and discoveries of new forms will also result in changes in the taxonomy of these groups in the near future. The taxonomy we present here is certainly not definitive, it is merely a working hypothesis, based on the information we can muster.

Space, unfortunately, does not allow for a detailed analysis of the history and issues concerning the systematic arrangements and the taxonomy of the species and subspecies we recognize here. We present some notes and mention the principal literature we have at hand for each of the genera that justifies or has influenced the taxonomy we present. More details and discussion can be found particularly in Rylands, Coimbra-Filho and Mittermeier (1993, Rylands et al. 2000, 2006, Rylands, Mittermeier and Coimbra-Filho in press) and Groves (2001, 2005), but there can certainly never be any justification for not referring to older literature and past major works such as those of Elliott (1913), Lawrence (1933), Cruz Lima (1945), Kellogg and Goldman (1944), Hershkovitz, Cabrera (1957), and Hill (1957, 1960, 1962).

2.3.1 Pygmy and Dwarf Marmosets, Cebuella and Callibella

Although Hershkovitz (1977) recognized no subspecific forms for *Cebuella*, Napier (1976) and Van Roosmalen and Van Roosmalen (1997) argued that a southerly

Table 2.2 A taxonomy of Neotropical primates

Species	Gen.	Sp.	Common name
Family Callitrichidae			
1. *Cebuella pygmaea pygmaea* (Spix, 1823)	1	1	Pygmy marmoset
2. *Cebuella pygmaea niveiventris* (Lönnberg, 1940)			
3. *Callibella humilis* (Van Roosmalen, Van Roosmalen, Mittermeier & Fonseca, 1998)	2	2	Black-crowned dwarf marmoset
4. *Mico argentatus* (Linnaeus, 1771)	3	3	Silvery marmoset
5. *Mico leucippe* (Thomas, 1922)		4	Golden-white bare-ear marmoset
6. *Mico melanurus* (É. Geoffroy, 1812)		5	Black-tailed marmoset
7. *Mico intermedius* (Hershkovitz, 1977)		6	Aripuanã marmoset
8. *Mico emiliae* (Thomas, 1920)		7	Snethlage's marmoset
9. *Mico* cf. *emiliae*		8	Rondônia marmoset
10. *Mico nigriceps* (Ferrari & Lopes, 1992)		9	Black-headed marmoset
11. *Mico marcai* (Alperin, 1993)		10	Marca's marmoset
12. *Mico humeralifer* (É. Geoffroy, 1812)		11	Black and white tassel-ear marmoset
13. *Mico chrysoleucus* (Wagner, 1842)		12	Golden-white tassel-ear marmoset
14. *Mico mauesi* (Mittermeier, Schwarz & Ayres, 1992)		13	Maués marmoset
15. *Mico saterei* (Silva Jr. & Noronha 1998)		14	Sateré marmoset
16. *Mico manicorensis* Van Roosmalen, Van Roosmalen, Mittermeier & Rylands, 2000		15	Manicoré marmoset
17. *Mico acariensis*. Van Roosmalen, Van Roosmalen, Mittermeier & Rylands, 2000		16	Rio Acarí marmoset
18. *Callithrix jacchus* (Linnaeus, 1758)	4	17	Common marmoset
19. *Callithrix penicillata* (É. Geoffroy, 1812)		18	Black-tufted-ear marmoset
20. *Callithrix kuhlii* Coimbra-Filho, 1985		19	Wied's black-tufted-ear marmoset
21. *Callithrix geoffroyi* (É. Geoffroy, 1812)		20	Geoffroy's tufted-ear marmoset
22. *Callithrix aurita* (É. Geoffroy, 1812)		21	Buffy-tufted-ear marmoset
23. *Callithrix flaviceps* (Thomas, 1903)		22	Buffy-headed marmoset
24. *Callimico goeldii* (Thomas, 1904)	5	23	Goeldi's monkey, callimico

Table 2.2 (continued)

Species	Gen.	Sp.	Common name
25. *Saguinus nigricollis nigricollis* (Spix, 1823)	6	24	Spix's black-mantle tamarin
26. *Saguinus nigricollis graellsi* (Jiménez de la Espada, 1870)			Graell's black-mantle tamarin
27. *Saguinus nigricollis hernandezi* Hershkovitz, 1982			Hernández-Camacho's black-mantle tamarin
28. *Saguinus fuscicollis fuscus* (Lesson, 1840)		25	Lesson's saddle-back tamarin
29. *Saguinus fuscicollis fuscicollis* (Spix, 1823)			Spix's saddle-back tamarin
30. *Saguinus fuscicollis avilapiresi* Hershkovitz, 1966			Ávila Pires' saddle-back tamarin
31. *Saguinus fuscicollis cruzlimai* Hershkovitz, 1966			Cruz Lima's saddle-back tamarin
32. *Saguinus fuscicollis leucogenys* (Gray, 1866)			Andean saddle-back tamarin
33. *Saguinus fuscicollis lagonotus* (Jiménez de la Espada, 1870)			Red-mantle saddle-back tamarin
34. *Saguinus fuscicollis primitivus* Hershkovitz, 1977			Hershkovitz's saddle-back tamarin
35. *Saguinus fuscicollis illigeri* (Pucheran, 1845)			Illiger's saddle-back tamarin
36. *Saguinus fuscicollis nigrifrons* (I. Geoffroy, 1850)			Geoffroy's saddle-back tamarin
37. *Saguinus fuscicollis weddelli* (Deville, 1849)			Weddell's saddle-back tamarin
38. *Saguinus melanoleucus melanoleucus* (Miranda Ribeiro, 1912)		26	White saddle-back tamarin
39. *Saguinus melanoleucus crandalli* Hershkovitz, 1966			Crandall's saddle-back tamarin
40. *Saguinus tripartitus* (Milne-Edwards, 1878)		27	Golden-mantle saddle-back tamarin
41. *Saguinus mystax mystax* (Spix, 1823)		28	Spix's mustached tamarin
42. *Saguinus mystax pileatus* (I. Geoffroy, 1848)			Red-cap mustached tamarin
43. *Saguinus mystax pluto* (Lonnberg, 1926)			White-rump mustached tamarin
44. *Saguinus labiatus labiatus* (É. Geoffroy, 1812)		29	Southern red-bellied tamarin
45. *Saguinus labiatus thomasi* (Goeldi, 1907)			Thomas's red-bellied tamarin
46. *Saguinus labiatus rufiventer* (Gray, 1843)			Northern red-bellied tamarin
47. *Saguinus imperator imperator* (Goeldi, 1907)		30	Black-chinned emperor tamarin
48. *Saguinus imperator subgrisescens* (Lönnberg, 1940)			Bearded emperor tamarin
49. *Saguinus midas* (Linnaeus, 1758)		31	Golden-handed tamarin
50. *Saguinus niger* (É. Geoffroy, 1803)		32	Black-handed tamarin
51. *Saguinus inustus* (Schwarz, 1951)		33	Mottled-face tamarin
52. *Saguinus bicolor* (Spix, 1823)		34	Pied bare-face tamarin
53. *Saguinus martinsi martinsi* (Thomas, 1912)		35	Martin's bare-face tamarin
54. *Saguinus martinsi ochraceus* Hershkovitz, 1966			Ochraceous bare-face tamarin
55. *Saguinus leucopus* (Günther, 1877)		36	Silvery-brown tamarin
56. *Saguinus oedipus* (Linnaeus, 1758)		37	Cotton-top tamarin

Table 2.2 (continued)

Species	Gen.	Sp.	Common name
57. *Saguinus geoffroyi* (Pucheran, 1845)	7	38	Geoffroy's tamarin
58. *Leontopithecus rosalia* (Linnaeus, 1766)		39	Golden lion tamarin
59. *Leontopithecus chrysomelas* (Kuhl, 1820)		40	Golden-headed lion tamarin
60. *Leontopithecus chrysopygus* (Mikan, 1823)		41	Black lion tamarin
61. *Leontopithecus caissara* Lorini & Persson, 1990		42	Black-faced lion tamarin
Family Cebidae			
62. *Saimiri oerstedii oerstedii* (Reinhardt, 1872)	8	43	Black-crowned Central American squirrel monkey
63. *Saimiri oerstedii citrinellus* Thomas, 1904			Grey-crowned Central American squirrel monkey
64. *Saimiri boliviensis boliviensis* (I. Geoffroy & de Blainiulle, 1834)		44	Bolivian squirrel monkey
65. *Saimiri boliviensis peruviensis* Hershkovitz, 1984			Peruvian squirrel monkey
66. *Saimiri vanzolinii* Ayres, 1981		45	Vanzolini's squirrel monkey
67. *Saimiri sciureus sciureus* (Linnaeus, 1758)		46	Common squirrel monkey
68. *Saimiri sciureus albigena* (Von Pusch, 1941)			Colombian squirrel monkey
69. *Saimiri sciureus cassiquiarensis* (Lesson, 1840)			Humboldt's squirrel monkey
70. *Saimiri sciureus macrodon* (Elliot, 1907)			Ecuadorian squirrel monkey
71. *Saimiri ustus* I. Geoffroy, 1843	9	47	Golden-backed squirrel monkey
72. *Cebus apella apella* (Linnaeus, 1758)		48	Guianan brown tufted capuchin
73. *Cebus apella margaritae* Hollister, 1914			Margarita Island tufted capuchin
74. *Cebus macrocephalus* Spix, 1823		49	Large-headed tufted capuchin
75. *Cebus libidinosus* Spix, 1823		50	Bearded capuchin
76. *Cebus nigritus* (Goldfuss, 1809)		51	Black-horned capuchin
77. *Cebus robustus* Kuhl, 1820		52	Crested capuchin
78. *Cebus cay* Illiger, 1815		53	Hooded capuchin
79. *Cebus flavius* (Schreber, 1774)		54	Marcgraf's capuchin, blond capuchin
80. *Cebus xanthosternos* Wied-Neuwied, 1826		55	Yellow-breasted capuchin
81. *Cebus albifrons albifrons* (Humboldt, 1812)		56	White-fronted capuchin
82. *Cebus albifrons cuscinus* Thomas, 1901			Shock-headed capuchin
83. *Cebus albifrons cesarae* Hershkovitz, 1949			César Valley white-fronted capuchin
84. *Cebus albifrons malitiosus* Elliot, 1909			Brown white-fronted capuchin
85. *Cebus albifrons versicolor* Pucheran, 1845			Varied capuchin
86. *Cebus albifrons trinitatis* Von Pusch, 1941			Trinidad white-fronted capuchin
87. *Cebus albifrons aequatorialis* Allen, 1914			Ecuadorian white-fronted capuchin

Table 2.2 (continued)

Species	Gen.	Sp.	Common name
88. *Cebus capucinus capucinus* (Linnaeus, 1758)		57	White-faced capuchin
89. *Cebus capucinus limitaneus* Hollister, 1914			Honduran white-throated capuchin
90. *Cebus capucinus imitator* Thomas, 1903			Panamanian white-throated capuchin
91. *Cebus capucinus curtus* Bangs, 1905			Gorgona Island white-throated capuchin
92. *Cebus olivaceus olivaceus* Schomburgk, 1848		58	Guianan wedge-capped capuchin
93. *Cebus olivaceus nigrivittatus* Wagner, 1848			Wedge-capped capuchin
94. *Cebus olivaceus apiculatus* Hershkovitz, 1949			Pale weeper capuchin
95. *Cebus olivaceus brunneus* Allen, 1914			Brown weeper capuchin
96. *Cebus olivaceus castaneus* I. Geoffroy, 1851			Chestnut wedge-capped capuchin
97. *Cebus kaapori* Queiroz, 1992		59	Ka'apor capuchin
Family Aotidae	10		
98. *Aotus lemurinus* (I. Geoffroy, 1843)		60	Colombian or lemurine night monkey
99. *Aotus griseimembra* Elliot, 1912		61	Grey-legged night monkey
100. *Aotus zonalis* Goldman, 1914		62	Panamanian night monkey
101. *Aotus brumbacki* Hershkovitz, 1983		63	Brumback's night monkey
102. *Aotus trivirgatus* (Humboldt, 1811)		64	Douroucouli, owl monkey, night monkey
103. *Aotus vociferans* (Spix, 1823)		65	Noisy night monkey
104. *Aotus miconax* Thomas, 1927		66	Andean night monkey
105. *Aotus nancymaae* Hershkovitz, 1983		67	Nancy Ma's night monkey
106. *Aotus nigriceps* Dollman, 1909		68	Black-headed or Peruvian night monkey
107. *Aotus azarae azarae* (Humboldt, 1811)		69	Azara's night monkey
108. *Aotus azarae boliviensis* Elliot, 1907			Bolivian night monkey
109. *Aotus azarae infulatus* (Kuhl, 1820)			Feline night monkey
Family Pitheciidae	11		
110. *Callicebus modestus* Lönnberg, 1939		70	Beni titi
111. *Callicebus donacophilus* (D'Orbigny, 1836)		71	Reed titi, D'Orbigny's titi
112. *Callicebus pallescens* Thomas, 1907		72	Paraguayan yellow titi
113. *Callicebus ollalae* Lönnberg, 1939		73	Ollala's titi
114. *Callicebus oenanthe* Thomas, 1924		74	Andean titi, Isabelline titi
115. *Callicebus cupreus* (Spix, 1823)		75	Red titi, coppery titi
116. *Callicebus discolor* (I. Geoffroy & Deville, 1848)		76	Red-crowned titi
117. *Callicebus ornatus* (Gray, 1866)		77	Ornate titi

Table 2.2 (continued)

Species	Gen.	Sp.	Common name
118. *Callicebus caligatus* (Wagner, 1842)		78	Chestnut-bellied titi
119. *Callicebus dubius* Hershkovitz, 1990		79	Doubtful titi
120. *Callicebus stephennashi* Van Roosmalen, Van Roosmalen & Mittermeier, 2002		80	Stephen Nash's titi
121. *Callicebus cinerascens* (Spix, 1823)		81	Ashy titi
122. *Callicebus hoffmannsi* Thomas, 1908		82	Hoffmann's titi
123. *Callicebus baptista* Lönnberg, 1939		83	Lago do Baptista titi
124. *Callicebus moloch* (Hoffmannsegg, 1807)		84	Orabassu titi
125. *Callicebus brunneus* (Wagner, 1842)		85	Brown titi
126. *C. bernhardi* Van Roosmalen, Van Roosmalen & Mittermeier, 2002		86	Prince Bernhard's titi
128. *C. aureipalatii* Wallace, Gómez, A. M. Felton & A. Felton, 2006		87	Madidi titi
127. *Callicebus medemi* Hershkovitz, 1963		88	Medem's collared titi
128. *Callicebus torquatus* (Hoffmannsegg, 1807)		89	White-collared titi
129. *Callicebus lugens* (Humboldt, 1811)		90	Widow monkey, white-chested titi
130. *Callicebus lucifer* Thomas, 1914		91	Rufous-tailed collared titi
131. *Callicebus purinus* Thomas, 1927		92	Red-bellied collared titi
132. *Callicebus regulus* Thomas, 1927		93	Juruá collared titi
133. *Callicebus personatus* (É. Geoffroy, 1812)		94	Northern masked titi
134. *Callicebus nigrifrons* (Spix, 1823)		95	Black-fronted masked titi
135. *Callicebus melanochir* Wied-Neuwied, 1820		96	Southern Bahian masked titi, black-handed masked titi
136. *Callicebus barbarabrownae* Hershkovitz, 1990		97	Northern Bahian blond titi
137. *Callicebus coimbrai* Kobayashi & Langguth, 1999	12	98	Coimbra's titi
138. *Pithecia pithecia pithecia* (Linnaeus, 1758)		99	White-faced saki
139. *Pithecia pithecia chrysocephala* I. Geoffroy, 1850			Golden-faced saki
140. *Pithecia monachus monachus* (É. Geoffroy, 1812)		100	Geoffroy's monk saki
141. *Pithecia monachus milleri* Allen, 1914			Miller's monk saki
142. *Pithecia monachus napensis* Lönnberg, 1938			Napo monk saki
143. *Pithecia irrorata irrorata* Gray, 1842		101	Gray's bald faced saki
144. *Pithecia irrorata vanzolinii* Hershkovitz, 1987			Vanzolini's bald-faced saki
145. *Pithecia albicans* Gray, 1860		102	Buffy saki

Table 2.2 (continued)

Species	Gen.	Sp.	Common name
146. *Pithecia aequatorialis* Hershkovitz, 1987	13	103	Equatorial saki
147. *Chiropotes albinasus* (I. Geoffroy & Deville, 1848)		104	White-nosed bearded saki
148. *Chiropotes satanas* (Hoffmannsegg, 1807)		105	Black bearded saki
149. *Chiropotes chiropotes* (Humboldt, 1811)		106	Guianan bearded saki
150. *Chiropotes utahickae* Hershkovitz, 1985		107	Uta Hick's bearded saki
151. *Chiropotes israelita* (Spix, 1823)		108	Rio Negro bearded saki
152. *Cacajao calvus calvus* (I. Geoffroy, 1847)	14	109	White bald-headed uacari
153. *Cacajao calvus ucayalii* (Thomas, 1928)			Ucayali bald-headed uacari
154. *Cacajao calvus novaesi* Hershkovitz, 1987			Novaes' bald-headed uacaris
155. *Cacajao calvus rubicundus* (I. Geoffroy and Deville, 1848)			Red bald-headed uacari
156. *Cacajao melanocephalus* (Humboldt, 1811)		110	Humboldt's black-headed uacari
157. *Cacajao ouakary* (Spix, 1823)		111	Spix's black-headed uacari
Family Atelidae			
158. *Alouatta seniculus* (Linnaeus, 1766)	15	112	Red howler monkey
159. *Alouatta arctoidea* Cabrera, 1940		113	Ursine howler monkey
160. *Alouatta macconnelli* Elliot, 1910		114	Guianan red howler monkey
161. *Alouatta juara* Elliot, 1910		115	Juruá red howler monkey
162. *Alouatta puruensis* Lönnberg, 1941		116	Purús red howler monkey
163. *Alouatta sara* Elliot, 1910		117	Bolivian red howler monkey
164. *Alouatta nigerrima* Lönnberg, 1941		118	Black howler monkey
165. *Alouatta belzebul* (Linnaeus, 1766)		119	Red-handed howler monkey
166. *Alouatta discolor* (Spix, 1823)		120	Spix's red-handed howler monkey
167. *Alouatta ululata* Elliot, 1912		121	Maranhão red-handed howler monkey
168. *Alouatta guariba guariba* (Humboldt, 1812)		122	Northern brown howler monkey
169. *Alouatta guariba clamitans* Cabrera, 1940			Southern brown howler monkey
170. *Alouatta caraya* (Humboldt, 1812)		123	South American black howler monkey
171. *Alouatta palliata palliata* (Gray, 1849)		124	Golden-mantled howler monkey
172. *Alouatta palliata mexicana* (Merriam, 1902)			Mexican howler monkey
173. *Alouatta palliata aequatorialis* (Festa, 1903)			Ecuadorian mantled howler monkey
174. *Alouatta palliata coibensis* Thomas, 1902			Coiba Island howler monkey

Table 2.2 (continued)

Species	Gen.	Sp.	Common name
175. *Alouatta palliata trabeata* Lawrence, 1933			Azuero howler monkey
176. *Alouatta pigra* Lawrence, 1933		125	Central American black howler monkey
177. *Ateles geoffroyi geoffroyi* Kuhl, 1820	16	126	Geoffroy's spider monkey
178. *Ateles geoffroyi azuerensis* (Bole, 1937)			Azuero spider monkey
179. *Ateles geoffroyi frontatus* (Gray, 1842)			Black-browed spider monkey
180. *Ateles geoffroyi grisescens* Gray, 1866			Hooded spider monkey
181. *Ateles geoffroyi ornatus* (Gray, 1870)			Ornate spider monkey
182. *Ateles geoffroyi vellerosus* (Gray, 1866)			Mexican spider monkey
183. *Ateles geoffroyi yucatanensis* Kellogg & Goldman, 1944			Yucatán spider monkey
184. *Ateles fusciceps fusciceps* Gray, 1866		127	Brown-headed spider monkey
185. *Ateles fusciceps rufiventris* Sclater, 1872			Colombian black spider monkey
186. *Ateles chamek* (Humboldt, 1812)		128	Black-faced black spider monkey
187. *Ateles paniscus* (Linnaeus, 1758)		129	Red-faced black spider monkey
188. *Ateles marginatus* (É. Geoffroy, 1809)		130	White-whiskered spider monkey
189. *Ateles belzebuth* (É. Geoffroy, 1806)		131	White-bellied spider monkey
190. *Ateles hybridus hybridus* (I. Geoffroy, 1829)		132	Variegated spider monkey
191. *Ateles hybridus brunneus* Gray, 1872			Brown spider monkey
192. *Lagothrix lagotricha* (Humboldt, 1812)	17	133	Humboldt's woolly monkey
193. *Lagothrix cana cana* (É. Geoffroy in Humboldt, 1812)		134	Geoffroy's woolly monkey
194. *Lagothrix cana tschudii* Pucheran, 1857			Peruvian woolly monkey
195. *Lagothrix poeppigii* Schinz, 1844		135	Poeppig's woolly monkey
196. *Lagothrix lugens* Elliot, 1907		136	Colombian woolly monkey
197. *Oreonax flavicauda* (Humboldt, 1812)	18	137	Peruvian yellow-tailed woolly monkey
198. *Brachyteles arachnoides* (É. Geoffroy, 1806)	19	138	Southern muriqui
199. *Brachyteles hypoxanthus* (Kuhl, 1820)		139	Northern muriqui

(south of the Rio Solimões) form *niveiventris* Lönnberg, 1940 was valid (see Groves 2001, Rylands, Mittermeier and Coimbra-Filho in press). A number of studies on the phylogenetic affinity of the pygmy marmoset (*Cebuella pygmaea*) to the Amazonian marmosets (*Callithrix*) (for example, Rosenberger 1981, Barroso et al. 1997, Porter et al. 1997, Tagliaro et al. 1997) have indicated that it could, even should, be considered congeneric. Groves (2001) listed *Cebuella* as a subgenus of *Callithrix* (embracing all of the marmosets).

The black-crowned dwarf marmoset was first described in the genus *Callithrix* by Van Roosmalen et al. (1998) but was subsequently placed in its own genus *Callibella* by Van Roosmalen and Van Roosmalen (2003, see also Aguiar and Lacher 2003). Groves (2001) listed *Callibella* as a subgenus of *Callithrix*. The pygmy and dwarf marmosets are quite distinct in their size, morphology and habits when compared to *Callithrix*, and we maintain them in separate genera.

2.3.2 *Marmosets, Callithrix* and *Mico, and Goeldi's Monkey, Callimico*

The Amazonian marmosets were formerly considered to be of the genus *Callithrix* (see Hershkovitz 1977). The argument that *Cebuella* should be included in the genus *Callithrix* centers on the conclusion, from both morphological and genetic studies, that the pygmy marmoset is more closely related to the Amazonian marmosets (the *argentata* group of Hershkovitz [1977]) than the latter are to the Atlantic forest (non-Amazonian) marmosets (the *jacchus* group of Hershkovitz [1977]). Schneider et al. (1993) and Schneider and Rosenberger (1996), however, also concluded that their molecular genetic data are compatible with *jacchus* and *pygmaea* being congeneric. Although closely related to the Amazonian marmosets, we believe that *Cebuella pygmaea* should be maintained in a separate genus (see Groves 2004; Rylands et al. 2000, Rylands, Mittermeier and Coimbra-Filho in press). The oldest generic name applicable to the Amazonian marmosets alone is *Mico* Lesson, 1840 (type species *Mico argentatus*).

Hapale emiliae was first described by Thomas (1904) from the Rio Iriri, southern Pará. Hershkovitz (1977) regarded it to be a dark form of *Callithrix argentata*. Vivo (1985, 1991) revalidated it on the basis of specimens from the state of Rondônia. The Rondônia marmoset described by Vivo should be considered distinct, however, because its distribution is disjunct—separated from that of *Hapale emiliae* by *M. melanurus* (see Rylands, Coimbra-Filho and Mittermeier 1993). Sena (1998) and Ferrari, Sena and Schneider (1999) found *M. emiliae* to be more similar to *M. argentatus* than the "*emiliae*" from Rondônia. Alperin (1995) argued that *Mico nigriceps* (Ferrari and Lopes 1992) and "*emiliae*" from Rondônia belong to the same species.

To date, *Callimico* is a monotypic genus, although speculation persists regarding the possibility of there being more than one species or subspecies. Vàsàrhelyi (2002) examined the genetic structure of the founder stock of captive callimicos, and concluded that more than one cryptic subspecies or species may be represented.

2.3.3 *Tamarins and Lion Tamarins,* Saguinus *and* Leontopithcus

The taxonomy of the tamarins and lion tamarins has been quite stable since the assessments of Hershkovitz (1966, 1977, 1979, 1982) with some few modifications we have adopted here. Except for the two subspecies of *Cebuella*, where a clear geographic distinction has yet to be established, all marmosets are now listed as species, and the question remains whether the numerous tamarin subspecies of Hershkovitz (1977, 1979, 1982) should also now be considered species.

Saguinus nigricollis graellsi is listed as a full species by Hernández-Camacho and Cooper (1976) on the basis of supposed sympatry with a population of *S. nigricollis* in the region of Puerto Leguízamo in southern Colombia. Defler (2004) discussed the evidence concerning this and found it to be inconclusive. He listed *graellsi* as a subspecies of *S. nigricollis*, although Groves (2001) listed it as a distinct species following Hernández-Camacho and Cooper (1976).

The taxonomy of *Saguinus fuscicollis* is based on Hershkovitz (1977; see also Cheverud and Moore, 1990), but there are some suggested modifications. In a molecular genetic study of the phylogeny of the genus, Cropp, Larson and Cheverud (1999) found that the form *fuscus* was closer to *S. nigricollis* than to *S. fuscicollis* and gave it species status. *Saguinus f. melanoleucus* and *S. f. crandalli* were listed as subspecies of *S. melanoleucus* by Coimbra-Filho (1990) and Groves (2001, 2005), although Tagliaro et al. (2005) found that differences between *melanoleucus* and *weddelli* were no larger than among the *weddelli* specimens. *Saguinus f. acrensis*, listed by Hershkovitz (1977), is not considered a valid form but a hybrid *S. f. fuscicollis* × *S. f. melanoleucus* from the upper Rio Juruá, following Peres, Patton and Silva (1996). *Saguinus f. crandalli* (listed here as a subspecies of *melanoleucus*) is of unknown provenance (Hershkovitz 1977), and may also be a hybrid. Hershkovitz (1977) listed the form *tripartitus* as a subspecies of *S. fuscicollis*, but Thorington (1988) argued for its species' status due to its supposed sympatry with *S. f. lagonotus*. Rylands, Coimbra-Filho and Mittermeier (1993, Rylands et al. 2000) and Groves (2001) listed it as a species, but a re-evaluation of the evidence for its distribution indicates that both Hershkovitz (1977) and Thorington (1988) may have been wrong (Heymann 2000; Rylands and Heymann in prep.), and sympatry between *S. f. lagonotus* and *S. tripartitus* has yet to be confirmed. M. G. M. van Roosmalen recorded a new form of saddleback tamarin in the interfluvium of the rios Madeira and Purus; a subspecies bounded to the south by the Rio Ipixuna and to the north by the *várzeas* of the Rio Solimões (Van Roosmalen 2003; described 16 August 2003).

Hershkovitz (1977) recognized three subspecies of the mustached tamarin, *S. mystax*. Groves (2001), however, found that while two, *mystax* and *pluto*, were quite similar, the form *pileatus* is quite distinct, and he listed it as a separate species, *S. pileatus*. The problem with this is that *pileatus* is sandwiched between the ranges of *mystax* and *pluto*, separating them geographically, and indicating that if *pileatus* is not a subspecies of *mystax*, then *pluto* too must be a distinct species. Groves (2001) also listed the form *S. labiatus rufiventer*, considered by Hershkovitz (1977) to be a synonym of *S. l. labiatus*.

The black-handed tamarin (*S. niger*) was considered by Hershkovitz (1977) to be a subspecies of *S. midas*. Melo et al. (1992) examined 20 blood genetic systems in *midas* and *niger* and obtained results compatible with their classification as subspecies, but Natori and Hanihara (1992) studying the postcanine dentition found *S. m. midas* to be more similar to *S. bicolor* than to *S. m. niger*, and argued that *niger* and *midas* should be considered distinct species. Tagliaro et al. (2005) came to the same conclusion, showing a grouping of *S. midas* – *S. bicolor* with the bare-faced tamarins of northwestern Colombia and Panama (*S. oedipus, S. leucopus*, and *S. geoffroyi*). Vallinoto et al. (2006) found that *S. midas* from the Rio Uatumã had a haplotype distinct from *S. midas* from the Rio Trombetas to the east, indicating a possibility of geographical races or distinct species. Vallinoto et al. (2006) also indicated that the Rio Tocantins may act as a barrier to gene flow for *S. niger* and, as such, that there may be two taxa of black-handed tamarins. This was presaged in the molecular genetic analysis by Tagliaro et al. (2005).

Whereas Hershkovitz (1977) placed the bare-faced tamarins *ochraceus* and *martinsi* as subspecies of *S. bicolor*, we here follow Groves (2001, 2005), in considering *martinsi* and *bicolor* as distinct species, with *ochraceus* being a subspecies of the former. Coimbra-Filho, Pissinatti and Rylands (1997) indicated the possibility that *ochraceus* may have arisen as a natural hybrid, intermediate between *bicolor* to its west and *martinsi* to the east. Although only one mottled-face tamarin, *S. inustus*, is recognized, Hernández-Camacho and Defler (1991) and Defler (2004) have indicated the probable existence of two subspecies in Colombia. Hershkovitz (1977) considered *S. geoffroyi* to be a subspecies of *S. oedipus*, but a number of studies have argued for them being separate species (see Rylands 1993, Groves 2001, 2005). Finally, the lion tamarins, *Leontopithecus*, are considered separate species following Rosenberger and Coimbra-Filho (1984) (see Rylands, Coimbra-Filho and Mittermeier, 1993). Coimbra-Filho (1990) considered *L. caissara* Lorini and Persson 1990 to be a subspecies of *L. chrysopygus*.

2.3.4 The Squirrel Monkeys, Saimiri

Saimiri taxonomy follows Hershkovitz (1984) and Groves (2001). An alternative taxonomy was presented by Thorington (1985). This for some reason never "caught on" but, being a solid and carefully considered assessment, should not be ignored in future studies and re-assessments of squirrel monkey systematics. Hernández-Camacho and Defler (1991) recognized *S. sciureus caquetensis* Allen, 1916, given as a junior synonym of *S. sciureus macrodon* by Hershkovitz (1984). Costello et al. (1993) argued for the recognition of just two species: *S. sciureus* in South America and *S. oerstedii* in Panama and Costa Rica. The findings of Boinski and Cropp (1999), however, strongly supported the Hershkovitz (1984) taxonomy, advocating four species: *Saimiri sciureus, S. boliviensis, S. oerstedii* and *S. ustus*. Hershkovitz (1987b, footnote p.22) indicated his recognition of *jaburuensis* Lönnberg, 1940 and *pluvialis* Lönnberg, 1940 as subspecies of *S. b. boliviensis*.

They are listed by Groves (2001) as synonyms. Hershkovitz (1987b) referred to *S. vanzolinii* as a subspecies of *S. boliviensis*. Rylands et al. (2006) discussed the taxonomy of the Central American squirrel monkeys.

2.3.5 The Tufted Capuchin Monkeys, Cebus

The taxonomy of the tufted capuchins (*sensu* Hershkovitz 1949, 1955) here follows Silva Jr. (2001), who did not recognize any subspecific forms. Groves (2001) presented an alternative taxonomy for the tufted capuchins as follows: *C. apella apella*; *C. a. fatuellus* (Linnaeus, 1766); *C. a. macrocephalus*; *C. a. peruanus* Thomas, 1901; *C. a. tocantinus* Lönnberg, 1939; *C. a. margaritae?*; *C. libidinosus libidinosus*; *C. l. pallidus* Gray, 1866; *C. l. paraguayanus* Fischer, 1829; *C. l. juruanus* Lönnberg, 1939; *C. nigritus nigritus*; *C. n. robustus*; *C. n. cucullatus* Spix, 1823; and *C. xanthosternos* (see Fragaszy et al. 2004; Rylands, Kierulff and Mittermeier 2005). Groves (2001) and Silva Jr. (2001) as such differ in their definitions of the forms *Cebus apella* and *C. macrocephalus*. *Cebus a. fatuellus, C. a. peruanus*, and *C. libidinosus juruanus* recognized by Groves (2001) are considered synonyms of *C. macrocephalus* by Silva Jr. (2001). Silva Jr. (2001) considered *C. apella tocantinus* a synonym of *C. apella*, and *C. l. juruanus* from the upper Rio Juruá a synonym of *C. macrocephalus*. *C. libidinosus pallidus* and *C. l. paraguayanus* are referred to as *Cebus cay* by Silva Jr. (2001). Both Groves (2001) and Silva Jr. (2001) were undecided about *C. a. margaritae* of the Venezuelan Island of Margarita.

Of the three subspecies of *C. nigritus* listed by Groves (2001), Silva Jr. (2001) considered *robustus* to be a separate species, and *cucullatus* a synonym of *C. nigritus*. *Cebus queirozi*, recently described in Pontes, Malta and Asfora (2006) is evidently a junior synonym of *C. flavius*, or, as argued by Oliveira and Langguth (2006), unavailable for lack of a registered type specimen. Distinct genetically (Seuánez et al. 1986), *C. xanthosternos* is considered a distinct species by both Groves (2001) and Silva Jr. (2001).

2.3.6 The Untufted Capuchin Monkeys, Cebus

Hershkovitz (1949) listed 13 subspecies of *C. albifrons*. Many of them were Colombian, and subsequently considered in some detail by Hernández-Camacho and Cooper (1976), Defler and Hernández-Camacho (2002), and Defler (2004). Hernández-Camacho and Cooper (1976) concluded that: 1. *C. a. malitiosus* is a well-defined subspecies of the northern slopes of the Santa Marta Mountains; 2. the light *C. a. cesarae* from the Río Cesar, Magdalena valley is a well-defined subspecies; 3. *C. a. versicolor* is a complex of forms from the Cauca-Magdalena interfluvium, including, besides, *C. a. versicolor* (intermediate phase), *C. a. leucocephalus* Gray, 1865) (dark phase) and *C. a. pleei* Hershkovitz, 1949 (light phase); 4. *C. a. adustus* Hershkovitz, 1949 probably occurs in piedmont forests of western

Arauca, the northern tip of Boyacá and north Santander, as well as the Lake Maracaibo region and upper Apure basin of Venezuela. 5. *C. a. unicolor* Spix, 1823, widespread in the upper Amazon, is very similar to the type species, and a junior synonym of *C. a. albifrons* Humboldt, 1812 (confirmed with further study [Defler and Hernández-Camacho 2002]).

We list here seven forms: *Cebus a. albifrons* and *C. a. versicolor* (recognized by Groves 2001 and Defler 2004); *C. a. cuscinus* (recognized by Groves 2001, and listed as *C. a. yuracus* Hershkovitz, 1949 by Defler 2004); *Cebus a. cesarae* and *C. a. malitiosus* (recognized by Defler 2004, but not Groves 2001); *C. a. trinitatis* (recognized by Groves 2001) and *C. a. aequatorialis* (recognized by Groves 2001).

Of the white-faced capuchins *C. capucinus*, Defler (2004) noted that three subspecies had been recorded for Colombia: *C. c. capucinus, C. c. nigripectus* Elliot, 1909; and *C. c. curtus* of the Pacific Island of Gorgona. Neither *nigripectus* nor *curtus* are recognized as valid by Hershkovitz (1949), Hernández-Camacho and Cooper (1976), Groves (2001) or Defler (2004). This listing maintains *C. c. curtus*, however, because it is an island population, and we believe that further studies are necessary. Fragaszy et al. (2004) and Rylands et al. (2006) discussed the taxonomy of the doubtfully valid Central American subspecies: *C. c. limitaneus* and *C. c. imitator*.

Hershkovitz (1949) listed five subspecies of the weeper or wedge-capped capuchin. Their ranges are not known; Hershkovitz plotted only their type localities (see, however, Bodini and Pérez-Hernández [1987] for ranges in Venezuela). Neither Silva Jr. (2001) nor Groves (2001) considered them valid. They continue to be listed here, however, because a detailed, modern, study (genetic/morphological/geographical) of the taxonomy of *Cebus olivaceus* has yet to be carried out. *Cebus o. brunneus* of Venezuela is definitely distinct from *olivaceus* in Suriname, for example. Queiroz (1992) described *Cebus kaapori* from south of the Amazon in Pará and Maranhão. It is of the weeper capuchin group in appearance, but is generally recognized as a distinct species (but see Harada and Ferrari 1996). Fragaszy et al. (2004) provided a more detailed discussion of the taxonomy of this group.

2.3.7 *The Night Monkeys,* Aotus

Reviewing the entire taxonomy and distributions of the night monkeys, *Aotus*, Ford (1994) carried out a multivariate analyses of craniodental measures and pelage patterns and color, and also took into consideration chromosomal data and blood protein variations. Ford (1994) concluded that there was "good support" for just two species north of the Río Amazonas: *A. trivirgatus* east and north of the Rio Negro, and the polymorphic *A. vociferans* to the west of the Rio Negro. *Aotus vociferans*, as such, would include all the forms north of the Río Amazonas/Solimões in Brazil (west of the Rio Negro), Peru, Colombia and Ecuador, and in the Chocó, northern Colombia and Colombian Andes, and Panama: *brumbacki, lemurinus, griseimembra,* and *zonalis*.

It is doubtful, however, that the current taxonomy provides a true picture of the diversity of the genus *Aotus*. Ruiz-Herrera et al. (2005) reported that cytogenetic studies have characterized 18 different karyotypes with diploid numbers ranging from 46 to 58 chromosomes. The taxonomy of the night monkeys here follows the revision by Hershkovitz (1983), with some modifications for the Colombian and Central American forms.

Hernández-Camacho and Cooper (1976) restricted both *lemurinus* and *griseimembra* to Colombia, while recognizing the form *zonalis* as the night monkey of northwestern Colombia (Chocó) and Panama. Hershkovitz (1983) made no mention of the name *zonalis*. Groves (2001) followed Hernández-Camacho and Cooper (1976) in recognizing *zonalis* as the form in Panama, and listed it as a subspecies of *lemurinus* along with *griseimembra* and *brumbacki*. Defler, Bueno and Hernández-Camacho (2001) concluded that the karyotype of *Aotus hershkovitzi* Ramirez-Cerquera, 1983 (from the upper Río Cusiana, Boyacá, Colombia; 2n = 58) was in fact that of true *lemurinus*, and that the karyotypes which Hershkovitz (1983) had considered to be those of *lemurinus* were in fact of *zonalis*. Defler, Bueno and Hernández-Camacho (2001, Defler and Bueno 2003, Defler 2004) concluded that *A. lemurinus* of Hershkovitz (1983) is in fact three karyotypically well-defined species, and that the night monkeys of the lowlands of Panama and the Chocó region of Colombia belong to the species *A. zonalis*, and those of the Magdalena valley to *A. griseimembra*, while those above altitudes of 1500 m should correctly be referred to as *A. lemurinus* (see Rylands et al. 2006). The form *infulatus* is listed here as a subspecies of *Aotus azarae* following Groves (2001), although Hershkovitz (1983) considered it a distinct species.

2.3.8 The Titi Monkeys, *Callicebus*

The taxonomy of *Callicebus* follows Hershkovitz (1988, 1990), Groves (2001) and Van Roosmalen, Van Roosmalen and Mittermeier (2002). Groves (2001) listed all the titi monkeys as full species except for those in the *torquatus* group, for which he recognized only two species—*C. torquatus* and *C. medemi*, and the *personatus* group of the Atlantic forest in which he again recognized just two species—*C. personatus* and *C. coimbrai*. Van Roosmalen, Van Roosmalen and Mittermeier (2002) listed all titi monkeys as full species. We continue to recognize *Callicebus dubius*, although Groves (2001) considered it a synonym of *C. cupreus*.

Defler (2004) considered all the Colombian titi monkeys to be subspecies of *torquatus* or *cupreus*, following the taxonomy of Hershkovitz (1990). In their karyological analysis Bueno et al. (2006), however, indicated that the form *ornatus* should be classified a distinct species. T. R Defler is currently investigating the long-ignored mention by Moynihan (1976) of an apparently undescribed titi between the ríos Orteguaza and Caquetá, in southern Caquetá Department, near Valparaíso, Colombia.

Groves (2001) maintained *C. medemi* of the *C. torquatus* group as a full species because of its geographical isolation. Defler (2004) pointed out that it was not in fact isolated; it is "found in continuous forest which harbors (to the east), *C. torquatus lucifer*" (p. 300). Defler (2004) argued that the members of the *C. torquatus* group should all be considered subspecies, following Hershkovitz (1990). Heymann, Encarnación and Soini (2002) have indicated that the taxonomy of this group requires further study. They reported that the diagnostic features for *C. lucifer* provided by Hershkovitz (1990) and repeated in Van Roosmalen, Van Roosmalen and Mittermeier (2002) (used to distinguish it from *C. medemi*, for example) were inconsistently represented in different localities in northeastern Peru.

Bonvicino et al. (2003b) argued that *Callicebus lugens* is a distinct species on the basis of its morphology and karyological and molecular analyses (compared with *C. purinus* and *C. torquatus*). Bueno and Defler (2007), agreeing that its karyotype argues for its elevation to a full species, concluded the possibility that *C. lugens* (*sensu stricto*) may be limited to the east of the Rio Negro, while *C. torquatus* (with three subspecies) may be restricted to the west. Their data indicated the need for further karyological studies and a re-assessment of the taxonomy of the group. Molecular genetic studies of *C. lugens* on the upper Rio Negro in Brazil by Casado, Bonvicino and Seuánez (2007) indicated an even more complex situation. The karyotypes are the same ($2n = 16$) either side of the Rio Negro, but they have distinct haplotypes, suggesting different "evolutionary lineages".

2.3.9 The Saki Monkeys, *Pithecia*

The taxonomy of the sakis, *Pithecia*, follows Hershkovitz (1987a), except in the recognition of *P. monachus napensis* from La Coca, Río Napo, Ecuador. It is evident, however, that further field, genetic, and museum studies will considerably increase the diversity in this genus (most particularly in Hershkovitz's [1987a] *Pithecia monachus* group). (L. Marsh pers. comm. 2006). Groves (2001) follows Hershkovitz's (1987a) taxonomy. Marsh (2004, 2006) reported that the saki at the Tiputini Biodiversity Station, on the Río Tiputini, Yasuní National Park, Ecuador, is not referable to *Pithecia aequatorialis* as was supposed by Hershkovitz (1987a), and may be a new form.

2.3.10 The Bearded Saki Monkeys, *Chiropotes*

The taxonomy of the bearded sakis, *Chiropotes*, is based on the revision of the genus by Hershkovitz (1985). The three subspecies of *C. satanas* (*satanas, chiropotes* and *utahicki*) are here listed as full species following the recommendation of Silva Jr. and Figueiredo (2002) and Figueiredo et al. (2006).

Bonvicino et al. (2003a) resurrected the name *israelita* for the bearded sakis discovered by J. P. Boubli to the west of the Rio Branco, tributary of the Rio Negro,

Amazonas, Brazil (Boubli 2002). Genetic and pelage differences distinguish it from the bearded sakis of the Guianas to the east of the Rio Branco. There is however, a possible confusion concerning the correct names. Silva Jr. and Figueiredo (2002) argue that the form west of the Rio Branco is, correctly C. *chiropotes*, according to its type locality, and the form to the east of the Rio Branco, extending through the states of Pará and Amapá, and into the Guianas (Guyana, Suriname, French Guiana) is C. *sagulatus* (Traill, 1821) from Demerara.

2.3.11 The Uacaris, *Cacajao*

The taxonomy of the uacaris (*Cacajao*) is based on the revision by Hershkovitz (1987b). Silva Jr. and Martins (1999) extended the range of C. *c. novaesi* to the upper Juruá, along the Rio Eiru, and reported on the occurrence of a white uacari along the Rio Jurupari, affluent of the Rio Envira, in Acre, that is distinct from *novaesi*. Recent observations in the middle Rio Juruá in the municipality of Carauari, have also revealed the presence of a white uacari, sympatric (but occupying different habitats) with what is presumed to be C. *c. novaesi* (A. Ravetta *in litt.* 31 August 2005). What relation this population might have to the Rio Jurupari population is not known. Genetic studies by W. Figueiredo (Museu Paraense Emílio Goeldi, Belém) are indicating that C. *c. calvus* (interfluvium of the rios Japurá and Solimões) and the white uacari of the Rio Juruá are genetically quite distinct and that the taxonomy of C. *calvus, sensu* Hershkovitz (1987b), requires revision (J. M. Cardoso da Silva pers. comm. 2005). Figueiredo et al. (2006) reported that their genetic analysis of the two black-headed uacaris—C. *m. melanocephalus* and C. *m. ouakary*—provided evidence that they should be considered different species.

2.3.12 The Howler Monkeys, *Alouatta*

Hill (1962) and Stanyon et al. (1995) listed nine subspecies of the red howling monkey, A. *seniculus*: A. *s. seniculus*, A. *s. arctoidea*, A. *s. stramineus*, A. *s. macconelli*, A. *s. insulanus*, A. *s. amazonica*, A. *s. juara*, A. *s. puruensis*, and A. *s. sara*. *Alouatta seniculus* from Cartagena, Bolívar, Colombia, is the red howling monkey from the northwestern Amazon. *Alouatta s. arctoidea* is recognized as a full species because Stanyon et al. (1995) concluded that the number of chromosomal differences between A. *s. sara* and A. *s. arctoidea* were on a similar scale to that found previously between A. *s. sara* and A. *s. seniculus* by Minezawa et al. (1986), which had resulted in A. *s. sara* being considered a full species. Groves (2001) neverthelesss maintained it as a subspecies of A. *seniculus*. *Alouatta s. straminea* is not a valid name for the red howling monkey because its type specimen is a female *Alouatta caraya*, as demonstrated by Rylands and Brandon-Jones (1998). Bonvicino, Fernandes and Seuánez (1995) believe that the red howler monkeys either side of the Rio Trombetas are karyotypically distinct and should be considered different species, and continue to use the name *stramineus* for the red howlers of the

Rio Negro basin west of the Rio Trombetas, while adopting the name *macconnelli* (type locality Demerara) for those east of the Rio Trombetas. In contrast to Bonvicino, Fernandes and Seuánez (1995), Sampaio, Schneider and Schneider (1996) concluded that the karyotypic differences between the howlers either side of the Rio Trombetas were inconsequential. In their molecular phylogenetic study, Bonvicino, Lemos and Seuánez (2001) found low levels of divergence between *A. macconnelli, A. stramineus* (howlers from the west of the Rio Trombetas) and *A. nigerrima*— lower than recorded within *A. caraya* and *A. belzebul*. A molecular genetic study by Figueiredo et al. (1998) failed to show a difference between the red howlers either side of the Rio Trombetas. Gregorin (2006) concluded that variation in pelage coloration provided no evidence for more than one species in the northeastern Amazon, in the Guianas, east of the Rio Negro and on both sides of the Rio Trombetas. *Alouatta macconnelli* may, therefore, be the right name for the red howler monkey of the entire northeastern Amazon and Guianas, although there are two other candidates for the name: *Mycetes auratus* Gray, 1845 and *M. laniger* Gray, 1845 (Rylands and Brandon-Jones 1998).

Alouatta s. insulanus of the Island of Trinidad, is considered to be a synonym of *A. macconnelli* (see Groves 2001). Groves (2001) considered *A. s. amazonica* Lönnberg, 1941 from Codajáz, north of the Rio Solimões, west of the Rio Negro, and *A. s. puruensis* from the Rio Purus to be junior synonyms of *A. s. juara*. Gregorin (2006) tentatively considered *amazonica* to be a junior synonym of *A. juara*, extending the range of this species to the north of the Rio Solimões, but Gregorin himself stated that this was mere speculation, and the identity of the howler monkeys west of the Negro in the north-central Amazon (the eastern range limits of *A. seniculus* from Colombia) remains unclear. *Alouatta s. juara* from the Rio Juruá, is listed here as a full species following Gregorin (2006). Groves (2001) recognized *juara* as a subspecies of *A. seniculus*, with *A. s. amazonica* and *A. s. puruensis* (here considered distinct) as synonyms. *Alouatta s. puruensis* from Jaburú, Rio Purus, is recognized by Gregorin (2006). *Alouatta s. sara* is recognized as a full species following Minezawa et al. (1986; see also Stanyon et al. 1995; Groves 2001).

Cruz Lima (1945), Langguth et al. (1987), and Bonvicino, Langguth and Mittermeier (1989) studied the howling monkeys that Hill (1962) had listed as subspecies of *A. belzebul*. Following Cruz Lima (1945), Groves (2001), and Gregorin (2006), the form *nigerrima* is here considered a full species. Cytogenetic studies have also indicated that *nigerrima* is sufficiently distinct as to warrant species status (Armada et al. 1987; Lima and Seuánez 1989), and, besides, that it is more closely related to *seniculus* than to *belzebul* (see Oliveira 1996). Groves (2001) did not recognize any subspecies for *A. belzebul*: the forms *discolor, mexianae*, and *ululata* were given as junior synonyms. Here we follow Gregorin (2006) who recognized the forms *belzebul, discolor*, and *ululata* as full species, and placed *mexianae* as a junior synonym of *A. discolor*. Nascimento et al. (2005) were unable to differentiate the disjunct populations of *A. belzebul* from the Atlantic forest (Paraíba) and the Amazon (southern Pará).

Rylands and Brandon-Jones (1998) and Gregorin (2006) discussed the validity or otherwise of the use of the alternative names of *fusca* and *guariba* for the brown howling monkey of the Atlantic forest of Brazil and Argentina. Rylands and Brandon-Jones (1998) indicated that *guariba* Humboldt, 1812 is the available name, Gregorin (2006) that *fusca* E. Geoffroy, 1812 is correct. Gregorin (2006), studying the morphology of the cranium and hyoid apparatus, considered that the two brown howlers listed by Rylands et al. (2000) and Groves (2001) as subspecies should be full species. We reserve judgment on this until genetic studies can be brought to bear. Harris et al. (2005) found differences between populations of *A. guariba clamitans* in southern Brazil—from Rio de Janeiro on the one hand and Santa Catarina on the other. They showed that these correspond to differences in karyotype recorded by Koiffman (1977) and Oliveira, Lima and Sbalqueiro (1995, Oliveira et al. 1998, 2002). Maximum genetic distances found by Harris et al. (2005) were considerably greater than those recorded for *A. caraya* and *A. belzebul* by Nascimento et al. (2005), and they argued that further research may result in the recognition of three species. We continue with the names and subspecific classification as used by Rylands et al. (2000) and Groves (2001, 2005) until the taxonomy becomes better defined.

Nascimento et al. (2005) showed that populations of *Alouatta caraya* from Santa Cruz, Bolivia (Chaco) are differentiated from those in various localities in the state of Mato Grosso and (one specimen) Goiás, further north, indicating the possibility of two rather than one taxa of the South American black howler. The taxonomy of the howlers of Mesoamerica and the Caribbean and Pacific coasts of Colombia and Ecuador is based on Lawrence (1933), Hill (1962), Hall (1981), Froehlich and Froehlich (1987), and Cortés-Ortiz et al. (2003). Groves (2001, 2005) recognized only *A. palliata* (no subspecies), *A. pigra*, and *A. coibensis* (no subspecies). Rylands et al. (2006) reviewed the taxonomy and distributions of *A. palliata*, *A. coibensis* and *A. pigra*.

Cortés-Ortiz et al. (2003) found that *A. palliata* and *A. coibensis* comprise a very closely related and monophyletic group of mtDNA lineages. Divergence between *A. palliata* and *A. coibensis* is similar to mtDNA distances observed between geographically-separated populations *within* each of these two species. Rylands et al. (2006) maintained the taxonomy suggested by Froehlich and Froehlich (1987) for the forms from the Azuero Peninsula (*A. coibensis coibensis*) and the Island of Coiba (Panama) (*A. c. trabeata*), but it is evident that the findings of the molecular genetic analyses of Cortés-Ortiz et al. (2003) would relegate them to synonyms of *A. palliata*. Groves (2001) listed *A. coibensis*, with *trabeata*, as a synonym. There is still some discussion concerning the correct name for the Central American black howler, *A. pigra*. Napier (1976) referred to it as *A. villosa* and, revisiting the issue, Brandon-Jones (2006) argued that *A. pigra* is a junior synonym.

2.3.13 *The Spider Monkeys,* Ateles

The taxonomy of the spider monkeys is based on Kellogg and Goldman (1944) and Hill (1962). The forms *hybridus, chamek* and *marginatus* are listed as distinct species, and *A. fusciceps robustus* Allen, 1914 is considered a junior synonym of *A. f. rufiventris*, following Heltne and Kunkel (1975) (see Rylands et al. 2000, 2006). Silva-López, Motta-Gill and Sánchez-Hernández (1996) argued that *Ateles geoffroyi pan* Schlegel, 1876 was not a valid taxon. Hernández-Camacho and Defler (1991) and Defler (2004) argued for the validity of the form *brunneus*, listed here as a subspecies of *A. hybridus*. The taxonomy and distributions of *A. geoffroyi* and *A. fusciceps* were reviewed by Rylands et al. (2006).

Van Roosmalen (2003) reported on the rediscovery of *A. longimembris* Allen, 1914, from Barão de Melgaço, headwaters of the Rio Gy-paraná, Mato Grosso, Brazil. Cruz Lima (1945) recognized this spider monkey as a subspecies of *A. paniscus*. The description of the range given by Cruz Lima (1945; p. 127) extends west into what is currently recognized to be that of *A. chamek*. Groves (2001) considered it to be a synonym of *A. chamek*. Van Roosmalen described it as differing from *A. chamek* by "cranial characters (i.e., proportionally heavy jaws and canines not enabling adults to fully close their mouth), triangular pink muzzle including chin, black face, and black triangular blaze of backwardly directed black hairs on the forehead."

Collins (1999) and Collins and Dubach (2000) argued strongly for the species status of the form *hybridus*. Their position was reinforced by Nieves et al. (2005). Hernández-Camacho and Defler (1991) and Defler (2004) argued for the validity of the form *brunneus* from the Departments of Bolivar, Antioquia and Caldas, between the lower Ríos Cauca and Magdalena in Colombia. It is listed here as a subspecies of *A. hybridus*, as recommended by Defler (2004).

2.3.14 The Woolly Monkeys, *Lagothrix* and *Oreonax*

The taxonomy of *Lagothrix* is based on Fooden (1963), but follows Groves (2001) in recognizing *cana, lugens,* and *poeppigii* as full species rather than subspecies of *Lagotricha*. Groves (2001) also recognized *tschudii* Pucheran, 1857 from Peru (see Cruz Lima 1945). An isolated population of *Lagothrix* was discovered by Wallace and Painter (1999) in Madidi National Park, Bolivia, at 1500 m. This may be a new taxon or *L. cana tschudii*, and is the only modern record of *Lagothrix* occurring in Bolivia. The provenance of a juvenile woolly monkey with orange-colored pelage illustrated by Cruz Lima (1945) was unknown until recently. A population was found by Carlos A. Peres on the upper Rio Jutaí, upper Amazon (within the supposed range of *L. poeppigii*) and reported by Van Roosmalen (2003).

As a result of his comparative studies of cranial morphology in the atelines, Groves (2001, 2005) concluded that the yellow-tailed woolly monkey should properly be considered a monotypic genus, *Oreonax* Thomas, 1927.

2.3.15 The Muriquis, *Brachyteles*

Vieira (1944) recognized two subspecies of *Brachyteles*. Evidence provided by Lemos de Sá et al. (1990), Fonseca et al. (1991) and Lemos de Sá and Glander (1993) indicated that Vieira's (1944) standing was valid, but that differentiation is even more extreme and justifies the classification of the two forms as separate species (see also Coimbra-Filho, Pissinatti and Rylands 1993). Groves (2001, 2005) listed the two muriquis as separate species.

2.4 Discussion

A full understanding of the diversity of primates in the Neotropical region is now an urgent task. The relentless ruination—fragmentation, degradation, and destruction—of the forests of South and Middle America is making it increasingly difficult to map the distributions of the primate species and subspecies currently recognized. In Mesoamerica and the eastern Brazilian Atlantic forest it is now impossible to detect clines and to fully document the natural variation that is so necessary to make judgements concerning the thresholds for determining that two populations are consistently different, and as such species—the unit we use to prioritize conservation investments. As we lose the population diversity of the neotropical primates we lose our capacity to detect phylogenetic processes and the true nature of the Neotropical primate radiation. It is quite possible that, what we would today judge to be species or subspecies have already been lost in the Atlantic forest and Mesoamerica, and Andean Colombia, Peru and Venezuela, where forest loss, fragmentation and hunting are most accentuated. Complex patterns of genetic and morphological variation are, in addition, confused irreparably by the constant transport of pets, and their release, often in large numbers, far from where they occur naturally. Marmosets and capuchin monkeys in the Atlantic forest are today quite severely mixed up due to interbreeding between geographically distant forms. The possibility that species have been, or will be, lost without us even detecting their existence is underlined by the fact that we are still finding species never before described—14 since 1990 (Table 2.3)—and even new genera (*Callibella* M. G. M van Roosmalen and T. van Roosmalen, 2003). Stranger still is the rediscovery of forms described many years ago but forgotten, such as the distinct bearded sakis west of the Rio Branco in the Brazilian Amazon (Boubli 2002; Bonvicino et al. 2003a), and Marcgrave's capuchin *Cebus flavius* (Schreber, 1774) on the coast of Pernambuco (see Oliveira and Langguth, 2006).

A neatly explained taxonomy with a well-drawn map unfortunately tends to inspire complacency. Both species' definitions and geographic distributions are hypotheses, which require continuous testing; evaluating the quality and quantity of information upon which they are based. Hershkovitz (1977), for example, indicated that *Saguinus mystax pluto* occurred between the Rios Purus and Madeira, from the Rio Solimões south into northern Bolivia. His hypothesis was based on a poorly defined type locality and just two other, dubious, locality records (Rylands, Coimbra-Filho and Mittermeier 1993). Although still poorly documented, it would seem that its true range is considerably smaller, and only to the west, not the east, of the (lower) Purus.

Table 2.3 Species and subspecies of Neotropical primates described since 1990

Callithrix nigriceps Ferrari and Lopes, 1992	Black-headed marmoset
Callithrix mauesi Mittermeier, Ayres & Schwarz, 1992	Maués marmoset
Callithrix argentata marcai Alperin, 1993	Marca's marmoset
Callithrix saterei Sousee Silva Jr & Noronha, 1998	Sateré marmoset
Callithrix humilis Van Roosmalen, Van Roosmalen, Mittermeier & Fonseca, 1998	Black-crowned dwarf marmoset
Callithrix manicorensis Van Roosmalen, Van Roosmalen, Mittermeier & Fonseca, 2000	Manicoré marmoset
Callithrix acariensis Van Roosmalen, Van Roosmalen, Mittermeier & Fonseca, 2000	Rio Acarí marmoset
Leontopithecus caissara Lorini and Persson, 1990	Black-faced lion tamarin
Callicebus bernhardi Van Roosmalen, Van Roosmalen and Mittermeier, 2002	Prince Bernhard's titi monkey
Callicebus stephennashi Van Roosmalen, Van Roosmalen and Mittermeier, 2002	Stephen Nash's titi monkey
Callicebus personatus barbarabrownae Hershkovitz, 1990	Blond titi
Callicebus coimbrai Kobayashi & Langguth, 1999	Coimbra-Filho's titi monkey
Callicebus aureipalatii Wallace, Gómez, A. Felton & A. M. Felton, 2006	Madidi titi monkey
Cebus kaapori Queiroz, 1992	Ka'apor capuchin

Further research is urgently needed to verify and detail the taxonomies and distributions of all of the Neotropical primates. Modern taxonomic and biogeographic revisions are especially needed for the woolly monkeys (*Lagothrix*) (the most recent was Fooden [1963]), the sakis (*Pithecia*), the collared titis of the *Callicebus torquatus* group, and the untufted capuchins, currently ascribed to just two species, *Cebus olivaceus* and *C. albifrons* (see Hershkovitz 1949, Bodini and Pérez-Hernández 1987, Defler 2004). It is becoming evident that the taxonomy of *Pithecia* published by Hershkovitz in 1987, although a major contribution, is still very far from reflecting the true diversity of the genus (L. Marsh, pers. comm.). Other genera which have received much attention over recent years but are still subject to severely divergent opinions and confusion are the night monkeys (*Aotus*), the tufted capuchin monkeys (*Cebus*), and the squirrel monkeys (*Saimiri*).

2.5 Summary

The modern taxonomy of the Infraorder Platyrrhini has been deeply influenced by the numerous publications of the late Philip Hershkovitz. This has meant that in many aspects platyrrhine taxonomy has been extraordinarily stable over the last three decades, while his work has at the same time provided the wherewithal for considerable refinement and adjustments. It may well be that the legacy of Hershkovitz is the cause of there being more species and subspecies of primates in the New World than in Africa and Asia, providing as he did the capacity to compare findings with what is known, both in terms of the physiognomy of the primates under scrutiny and their supposed distributions. The latest taxonomies of the non-human primates indicate approximately 630 species and subspecies in 71 genera and 17 families. Of these, five families, 19 genera and 199 species and subspecies are Neotropical— 31% of the primates. Africa has 169 species and subspecies and Asia 186. Notably, of the 53 "new" primates described since 1990, 15 were from the Neotropics of which nearly half (seven) are marmosets and five are titi monkeys (two are capuchin monkeys, and the black-faced lion tamarin, *Leontopithecus caissara* completes the set). Both the titi monkeys, *Callicebus*, and the marmosets were extensively revised by Hershkovitz prior to 1990. The groups that he did not revise prior to his death were the atelines and capuchin monkeys, both of which are still confused in their taxonomy, although the variability in the latter group particularly makes a better understanding of their systematics the biggest challenge among all the platyrrhine genera. Two further tendencies deserve mention. The first is associated with the desire to conserve the full diversity of primates, an aspect which drags taxonomy from the realm of cataloguing and academic pursuit into the applied sciences. Precaution allows people to unashamedly "split." The second is related to our increased knowledge of the geography of the phenotypes we observe in situ, which has made it increasingly difficult to accept single definitions or dichotomies of species and subspecies. This and the new insights resulting from molecular genetics and chromosome studies, have promoted the adoption of the Phylogenetic Species Concept and the gradual rejection of interpretations of variation using the category of subspecies. In this chapter, I discuss these aspects as related to the taxonomy of the New World monkeys, indicating particularly where change has been prevalent, and the challenges we still face in achieving an understanding of their full diversity, not least of which are the widespread loss of their forests and the introduction and spread of species outside of their natural ranges.

References

Aguiar, J. M., and Lacher Jr., T. E. 2003. On the morphological distinctiveness of *Callithrix humilis* Van Roosmalen et al. 1998. Neotrop. Primates 11(1):11–18.

Alperin, R. 1995. *Callithrix nigriceps* Ferrari e Lopes, 1992: Um sinônimo júnior de *Callithrix emiliae* Thomas, 1920. In Programa e Resumos: VII Congresso Brasileiro de Primatologia, p.79, Universidade Federal do Rio Grande do Norte, Natal, 23–28 July, 1995. Natal: Sociedade Brasileira de Primatologia.

Armada, J. L. A., Barroso, C. M. L., Lima, M. M. C., Muniz, J. A. P. C., and Seuánez, H. N. 1987. Chromosome studies in *Alouatta belzebul*. Am. J. Primatol. 13:283–296.

Barroso, C. M. L., Schneider, H., Schneider, M. P. C., Sampaio, I., Harada, M. L., Czelusniak, J., and Goodman, M. 1997. Update on the phylogenetic systematics of New World monkeys: Further DNA evidence for placing the pygmy marmoset (*Cebuella*) within the genus *Callithrix*. Int. J. Primatol. 18(4):651–674.

Bodini, R., and Pérez-Hernández, R. 1987. Distribution of the species and subspecies of cebids in Venezuela. Fieldiana, Zool., New Series (39):231–244.

Boinski, S., and Cropp, S. J. 1999. Disparate data sets resolve squirrel monkey (*Saimiri*) taxonomy: Implications for behavioral ecology and biomedical usage. Int. J. Primatol. 20(2):237–256.

Bonvicino, C. R., Langguth, A., and Mittermeier, R. A. 1989. A study of the pelage color and geographic distribution in *Alouatta belzebul* (Primates: Cebidae). Rev. Nordestina Biol. 6(2): 139–148.

Bonvicino, C. R., Fernandes, M. E. B., and Seuánez, H. N. 1995. Morphological analysis of *Alouatta seniculus* species group (Primates, Cebidae): A comparison with biochemical and karyological data. Hum. Evol. 10(2):169–176.

Bonvicino, C. R., Lemos, B., and Seuánez, H. N. 2001. Molecular phylogenetics of howler monkeys (*Alouatta*, Platyrrhini): A comparison with karyotypic data. Chromosoma 110:241–246.

Bonvicino, C. R., Boubli, J. P., Otazu, I. B., Almeida, F. C., Nascimento, F. F., Coura, J. R., and Seuánez, H. N. 2003a. Morphologic, karyotypic, and molecular evidence of a new form of *Chiropotes* (Primates, Pitheciinae). Am. J. Primatol. 61(3):123–133.

Bonvicino, C. R., Penna-Fieme, H. N., Nascimento, F. do, Lemos, B., Stanton, R., and Seuánez, H. N. 2003b. The lowest diploid number (2n = 16) yet found in any of the primates: *Callicebus lugens* (Humboldt, 1819). Folia Primatol. 74:141–149.

Boubli, J. P. 2002. Western extension of the range of bearded sakis: A possible new taxon of *Chiropotes* sympatric with *Cacajao* in the Pico da Neblina National Park. Neotrop. Primates 19(1):1–4.

Brandon-Jones, D. 2006. Apparent confirmation that *Alouatta villosa* (Gray, 1845) is a senior synonym of *A. pigra* Lawrence, 1933 as the species-group name for the black howler monkey of Belize, Guatemala and Mexico. Primate Conserv. (21):41–43.

Brandon-Jones, D., and Groves, C. P. 2002. Neotropical primate family-group names replaced by Groves (2001) in contravention of Article 40 of the International Code of Zoological Nomenclature. Neotrop. Primates 10(3):113–115.

Brandon-Jones, D., Eudey, A. A., Geissmann, T., Groves, C. P., Melnick, D. J., Morales, J. C., Shekelle, M., and Stewart C.-B. 2004. Asian primate classification. Int. J. Primatol. 25(1): 97–164.

Bueno, M. L., and Defler, T. R. 2007. Esta presente *Callicebus lugens* en Colombia? Unpublished manuscript.

Bueno, M. L., Ramírez-Orjuela, C., Leibovici, M., and Torres, O. M. 2006. Información cariológica del género *Callicebus* en Colombia. Rev. Acad. Colomb. Cienc. 30(114):109–115.

Cabrera, A. 1957. Catalogo de los mamíferos de América del Sur. Rev. Mus. Argentino de Cienc. Nat. "Bernardino Rivadavia" 4(1):1–307.

Canavez, F. C., Moreira, M. A. M., Simon, F., Parham, P., and Seuánez, H. N. 1999. Phylogenetic relationships of the Callitrichinae (Platyrrhini, Primates) based on beta2-microglobulin DNA sequences. Am. J. Primatol. 48(3):225–236.

Casado, F., Bonvicino, C. R., and Seuánez, H. N. 2007. Phylogeographic analyses of *Callicebus lugens* (Platyrrhini, Primates). J. Hered. 98(1):88–92.

Chaves, R., Sampaio, I., Schneider, M. P. C., Schneider, H., Page, S. L., and Goodman, M. 1999. The place of *Callimico goeldii* in the callitrichine phylogenetic tree: Evidence from von Willebrand Factor Gene Intron II sequences. Molec. Phylogenet. Evol. 13:392–404.

Cheverud, J. M., and Moore, A. J. 1990. Subspecific variation in the saddle-back tamarin (*Saguinus fuscicollis*). Am. J. Primatol. 21:1–15.

Coimbra-Filho, A. F. 1990. Sistemática, distribuição geográfica e situação atual dos símios brasileiros (Platyrrhini – Primates). Rev. Brasil. Biol. 50:1063–1079.

Coimbra-Filho, A. F., Pissinatti, A., and Rylands, A. B. 1993. Breeding muriquis (*Brachyteles arachnoides*) in captivity: The experience of the Rio de Janeiro Primate Centre (CPRJ-FEEMA) (Ceboidea, Primates). Dodo, J. Wildl. Preserv. Trusts 29:66–77.

Coimbra-Filho, A. F., Pissinatti, A., and Rylands, A. B. 1997. A simulacrum of *Saguinus bicolor ochraceus* Hershkovitz, 1966, obtained through hybridising *S. b. martinsi* and *S. b. bicolor* (Callitrichidae, Primates). In A Primatologia no Brasil – 5, eds. S. F. Ferrari, and H. Schneider, pp. 179–184. Belém: Sociedade Brasileira de Primatologia.

Collins, A. C. 1999. Species status of the Colombian spider monkey, *Ateles belzebuth hybridus*. Neotrop. Primates 7(2):39–43.

Collins, A. C., and Dubach, J. 2000. Phylogenetic relationships of spider monkeys (*Ateles*) based on mitochondrial DNA variation. Int. J. Primatol. 21(3):381–420.

Cortés-Ortiz, L., Bermingham, E., Rico, C., Rodriguez-Luna, E., Sampaio, I., and Ruiz-Garcia, M. 2003. Molecular systematics and biogeography of the Neotropical monkey genus, *Alouatta*. Molec. Phylogenet. Evol. 26:64–81.

Costello, R. K., Dickinson, C., Rosenberger, A. L., Boinski, S., and Szalay, F. S. 1993. Squirrel monkey (genus *Saimiri*) taxonomy: A multidisciplinary study of the biology of the species. In Species, Species Concepts, and Primate Evolution, eds. W. H. Kimbel, and L. B. Martin, pp.177–210. New York: Plenum Press.

Cronin, J. E., and Sarich, V. M. 1978. Marmoset evolution: The molecular evidence. In Primates in Medicine, Vol. 10, eds. N. Gengozian, and F. Deinhardt, pp.12–19. Basel: S. Karger.

Cropp, S. J., Larson, A., and Cheverud, J. M. 1999. Historical biogeography of tamarins, genus *Saguinus*: The molecular phylogenetic evidence. Am. J. Phys. Anthropol. 108:65–89.

Cruz Lima, E. da. 1945. Mammals of Amazônia, Vol. 1. General Introduction and Primates. Belém do Pará: Contribuições do Museu Paraense Emílio Goeldi de História Natural e Etnografia.

Defler, T. R. 2004. Primates of Colombia. Washington, DC: Tropical Field Guide Series, Conservation International.

Defler, T. R., and Bueno, M. L. 2003. Karyological guidelines for *Aotus* taxonomy. Am. J. Primatol. 60(suppl. 1):134–135.

Defler, T. R., and Hernández-Camacho, J. I. 2002. The true identity and characteristics of *Simia albifrons* Humboldt, 1812: Description of neotype. Neotrop. Primates 10:49–64.

Defler, T. R., Bueno, M. L., and Hernández-Camacho, J. I. 2001. Taxonomic status of *Aotus hershkovitzi*: Its relationship to *Aotus lemurinus lemurinus*. Neotrop. Primates 9:37–52.

Elliot, D. G. 1913. A Review of Primates, Monograph Series, American Museum of Natural History, New York.

Ferrari, S. F., and Lopes, M. A. 1992. A new species of marmoset, genus *Callithrix* Erxleben, 1777 (Callitrichidae, Primates) from western Brazilian Amazonia. Goeldiana Zoologia (12):1–3.

Ferrari, S. F., Sena, L., and Schneider, M. P. C. 1999. Definition of a new species of marmoset (Primates: Callithrichinae) from southwestern Amazonia based on molecular, ecological, and zoogeographic evidence. In Livro de Resumos, IX Congresso Brasileiro de Primatologia, pp. 80–81, Santa Teresa, Espírito Santo, 25–30 July 1999. Santa Teresa, Espírito Santo: Sociedade Brasileira de Primatologia.

Figueiredo, W. B., Carvalho-Filho, N. M., Schneider, H., and Sampaio, I. 1998. Mitochondrial DNA sequences and the taxonomic status of *Alouatta seniculus* populations in northeastern Amazonia. Neotrop. Primates 6:73–77.

Figueiredo, W. B., Silva, J. M., Bates, J. M., Harada, M. L, and Silva Jr., J. S. 2006. Conservation genetics and biogeography of pitheciins. Int. J. Primatol. 27(suppl. 1):Abstract #510.

Fonseca, G. A. B. da, Lemos de Sá, R. M., Pope, T. R., Glander, K. E., and Struhsaker, T. T. 1991. A pilot study of genetic and morphological variation in the muriqui (*Brachyteles arachnoides*) as a contribution to a long-term conservation management plan. Report, World Wildlife Fund – US, Washington, DC. 58pp.

Fooden, J. 1963. A revision of the woolly monkeys (genus *Lagothrix*). J. Mammal. 44(2):213–247.

Ford, S. M. 1994. Taxonomy and distribution of the owl monkey. In *Aotus*: The Owl Monkey, eds. J. F. Baer, R. E. Weller, and I. Kakoma, pp. 1–57. New York: Alan R. Liss.

Fragaszy, D. M., Visalberghi, E., Fedigan. L., and Rylands A. B. 2004. Taxonomy, distribution and conservation: Where and what are they, and how did they get there? In The Complete Capuchin: The Biology of the Genus *Cebus*, D. Fragaszy, L. Fedigan, and E. Visalberghi, pp. 13–35. Cambridge, UK: Cambridge University Press.

Froehlich, J. W., and Froehlich, P. H. 1987. The status of Panama's endemic howling monkeys. Primate Conserv. (8):58–62.

Gregorin, R. 2006. Taxonomy and geographic variation of species of the genus *Alouatta* Lacépède (Primates, Atelidae) in Brazil. Rev. Brasil. Zool. 23(1):64–144.

Groves, C. P. 2001. Primate Taxonomy. Washington, DC: Smithsonian Institution Press.

Groves, C. P. 2004. The what, why and how of primate taxonomy. Int. J. Primatol. 25(5): 1105–1126.

Groves, C. P. 2005. Order Primates. In Mammal Species of the World: A Taxonomic and Geographic Reference, 3rd Edition, Vol. 1, eds. D. E. Wilson, and D. M. Reeder, pp. 111–184. Baltimore: Johns Hopkins University Press.

Grubb, P., Butynski, T. M., Oates, J. F., Bearder, S. K., Disotell, T. R., Groves, C. P., and Struhsaker, T. T. 2003. Assessment of the diversity of African primates. Int J. Primatol. 24(6):1301–1357.

Hall, E. R. 1981. The Mammals of North America. Vol. 1. New York: John Wiley and Sons.

Harada, M. L., and Ferrari, S. F. 1996. Reclassification of *Cebus kaapori* Queiroz 1992 based on new specimens from eastern Pará, Brazil. In Abstracts. XVIth Congress of the International Primatological Society, XIXth Congress of the American Society of Primatologists, #729, August 11–16, 1996. Madison, Wisconsin.

Harris, E. E., Gifalli-Inghetti, C., Braga, Z. H., and Koiffman, C. P. 2005. Cytochrome b sequences show subdivision between populations of the brown howler monkey *Alouatta guariba* from Rio de Janeiro and Santa Catarina, Brazil. Neotrop. Primates 13(2):16–17.

Heltne, P. G., and Kunkel L. M. 1975. Taxonomic notes on the pelage of *Ateles paniscus paniscus, A. p. chamek* (*sensu* Kellogg and Goldman, 1944) and *A. fusciceps rufiventris* (= *A. f. robustus* Kellogg and Goldman, 1944). J. Med. Primatol. 4:83–102.

Hernández-Camacho, J., and Cooper, R. W. 1976. The non-human primates of Colombia. In Neotropical Primates: Field Studies and Conservation, eds. R. W. Thorington Jr., and P. G. Heltne, pp. 35–69. Washington, DC: National Academy of Sciences.

Hernández-Camacho, J., and Defler, T. R. 1991. Algunos aspectos de la conservación de primates no-humanos en Colombia. In La Primatología en Latinoamérica, eds. C. J. Saavedra, R. A. Mittermeier, and I. B. Santos, pp. 67–100. Washington, DC: World Wildlife Fund.

Hershkovitz, P. 1949. Mammals of northern Colombia. Preliminary report No. 4: Monkeys (Primates), with taxonomic revisions of some forms. Proc. U. S. Natl. Mus. 98:323–427.

Hershkovitz, P. 1955. Notes on the American monkeys of the genus *Cebus*. J. Mammal. 36: 449–452.

Hershkovitz, P. 1966. Taxonomic notes on tamarins, genus *Saguinus* (Callithricidae, Primates) with descriptions of four new forms. Folia Primatol. 4:381–395.

Hershkovitz, P. 1977. Living New World Monkeys (Platyrrhini) with an Introduction to Primates, Vol. 1. Chicago: Chicago University Press.

Hershkovitz, P. 1979. Races of the emperor tamarin, *Saguinus imperator* Goeldi (Callitrichidae, Primates). Primates 20(2):277–287.

Hershkovitz, P. 1982. Subspecies and geographic distribution of black-mantle tamarins *Saguinus nigricollis* Spix (Primates: Callitrichidae). Proc. Biol. Soc. Wash. 95(4):647–656.

Hershkovitz, P. 1983. Two new species of night monkeys, genus *Aotus* (Cebidae, Platyrrhini): A preliminary report on *Aotus* taxonomy. Am. J. Primatol. 4:209–243.

Hershkovitz, P. 1984. Taxonomy of squirrel monkeys, genus *Saimiri* (Cebidae, Platyrrhini): A preliminary report with description of a hitherto unnamed form. Am. J. Primatol. 7:155–210.

Hershkovitz, P. 1985. A preliminary taxonomic review of the South American bearded saki monkeys genus *Chiropotes* (Cebidae, Platyrrhini), with the description of a new subspecies. Fieldiana, Zool., New Series (27):iii +46.

Hershkovitz, P. 1987a. The taxonomy of South American sakis, genus *Pithecia* (Cebidae, Platyrrhini): A preliminary report and critical review with the description of a new species and new subspecies. Am. J. Primatol. 12:387–468.

Hershkovitz, P. 1987b. Uacaries, New World monkeys of the genus *Cacajao* (Cebidae, Platyrrhini): A preliminary taxonomic review with the description of a new subspecies. Am. J. Primatol. 12:1–53.

Hershkovitz, P. 1988. Origin, speciation, and distribution of South American titi monkeys, genus *Callicebus* (Family Cebidae, Platyrrhini). Proc. Acad. Nat. Sci. Philadelphia 140(1): 240–272.

Hershkovitz, P. 1990. Titis, New World monkeys of the genus *Callicebus* (Cebidae, Platyrrhini): A preliminary taxonomic review. Fieldiana, Zool., New Series (55):1–109.

Heymann, E. W. 2000. Field observations of the golden-mantled tamarin, *Saguinus tripartitus*, on the Rio Curaray, Peruvian Amazonia. Folia Primatol. 71(6):392–398.

Heymann, E. W., Encarnación C. F., and Soini, P. 2002. On the diagnostic characters and geographic distribution of the "yellow-handed" titi monkey, *Callicebus lucifer*, in Peru. Neotrop. Primates 10(3):124–126.

Hill, W. C. O. 1957. Primates Comparative Anatomy and Taxonomy III. Pithecoidea Platyrrhini (Families Hapalidae and Callimiconidae). Edinburgh: Edinburgh University Press.

Hill, W. C. O. 1960. Primates Comparative Anatomy and Taxonomy IV. Cebidae Part A. Edinburgh: Edinburgh University Press.

Hill, W. C. O. 1962. Primates Comparative Anatomy and Taxonomy V. Cebidae Part B. Edinburgh: Edinburgh University Press.

Kellogg R., and Goldman, E. A. 1944. Review of the spider monkeys. Proc. U. S. Natl. Mus. 96:1–45.

Koiffman, C. P. 1977. Variabilidade Cromossômica na Família Cebidae (Primates, Platyrrhini). Doctoral dissertation, Universidade de São Paulo, São Paulo.

Langguth, A., Teixeira, D. M., Mittermeier, R. A., and Bonvicino, C. 1987. The red-handed howler monkey in northeastern Brazil. Primate Conserv. (8):36–39.

Lawrence, B. 1933. Howler monkeys of the *palliata* group. Bull. Mus. Comp. Zool. (Harvard University) 75:314–354.

Lemos de Sá, R. M., and Glander, K. E. 1993. Capture techniques and morphometrics for the woolly spider monkey, or muriqui (*Brachyteles arachnoides* É. Geoffroy, 1806). Am. J. Primatol. 29:145–153.

Lemos de Sá, R. M., Pope, T. R., Glander, K. E., Struhsaker, T. T., and Fonseca, G. A. B. da. 1990. A pilot study of genetic and morphological variation in the muriqui (*Brachyteles arachnoides*). Primate Conserv. (11):26–30.

Lima, M. M. C., and Seuánez, H. N. 1989. Cytogenetic characterization of *Alouatta belzebul* with atypical pelage coloration. Folia Primatol. 52:97–101.

Lönnberg, E. 1938. Remarks on some members of the genera *Pithecia* and *Cacajao* from Brazil. Ark. Zool. 30A(18):1–25.

Lorini, V. G., and M. L. Persson. 1990. Uma nova espécie de *Leontopithecus* Lesson, 1840, do sul do Brasil (Primates, Callitrichidae). Bol. Mus. Nac., Rio de Janeiro, nova sér. Zoologia (338):1–14.

Marsh, L. K. 2004. Primate species at the Tiputini Biodiversity Station, Ecuador. Neotrop. Primates 12(2):75–78.

Marsh, L. K. 2006. Identification oand conservation of a new species of *Pithecia* in Amazonian Ecuador. Int. J. Primatol. 27(suppl. 1):Abstract #508.

Melo, A. C. A., Sampaio, M. I., Schneider, M. P. C., and Schneider, H. 1992. Biochemical diversity and genetic distance in two species of the genus *Saguinus*. Primates 33(2):217–225.

Minezawa. M., Harada, M., Jordan, O. C., and Valdivia Borda, C. J. 1986. Cytogenetics of the Bolivian endemic red howler monkeys (*Alouatta seniculus sara*): Accessory chromosomes and Y-autosome translocation related numerical variations. Kyoto Univ. Overseas Res. Rep. New World Monkeys 5:7–16.

Mittermeier, R. A., Ganzhorn, J. U., Konstant, W. R., Glander, K., Tattersall, I., Groves, C. P., Rylands, A. B., Hapke, A., Ratsimbazafy, J., Mayor, M. I., Louis Jr., E. E., Rumpler, Y., Schwitzer, C., and Rasoloarison, R. M. In press. Lemur Diversity in Madagascar. Int. J. Primatol.

Moynihan, M. 1976. The New World Primates: Adaptive Radiation and the Evolution of Social Behavior, Languages, and Intelligence. Princeton, NJ: Princeton University Press.

Napier, P. H. 1976. Catalogue of the Primates in the British Museum (Natural History). Part I. Families Callitrichidae and Cebidae. London: British Museum (Natural History).

Nascimento, F. F., Bonvicino, C. R.,. Silva, F. C. D. da, Schneider, M. P. C., and Seuánez, H. N. 2005. Cytochrome b polymorphisms and population structure of two species of *Alouatta* (Primates). Cytogenet. Genome Res. 108:106–111.

Natori, M., and Hanihara, T. 1992. Variations in dental measurements between *Saguinus* species and their systematic relationships. Folia Primatol. 58:84–92.

Nieves, M., Ascunce, M. S., Rahn, M. I., and Mudry, M. D. 2005. Phylogenetic relationships among some *Ateles* species: The use of chromosome and molecular markers. Primates 46: 155–164.

Oliveira, E. H. C. 1996. Estudos Citogenéticos e Evolutivos nas Espécies Brasileiras e Argentinas do Gênero *Alouatta* Lacépède, 1799 (Primates, Atelidae). Master's thesis, Universidade Federal do Paraná, Curitiba.

Oliveira, E. H. de, Lima M. M. C. de, and Sbalqueiro, I. J. 1995. Chromosomal variation in *Alouatta fusca*. Neotrop. Primates 3:181–182.

Oliveira, E. H. de, Lima M. M. C. de, Sbalqueiro, I. J., and Pissinatti, A. 1998. The karyotype of *Alouatta fusca clamitans* from Rio de Janeiro, Brazil: Evidence for a Y chromosome translocation. Genet. Molec. Biol. 21:361–364.

Oliveira, E. H. de, Neusser, M., Figueiredo, W. B., Nagamachi, C., Pieczarka, J. C., Sbalqueiro, I. J., Wienberg, J., and Müller, S. 2002. The phylogeny of howler monkeys (*Alouatta*, Platyrrhini) Reconstruction by multi-color cross-species chromosome painting. *Chromosome Res.* 10: 669–683.

Oliveira, M. M. de, and Langguth, A. 2006. Rediscovery of Marcgrave's capuchin monkey and designation of a neotype for *Simia flavia* Schreber, 1774 (Primates, Cebidae). Bol. Mus. Nac., nova sér., Rio de Janeiro (523):1–16.

Pastorini, J., Forstner, M. R. J., Martin, R. D., and Melnick, D. J. 1998. A reexamination of the phylogenetic position of *Callimico* (Primates) incorporating new mitochondrial DNA sequence data. J. Molec. Evol. 47(1):32–41.

Peres, C. A., Patton J. L., and Silva, M. N. F. da. 1996. Riverine barriers and gene flow in Amazonian saddle-back tamarins. Folia Primatol. 67(3):113–124.

Pontes, A. R. M., Malta, A., and Asfora, P. H. 2006. A new species of capuchin monkey, genus *Cebus* Erxleben (Cebidae, Primates): Found at the very brink of extinction in the Pernambuco Endemism Centre. *Zootaxa* 1200:1–12. Website: http://www.mapress.com/zootaxa/. Accessed 11 May 2006.

Porter, C. A., Czelusniak, J., Schneider, H., Schneider, M. P. C., Sampaio, I., and Goodman, M. 1997. Sequences of the primate epsilon-globin gene: Implications for systematics of the marmosets and other New World primates. Gene 205(1–2):59–71.

Queiroz, H. L. 1992. A new species of capuchin monkey, genus *Cebus* Erxleben 1977 (Cebidae, Primates), from eastern Brazilian Amazonia. Goeldiana Zoologia (15):1–3.

Rosenberger, A. L. 1980. Gradistic views and adaptive radiation of platyrrhine primates. Z. Morph. Anthrop. 71(2):157–163.

Rosenberger, A. L. 1981. Systematics: The higher taxa. In Ecology and Behavior of Neotropical Primates, Vol. 1, eds. A. F. Coimbra-Filho, and R. A. Mittermeier, pp.9–27. Rio de Janeiro: Academia Brasileira de Ciências.

Rosenberger, A. L., and Coimbra-Filho, A. F. 1984. Morphology, taxonomic status and affinities of the lion tamarins, *Leontopithecus* (Callitrichinae, Cebidae). Folia Primatol. 42: 149–179.

Ruiz-Herrera, A., García, F., Aguilera, M., Garcia, M., and Fontanals, M. P. 2005. Comparative chromosome painting in *Aotus* reveals a highly derived evolution. Am. J. Primatol. 65:73–85.

Rylands, A. B. 1993. The bare-face tamarins *Saguinus oedipus oedipus* and *Saguinus oedipus geoffroyi*: Subspecies or species? Neotrop. Primates 1(2):4–5.

Rylands, A. B., and Brandon-Jones, D. 1998. Scientific nomenclature of the red howlers from the northeastern Amazon in Brazil, Venezuela, and the Guianas. Int. J. Primatol. 19(5): 879–905.

Rylands, A. B., Coimbra-Filho, A. F., and Mittermeier, R. A. 1993. Systematics, distributions, and some notes on the conservation status of the Callitrichidae. In Marmosets and Tamarins: Systematics, Behaviour and Ecology, ed. A. B. Rylands, pp. 11–77. Oxford: Oxford University Press.

Rylands, A. B, Kierulff, M. C. M., and Mittermeier, R. A. 2005. Some notes on the taxonomy and distributions of the tufted capuchin monkeys (*Cebus*, Cebidae) of South America. Lundiana 6(supl.):97–110.

Rylands, A. B., Mittermeier, R. A., and Coimbra-Filho, A. F. in press. The systematics and distributions of the marmosets (*Callithrix, Callibella, Cebuella*, and *Mico*) and callimico (*Callimico*) (Callitrichidae, Primates). In The Smallest Anthropoids: The Marmoset/Callimico Radiation, eds. S. M. Ford, L. C. Davis, and L. Porter. New York: Springer.

Rylands, A. B., Schneider, H., Langguth, A., Mittermeier, R. A., Groves, C. P., and Rodríguez-Luna, E. 2000. An assessment of the diversity of New World primates. Neotrop. Primates 8(2):61–93.

Rylands, A. B., Groves, C. P., Mittermeier, R. A., Cortés-Ortiz, L., and Hines, J. J. 2006. Taxonomy and distributions of Mesoamerican primates. In New Perspectives in the Study of Mesoamerican Primates: Distribution, Ecology, Behavior and Conservation, eds. A. Estrada, P. Garber, M. Pavelka, and L. Luecke, pp. 29–79. New York: Springer.

Sampaio, M. I. da C., Schneider, M. P. C., and Schneider, H. 1996. Taxonomy of the *Alouatta seniculus* group: Biochemical and chromosome data. Primates 37(1):67–73.

Schneider, H., and Rosenberger A. L. 1996. Molecules, morphology, and platyrrhine systematics. In Adaptive Radiations of Neotropical Primates, eds. M. A. Norconk, A. L. Rosenberger, and P. A. Garber, pp. 3–19. New York: Plenum Press.

Schneider, H., Schneider, M. P. C., Sampaio, I., Harada, M. L., Stanhope, M., Czelusniak, J., and Goodman, M. 1993. Molecular phylogeny of the New World monkeys (Platyrrhini, Primates). Molec. Phylogenet. Evol. 2(3):225–242.

Sena, L. dos S. 1998. Filogenia do gênero *Callithrix* Erxleben 1777 (Callitrichinae, Platyrrhini) Baseada em Sequências do Gene Mitocondrial da Citocromo Oxidase II (COII). Master's thesis, Universidade Federal do Pará, Belém.

Seuánez, H. N., Armada, J. L., Freitas, L., Rocha e Silva, R. da, Pissinatti, A., and Coimbra-Filho, A. F. 1986. Intraspecific chromosome variation in *Cebus apella* (Cebidae, Platyrrhini): the chromosomes of the yellow-breasted capuchin *Cebus apella xanthosternos* Wied, 1820. Am. J. Primatol. 10:237–247.

Seuánez, H. N., Forman, L., Matayoshi, T., and Fanning, T. G. 1989. The *Callimico goeldii* (Primates, Platyrrhini) genome: Karyology and middle repetitive (LINE-1) DNA sequences. Chromosoma 98:389–395.

Silva Jr., J. de S. 2001. Especiação nos Macacos-prego e Caiararas, Gênero *Cebus* Erxleben, 1777 (Primates, Cebidae). Doctoral thesis, Universidade Federal do Rio de Janeiro, Rio de Janeiro. 377pp.

Silva Jr., J. de S. 2002. Sistemática dos macacos–prego e caiararas, gênero *Cebus* Erxleben, 1777 (Primates, Cebidae). In Livro de Resumos, X Congresso Brasileiro de Primatologia: Amazônia – A Última Fronteira, p. 35, 10–15 November 2002. Bélém: Sociedade Brasileira de Primatologia.

Silva Jr., J. S., and Figueiredo, W. M. B. 2002. Previsão sistemática dos cuxiús, genêro *Chiropotes* Lesson, 1840 (Primates, Pitheciidae). In Livro de Resumos: X° Congresso Brasileiro de Primatologia "Amazônia: A Última Fronteira", p. 21, 10–15 de novembro de 2002. Belém: Sociedade Brasileira de Primatologia.

Silva, Jr., J. de S., and Martins, E. de S. 1999. On a new white bald uakari population in southwestern Brazilian Amazonia. Neotrop. Primates 7(4):119–121.

Silva-López, G., Motta-Gill, J., and Sánchez-Hernández, A. I. 1996. Taxonomic notes on *Ateles geoffroyi*. Neotrop. Primates 4(2):41–44.

Stanyon, R., Tofanelli, S., Morescalchi, M. A., Agoramoorthy, G., Ryder, O. A., and Wienberg, J. 1995. Cytogenetic analysis shows extensive genomic rearrangements between red howler (*Alouatta seniculus* Linnaeus) subspecies. Am. J. Primatol. 35:171–183.

Tagliaro, C. H, Schneider, M. P. C., Schneider, H., Sampaio, I. C., and Stanhope, M. J. 1997. Marmoset phylogenetics, conservation perspectives, and evolution of the mtDNA control region. Molec. Biol. Evol. 14(6):674–684.

Tagliaro, C. H., Schneider, H., Sampaio, I., Schneider M. P. C., Vallinoto, M., and Stanhope, M. 2005. Molecular phylogeny of the genus *Saguinus* (Platyrrhini, Primates) based on the ND1 mitochondrial gene and implications for conservation. Genet. Molec. Biol. 28(1): 46–53.

Thomas, O. 1904. New *Callithrix, Midas, Felis, Rhipidomys*, and *Proechimys* from Brazil and Ecuador. Ann. Mag. Nat. Hist. 14(7):188–196.

Thorington Jr., R. W. 1985. The taxonomy and distribution of squirrel monkeys (*Saimiri*). In Handbook of Squirrel Monkey Research, eds. L. A. Rosenblum, and C. L. Coe, pp. 1–33. New York: Plenum Press,

Thorington Jr., R. W. 1988. Taxonomic status of *Saguinus tripartitus* (Milne-Edwards, 1878). Am. J. Primatol. 15:367–371.

Vallinoto, M., Araripe, J., Rego. P. S., Tagliaro, C. H., Sampaio, I., and Schneider, H. 2006. Tocantins River as an effective barrier to gene flow in *Saguinus niger* populations. Genet. Molec. Biol. 12:823–833.

Van Roosmalen, M. G. M. 2003. New species from Amazonia. Website <http:amazonnewspecies. com>, 5 August 2003, updated 20 September 2003. Accessed 26 September 2003.

Van Roosmalen, M. G. M., and Van Roosmalen, T. 1997. An eastern extension of the geographical range of the pygmy marmoset, *Cebuella pygmaea*. Neotrop. Primates 5(1):3–6.

Van Roosmalen, M G. M., and Van Roosmalen, T. 2003. The description of a new marmoset genus, *Callibella* (Callitrichinae, Primates), including its molecular phylogenetic status. Neotrop. Primates 11(1):1–10.

Van Roosmalen, M. G. M., Van Roosmalen, T., and Mittermeier, R. A. 2002. A taxonomic review of the titi monkeys, genus *Callicebus* Thomas, 1903, with the description of two new species, *Callicebus bernhardi* and *Callicebus stephennashi*, from Brazilian Amazonia. Neotrop. Primates 10(Suppl.):1–52.

Van Roosmalen, M. G. M., Van Roosmalen, T., Mittermeier, R. A., and Fonseca, G. A. B. da. 1998. A new and distinctive species of marmoset (Callitrichidae, Primates) from the lower Rio Aripuanã, state of Amazonas, central Brazilian Amazonia. Goeldiana Zoologia (22)s:1–27.

Vàsàrhelyi, K. 2002. The nature of relationships among founders in the captive population of Goeldi's monkey (*Callimico goeldii*). Evol. Anthropol. (suppl. 1):155–158.

Vieira, C. da C. 1944. Os símios do Estado de São Paulo. Pap. Avuls. Zool., São Paulo 4:1–31.

Vivo, M. de. 1985. On some monkeys from Rondônia, Brasil (Primates: Callitrichidae, Cebidae). Pap. Avuls. Zool., São Paulo (4):1–31.

Vivo, M. de. 1991. Taxonomia de *Callithrix* Erxleben, 1777 (Callitrichidae, Primates). Belo Horizonte: Fundação Biodiversitas.

Wallace, R. B., and Painter. R. L. E. 1999. A new primate record for Bolivia: An apparently isolated population of common woolly monkeys representing a southern range extension for the genus *Lagothrix*. Neotrop. Primates 7(4):111–112.

Chapter 3
Paleogeography of the South Atlantic: a Route for Primates and Rodents into the New World?

Felipe Bandoni de Oliveira, Eder Cassola Molina, and Gabriel Marroig

3.1 Introduction

The history of primates and rodents in South America started in the Oligocene, around 30 million years ago (Ma) (Hoffstetter 1969; Simpson 1980; Wyss et al. 1993; Takai et al. 2000), with the possibility of an even earlier Eocene occurrence for rodents (Frailey and Campbell 2004). By that time, South America was already separated from Africa and not yet connected to North America via the Isthmus of Panama (Scotese 2004). If primates and rodents arrived between 50 and 20 Ma, two critical questions arise: where did they come from, and how did they reach South America? The question "where" generated great controversy in the past (Ciochon and Chiarelli 1980; George and Lavocat 1993; Goldblatt 1993). During the twentieth century, the most widely accepted opinion was that the New World monkeys (Platyrrhini) and the Old World monkeys (Catarrhini) evolved their higher primate features in parallel in Africa and South America from different prosimian ancestors (Gazin 1958; Simons 1961; Fleagle and Gilbert 2006). No prosimian fossil is known from South America, but given their Eocene abundance in North America, the possibility of a migration across the Caribbean Sea was entertained in the past (Wood 1980, 1993). An equivalent hypothesis was proposed for rodents, implying convergent evolution of a specialized jaw morphology (hystricognathy) in South American caviomorphs and African phiomorph rodents (Ciochon and Chiarelli 1980; Wood 1993).

Nonetheless, with the increasing acceptance of phylogenetic methods, plenty of evidence that platyrrhines and catarrhines are sister taxa and share a common ancestry became available, rendering convergent evolution of anthropoid features from prosimians an improbable alternative. The same holds for South American caviomorphs and African phiomorphs. Recent molecular and fossil analyses clearly indicate that these South American lineages each represent monophyletic groups that are most closely related to African forms (Nedbal et al. 1994; Kay

G. Marroig (✉)

Departamento de Genética e Biologia Evolutiva, Instituto de Biociências, Universidade de São Paulo, Rua do Matão, 277, 05508-900, São Paulo, Brazil
e-mail: gmarroig@usp.br

P.A. Garber et al. (eds.), *South American Primates,* Developments in Primatology: Progress and Prospects, DOI 10.1007/978-0-387-78705-3_3,
© Springer Science+Business Media, LLC 2009

et al. 1997; Flynn and Wyss 1998; Goodman et al. 1998; Takai et al. 2000; Huchon and Douzery 2001; Schrago and Russo 2003; Poux et al. 2006). In contrast to the absence of likely platyrrhine and caviomorph ancestors in North America, fossils from the Eocene of Fayum, Egypt, exhibit numerous traits similar to living and fossil platyrrhines and caviomorphs from South America (Lavocat 1980; Van Couvering and Harris 1991; Kay et al. 1997; Takai et al. 2000; Fleagle and Gilbert 2006). *Proteopithecus*, for instance, has no features that could distinguish it from basal platyrrhines, leading some authors to propose that it might be part of their early radiation (Takai et al. 2000). Altogether, phylogenetic fossil and molecular evidence favor the hypothesis that platyrrhines and caviomorphs originated from groups that migrated from Africa to South America. The other biogeographical question, however, remains unanswered: how did monkeys and rodents manage to travel across the Atlantic Ocean?

South America is separated from Africa by at least 2600 km of ocean, and primates and rodents first appeared well after the onset of the Gondwana break-up, around 100 Ma (Scotese 2004; Fleagle and Gilbert 2006). Three hypotheses have been formulated to explain their possible transatlantic migration: land bridges, volcanic island hopping, and floating island rafting (Table 3.1; Hoffstetter and Lavocat 1970; Simpson 1980; Ciochon and Chiarelli 1980; Houle 1999). Paleo-geographic reconstructions and geophysical evidence clearly dismiss the existence of a complete land bridge between Africa and South America during the Cenozoic (Sclater et al. 1977; Markwick and Valdes 2004; Eagles 2007). Nonetheless, drilling studies in the South Atlantic provided evidence of subaerial exposure as late as 25 Ma for some points which are now at more than 1 km below sea level (Barker 1983; Parrish 1993). Unfortunately, due to scattered nature of these data, it is not possible to determine if the distribution of these islands in time and space would have been sufficient to enable mammals to migrate by island hopping. The floating island model remains a plausible alternative and is compatible with paleocurrent directions from 60 Ma to the present (Haq 1981; Parrish and Curtis 1982). However, a critical condition for floating island migration is that distances should be small enough to allow animals to survive until they have successfully reached a larger land mass. Other studies have modeled this kind of migration using paleogeographic reconstructions (Houle 1999, based on Nürnberg and Müller 1991). Nonetheless, recent data on sea level changes (Miller et al. 2005), more precise dating of the ocean floor (Müller et al. 1997), and new map manipulation techniques (Wessel and Smith 1998; Markwick and Valdes 2004) offer new bases to re-estimate more accurately migration distances, and the feasibility of a proposed rafting migration can be more critically examined.

In this chapter, we studied the paleogeography of the South Atlantic during the probable period of crossing of caviomorphs and platyrrhines to re-evaluate the possible role of island hopping and floating islands in their proposed migration. Both the contour and position of South America and Africa are known not to have been the same during the Cenozoic. Due to continental drift, the African and South American tectonic plates have been separating for at least 100 million years (Sclater et al. 1977; Ford and Golonka 2003; Scotese 2004). Additionally, ocean bathymetry

Table 3.1 A glossary of the terms used in this chapter

Bathymetry	The measurement of the depth of water bodies.
Floating island model	Mode of dispersal in which organisms are passively transported in an island across wide water bodies. These islands are typically formed by pieces of land and plants detached from the margins of large rivers. See Houle (1998 and 1999) for details on size and wind effects on these islands.
Island hopping model	Mode of dispersal in which organisms migrate across large water bodies through sets of islands (also called "stepping-stones"). In this scenario, all islands do not persist during the entire migration between land masses, but adjacent islands are successively connected along geological time.
Land bridge model	Mode of dispersal in which two land masses were connected in the past, but not anymore. This model was widely used to explain disjunct biogeographical patterns between continents before continental drift became accepted.
Thermal subsidence	Relative subsidence of the lithosphere due to heat loss and subsequent contraction. Empirical data show that older oceanic lithosphere lie deeper than the more recently formed, and mathematical models of thermal subsidence rates were developed to predict depth from age (e.g., Parsons and Sclater 1977).

also has changed due to thermal subsidence of the oceanic floor, with depth increasing with age (Sclater and Mckenzie 1973), and due to changes in sea level over the last 65 Ma (Miller et al. 2005). We modeled these three factors, i.e., horizontal plate motion, thermal subsidence of the oceanic lithosphere, and global sea level fluctuations at four time-slices along the Cenozoic (20, 30, 40 and 50 Ma) in order to reconstruct a plausible scenario in which the migration of primates and rodents to South America could have taken place.

3.2 Material and Methods

We reconstructed the position of the continents in the period when migration of primates and rodents presumably occurred (between 20 and 50 Ma) based on magnetic anomalies. The periodic reversal of the Earth's magnetic field can be used to date the oceanic floor and many age maps were developed using this information (e.g., Müller et al. 1997). The past position of African and South American plates can be reconstructed by fitting together magnetic lineations of a certain age from opposite sides of the Mid-Atlantic Ridge axis (Pitman et al. 1993; Scotese 2004). In this study, we used the digital age grid of the ocean floor provided by Müller et al. (1997).

We superimposed the past position of Africa and South America on paleobathymetric reconstructions of the Atlantic Ocean based on the thermal subsidence of the oceanic lithosphere. For terrains younger than 80 Ma, theoretical models predict that subsidence rates follow the relationship (Parsons and Sclater 1977; Markwick and Valdes 2004):

$$d = 2500 + 350\,\sqrt{t}$$

For areas older than 80 Ma, we used the following equation (Parsons and McKenzie 1978; Kearey and Vine 1996):

$$d = 6400 - 3200\exp(-t/62.8)$$

in both cases, t is the age of the rocks in million years and d their depth in meters. Empirical drilling reveals a strong predicting capacity for these models, with an associated error of about 300 m (Sclater and Mckenzie 1973; Parsons and Sclater 1977). We applied these models assuming symmetry across spreading centers (Fairhead and Maus 2003). Where complex features exist, like in hotspots or ocean plateaus (e.g., guyots, seamounts), we superimposed them on the age-depth curves as positive features, and their past depths were calculated accordingly (Markwick and Valdes 2004).

Present deep-ocean topographic information was obtained from the datasets of the Land Processes Distributed Active Archive Center, at the United States Geological Survey (available at http://lpdaac.usgs.gov); this information combines direct drilling with satellite and sonar data. Maps were generated with Generic Mapping Tools (Wessel and Smith 1998).

Sea level fluctuations were incorporated in the analysis by adding the effects of the lowest sea level stand since 50 Ma. The most recent studies estimate a minimum of 150 m below the present level in the 10^6-year scale (Miller et al. 2005). Given the uncertainty of this time scale and the possibility that the migration event could have occurred in an extreme situation, it seems reasonable to assume a minimum 150 m sea level regression, which was added to our reconstructions. Local effects such as particular subsidence of coastal areas are generally limited in space and therefore should not affect our general results substantially.

Any attempt to reconstruct paleobathymetry should consider increased subsidence rates of oceanic crust due to sediment loading and reduced depth due to sediment thickness. Other studies predict relatively thin sediment layers (less than 200 m) for oceanic crust younger than 90 Ma, which are the majority reconstructed here (Brown et al. 2006); these effects would not modify our main results, as they are probably of one order of magnitude smaller than the tectonic effects modeled. However, sediment layers could be thicker for older crust. A more detailed analysis, accounting for latitudinal variation in sediment thickness and integrating direct drilling data, would be a more complete approach to correct for sedimentation effects, but we do not attempt such precise reconstructions here. Another significant source of error is the vertical tectonic movements in fracture zones and aseismic ridges (Bonatti 1978; Barker 1983; Gasperini et al. 2001), in which subsidence rates are faster than predicted by the depth vs. age curves. Given that our paleobathymetric reconstructions are based only on paleobasement depths, and considering that in fracture zones in the South Atlantic subsidence is generally faster than predicted by the depth vs. age models used in our analyses (e.g., Bonatti 1978; Barker 1983; Gasperini et al. 2001), both effects (i.e., sediment accumulation and vertical tectonic movements) would reduce depth. Therefore, the reconstructions presented here should be considered a "maximum depth", which is a conservative approach to estimate paleobathymetry, and any exposed land resulting from such reconstructions could be potentially larger due to these unaccounted factors.

3.3 Results

Our reconstructions agree with previous studies in that there was no complete land connection between Africa and South America after 50 Ma (Sclater et al. 1977; Nürnberg and Müller 1991; Ford and Golonka 2003; Scotese 2004; Eagles 2007). However, they suggest the existence of considerable extensions of dry land in the South Atlantic during part of the Tertiary, especially before 40 Ma (Fig. 3.1). At 50 Ma, the shortest distance between Africa and South America is around 1000 km in a straight line (from present day Sierra Leone to Paraíba state, in Brazil); that is probably the minimum distance that intercontinental migrants would need to cover. The ocean is wider further south, but several islands of considerable size (more than 200 km in length) persisted along the present-day submerged Rio Grande Rise and Walvis Ridge. Between 20 and 30° S, at 50 Ma, a long series of close islands stretched from the African shore, and at least one large island (around 500 km in length) was formed by the emergent top of the Rio Grande Rise. The set of islands and shallow waters between 20 and 30° S is interrupted west of the Rio Grande Rise by the Pelotas Basin, a wide area where deeper waters already existed. At 40 Ma, our reconstruction exhibits some disruptions of the islands present at 50 Ma, but the general situation remained the same, with a combination of islands and shallow terrain (less than 1000 m) forming a long strip in the South Atlantic. Another noticeable feature at 40 and 50 Ma is the long set of islands (at least 800 km long) stretching from the Brazilian coast at 20° S (at the present day Martin Vaz Archipelago; Fig. 3.1).

Our data suggest that most of the islands that existed before 40 Ma did not persist after 30 Ma (Fig. 3.1). Although terrain shallower than 1000 m probably existed in the South Atlantic between 20 and 30° S until 20 Ma, by this time the number of islands is dramatically reduced and only small areas (less than 200 km in length) of Rio Grande Rise and Walvis Ridge were emergent. At 20 Ma these islands are virtually absent and the closest distance between Africa and South America is around 2000 km. At this period, the chances of a transcontinental migration for terrestrial animals seem much less probable than before 40 Ma.

3.4 Discussion

Our reconstructions suggest the existence of a series of islands and shallow terrain in the South Atlantic during the mid-Cenozoic, particularly between 40 and 50 Ma. These paleogeographic features, which are underwater today, might have reduced

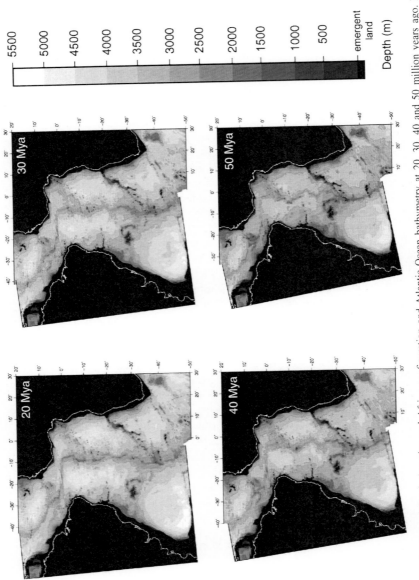

Fig. 3.1 Reconstruction of South America and Africa configuration and Atlantic Ocean bathymetry at 20, 30, 40 and 50 million years ago. Present day coastlines are represented by a white continuous line. At the equator, 10° are approximately 1,100 km

considerably the distance of a possible migration of primates and caviomorph rodents from Africa to South America. It is unlikely that an uninterrupted land bridge between the two continents existed after 80 Ma (Scotese 2004), but our data suggest the existence of large islands in the South Atlantic up to 40 Ma. These islands were probably present since the separation of Africa and South America, as they also appear in late Maastrichtian (70 Ma – Markwick and Valdes 2004) and late Paleocene reconstructions (55 Ma – Lawver and Gahagan 2003). Direct drilling data also corroborate their existence, as red algae remains (which need light to grow), shallow water animals, and rocks formed under aerial exposure were found on samples from the Rio Grande Rise. The youngest sample, dated from the late Oligocene, was drilled in a spot distant from the top of the Rise, at present day 1600 m depth, and contains rocks typically formed in shallow water. This means the crest of the Rio Grande Rise could have been as much as 600 m above sea level (Deep Sea Drilling Project Leg 72; Barker 1983; Parrish 1993). Evidence of subsided islands also exists east of the Mid-Atlantic Ridge. Middle Eocene volcanic rocks, probably extruded above sea level, were drilled around 1000 km away from Africa on the western Walvis Ridge (Ocean Drilling Project Leg 208; Parrish 1993), showing that at least part of its crest was exposed by 40 Ma. The Vema Transverse Ridge, which offsets the Mid-Atlantic Ridge by 320 km and is presently 600 m below sea level, has been found to be capped by carbonate platforms (reef limestone) that formed around 3–4 Ma (Kastens et al. 1998). The size of these subaerially exposed features during the Eocene has not been clearly determined yet, but our data suggest that islands formed on top of Rio Grande Rise could have been as long as 500 km at 50 Ma (Fig. 3.1).

Some studies interpret the Rio Grande Rise and the Walvis Ridge as a part of a hotspot track generated during the late Cretaceous and the Paleogene, which was initially focused below the Paraná-Etendeka large igneous province (around 135 Ma), and is now below Tristan da Cunha and the Gough islands (O'Connor and Duncan 1990; Schettino and Scotese 2005). Our reconstructions do not account for this anomalous lithospheric structure, which probably has a different subsidence history compared to the surrounding seafloor (Barker 1983; Eagles 2007). A more accurate approach should use more complicated models, considering mantle plume temperature, magma supply rate and lithosphere loading by the extra volcanics of the hotspot. Similarly, fracture zones seem to have particular subsidence histories (Bonatti 1978; Gasperini et al. 2001). Nonetheless, regarding the Walvis Ridge and Rio Grande Rise, it seems reasonable to suppose that these features had a faster subsidence than predicted by the age vs. depth curves used here (Barker 1983). Given that tectonic processes in these anomalous lithospheric features could be much more complex and do not allow direct modeling at this moment (Fairhead and Wilson 2005), the scenario exhibited by our reconstructions should be considered a "maximum depth" estimate for the South Atlantic at 20, 30, 40 and 50 Ma; in other words, this should be viewed as the worst possible scenario for a transoceanic animal migration. Despite not accounting for the particularities stated above, our data have important implications for potential routes for mammal dispersal between Africa and South America.

Based only on our data, one can effectively discard the land bridge hypothesis and make a stronger case for the floating island model. Even in the earliest reconstruction, at 50 Ma, there is no complete connection between Africa and South America, implying that if land animals migrated from one continent to the other after this period, some kind of oceanic dispersal must have occurred. Additionally, the absence of mammals originally from South America in the African fossil record indicates a selective dispersal route, compatible with a hypothesis in which oceanic currents played a prominent role. Other studies confirm that paleocurrents and paleowinds favored a westward crossing of the Atlantic from Africa. Since the formation of deep water connection between South and Central Atlantic, currents have flowed from the southern tip of Africa, turned westwards near the equator across the Atlantic, and then southwards at the South American coastline, generating a wide counterclockwise pattern of water circulation (Haq 1981; Parrish and Curtis 1982; Parrish 1993). Models based on present-day wind speeds and in paleodistances similar to the ones presented here (around 1000 km between Africa and South America at 50 Ma, 1500 km at 40 Ma, and 2000 km at 30 Ma) predict that an eventual floating island would take 5–15 days to cross the Atlantic Ocean from Africa, making it a feasible mode of dispersal for small or medium-sized mammals (Houle 1998 and 1999). Although there was no complete land connection across the Atlantic, our data suggest that the presence of islands close to each other on the Martin Vaz hotspot track (Fig. 3.1, latitude 20°S) could have facilitated a possible crossing from Africa. Using our most migration-friendly reconstruction, at 50 Ma, these islands could have formed a peninsula stretching at least 500 km into the Atlantic, potentially reducing migration distance. However, as we did not model specific hotspot effects in the reconstructions, it is not certain how far these islands stretched, given that hotspot activity could vary in time (Barker 1983; Fairhead and Wilson 2005).

Our data do not provide enough resolution to decide between the island hopping and the floating island modes of dispersal, as both have arguments for and against. It is unlikely that a migration across the entire Atlantic would be feasible by hopping from one island to another, as their distribution in time and space does not seem to form a continuous emergent feature. However, a scenario in which part of the way was covered rapidly in a floating island and part by slow island hopping (e.g., at the Martin Vaz hotspot track) cannot be discarded. Although one of the best candidates for a source of floating islands, the Congo River, failed to flow to the Atlantic Ocean before 30 Ma (Stankiewicz and de Wit 2006), the floating island model, or a combination with the island hopping model, are the ones that best fit the paleogeographic data.

Our reconstructions also could shed light on the timing of the possible migration event of caviomorphs and platyrrhines. The mean distance to be traveled increased with time since the split between Africa and South America (Scotese 2004; Eagles 2007), and the same reasoning applies to the thermal subsidence of oceanic lithosphere, as ocean depths increased with time. Our data suggest that paleogeographic conditions remained most favorable for a transatlantic migration until 40 Ma. This is at least 10 Ma earlier than the oldest fossil occurrences for both Platyrrhini and Caviomorpha in South America. Considerable discussion exists on their oldest fossil relatives from the Old World (Marivaux et al. 2002; Fleagle and Gilbert 2006). Regarding primates, the first undisputed anthropoids are from the early Oligocene of Fayum, Egypt (around 30 Ma; Simons and Rasmussen 1994; Seiffert 2006), suggesting that migration to South America could only happen after this period. Nevertheless, there is considerable evidence supporting an earlier anthropoid origin. Some authors defend that earlier fossils from Africa and Asia do have anthropoid features, pushing the origin of the group back at least to the middle Eocene (ca. 45 Ma; Beard et al. 1996; Kay et al. 1997; Beard 2006), and even to the Late Paleocene (ca. 55 Ma; Godinot 1994). The situation is similar with respect to rodents, given that the earliest undisputed hystricognaths are from the Oligocene of Pakistan and Egypt (ca. 35 Ma), with the possibility of a Middle Eocene origin for the group (Bryant and McKenna 1995; Marivaux et al. 2002).

Molecular studies also suggest earlier origins for both platyrrhines and caviomorphs. Coalescence analyses calibrated by fossils indicate that the Old and New World monkeys lineages split around 40 Ma (Goodman et al. 1998; Schrago and Russo 2003); a study of nuclear genes suggested that caviomorphs separated from their African counterparts (phiomorphs) sometime between 45 and 35 Ma (Poux et al. 2006). Moreover, recently described caviomorph fossils from the Santa Rosa Formation in Peru may be Eocene in age and exhibit considerable morphological variation (Frailey and Campbell 2004). This piece of evidence demonstrates that caviomorphs were already a very diversified group at this period, a finding corroborated by molecular data (Mouchaty et al. 2001; Schrago and Russo 2003; Poux et al. 2006). Coupled with the presence of hystricognaths in Asia, Africa and South America in the earliest Oligocene (ca. 35 Ma), these findings strongly point to an Eocene origin for caviomorphs (Marivaux et al. 2002). Overall, the available evidence suggests that platyrrhines and caviomorph rodents may have arisen well before than what the current fossil record indicates. If that is correct, the timing of their proposed crossing from Africa to South America is in greater agreement with our findings, and could have happened between 40 and 50 Ma, when paleogeographic conditions were most favorable.

Migration scenarios involving North America and Antarctica were also proposed in the past (Wood 1993; Houle 1999), and deserve attention. The oldest anthropoid fossils were excavated in the Old World, and phylogenetic analyses of both hystricognath rodents and anthropoid primates strongly suggest that African and South American forms are derived from a common ancestor (Nedbal et al. 1994; Kay et al. 1997; Flynn and Wyss 1998; Goodman et al. 1998; Takai et al. 2000; Huchon and Douzery 2001; Schrago and Russo 2003; Poux et al. 2006). In this context, the most probable hypothesis is that both groups originated in Africa or Asia and migrated after the Paleocene to the New World. Antarctica separated from Africa at least 130 Ma, and from South America around 30 Ma (Scotese 2004; Lawver and Gahagan 2003). Thus, if land mammals have used this route, they would have needed a transoceanic crossing from Africa or Asia to Antarctica, and a posterior land migration (if before 30 Ma), or another oceanic crossing (if later than 30 Ma) to South America. The case is similar with respect to North America, as we would expect a migration from Asia through the Bering Strait and a posterior Caribbean Sea crossing to South America. Although distances between North and South America were smaller than between Africa and South America during most of the Cenozoic, paleocurrents and paleowinds were more favorable for a migration from Africa (Haq 1981; Parrish and Curtis 1982; Parrish 1993). Nevertheless, the strongest argument against scenarios involving North America and Antarctica is the complete absence of fossils of likely ancestors of platyrrhines and caviomorphs. If the migration route involved North America or Antarctica, one should assume that platyrrhine and caviomorph ancestors did not leave any fossils in these continents, or that they have not yet been found. Considering the abundant record of Paleocene and Eocene mammal fossils in North America, including primates, it seems unlikely to assume that only anthropoids were not preserved, especially if they had to disperse all the way from Bering Strait to Central America. Even in the relatively scarce fossil record of Antarctica, land mammals in the Paleogene are documented, but no primates or rodents (Houle 1999; Briggs 2003). Thus, the oceanic dispersal of African groups to South America sometime between 50 and 30 Ma stands as the most likely explanation to the distribution of fossil and present day caviomorph rodents and platyrrhines.

It is necessary to look carefully to the limits of the reconstructions presented here. We did not consider the effects of sedimentation and the particular tectonic behavior of anomalous areas, such as hotspot tracks. Sedimentation could cause faster subsidence due to sediment loading, or reduced depth by sediment accumulation. Brown, Gaina and Müller (2006) noted that for oceanic crust younger than 90 Ma, deviations in reconstructed bathymetry from the depth vs. age models would not be larger than 200 m. Greater discrepancies, however, are to be expected in crust older than 90 Ma: they could be up to 1000 m shallower than exhibited by our reconstructions. That would increase the number of islands close to continents and potentially favor an eventual migration. Similarly, the particular subsidence rates of the Rio Grande Rise-Walvis Ridge features are probably faster than the surrounding seafloor (Barker 1983). This makes the approach presented here a conservative way to look at the paleogeography of the South Atlantic within the context of primate and rodent migration from Africa to South America.

One interesting feature of our reconstructions is that it could provide a new background for the interpretation of the distributional patterns of other animal groups in addition to caviomorphs and platyrrhines. Some plants, freshwater fishes (cichlids and aplocheiloids), birds (parrots) and lizards (geckos) appear to have followed a post-Gondwanan dispersal pattern across the South Atlantic (Briggs 2003; Renner 2004; de Queiroz 2005). That could be explained by the favorable paleocurrents and island-punctuated scenario presented here, especially between 50 and 40 Ma. Perhaps the long lasting puzzle of the origin of South American monkeys and caviomorphs is not as unique as once thought.

3.5 Summary

The sudden appearance of platyrrhine primates and caviomorph rodents in the late Oligocene fossil record comprises an old puzzle for biologists and paleontologists, since South America was an isolated continent for most of the Tertiary.

The well-established phylogenetic relationships between these groups and African forms force acceptance of some kind of migration across the Atlantic Ocean. Many hypotheses have been put forward to account for this crossing, including floating island rafting, volcanic stepping-stone islands, and land bridges. Here we present paleogeographic reconstructions of the South Atlantic in order to re-evaluate the scenario in which such migration took place, modeling both continental drift and sea-floor thermal subsidence movements, while accounting for sea level changes. We analyse these data by bringing together evidence from the fossil record, estimated dates of phylogenetic divergence based on molecular data, geophysical modeling and paleocurrent estimates. Our reconstructions confirmed previous findings that reject complete land bridges between Africa and South America during the Cenozoic, but suggested the presence of islands of considerable size (>200 km in length) in the South Atlantic. Other paleogeographic features that could eventually reduce migration distance are discussed. Our data indicated that the most favorable period for a possible migration was between 40 and 50 million years ago. This evidence, coupled with favorable westward paleocurrents and paleowinds from Africa could have facilitated a transatlantic crossing via floating islands. Other organisms that seem to share the distributional patterns of platyrrhines and caviomorphs could also have dispersed between Africa and South America in this scenario.

References

Barker, P. F. 1983. Tectonic evolution and subsidence history of the Rio Grande Rise. In P. F. Barker (ed.), *Initial Reports of Deep Sea Drilling Project 72* (pp. 953–976). Washington: U.S. Government Printing Office.

Beard, K. C. 2006. Mammalian biogeography and anthropoid origins. In S. M. Lehman and J. G. Fleagle, (eds.), *Primate Biogeography* (pp. 439–468). New York: Springer.

Beard, K. C., Tong, Y. S., Dawson, M. R., Wang, J. W. and Huang, X. S. 1996. Earliest complete dentition of an anthropoid primate from the late middle Eocene of Shanxi Province, China. Science 272: 82–85.

Bonatti, E. 1978. Vertical tectonism in oceanic fracture zones. Earth Planet. Sci. Lett. 37: 369–79.

Briggs, J. C. 2003. Fishes and birds: Gondwana life rafts reconsidered. Syst. Biol. 52(4): 548–553.

Bryant, J. D. and McKenna, M. C. 1995. Cranial anatomy and phylogenetic position of *Tsaganomys altaicus* (Mammalia:Rodentia) from the Hsanda Gol Formation (Oligocene), Mongolia. Am. Mus. Novit. 3156: 1–42.

Brown, B., Gaina, C. and Müller, R. D. 2006. Circum-Antarctic palaeobathymetry: Illustrated examples from Cenozoic to recent times. Palaeogeogr., Palaeoclimatol., Palaeoecol. 231: 158–168.

Ciochon, R. L. and Chiarelli, A. B. 1980. Paleobiogeographic perspectives on the origin of the Platyrrhini. In R. L. Ciochon and A. B. Chiarelli (eds.), *Evolutionary Biology of New World Monkeys and Continental Drift* (pp. 459–493). New York: Plenum Press.

De Queiroz, A. 2005. The resurrection of oceanic dispersal in historical biogeography. Trends Ecol. Evol. 20(2): 68–73.

Eagles, G. 2007. New angles on South Atlantic opening. Geophys. J. Intl. 166: 353–361.

Fairhead, J. D. and Wilson, B. M. 2005. Plate tectonic processes in the South Atlantic Ocean: do we need deep mantle plumes? In G.R. Foulger, J.H. Natland, D.C. Presnall, and D. L. Anderson (eds.), *Plates, plumes and paradigms (Geological Society of America Special Paper 388)* (pp.537–553). Boulder: Geological Society of America.

Fairhead, J. D. and Maus, S. 2003. CHAMP satellite and terrestrial magnetic data help define the tectonic model for South America and resolve the lingering problem of the pre-break-up fit of the South Atlantic Ocean. Leading Edge 22(8): 779–783.

Fleagle, J. G. and Gilbert, C. P. 2006. The biogeography of primate evolution: the role of plate tectonics, climate and chance. In S. M. Lehman and J. G. Fleagle (eds.), *Primate Biogeography, Primate Biogeography* (pp. 375–418). New York: Springer.

Flynn, J. J. and Wyss, A. R. 1998. Recent advances in South American mammalian paleontology. Trends Ecol. Evol. 13: 449–454.

Frailey, C. D. and Campbell, K. E. 2004. The Rodents of the Santa Rosa Local Fauna. In K. E. Campbell Jr. (ed.), *The Paleogene mammalian fauna of Santa Rosa, Amazonian Peru (Natural History Museum of Los Angeles County, Science Series 40)* (pp. 71–130). Los Angeles: Natural History Museum of Los Angeles County.

Ford, D. and Golonka, J. 2003. Phanerozoic paleogeography, paleoenvironment and lithofacies maps of the circum-Atlantic margins. Mar. Petroleum Geol. 20: 249–285.

Gasperini, L., Bernoulli, D., Bonatti, E., Borsetti, A. M., Ligi, M., Sartori, N. A. R. and von Salis, K. 2001. Lower Cretaceous to Eocene sedimentary transverse ridge at the Romanche Fracture Zone and the opening of the equatorial Atlantic. Mar. Geol. 176: 101–119.

Gazin, C. L. 1958. A review of the Middle and Upper Eocene primates of North America. Smithsonian Misc. Collect.136: 1–112.

George, W. B. and Lavocat, R. 1993. The Africa-South America connection. New York and Oxford: Clarendon Press and Oxford University Press.

Godinot, M. 1994. Early North African primates and their significance for the origin of Simiiformes (=Anthropoidea). In J. G. Fleagle and R. F. Kay (eds.), *Anthropoid Origins* (pp.235–295). Plenum Press: New York.

Goodman, M., Porter, C. A., Czelusniak, J., Page, S. L., Schneider, H., Shoshani, J., Gunnell, G. and Groves, C. P. 1998. Toward a phylogenetic classification of primates based on DNA evidence complemented by fossil evidence. Mol. Phylogenet. Evol. 9: 585–598.

Goldblatt, P. 1993. Biological Relationships between Africa and South America. New Haven: Yale University Press.

Haq, B. U. 1981. Paleogene paleoceanography: Early Cenozoic oceans revisited. Oceanol. Acta 4(suppl.): 71–82.

Hoffstetter, M. R. 1969. Un primate de l'Oligocène inférieur sudamericain: *Branisella boliviana* gen. et sp. nov. C. R. Acad. Sci. Paris 269: 434–437.

Hoffstetter, M. R. and Lavocat, R. 1970. Decouverie dans le Deseadien de Bolivie de genres pentalophodenies appuyant les affinities africaines des Rongeurs Caviornorphes. C. R. Acad. Sci. Paris D 271: 172–175.

Houle, A. 1998. Floating islands: a mode of long-distance dispersal for small and medium-sized terrestrial vertebrates. Divers. Distrib. 4: 201–216.

Houle, A. 1999. The origin of platyrrhines: An evaluation of the Antarctic scenario and the floating island model. Am. J. Phys. Anthropol. 109: 541–559.

Huchon, D. and Douzery, E. J. P. 2001. From the old world to the new world: a molecular chronicle of the phylogeny and biogeograpy of hystricognath rodents. Mol. Phylogenet. Evol. 20: 238–251.

Kastens, K., Bonatti, E., Caress, D., Carrara, G., Dauteuil, O., Frueh-Green, G., Tartarotti, P. and Ligi, M. 1998. The Vema Transverse Ridge (Central Atlantic). Mar. Geophys. Res. 20: 533–556.

Kay, R. F., Ross, C. and Williams, B. A. 1997. Anthropoid origins. Science 275: 797–804.

Kearey, P. and Vine, F. J. 1996. Global Tectonics. Oxford: Blackwell Science.

Lavocat, R. 1980. The implications of rodent paleontology and biogeography to the geographical sources and origin of platyrrhine primates. In R. L. Ciochon and A. B. Chiarelli, (eds.), *Evolutionary Biology of New World Monkeys and Continental Drift* (pp. 93–103). New York: Plenum Press.

Lawver, L. A. and Gahagan, L. M. 2003. Evolution of Cenozoic seaways in the circum-Antarctic region. Palaeogeogr., Palaeoclimatol., Palaeoecol. 198: 11–37.

Marivaux, L., Welcomme, J. L., Vianey-Liaud, M. and Jaeger, J. J. 2002. The role of Asia in the origin and diversification of hystricognathous rodents. Zool. Scripta, 31: 225–239.

Markwick, P. J. and Valdes, P. J. 2004. Palaeo-digital elevation models for use as boundary conditions in coupled ocean–atmosphere GCM experiments: a Maastrichtian (late Cretaceous) example. Palaeogeogr., Palaeoclimatol., Palaeoecol. 213: 37–63.

Miller, K. G., Kominz, M. A., Browning, M. A., Wright, J. D., Mountain, G. S., Katz, M. E., Sugarman, P. J., Cramer, B. S., Christie-Blick, N. and Pekar, S. F. 2005. The Phanerozoic record of global sea-level change. Science 310: 1293–1298.

Mouchaty, S. K., Catzeflis, F., Janke, A. and Arnason, U. 2001. Molecular evidence of an African phiomorpha-South American caviomorpha clade and support for hystricognathi based on the complete mitochondrial genome of the cane rat (*Thryonomys swinderianus*). Mol. Phylogenet. Evol. 18: 127–135.

Müller, R. D., Roest, W. R., Royer, J. Y., Gahagan, L. M. and Sclater, J. G. 1997. Digital isochrons of the world's ocean floor. J. Geophys. Res. 102: 3211–3214.

Nedbal, M. A., Allard, M. W. and Honeycutt, R. L. 1994. Molecular systematics of hystricognath rodents: evidence from the mitochondrial 12S rRNA gene. Mol. Phylogenet. Evol. 3: 206–220.

Nürnberg, D. and Müller, R. D. 1991. The tectonic evolution of the South Atlantic from Late Jurassic to Present. Tectonophys. 191: 27–53.

O'Connor, J. M., and Duncan, R. A. 1990. Evolution of the Walvis Ridge –Rio Grande Rise hot spot system: Implications for African and South American plate motions over plumes. J. Geophys. Res. 95: 17475–17502.

Parrish, J. T. 1993. The palaeogeography of the opening South Atlantic. In W. George and R. Lavocat (eds.), The Africa-South America Connection (pp. 8–27). Oxford: Clarendon Press.

Parrish, J. T. and Curtis, R. L. 1982. Atmospheric circulation, upwelling and organic-rich rocks in the Mesozoic and Cenozoic. Palaeogeogr., Palaeoclimatol., Palaeoecol. 40: 31–36.

Parsons, B. and McKenzie, D. P., 1978. Mantle convection and the thermal structure of the plates. J. of Geophys. Res. 83: 4485–4496.

Parsons, B. and Sclater, J. G. 1977. Analysis of variation of ocean-floor bathymetry and heat-flow with age. J. Geophys. Res. 82: 803–827.

Pitman, W. C. I., Cande, S., LaBrecque, J. and Pindell, J. 1993. Fragmentation of Gondwana: the separation of Africa from South America. In P. Goldblatt (ed.), Biological Relationships Between Africa and South America (pp. 15–34). New Haven: Yale University Press.

Poux, C., Chevret, P., Huchon, D., de Jong, W. W. and Douzery, E. J. P. 2006. Arrival and diversification of caviomorph rodents and platyrrhine primates in South America. Syst. Biol. 55: 228–244.

Renner, S. 2004. Plant dispersal across the Tropical Atlantic by wind and sea currents. Int. J. Plant Sci. 165(4 Suppl.): S23–S33.

Schettino, A., and Scotese, C. R. 2005. Apparent polar wander paths for the major continents (200 Ma – present day): A paleomagnetic reference frame for global plate tectonic reconstructions. Geophys. J. Intl 163(2): 727–759.

Schrago, C. G. and Russo, C. A. M. 2003. Timing the origin of New World monkeys. Mol. Phylogenet. Evol. 20: 1620–1625.

Sclater, J. G., Hellinger, S. and Tapscott, C. 1977. Paleobathymetry of Atlantic Ocean from Jurassic to Present. J. Geol. 85: 509–552.

Sclater, J. G. and Mckenzie, D. P. 1973. Paleobathymetry of South-Atlantic. Geol. Soc. America Bull. 84: 3203–3216.

Scotese, C. R. 2004. A continental drift flipbook. J. Geol. 112: 729–741.

Simons, E. L.1961. The dentition of *Ourayia*: its bearing on relationships of omomyid primates. Postilla 54: 1–20.

Seiffert, E. R. 2006. Revised age estimates for the later Paleogene mammal faunas of Egypt and Oman. Proc. Natl. Acad. Sci. 103: 5000–5005.

Simons, E. L. and Rasmussen, D. T. 1994. A whole new world of ancestors: Eocene anthropoideans from Africa. Evol. Anthropol. 3: 128–139.

Simpson, G. G. 1980. Splendid isolation: the curious history of South American mammals. New Haven: Yale University Press.

Stankiewicz, J. and de Wit, M. J. 2006. A proposed drainage evolution model for Central Africa—Did the Congo flow east? J. Afr. Earth Sci. 44: 75–84.

Takai, M., Anaya, F., Shigehara, N. and Setoguchi, T. 2000. New fossil materials of the earliest new world monkey, *Branisella boliviana*, and the problem of platyrrhine origins. Am. J. Phys. Anthropol. 111: 263–281.

Van Couvering, J. A. and Harris, J. A. 1991. Late Eocene age of the Fayum mammal fauna. J. H. Evol. 21: 241–260.

Wessel, P. and Smith, W. H. F. 1998. New, Improved Version of Generic Mapping Tools Released. EOS Trans. AGU 79: 579.

Wood, A. E. 1980. The origin of the caviomorph rodents from a source in Middle America: a clue to the area of origin of the platyrrhine primates. In R. L. Ciochon and A. B. Chiarelli (eds.), Evolutionary Biology of New World Monkeys and Continental Drift (pp. 79–92). New York: Plenum Press.

Wood, A. E. 1993. The history of the problem. In W. George and R. Lavocat (eds.), The Africa-South America connection (pp. 1–7). Oxford: Clarendon Press.

Wyss, A. R., Flynn, J. J., Norell, M. A., Swisher, C. C., Charrier, R., Novacek, M. J. and Mckenna, M. C. 1993. South America earliest rodent and recognition of a new interval of mammalian evolution. Nature 365: 434–437.

Chapter 4
Platyrrhine Ecophylogenetics in Space and Time

Alfred L. Rosenberger, Marcelo F. Tejedor, Siobhán B. Cooke,
and Stephen Pekar

4.1 Introduction

We are far from developing an informed synthesis regarding the evolution of New
World Monkeys – probably decades away. For even with the important strides made
over the past 30–40 years regarding platyrrhine ecology and behavior, there are large
gaps in our knowledge of the evolutionary and historical context. The scarceness
of fossils is but one factor. Equally critical is our incomplete knowledge of large-
scale changes to the continent of South America (SAM), pertinent to the evolution
of its fauna. An objective of this paper is to review some of this information as
a basis for interpreting the platyrrhines from an ecophylogenetic point of view in
space and time. Our goal is to integrate information on living and extinct forms in
order to identify community or regional patterns of platyrrhine evolution, rather than
examining the moderns and fossils as distinct entities or evolutionary problems. In
keeping with the South American emphasis of this volume, we do not consider the
primate fauna of the Middle American mainland but have elected to examine the
Caribbean subfossil monkeys for reasons that will become clear below.

We suggest that the casual way of thinking about New World Monkeys (NWM)
as a monolithic radiation inhabiting a rainforest wonderland – South America –
is a model that needs to be changed. The continent is about 2.5 times the size
of today's Amazonian rainforest in area, it contains diverse landscapes and habitats, and the Amazonian region changed vastly during the Cenozoic (e.g., Bigarella
and Ferreira 1985). At present, more of the continent is grassland than rainforest
(Fig. 4.1a), and the grasslands have been flourishing for 20–30 million years (*see*
below). The first primates to arrive did not encounter the Amazonia we know, for
it may have begun to take on its present character only about 15 Ma (Campbell
et al. 2006). Thus, even though the NWM have a monophyletic, unitary origin, their

A.L. Rosenberger (✉)
Brooklyn College, The City University of New York, Department of Anthropology and Archaeology; The Graduate Center, The City University of New York; The American Museum of Natural History, Department of Mammalogy; New York Consortium in Evolutionary Primatology
(NYCEP), New York, NY, USA
e-mail: alfredr@brooklyn.cuny.edu

P.A. Garber et al. (eds.), *South American Primates,* Developments in Primatology:
Progress and Prospects, DOI 10.1007/978-0-387-78705-3_4,
© Springer Science+Business Media, LLC 2009

Fig. 4.1a Grasslands of South America. *See* Color Insert.

evolution on the large landmass of SAM probably did not occur as a stately *in situ* unfolding.

Another adjustment in our thinking is bound up with the contrasts we tend to draw between Old and New World primates. We tend to see the platyrrhines as living in the largest, richest rain forest habitat in the world where the ecological dichotomies so evident among Old World anthropoids and strepsirhines – terrestriality and arboreality, diurnality and nocturnality, expansion of arid and humid habitats – seem like unnecessary evolutionary strategies. But it may be too facile to assume that a one-dimensional model of NWM evolving in an Edenic

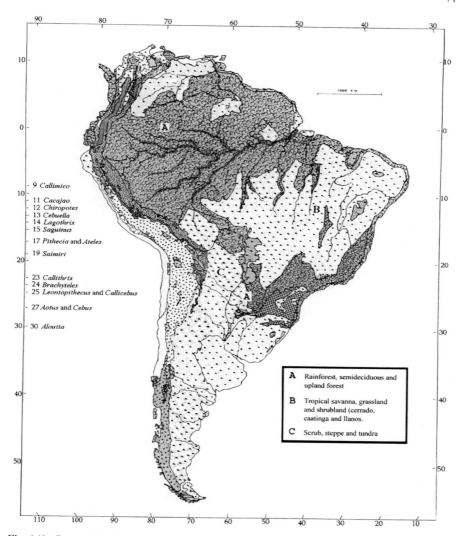

Fig. 4.1b Current distribution of tropical and subtropical forests, in South America (modified from Hershkovitz 1977). Approximate lowest latitudes of the distributions of living genera are shown on the left, compiled from Hershkovitz (1977), Kinzey (1997) and the BDGEOPRIM project, University Federal do Minas Gerais

arboreal milieu can aptly describe their full evolutionary history. To emphasize this point, consider the potential climatic implications of the important series of fossil platyrrhines from the Miocene of Patagonia. One locality in Southern Argentina, Killik Aike Norte, at about 51°S is situated more than 15° nearer to the South Pole than the southern edge of Africa. The Killik Aike Norte primates lived at the lowest latitude of any primates ever known.

The habitats occupied by today's platyrrhines are also less uniform than aerial views of Amazonia's monotonous canopy cover suggests. They are spread across a variety of biomes in Central and SAM (Hershkovitz 1977; *see* Fig. 4.1b). Calling NWM "neotropical" is something of a misnomer for while the geography of many NWM is bounded by the Tropics of Cancer and Capricorn, not all of this terrain consists of hot humid forest. Modern monkeys flourish in their own fashion as inhabitants of the extra-Amazonian, semi-deciduous domain of the Atlantic Coastal Forest (Mata Atlantica) of southeastern Brazil, for example. They also occur with success and regularity, but absent taxonomic diversity, in drier, sparsely treed, savannah-like habitats situated between the distant margins of Amazonia and the Mata Atlantica. These are tropical savannas, grasslands and shrubland habitats that comprise a large fraction of the landscape in Brazil and Venezuela. These species-poor primate communities in areas such as the Cerrado, Caatinga and Llanos regions (Fig. 4.1a), which are just as mature as rich, wet, tall-forest communities, also have a special importance for interpreting the NWM fossil record for, historically, the southern fossil NWM were anything but geographically tropical.

4.1.1 An Ecophylegenetic Approach

Why ecophylogenetic? As the foundational ideas of NWM systematics began to gel around phylogenetics (*see* Rosenberger 2002), evolutionary models have come to fuse phylogenetic and adaptational thinking (Rosenberger 1980, 1992). The result is a deeper appreciation of coherence, that platyrrhines diversified as interlinked arrays of taxa sharing unique patterns of behavior and ecology as much as they share clado-genic branches. There is a patterned structure to the platyrrhine ecological radiation that is its phylogeny. As a consequence, there are today four major guilds which we feel represent monophyletic groups, each occupying different adaptive zones. And, within each of these monophyletic groups, sublineages or genera have differentiated by evolving adaptive variations or alternative avenues of resource exploitation. To summarize, the four groups in general terms:

Callitrichines – Small-bodied predaceous frugivores specialized for locomotion and posture below the canopy of giant, large-crowned trees as well as within the fine terminal branches. They mostly exploit arthropods and insects as protein sources unless specially adapted via body size (and positional or dental specialization) for seizing large prey items (*Leontopithecus*, some *Saguinus*) or gums (especially *Callithrix, Cebuella*). Callitrichines are mostly Amazonian, but genera with derived character complexes related to specialized feeding patterns, *Leontopithecus* and *Callithrix*, are able to live successfully outside of Amazonia.

Cebines – Small-and medium-sized predaceous frugivores (*Saimiri*) and *bona fide* non-herbivorous, but also variously specialized omnivores (*Cebus*) who use the canopy as a resource base for exploiting arthropods and insects, which frequently may be embedded. *Saimiri* is decidedly Amazonian. *Cebus*, one of the order's champion generalists, is hardly limited geographically. The craniodentally specialized

C. apella group, adapted for destructive foraging, is highly successful outside of Amazonia even in the most sparse forest types.

Pitheciines – Small- and medium-sized hard-fruit and unripe-fruit eaters who live in the canopy and rely on a variety of protein sources, such as seeds (*Pithecia, Chiropotes, Cacajao*), insects (*Aotus*) and insects and leaves (*Callicebus*). Specializations on food types cluster pitheciin seed-predators such as *Chiropotes* and *Cacajao* in habitats near the Rio Amazonas, while the capacity to use low-value leaves and immature fruit allows *Callicebus* to be the most successful SAM pitheciine living at great distance from the Amazon basin, in the Mata Atlantica of Brazil and in the Chaco of western Paraguay. [We acknowledge that many researchers do not include *Aotus* in this group, but remain unconvinced that the molecular evidence outweighs a strong suite of derived morphological features shared exclusively by *Callicebus* and *Aotus* (Rosenberger and Tejedor, in press).]

Atelines – Large-bodied frugivores and semi-folivores who rely on large trees to forage and exploit leaves for protein (*Alouatta, Lagothrix, Brachyteles*), or are able to subsist almost entirely on fruit (*Ateles*). The capacity to exploit leaves for extensive periods allows two convergently derived members, *Brachyteles* and *Alouatta*, to live in less diverse, less fertile, semi-deciduous forests in the Mata Atlantica. *Alouatta*, the most folivorous platyrrhine, like the omnivorous *Cebus*, is quite successful in living under riverine conditions in sparsely treed, woodland and savannah habitats far from humid rain forests.

An obvious rationale for adopting an ecophylogenetic approach relates to our focus on the fossil record, as paleontology is naturally given to the study of interrelationships. Unfortunately, only a few scattered locations on the continent of SAM and in the Caribbean have produced platyrrhine fossils thus far; nothing is known from Middle America. What these fossils offer to better understand the nature and histories of the living species as ecological communities is a question we address. And the reverse: What do the living communities teach us about the paleobiology of fossil taxa when they co-occur? Finally, we ask if the Long Lineage Hypothesis, namely the argument that much of NWM evolution has been shaped by an impressive number of genera and lineages that have deep phylogenetic origins and little changed adaptive histories (*see* Rosenberger 1979; Delson and Rosenberger 1984; Setoguchi and Rosenberger 1987) is a robust explanation of the platyrrhines?

Although our interests in this paper can be satisfactorily addressed by examining genera and lineages, we voice a note of caution regarding an underappreciated challenge to community thinking, the poor state of alpha-level platyrrhine systematics. Without knowing how many *real biological taxa* – species – exist in any one place or region, how are we to articulate scientific questions about them? As readers of this chapter will know, while the study of NWM, historically, has rarely given energy to problems of alpha taxonomy, there has been an unprecedented proliferation of NWM species names put into use during the past 10–15 years. Comparing taxonomic works such as Napier (1976) and Groves (2001) presents a profound, de-constructivist image of this radical phenomenon. Almost always, these recent judgments have been made in the absence of original empirical research, new data or strong justification. Rather, this trend appears to stem from a shift in philosophies, thinking about what a species is and how it can be studied, and about how upward counts of biodiversity can be used, in theory, to promote conservation. We lament this diversion of purpose in the use of taxonomic terms for it can destabilize systematics, the core of all forms of biological research.

4.2 Before Platyrrhines: Paleoecology of SAM

Euprimates were achieving a pinnacle of success during the Eocene on the northern continents as adapiforms and tarsiiforms became widespread and diverse (e.g., Simons 1972; Szalay and Delson 1979; Gebo 2002; Gunnell and Rose 2002). The ultimate cause for this is related to tectonic events, but global climate may be a more direct, proximate factor. Primates appear to have been very adept at dispersal during the Eocene, presumably riding a wave of humid forest habitat expansion during a period that has been described as a "greenhouse world." Global temperatures during the Eocene were elevated, making this the ideal time for tropical flora and fauna to spread. Since the Late Paleocene Thermal Maximum (LPTM) and during the Early Eocene Climatic Optimum (EECO), between 50 and 55 Ma (Zachos et al. 2001), wet, botanically diverse forests were even in place thousands of kilometers south of the equator, at latitudes of 47–48°S (Wilf et al. 2003) in Patagonian Argentina. Thus, primates appeared abruptly in North America at the Paleo-Eocene boundary, 55–57 Ma, amidst a succession of moves from Asia (Beard 2002), including the migration of *Teilhardina* from China through Europe and into North America, a trip which may have occurred in as little as 20,000 years (Smith, Rose and Gingerich 2006). For the southern continents, we know less about diversity and biogeography. There is no evidence of their presence during the LPTM and the EECO in SAM. However, in equatorial Africa, by the end of the Eocene anthropoids were prolific in the Fayum, Egypt. There are no near-equator localities with mammals known in the Eocene of SAM but there are many concentrated in the Patagonian region. None have yielded any primates so far.

The absence of primates from the Patagonian Eocene is an important issue, but the existence of abundant fossils from other orders informs us about the conditions primates would eventually meet on the continent. Before they arrived, more than 20 different families representing over 12 orders of mammals were already established, apart from the rodents. These have been called the first phase, or Stratum 1, of the history of mammals in SAM, and include forms that were all indigenous to the continent (Fig. 4.2). The second stratum is characterized by two features, primates and caviomorph rodents, who are generally believed to be co-immigrants coming from the same source continent, and native faunal turnover. This phase brackets the late Eocene and early Miocene, when the archaic Stratum 1 mammals were replaced by their more advanced relatives. Many of the archaics were herbivorous, and they were becoming more adapted to open-country herbivory over time (Patterson and Pascual 1972; Pascual and Ortíz Jaureguizar 1990; Pascual 2006). By the late Oligocene-earliest Miocene, most of the Stratum 1 mammals had disappeared. Few could be found in the localities that have produced primates during the Early

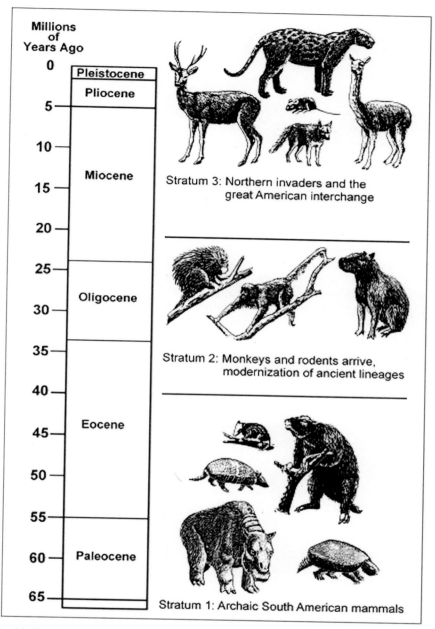

Fig. 4.2 The three major faunal strata in Cenozoic South America (from Flynn and Wyss 1998)

and Middle Miocene, 15–20 Ma. The final phase, Stratum 3, is when North American mammals began their sally into SAM to become an inexorable ecological force, while southern animals also filtered into the north. This is the Great American Biotic Interchange (Stehli and Webb 1985).

Although we assume that primates must have landed in SAM as rain forest-adapted animals, they arrived on a continent which was already becoming dominated by open country herbivorous mammals. Grasslands began to emerge during the Oligocene in the far south in Patagonia. While this may roughly coincide with the first appearance of primates in the fossil record, the oldest primate bearing locality is much nearer the equator, in Bolivia. Nevertheless, multiple lines of evidence (MacFadden 2000) show the shift from browsing to grazing occurred over a vast area, reaching from Patagonia to Bolivia, so that by Late Oligocene times the major mammalian faunas were adapted to and co-evolving with emergent grasslands. The native notoungulates provide a stunning example of this transition. This hoofed order of mammals was always abundant and taxonomically diverse in the Paleogene record (Cifelli 1985; Bond 1986; Pascual et al. 1996), with brachyodont (low-crowned) and bunodont teeth well suited for an herbivorous diet. But by the Miocene, notoungulates were characterized by high-crowned (hypsodont) and selenodont (shearing) cheek teeth, grazing adaptations. Jacobs et al. (1999) also note the presence of grasses in SAM even during the Paleocene and the dominance of mammalian herbivores in the early Oligocene Tinguiririrican fauna of Chile. Today, grasslands cover about 60% of the continent (Fig. 4.1a).

Still, there are indications of non-grazing niches among Stratum 1 and Stratum 2 mammals as well. Several SAM orders and families evolved bunodont (round, low cusps) molars during the Paleogene, some with specialized morphologies designed for a variety of diets. For example, polydolopines, caroloameghiniids and proto-didelphids were probably frugivorous marsupials with procumbent incisors and very low-crowned molars (see Goin 2003, and references therein). Didolodontids and protolipterniids were primitive ungulates ("condylarths") that also included small, primate-sized animals that were diverse and abundant (Cifelli 1985; Muizon and Cifelli 2000; Gelfo 2006). This varied fauna probably was widespread during the EECO at its peak around 52–50 Ma. Afterwards, the Eocene turned cooler for a long period of time, up until the early Oligocene (see Zachos et al. 2001). In the late Oligocene, during the Deseadan South American Land Mammal Age (SALMA), platyrrhines first appear in the record, in Bolivia (Table 4.1). One would expect this to have been the critical period for primates to appear in Patagonia as well, since we know that SAM caviomorph rodents existed then in Chile and Argentina (Flynn et al. 2003; Vucetich et al. 2005). However, no Patagonian localities of this age have produced primates thus far.

The Tinguiririrican SALMA, which was recently identified in Chile and is probably about 31–33 Ma, has an interesting fauna because it documents the simultaneous occurrence of dentally primitive herbivorous mammals and more advanced, hypsodont forms (Flynn et al. 2003: Hitz et al. 2006), thus probably reflecting major environmental changes near the base of the Oligocene. The transition toward high-crowned teeth in several groups of herbivorous mammals is also well documented

Table 4.1 The fossil record of platyrrhine primates. Pre-Pleistocene genera marked with an asterisk are possibly "living fossils," forms closely related to modern genera as actual or possible congeners, sister-taxa or clade members classified within the same tribe as the moderns

Epoch	Platyrrhine fossils	Geographic source	Fossil locality or formation	Age
Holocene	Antillothrix	Hispaniola	Samaná Bay, DR; Cueva de Berna, DR; Trouing Jeremie #5, Haiti; Trouing Marassa, Haiti; Trou Jean Paul, Haiti; Trouing Carfineyis, Haiti	ca. 3,580 BP
Pleistocene	Xenothrix	Jamaica	Long Mile Cave; Skeleton Cave	ca. 6,730 BP
	Paralouatta varonai	Cuba	Cueva del Mono fósil; Cueva Alta	Holocene – Miocene
	Protopithecus	Brazil	Minas Gerais; Toca da Boa Vista, Bahia	ca. 20,000 BP
	Caipora	Brazil	Toca da Boa Vista, Bahia	ca. 20,000 BP
	Alouatta	Brazil	Gruta dos Brejoes, Bahia	ca. 20,000 BP
Late Miocene	Acrecebus	Brazil	Solimoes Fm, Acre	9–6.8 Ma, Huayquerian
	Solimoea	Brazil	Solimoes Fm, Acre	9–6.8 Ma, Huayquerian
Middle Miocene	Neosaimiri*	Colombia	La Venta, Upper Magdalena Valley	13.5–11.8 Ma, Laventan
	Laventiana	Colombia	La Venta, Upper Magdalena Valley	13.5–11.8 Ma, Laventan
	Mohanamico*	Colombia	La Venta, Upper Magdalena Valley	13.5–11.8 Ma, Lamentan
	Patasola	Colombia	La Venta, Upper Magdalena Valley	13.5–11.8 Ma, Laventan
	Lagonimico	Colombia	La Venta, Upper Magdalena Valley	13.5–11.8 Ma, Lamentan
	Aotus*	Colombia	La Venta, Upper Magdalena Valley	13.5–11.8 Ma, Laventan
	Cebupithecia*	Colombia	La Venta, Upper Magdalena Valley	13.5–11.8 Ma, Lamentan
	Nuciruptor	Colombia	La Venta, Upper Magdalena Valley	13.5–11.8 Ma, Laventan

Table 4.1 (continued)

	*Miocallicebus**	Colombia	La Venta, Upper Magdalena Valley	13.5–11.8 Ma, Laventan
	*Stirtonia**	Colombia	La Venta, Upper Magdalena Valley	13.5–11.8 Ma, Laventan
	*Proteropithecia**	Argentina	Collon Cura Fm, Neuquen Province	15.8 Ma, Colloncuran
	Paralouatta marianae	Cuba	Domo de Zaza, Lagunitas Formation	~14.68–18.5 Ma
Early Miocene	*Homunculus*	Argentina	Santa Cruz Fm, Santa Cruz Province	16.5 Ma, Santacrucian
	Killikaike	Argentina	Santa Cruz Fm, Santa Cruz Province	16.5 Ma, Santacrucian
	Soriacebus	Argentina	Pinturas Fm, Santa Cruz Province	17.5–16.5 Ma, Santacrucian
	Carlocebus	Argentina	Pinturas Fm, Santa Cruz Province	17.5–16.5 Ma, Santacrucian
	*Dolichocebus**	Argentina	Sarmiento Fm, Chubut Province	~20 Ma, Colhuehuapian
	*Tremacebus**	Argentina	Sarmiento Fm?, Chubut Province	~20 Ma, Colhuehuapian
	Chilecebus	Chile	Abanico Fm, central Chile	20 Ma, Colhuehuapian
Late Oligocene	*Branisella*	Bolivia	Salla	26 Ma, Deseadan
	Szalatavus	Bolivia	Salla	26 Ma, Deseadan

at the locality of Gran Barranca, in central Patagonia, which presents a relatively continuous stratigraphic sequence from the Middle Eocene through the Eocene-Oligocene transition and the Early Miocene (Madden et al. 2005). Thus the Gran Barranca spans the EECO, the cooling phase that followed it, and also a brief period of warming, the Middle Miocene Climatic Optimum (MMCO) (Zachos et al. 2001, and references therein). Mammals are most abundant at Gran Barranca around the Eocene-Oligocene boundary, including caviomorphs, but primates are not possibly because they were rare if at all present.

The non-herbivorous mammals may have used woodland or forested habitats of some kind, interspersed within the grasslands. In fact, there is abundant paleobotanical evidence that forests of enormous species diversity existed deep in Patagonia in the early Eocene, coincident with the EECO (Wilf et al. 2003; Hinojosa 2005; Wilf et al. 2005), after which climatic cooling accompanied the extinction of some of their tropical elements (Zamaloa et al. 2006). In the north, in Colombia and Venezuela, Jaramillo et al. (2006) show a rapid increase in plant diversity during the EECO which, by early Middle Eocene, culminates in greater diversity than is found in the Holocene neotropics. Diversity drops thereafter until the early Oligocene. Thus SAM must have been a complex and changing mosaic of grasslands and forests during the Cenozoic, long before and all throughout the reign of primates in the north.

4.2.1 Before Platyrrhines: Climatic Influence of Antarctica

In the southern cone of SAM in particular, temperature and habitat during the Paleogene and early Neogene were influenced profoundly by proximity to Antarctica, by the climate and cryospheric history of that continent in what was then a uni-polar world. Early Eocene plant and pollen records from Antarctica, although sparse and fragmentary, provide a consistent view of a warm equable climate with a temperate forested continent at least along the coastline (e.g., Askin 1997; Francis and Poole 2002). Mean annual temperatures are estimated to have ranged up to 7–15°C during the PETM and early EECO on the Antarctic Peninsula (Francis and Poole 2002), oxygen isotope data from calcite cements provide paleotemperature estimates of 5–13° C (Pirrie et al. 1998), and deep-sea oxygen isotope records suggest that the warmest bottom water temperatures of the Cenozoic occurred during the early Eocene (up to 13°C) (Zachos et al. 2001).

After the EECO, Antarctica experienced a long-term cooling trend (Fig. 4.3), culminating with the expansion of continental sized ice sheets at the start of the Oligocene (e.g., Miller, Fairbanks, and Mountain 1987; Zachos et al. 2001; Francis and Poole 2002). During the middle and late Eocene, pollen and plant studies, while scant, indicate a deterioration of the environment from warm temperate to cool temperate climates (Francis and Poole 2002). Deep-sea isotopic records also suggest a cooling of bottom water from 12°C to 13°C during the early Eocene to perhaps 2–6° C by the late Eocene (e.g., Zachos et al. 2001). There is stratigraphic evidence for glaciers reaching the coast in the western Ross Sea during the late Eocene,

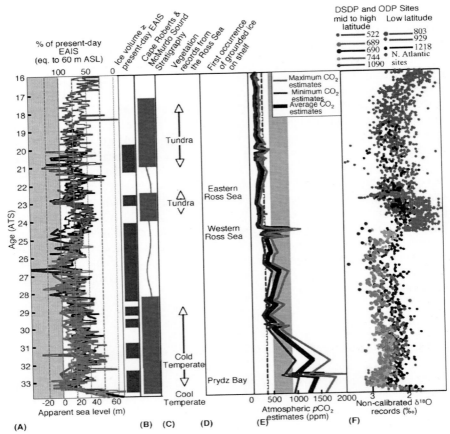

Fig. 4.3 Antarctica paleoenvironmental reconstruction from Early Oligocene to Middle Miocene (modified from Pekar and Christie-Blick, 2008). (**A**) Apparent sea-level (ASL) is global sea level plus the effects of water loading on the crust. The upper x-axis is the percent of the present-day East Antarctic Ice Sheet (EAIS) (equivalent to ~60 m ASL). The lower x-axis is ASL change: zero represents sea level resulting from ice volume lent to the present-day EAIS; increasing values represents sea-level rise; negative values represent ice volume greater than the present-day EAIS. Thick blue lines represent times when ice volume was ≥ than present-day EAIS. (**B**) Composite stratigraphy of cores drilled in the western Ross Sea. Thick brown lines represent times when sediment was preserved; red wavy lines represent hiatuses identified in cores. Note the excellent agreement when ice volume was ≥ than present-day EAIS and the timings of the hiatuses. (**C**) Reconstructed vegetation. (**D**) First occurrence of grounded ice based on core and seismic data around Antarctica. (**E**) Partial Pressure of Carbon Dioxide (pCO_2) estimates show decreasing values during the Oligocene reaching pre-industrial levels by the latest Oligocene and continuing into the early Miocene. Dashed line represents pre-industrial values (280 ppm); shaded box represents values predicted to occur this century. (**F**) Deep-sea oxygen isotope ($\delta^{18}O$) values. The abrupt decrease circa 24.5 Ma is due to a change in the source of data from high latitude to low latitude sites, with Southern Ocean sites below and mainly western equatorial Atlantic Site 929 above. (modified from Pekar and Christie-Blick, 2008.) *See* Color Insert.

although it is uncertain whether they were smaller tidewater glaciers or continental in scale (Barrett 1989). However, a growing consensus from a number of different data sources indicate that small to moderately sized ice sheets existed in Antarctica (presumably on the interior plateaus) during the middle and late Eocene (Browning et al. 1996; DeConto and Pollard 2003; Pekar et al. 2005; Miller et al. 2005).

The most significant climate shift of the Cenozoic occurred across the Eocene and Oligocene boundary (Miller et al. 1991; Zachos et al. 2001; Pekar et al., 2002), with relatively small ephemeral ice sheets expanding in the latest Eocene to large permanent ice sheets at the base of the Oligocene. In some instances, they expanded out onto the shelf during glacial maxima (e.g., Cooper et al. 1991), extending up to 500 km outboard of the present-day ice grounding lines in Prdyz Bay (Hambrey et al. 1991). This is confirmed by cores obtained from around Antarctica and a variety of other studies, which support the notion that the Oligocene ice sheets may have been 30–40% larger than the present-day East Antarctic Ice Sheet (EAIS) (Kominz and Pekar 2001; Lear et al. 2004; Pekar et al. 2002; Miller et al. 2005; Pekar and Christie-Blick 2008). In contrast, ice volume decreased to as little as 30–40 % of the present-day EAIS during glacial minima, suggesting a dynamic ice sheet during the Oligocene (Pekar et al. 2002; Lear et al. 2004; Pekar et al. 2006; Pekar and Christie-Blick 2008). Additionally, palynology records suggest that during glacial minima, climate changed from a cool temperate in the latest Eocene to a cold temperate climate in the earliest Oligocene to near tundra-like conditions by the late Oligocene (Francis and Poole 2002; Prebble et al. 2006).

The opening of the Drake Passage, which provided separation between SAM and the remnants of Gondwana, is believed to have been the final barrier to circum-Antarctic circulation. However, the timing of the opening is still uncertain with estimates of the opening to shallow waters ranging from the late middle Eocene (~41 Ma) to early Oligocene (Lawver and Gahagan 1998; Scher and Martin 2004). A deep-water passage developed somewhat later in the Oligocene (Livermore et al. 2004; Pekar et al. 2006). Numerical modeling studies suggest the deep water Drake Passage significantly altered oceanic circulation patterns and in turn may have affected the climate on Antarctica and SAM (Mikolajewicz et al. 1993; Nong et al. 2000; Toggweiler and Bjornsson 2000; Sijp and England 2004).

During the early Miocene (23–16 Ma), Antarctica remained heavily glaciated, with ice volume being up to 30% larger than the present-day EAIS during glacial maxima, shrinking to about 40–50% of the present-day EAIS during glacial minima (Lear et al. 2004; Pekar and DeConto 2006). The coastline was sparsely vegetated, with near tundra like conditions during glacial minima (Francis and Poole 2002; Prebble et al. 2006). This is consistent with the evidence that grasslands prevailed at this time in nearby Patagonia (*see* above). Interpretations of Antarctic climate and cryosphere conditions during the middle and late Miocene through the Pliocene (15–2 Ma) are more controversial. Data from outcrops located in East Antarctica suggest a warmer, more dynamic ice sheet during the late Neogene (Webb and Harwood 1991, 1993; Francis and Hill 1996; Ashworth et al. 1997), while evidence from the geomorphologic landscape evolution of the ice-free Dry Valleys region of the continent suggests persistent cold polar condition since the Middle Miocene

(Marchant et al. 1993; Marchant et al. 1996; Sudgen et al. 1993) Fossil plants as well as insects, invertebrates and palaeosols dated as late Neogene, indicate the presence of tundra-like conditions with estimates of mean annual temperature of −12°C, and a short summer season with temperatures up to +5°C (Wilson et al. 1998; Francis et al. 2007). The relevance of a more dynamic ice sheet is that warming and cooling phases may be relatable to Patagonian climate, possibly corresponding to the flux of warmer and more forested habitats that primates prefer.

4.3 NWM: the Temporospatial and Ecophylogenetic Setting

Sites producing fossil platyrrhines (Fig. 4.4a) occur across the entire length of the Andean backbone of SAM. The outstanding fact here is their occurrence at very low latitudes in Argentina and Chile, with one locality, Killik Aike Norte, situated about 1500 km away from the tip of the Antarctic Peninsula. Paleontologists first discovered fossil NWM in 1836 (Lund 1840) but concerted efforts to recover them by mounting frequent, systematic expeditions began only in the 1980s. Thus, the New World fossil record cannot be expected to match the abundance of fossil catarrhines. What we can safely assume about the SAM record is that we have only a few clues to a small portion of the story. Most critically, we are probably missing the very beginning. The oldest fossil NWM are the Bolivian *Branisella boliviana* and *Szalatavus attricuspis* (*see* Fleagle and Tejedor 2002), about 26 Ma from the Deseadan SALMA, but this date may be long after the time when anthropoids first entered the continent.

It is generally thought that platyrrhines arrived in SAM along with caviomorph rodents, for neither order is present early in the Paleogene, both seem to first appear at about the same time, and many believe both originated in Africa and dispersed across the Atlantic Ocean. The earliest radiometric ages of the mammal assemblage where caviomorph fossils are found is ∼31–32 Ma for the Chilean fossils (Flynn et al. 2003) and no more than 33.4 Ma for the Patagonian forms (Vucetich et al. 2005). But in both cases, the taxonomic diversity of the rodents is very low, and they are not abundant among the various fossil specimens collected. Even more, the Patagonian caviomorphs present a very primitive pattern of enamel microstructure (Vucetich et al. 2005, and references therein), leading to the suggestion that their SAM origins are not much older than the early Oligocene or latest Eocene. Since rodents are always by far more abundant than primates later in the SAM record, monkeys could have been even more rare proportionately early in their history. The absence of platyrrhines from some pre-Deseadan faunas may thus be an artifact of sampling, but it may also reflect different ecological requirements.

However, Poux et al. (2006) use molecular data to place the date of the caviomorph's origin at 37–45 Ma, in the Eocene. The latest date they give for platyrrhine origins is ∼39 Ma, also Eocene, about 13 million years before the fossils of Salla. In contrast, in Africa, archaic fossil anthropoids are known in abundant diversity from the upper Eocene, 36 Ma (Seiffert et al. 2005) and onwards,

Fig. 4.4a Geographic distribution of continental South American fossil and subfossil platyrrhines

but modern-looking catarrhines are only well established in the early Miocene (Harrison 2002). So, it is no surprise that many platyrrhine fossils, relatively young, are of a modern aspect and relatable to the living forms. Except for questions about the earliest genera, *Branisella* and *Szalatavus*, whose affinities are not at all clear, what we have sampled thus far appears to be earlier forms of the one radiation that produced modern lineages (e.g., Rosenberger 2002). Once the NWM record becomes readable in the early and middle Miocene in terms of morphological

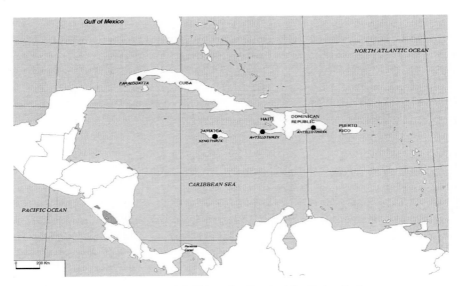

Fig. 4.4b Geographic distribution of Caribbean fossil and subfossil platyCarib map

diversity that is interpretable with some degree of confidence, controversy notwithstanding, the fossils appear to be quite closely related and adaptively similar to
modern counterparts (e.g., Stirton 1951; Rosenberger 1979; Delson and Rosenberger 1984; Setoguchi and Rosenberger 1987). A preponderance of long lived
genera, generic lineages and morphological stasis may be a defining feature of
the platyrrhine radiation over the past 15–20 million years – the Long Lineage
Hypothesis. This idea has obvious significance when considering the evolution of
communities.

Rosenberger, Delson and colleagues (Rosenberger 1979; Delson and Rosenberger 1984; Setoguchi and Rosenberger 1987; Rosenberger 1992) suggested that the
platyrrhine and catarrhine radiations may be fundamentally different in this regard, for
Old World anthropoids have evolved as a succession of replacing adaptive radiations
and few genera or lineages reveal the relatively great time depth seen in SAM. In
contrast, in the New World one can identify several forms that may have congeners,
sister-taxa or potential ancestors in the Miocene fossil record between 12 Ma and
20 Ma, or are so closely related that they are classified in the same tribe (Table 4.1).

By the same token, sampling error is a difficulty when interpreting the history of
living communities. For example, the semi-arid Cerrado grasslands of central east
Brazil supports small assemblages of two or three monkey species, often living as
riverine pioneers. The composition of small mammals in these communities indicate
that they are extensions of both the Amazonian forests and the Mata Atlantica (Mares
et al. 1986); either may be a potential source for the primates. The adaptable *Callithrix*,
Cebus and *Alouatta* often occur in sympatric species dyads, which is consistent with
the low biological productivity of these areas. But the subfossils discovered at sites

squarely within the present day Cerrado, or at the interface between Cerrado and the Mata Atlantica, the mega-sized *Protopithecus* and *Caipora* (? ?; Cartelle and Hartwig 1996; Hartwig and Cartelle 1996), strain the imagination: How they, too, could have coexisted with *Cebus* and *Alouatta* as the evidence indicates they did (*see* Tejedor, Rosenberger and Cartelle 2008), in the sparsely forested environs suggested by other contemporaneous mammals, such as giant ground sloths and grazing, fleet footed notoungulates (*see* Cartelle 1994)? At the same time, it is hard to fathom how *Protopithecus* and *Caipora* could have lived in a proto-Mata Atlantica if its structure closely resembled the present day forest.

Our final caveat relates to the fact that we give scant attention to the smallest platyrrhines, the callitrichines. While there are a few points of interest that arise below concerning them, marmosets and tamarins are minimally represented in the platyrrhine record and we suspect that this situation, the bias against finding small-sized specimens, will continue until there is an extensive, concerted effort to find them by adopting screen washing techniques at appropriate localities.

4.4 The Platyrrhine Provinces in Space and Time

With fossils occurring far beyond the domain of the Amazonian basin, which represents only 40% of the continent in area, rather than presuming a "center of evolution" around a monolithic Amazonia, we divide NWM into four regional groups or provinces based on fauna, geography and habitat: Amazonian, Atlantic, Caribbean and Patagonian. Each has interesting ecophylogenetic connections to fossil and sub-fossil species. Absent living non-human primates, the Caribbean province includes animals that are entirely extinct. Evidence indicates they have affinities with fossil SAM as opposed to living Central American species. The Patagonian province is also represented exclusively by fossils, ranging from the late Oligocene through the middle Miocene. Our concept of Patagonia is an extension of conventional usage. In addition to the common characterization of it as the region of Argentina east of the Andes and south of the Río Colorado, we also include places that encompass primate bearing sites within the southern Andes, such as the Abanico Formation of central Chile that yielded *Chilecebus carrascoensis* (Flynn et al. 1995). The growing Andean cordillera was certainly not a difficult barrier for the primates during the early Miocene, and there is no geographical or ecological reason to separate central Chile from Patagonia as different provinces.

We have not included Salla, Bolivia, with its two (or three, *see* below) important primate genera, in either the Amazonian or Patagonian provinces. Despite its geographical proximity to the former, the mammal fauna is distinctly different and there are indications of a semi-arid environment (MacFadden 1990). Comparisons of the Deseadan mammal faunas of Patagonia and Salla also highlight important ecological differences, especially among the rodents (Vucetich 1986; Pascual and Ortíz Jaureguizar 1990; Vucetich et al. 2005). Greater aridity is indicated in the south. Additionally, the pre-Deseadan rodents of Patagonia differ from Salla and from the

Oligocene rodents from Peru (Vucetich et al. 2005; Frailey and Campbell 2004). Thus while this Bolivian locale likely represents another ecological community, we still know too little about its primates to warrant much elaboration. Similarly, the various regions on the continent with distinctive habitats that currently support primates as low-diversity/density faunas, notably the Cerrado and Caatinga of east and central Brazil and the Llanos of Venezuela, are not distinguished. They are important to our thesis as analogues of several fossil assemblages, but do not merit recognition as separate provinces.

4.4.1 The Amazonian Province

The Amazon rainforest is the most dynamic terrestrial ecological region in the world, as it annually transforms with rainfall patterns, from being a river and then a lake, as seasons cycle from dry to wet (*see* Hess et al. 2003). More than a dozen primate species and as many as a dozen genera can easily be found at many lowland locations within the province. How these primates operate within a structured community has been the subject of excellent discussions by Terborgh (1983, 1985) who explains the general ecological factors and species specific feeding, locomotor and behavioral strategies that enable local species packing. His studies shed light on the overall picture of Amazonia that has meaning for our ecophylogenetic assessment of fossils throughout SAM.

Six of the 16 extant primate genera we recognize taxonomically are endemic to the region: *Cebuella, Callimico, Pithecia, Chiropotes, Cacajao* and *Lagothrix*. Others found within the lowland forests that would be considered endemic had they not dispersed into Central America (relatively recently, after uplift of the isthmus ca. 3 Ma (Stehli and Webb 1985)) include *Saguinus, Saimiri, Aotus*, and *Ateles*. None of the latter genera occur in the other major tropical forest of SAM, the Mata Atlantica. Of the four northern out-dispersers, *Ateles* has been able to penetrate deep into Mexico; *Saimiri* only goes as far as Costa Rica; *Saguinus* and *Aotus* are limited to Panama. This pattern means that a static, present-day definition of Amazonian endemism is rather arbitrary. The effective pre-Pleistocene measure of primate endemism in the Amazonian Province is higher than a literal count implies, closer to 10 genera. There is one endemic tribe, Pitheciini (*Pithecia, Chiropotes* and *Cacajao*). The latter's geographic exclusivity (they even cluster close to the river, or are sparsely distributed away from it) is instructive as to the ecological basis of their endemism, for they are not present in Central America or the Mata Atlantica even though habitat (Central America) or close relatives (*Callicebus* in the Mata Atlantica) would suggest this is an anomaly. Instead, this pattern probably means that seed-eaters cannot make it in the Mata Atlantica, just as extreme soft-fruit feeders, i.e., *Ateles*, also cannot be supported there or in sympatry with other large-bodied atelines, *Alouatta, Brachyteles*, or their local predecessors (e.g., *Protopithecus, Caipora*).

Three other modern platyrrhines that live in Amazonia have distributions over-lapping into other tropical forests and are clearly non-endemics. *Callicebus, Cebus* and *Alouatta* occur in the Mata Atlantica, and *Cebus* and *Alouatta* occur in Central America. *Alouatta*, like *Ateles*, occupies Mexico in its northernmost limit, and it also co-occurs with *Cebus* in many marginal, sparsely forested habitats in the tropical grasslands of the Cerrado and Llanos. These taxa are critical to understanding the low-diversity communities of Patagonia, as discussed below.

Fossils are known from a few localities that are actually within Amazonia today, but the most important source of paleontological information comes from the La Venta fauna of Colombia, which is now situated in the arid Magdalena Valley. There is a consensus that before the uplift of the Cordillera Oriental, which now walls off the Magdalena river from the Amazon basin, this region was part of the latter's drainage system and biotically part of the Amazonian province. The evidence for this has been growing since primates were first discovered there the 1940s. Kay (1997) cogently review much of this information. La Venta is a sub-sample of the cohesive, ~12–14 Ma Amazonian community, the best fossil repre-sentations of that biome that we know. In the fossiliferous Acre basin of western Brazil is a second, younger area, approximately 8 Ma, which has produced primates (Kay and Cozzuol 2006) and other tropical forms. It is also part of the Amazonian province.

4.4.2 The Atlantic Province

Outside the lowland Amazon basin there is a second major tropical forest now situ-ated along the eastern margin of the continent, within Brazil, the Mata Atlantica. Our primate-centric use of the term is restricted to the relatively humid semi-deciduous forests in southeastern Brazil, bordering the margin of Atlantic Ocean and extending plume-like in the south to the vicinity of Paraguay. This habitat is quite different from Amazonia. It has a radically different physiography, lacking the basin and river-network system that defines the Amazon, for example, and so large volumes of moisture are not held within it for long periods of time. In addition to rainfall, these humid forests rely on trade winds off the Atlantic Ocean for moisture, but they are also given to more intense dry seasons.

For all its biodiversity, the Atlantic province is by comparison with Amazonia a low-productivity habitat. Only six primate genera live there: *Brachyteles, Leon-topithecus, Callicebus, Callithrix, Alouatta* and *Cebus*. The first two are endemic to the region, very highly endangered and now confined to small tracts of forest, but records indicate that *Brachyteles* was once almost as widely distributed in the Mata Atlantica as *Alouatta* and *Cebus* are today. *Callicebus* is also very widely distributed. Furthermore, not all of the six platyrrhine genera occur at any one place, another indication of the lower carrying capacity of the environment. The most com-mon sympatric combinations involve one (rarely two together) of the callitrichines, plus *Cebus, Alouatta* and, in areas where they have escaped the holocaust of human habitation, *Brachyteles, Callithrix* and *Leontopithecus* tend to replace one another. Overall, this assessment probably underestimates primate biodiversity somewhat because the present composition of the Mata Atlantica flora and fauna was also strongly influenced by the Pleistocene refugial phenomenon (e.g., Cerqueira 1982), which may have pushed some forms to extinction recently without recruiting others.

The character of the Atlantic fauna is defined, in a sense, by *Cebus, Alouatta* and *Brachyteles*. The ecological adaptability of *Cebus* and *Alouatta* has already been stressed, and *Brachyteles* appears to fall into a similar category. The convergence of the endemic *Brachyteles* on howler-like semi-folivory (Rosenberger and Strier 1989; Rosenberger 1992; *see* also Anthony and Kay 1993) is a measure of the inability of the Mata Atlantica to support even one large-bodied, committed soft-fruit frugivore like the closely related *Ateles*, with which *Brachyteles* shares many positional and locomotor specializations (e.g., Rosenberger and Strier 1989). Amazonian localities often support two large frugivores, *Ateles* and *Lagothrix*.

The presence of only one pitheciine in the Atlantic province, the non-seed-eating *Callicebus*, also attests to its non-Amazonian floristic character. The success of the widespread *Callicebus* may mean that in the Mata Atlantica primates occupying the middle body size range do not have seeds as a protein option. Evolving sympatrically with *Cebus* monkeys who target deftly proteinaceous prey, leaves offer a viable, necessary option, for *Callicebus*, here and in Amazonia (*see* Norconk 2007). The smaller cebids of the Mata Atlantica, phylogenetically canalized to rely on arthropod and insects for protein, are also distinguished by derived feeding and foraging specializations relative to their Amazonian counterparts. Thus *Leontopithecus* is derivedly large in body size and preys more on vertebrates than other callitrichines. And, *Callithrix* is gumivorous. In a sense, *Leontopithecus* and *Callithrix* combine to exemplify the *Saguinus*-like niche.

Generally, the Atlantic region assembles a relatively small collection of primates that are either pioneers, adept generalists or targeted dietary specialists when it comes to exercising the protein needs – animals or leaves – of their diets, but neither seed-eating, nor hyper-frugivory seems to be available as a strategy. Each of the four major monophyletic groups of platyrrhines are represented by at least one genus, thus, making it easy to envision the fauna as a product of vicariance, the splitting of a once continuous connection between Mata Atlantica and Amazonas. *Cebus, Alouatta* and *Callicebus* are most prone to be successful in such a scenario, being ecologically the most generalized members of their subfamilies or tribes; their anatomical specializations (autapomorphies) make them so. But they each also require different resources and get them by using different feeding strategies. *Cebus* is the highly omnivorous predaceous frugivore; *Alouatta* is able to rely heavily on leaves; *Callicebus* can eat relatively hard fruits as well as insects and also a fair proportion of leaves. It also is interesting that the subfamilies that have succeeded in establishing two new genera, callitrichines and atelines, have in each case produced distinctly different forms from those living in Amazonia. *Leontopithecus*, the largest living callitrichine, is highly predatory and specialized for foraging in microhabitats such as the large bromeliad bowels that are typical of the Mata Atlantica habitat. *Brachyteles*, phyletically an atelin and postcranially adapted to a high-energy lifestyle that differs radically from the low-energy existence of an *Alouatta*, has secondarily become a semi-folivore.

The Brazilian subfossils mentioned above also have bearing on the evolution of the Atlantic province. Two genera are even larger than extant atelines, *Protopithecus* and *Caipora*, both one and a half to twice the weight of any extant ateline. These come from the states of Minas Gerais and Bahia (where the Mata Atlantica is almost now entirely extinct). The ages of these fossils cannot be determined; they are considered to be late Pleistocene or Holocene (MacPhee and Woods 1982). In addition to their surprisingly large size, one other factor is pertinent here. Both appear to be decidedly frugivorous (e.g., Cooke et al. 2007), and neither has a dental morphology that suggests a howler-like compromise between fruit and leaves (*see* Rosenberger and Kinzey 1976). Other subfossil mammals (megatheres, ungulates) that have come from the same caverns as *Protopithecus* and *Caipora* indicate presence of a more open country environment than humid tropical forest. The proposition that terrestrially is a possible form of locomotion in one or both of these mega-platyrrhines (Heymann 1998) needs to be investigated, but preliminary morphological studies of the postcranium (Cartelle and Hartwig 1996; Hartwig and Cartelle 1996) have failed to find evidence for this.

What the subfossil primates point to is the existence of a former habitat in the Atlantic province that was probably more productive than the current forest, one that could support one or more large bodied, highly active frugivores that might have generally resembled *Lagothrix* in diet and locomotion. This is consistent with the evolutionary model that a continuous connection between the Amazon and the Atlantic provinces existed in the past. *Protopithecus* and *Caipora* may have been users of large trees in mature, continuous-canopy forests that preceded the drier, semi-deciduous Mata Atlantica.

4.4.3 The Caribbean Province

Three genera of subfossil primates have been described from the Caribbean, and a fourth new form is now under study. One of the most prominent facts about the Caribbean primates is that all are endemic. There is also pointed debate about the relationships of these forms and we caution that little is known regarding their functional morphology. Thus, our interpretations regarding the lifestyles of these NWM are meant to be provisional. A second crucial point is that so far we are only certain that one monkey genus occurs on each of the islands bearing primates. If an artifact, this means more exciting forms await discovery. If real, it opens up new lines of questioning regarding insularity.

Three of the islands of the Greater Antilles, Cuba, Hispaniola, and Jamaica have fossil and subfossil primates. These islands probably were not modern in appearance until the late Miocene at the earliest, but there were certainly subaerial portions of the major islands exposed as early as the Eocene (Iturralde-Vinent and MacPhee 1999). Unfortunately, few mammalian fossils exist from this early period, but there is a considerable fossil record of the sloths, rodents, insectivores, bats, and (a few) primates from the Miocene through the Holocene.

There are alternative hypotheses explaining how the Caribbean was colonized by terrestrial mammals and other vertebrates. For example, Hedges and colleagues (e.g., Hedges 1996, 2001; Hass et al. 2001) have largely supported an overwater dispersal model, where many separate propogules arrived from the South American mainland onto the islands since the early Oligocene. Another view is that of Iturralde-Vinent and MacPhee (1999; MacPhee and Iturralde-Vinent 1994, 1995), who posit that a landspan once existed where the Greater Aves Ridge now lies, which connected the mainland of South America and Cuba, Hispaniola and Puerto Rico as well as the islands of the Lesser Antilles. This landspan, termed GAARlandia, was subaerial only briefly during the Eocene-Oligocene transition and allowed in at least two lineages of sloths (White and MacPhee 2001), a primate, bats, and numerous rodents. MacPhee and colleagues believe a single primate species was ancestral to the entire primate fauna, a notion contested by Rosenberger (2002). Either way, the mammals experienced a significant adaptive radiation, and today we know of 60 extant and 75 extinct non-introduced land mammal species (Daválos 2004). There is a striking degree of endemism in this fauna, with fifty percent of (non-introduced) bat species, and one hundred percent of the non-volant mammals being native (Hedges 2001). These newcomers would have encountered an earlier fauna of likely North American derivation, including the relatives of the insectivores *Solenodon* and *Nesophontes* that may have entered the Antilles sometime during the Eocene (e.g, MacDowell 1958).

The earliest evidence for primates in the Greater Antilles is from the Cuban Miocene (14.68–18.5 Ma) site of Doma de Zaza, where a lone primate astragalus was recovered in the early 1990s and described as *Paralouatta marianae* (MacPhee et al. 2003). Other terrestrial mammals known from Domo de Zaza are a megalonychid sloth and a capromyid rodent. The geological evidence, and the presence of marine fauna, indicates that the site lay along the banks of the sea where several depositional environments were present (MacPhee et al. 2003). Afterwards, the primate fossil record is silent until the late Pleistocene or Holocene when at least four species of primates were present in the Greater Antilles. These include *Xenothrix mcgregori*, *Antillothrix bernensis*, *Paralouatta varonai* and the new form.

The Jamaican monkey *Xenothrix* has a peculiar dentition (*see* Rosenberger 1977; MacPhee and Horovitz 2004) characterized by large, bunodont, thick-enameled cheek teeth and, most likely, broad upper incisors (Rosenberger 2002). Postcranially, its morphology is consistent with a slow-climbing positional repertoire (MacPhee and Fleagle 1991; MacPhee and Meldrum 2006). *Paralouatta* also has cheek teeth that are crested and wear down in a conspicuously flat fashion, indicating thick enamel. Its postcranials have been interpreted as possibly indicating semi-terrestriality (MacPhee and Meldrum 2006). Early studies of *Antillothrix* have also emphasized the bunodont shape of its crowns (MacPhee and Woods 1982), and postcranial remains are indicative of arboreal quadrupedalism with some leaping (*see* Ford 1990; MacPhee and Meldrum 2006). A common functional thread here appears to be protection against exogenous dietary abrasives. This may imply diets including fibrous and/or woody materials, such as pith, hard-shelled fruits or tough seeds, or a *Cebus*-like omnivory. More exotic adaptations, such as an embedded grub eating diet like Aye-Ayes, can also be entertained, e.g., *Xenothrix*. Wear resistance would also be consistent with the suggestion of a semi-terrestrial habit in *Paralouatta*. Basically, in the hypersensitive selective environment of enforced insularity on a small landmass, each of the Caribbean primates may have been adapting to the ultimate fallback, critical-function foods (*see* Rosenberger and Kinzey 1976), which may have meant omnivory.

4.4.4 Patagonian Province

As discussed above, during the Cenozoic the relatively flat eastern Patagonian lowlands would have been gradually evolving into its modern form – arid, cold and windy, strongly influenced by the nearby Antarctic, recipient of its cold winds and bottom waters, rain-blocked on the Pacific side by the uplifting Andes. Patagonia would have lacked the humid eastern trade winds that feed the higher subequatorial latitudes and the semi-deciduous Atlantic coastal forest. The environment has become most suitable to grasslands, which began to proliferate 30 million years ago, almost 10–15 million years before they rooted in North America (Flynn and Wyss 1998; MacFadden 2000; Jacobs et al. 1999). Even the localities that have produced primates evidence grazing and/or browsing notoungulates and rodents, and fleet predatory phorusrhacid birds that also signal open environments. Savanna, with sparse, spaced trees and an established herbaceous ground layer, is a possibility.

The phases of warmer and colder periods suggested by the dynamism of ice sheets in the Antarctic during the Miocene (Fig. 4.3) likely drove episodic expansion and contraction of more treed habitats where local conditions permitted. The source of warm-adapted flora that perhaps crept in from the north may have included proto-Amazonian or proto-Atlantic elements, but this would have involved habitat expansion over vast distances, accompanied by change reflecting local adaptation. Evolving native vegetation may have been even more important. The forests that developed during the EECO in Patagonia (Wilf et al. 2003; 2005; Hinojosa 2005) probably was a resource base for the later emergence of treed habitats that occurred in Patagonia during the Miocene. This could have included large swaths of forested terrain, sparse savanna and also riverine corridors where small communities of primates could have easily lived as they do now in the Atlantic province and in the Cerrado, Llanos and Chaco grasslands of Brazil, Venezuela and Bolivia.

The local Patagonia flora may have been a sort that was not conducive to high primate diversity. Two lines of evidence suggest that it was unlikely to have been a transplanted "Amazonia of the South", with similarly high productivity. First, new syntheses of the geological evidence pertaining to the topographic and rainfall features forming the foundation of Amazonia suggest the basin had a relatively late origin, coming long after primates are known to have occupied Patagonia (Campbell et al. 2006). Thus, if far north botanical elements were drawn to the southern cone, they were likely to have been different from the substance that makes today's Amazonian province what it is. Additionally, there may have been a physical barrier limiting southward dispersal. During the middle and late Miocene there is evidence of a large inland sea, the Paranaian Sea, situated north of Patagonia. It reached as far north as Bolivia and stretched from the Atlantic Ocean in the vicinity of northern Argentina and Uruguay to the base of the Andes (Aceñolaza 2000; Alonso 2000; Cozzuol 2006). This could have acted as a strong barrier for dispersal of terrestrial plants and animals either from or to the north.

In a mixed regional environment that favored grasslands but also allowed forests to exist, the primates that came to live in the Patagonian province during the late Oligocene and early Miocene were likely to have been adaptively similar to the pioneers and colonists that today inhabit second tier and marginal primate habitats in SAM. This means we might expect animals having lifestyles consistent with the Mata Atlantica and even the Cerrado, adaptable taxa like *Alouatta*, *Cebus* and *Callicebus*. But gum-eating marmoset-like species, quite successful in Cerrado and even in the drier Caatinga, are able to co-exist with other primates and would probably have done well also.

How do these predictions fit with the empirical evidence? An interesting eco-phylogenetic feature of the Patagonian primate faunas as it is currently known is the dominance of pitheciines, and the general resemblances of some to the inferred formative pitheciine feeding strategies (*see* Rosenberger 1992). These include *Homunculus*, *Tremacebus*, *Soriacebus* and *Carlocebus*. First, they are all small- to middle-sized NWM. In *Homunculus* and *Soriacebus*, prognathic incisors and premolar morphology indicate hard-fruit harvesting. *Homunculus* cheek teeth, of which we only have several, are also interesting in that they are worn but in a pattern that exposes rows of molar shearing crests on the crown surface. Their cheek teeth also seem to be relatively large. This is suggestive of at least some folivory, and a lifestyle in which *Homunculus* may resemble *Callicebus* living in the Mata Atlantica. In her summary, Norconk (2007) notes that the percentage of leaves in the diet of *Callicebus* in four field studies ranges from 4% to 23%– 66%. The spread at the end of the range reflects seasonal variation. Like *Callicebus*, *Homunculus* may have relied on leaves more than insects to fulfill its protein needs.

Homunculus also has small, non-projecting canines. As is well known, this is associated with monogamy and territorial defense in *Callicebus* and *Aotus* (e.g., Fernandez-Duque 2007). Additionally, *Homunculus* has jaws that deepen posteriorly, as in most *Aotus* but less than the extent seen in *Callicebus*. It may be that this feature is connected with the development of a vocal sac that enables the animals to produce powerful calls as a component of ritualized territorial behavior. One of the advantages of this syndrome is that it makes it possible for group sizes to remain small and efficiently spaced apart, which would be an advantage in low-productivity environments.

Soriacebus clearly had advanced, pitheciin-like prising incisors and the massive canines and anterior premolars associated with cracking hard-covered fruits, but it had unique cheek teeth that do not resemble pitheciins morphologically. So, while possible, there is insufficient anatomical evidence to argue by way of detailed analogy that *Soriacebus* was already an advanced seed-eater comparable to sakis and uakaris, where seeds comprise two-thirds or more of the diet (Norconk 2007). Rosenberger (1992) argued that *Soriacebus* was a primitive pitheciine, adaptively, able to ingest unripe, woody fruit, but without specializing on the seed for protein. Bown and Larriestra (1990) saw evidence of a varied environment in the Pinturas Formation where *Soriacebus* occurs, including tropical forest, areas that were perhaps partly forested and adequately watered, and areas sufficiently arid to allow sand dunes to form. They say (pg. 108), ". . .if appreciable climatic drying did accompany dune formation, some forest-dwelling mammals (including monkeys) were either unaffected by it or quickly reestablished themselves following dune formation." This description bears little resemblance to Amazonia.

Tremacebus molar teeth, known only from fragments of very badly worn teeth, are nonetheless grossly similar to some molars allocated to *Homunculus*, which show deep pockets of gross crown wear. Skull shape and measures of orbit size indicates eyeball proportions intermediate between small-eyed *Callicebus* and large-eyed modern *Aotus* (Fleagle and Rosenberger 1983; Kay et al. 2004). There is little reason to doubt that *Tremacebus* was nocturnal or crepuscular or cathemeral, roughly like *Aotus*, even though its eyes may have been slightly smaller and its olfactory lobe may not have been as enlarged relatively as *Aotus* (*see* also Kay et al. 2004). There is no reason to expect a 20 million year old owl monkey to be identical to an extant owl monkey. *Tremacebus* may have been quite similar in its use of time and light conditions to *Aotus* populations that live in the southern, more temperate limits of its distribution in Argentina and Paraguay, where their activity is conditioned by ambient temperature and moon-light phases (Fernandez-Duque 2007). They are more active during daytime hours when the nighttime temperature is low and/or nightlight is limited.

Only two non-pitheciine genera are known in Patagonia, *Dolichocebus* and *Killikaike*, although we fully expect more of the adaptable cebines and alouattins will eventually be discovered The skull shapes of both (Rosenberger 1992; Tejedor et al. 2006) suggest they had typically large cebine brains, thus perhaps similar foraging systems. The dentition of *Dolichocebus* is still hardly known.

Recent paleoecological studies are consistent with this scenario. Among the subregions where fossil primates are found in Argentina, the Santa Cruz Formation is an early Miocene stratigraphic unit deposited in Santa Cruz province (Ameghino 1906; Russo and Flores 1972). The duration of the Santa Cruz was probably under a million years (Tejedor et al. 2006). Yet it is one of the most representative, diverse, and rich vertebrate assemblages in South America, and is exposed from the Atlantic coast west to the Andean foothills. The most complete and best preserved fossils come from the coast, especially at sites between the rivers Coyle and Gallegos. Primates are not anomalies at these localities because they have also produced other mammals that tend to inhabit humid forests, such as sloths and some caenolestoid marsupials.

Tauber (1994, 1997a,b), who recorded hundreds of specimens of vertebrates, including mammals and a primate skull (Tauber 1991), developed a broad basis for biostratigraphic and paleoecologic studies of this area. He found several indications that from the latest, early Miocene onward climate changed from being warm and humid toward becoming drier and more seasonal. This was associated with a reduction in arboreal habitats and a shift toward open grasslands. Tauber showed there was a decrease in mammalian diversity over time, especially of the smaller forms such as microbiotheres and palaeothentid marsupials, and the small rodents. In contrast, armored glyptodontids and cursorial toxodontids, both herbivores, became larger and more diverse; hypsodont dentitions became more common among notoungulates; and, brachydont ungulates showed a progressive shift in body size frequencies.

That Patagonia was not uniformly tropical across the land mammal ages and geographical areas occupied by primates is shown in other ways as well, including palynological studies in the Pinturas Formation (Zamaloa 1993) that found temperate families of trees, and sedimentology (Bown and Larriestra 1990). Palinostratigraphic work in coastal Patagonia (Barreda and Palamarczuck 2000) showed that xerophytic plants were replaced by more tropical elements during the latest Early Miocene-earliest Middle Miocene. From the mammals, Vucetich (1994; *see* also Vucetich and Verzi 1994) concluded that the paleoenvironment of the western Santacrucian exposures could have been similar to the coastal Santa Cruz Formation in the east; probably woodland savannahs, or possibly more arid than that.

4.5 Discussion

4.5.1 Landscapes and Early History

The Cenozoic geographic isolation of SAM has been an important factor in shaping the evolution of platyrrhines. As we have proposed elsewhere (Delson and Rosenberger 1984), this situation enforced intra-community interactions among the NWM – potentially for 26 million years or more – while eliminating the sort of competitive faunal mixing that happened in the Old World among primate communities that met when anthropoids moved between continents. Continental scale isolation is possibly one of the correlates of the long lineage syndrome; strongly differentiated niches evolving early in the platyrrhine radiation followed by the establishment of an ecophylogenetic balance among the differentiating sublineages. Because no competing guilds were injected wholesale into the continent, the greater platyrrhine community may have arrived at a relatively steady-state ecological framework that allowed various lineages to survive for long periods in conditions of relative stasis, without being subject to massive faunal turnover.

Other important factors of continental scale that shaped the evolution of NWM involved the size and contour of the landmass, its deep extension into southern latitudes, and its tectonic history. These elements had profound consequences for regional communities and the radiation as a whole. It meant that the widest segment of the continent, in the north, was strongly influenced by a hot house equatorial climatic regime, while the long, narrow southern cone was influenced by its proximity to the ice box climatic engine of Antarctica, more powerful during much of the Cenozoic. Three major tectonic events also had crucial consequences. In the south, separation of the Fuegan tip of SAM from Antarctica and the deepening of the Drake Passage strongly influenced climate for more than 30 million years. It molded and enforced cold-adapted Patagonian habitats that may only have been interrupted locally and intermittently. In the north and west, episodes of Andean uplift during the past 15 million years may have effectively created the physiographic foundations of the Amazon basin. In the Caribbean, crustal movements opened a pre-isthmian exit for primates out of the continent and into the Greater Antilles.

Of course, the truly seminal event in NWM history was the coming of anthropoids to the continent. In the absence of any resident ecological cognates, upon encountering a conducive tropical habitat the newly arrived anthropoids probably underwent an explosive adaptive radiation driven by competitive release, comparable to what happened in Madagascar long before. There are still no fossils that bear witness to this process for the record probably does not extend deeply enough in time, and the oldest site, Salla, Bolivia, 26 Ma, has produced only 2–3 genera (by our count) so far, all difficult to interpret. The caviomorph rodents, better represented as early fossils, were possibly co-immigrants although here, too, there is still no corroborating fossil evidence of contemporaneity.

Early caviomorphs, known from Chile and Argentina, have a long history in Patagonia. They occur for perhaps 5–7 million years before any evidence for NWM on the continent. The near absence of NWM from Argentina's Gran Barranca, from localities older than the Colhuehuapian that produce rodents, is interesting. Gran Barranca has been well sampled (using screen washing) and spans several millions of years in time. The rarity of platyrrhines there may be another indication that the far southern latitudes were inhospitable to primates for long periods until the early Miocene, when the earliest NWM begin to appear at the 20 Ma Abanico Formation of Chile (Wyss et al. 1994). Or, it may mean that the rate of dispersal of caviomorphs, much more cosmopolitan in their distribution and more labile in their habitat requirements, was much higher than the primate rate. But none of the Colhuehuapian primates (Table 4.1) can be considered stem platyrrhines, contra Flynn et al. (1995) and Kay et al. (2008). Quite the opposite, *Dolichocebus* and *Tremacebus* are assignable to extant lineages, cebines and aotins, respectively. *Chilecebus* is still insufficiently described, but it appears to be phylogenetically nested among the cebids, possibly within a derived clade. Therefore, the Patagonian forms are neither near the ancestry of the radiation nor is there evidence that they arose in the southern cone.

It is tempting to position Salla, Bolivia, as a central datum in the early history of NWM for it is the oldest site that has produced monkeys thus far. MacFadden (1990:19) did so in speculating that his reconstruction of Salla's habitat as semi-arid was evidence that "...the earliest South American primates may have lived in a non-rainforest environment and is therefore in contrast to the dominant adaptive/environmental setting for platyrrhines today.... Perhaps this was the original environment into which platyrrhines radiated and the tropical, rain-forest environments of today represents secondary evolution into a new adaptive zone." We consider this a stretch. A more tempered explanation is that Salla primates, like various modern NWM and many extinct Patagonian forms with close affinity to the moderns, were able to exist in marginal (gallery forest?) habitats as part of their arborealist heritage. We still have no indications from the postcranium, except for one possibility in the Caribbean, of terrestrially adapted platyrrhines, which should have thrived in the south if they ever passed the arbo-terrestrial Rubicon. Even though Salla primates exhibit high-crowned wear resistant teeth (Kay et al. 2001), we believe these would have been advantageous in a forested habitat adjacent to dust-producing grasslands. There is also one Salla specimen, a mandible, whose allocation has been largely ignored (but *see* Rosenberger 1981; Rosenberger et al. 1991), that has a bunodont *Cebus*-like molar and a thick, shallow jaw. It may be a cebine, i.e., a part of the most adaptable, derived platyrrhine guilds. Thus we see it more profitable to regard Salla not as a mirror of the adaptive profile of ancestral platyrrhines; rather, as another example of the meandering evolutionary pathways taken by splinters of the NWM adaptive radiation as they evolved locally under shifting climatic and environmental conditions, including milieus that were non-Amazonian in aspect.

4.5.2 *Evolving in the SAM Provinces*

Eventually, the platyrrhine radiation "settled" in the Amazonian and Atlantic provinces. From our modern perspective, the immense diversity of the Amazonian community became the centerpiece of the evolving platyrrhine fauna, starting at least in the middle Miocene. However, it would be wrong to assume all the taxa and adaptive profiles represented there evolved *in situ* as responses to the ecology of that great lowland forest. South America is a large continent with a complex history, and monkeys have colonized different regions across time and space. The Amazonian primate community may be a composite fauna that includes forms drawn from other provinces or areas of SAM.

An integration of paleontology, phylogenetics and the phylogeography (i.e., of important primate fruit foods) may be employed in testing this composite fauna hypothesis. Although the evidence is probably not sufficiently robust at this time to address the problem in depth, we offer several ideas for exploration. Among the atelid endemics, it seems likely that the most committed seed-eaters, *Chiropotes* and *Cacajao*, and fruit-eaters, *Ateles*, originated in Amazonia. No fossils particularly close to any of these have been found yet but molecular evidence suggests relatively recent origins for them (Opazo et al. 2006), at about 9 Ma and 13 Ma respectively, which postdates the time when mountain building in the west initiated the transformational geological processes that produced the basin (*see* below).

As examples of taxa that may have had an extra-Amazonian origin, *Alouatta* and *Cebus* are good candidates due to their highly generalized ecological nature. Paleontology (e.g., Rosenberger 1979) and molecules (e.g., Opazo et al. 2006) indicate a pre-20 Ma origin for the *Cebus* lineage and, as discussed above, alouattins may have had a long pre-Laventan history as well. In a sense, *Alouatta* and

Cebus exhibit derived feeding and foraging adaptations that enabled them to become successful ecological generalists, seasonally adaptable, flexible in diet and substrate requirements, and comparatively undeterred by ecological barriers that geographically limit other monkeys. They are also intermediate in body size relative to the full range of Amazonian primates, neither too small nor too large to mandate a narrow zone of ecological tolerance. *Cebus*, a phyletic giant among modern cebids (Rosenberger 1992; but *see* Kay and Cozzuol 2006), has eluded the small-body size energetic and foraging constraints imposed by the ecophyletic inertia of its ancestry within the fundamentally insectivorous, cebid guild. *Alouatta* is the smallest living ateline genus, contrary to what might be expected as the member most prone to leaf eating, and its smallest living species occur in the Mata Atlantica (Table 4.2). This means howlers have fewer substrate constraints than heavier forms which need large canopies to support foraging and feeding requirements. Thus *Alouatta* and *Cebus*, well known across the continents as ecological pioneers, could easily have arisen in habitats defined by lower productivity, more seasonality, and less physical majesty than the big-canopy, multi-tiered, lush forests of Amazonia.

A second set of forms that may have arisen outside Amazonia are *Aotus* and *Callicebus*. As a non-endemic, living also in the Atlantic province, it is easy to see *Callicebus* evolving in connection with a low-productivity habitat, able to use leaves as a protein source even at a body size much smaller than *Alouatta*. Even more interesting is the point that both may have close evolutionary ties to extinct early middle Miocene forms from the southern cone. *Callicebus* has been potentially linked with *Homunculus*, and *Tremacebus* is closely related to *Aotus* (including the Laventan *A. dindensis* (*see* Fleagle and Tejedor 2002 for a review). While neither of these fossils is sufficiently well known anatomically to reveal characters barring them from a direct ancestry of the moderns, the possibility that they may be sister-taxa to the

Table 4.2 Body weights of extinct and extant atelines

Species	Weight (kg)
Alouatta fusca	4.418[1]
Alouatta caraya	5.206[1]
Alouatta belzebul	5.585[1]
Alouatta palliata	6.015[1]
Alouatta seniculus	6.228[1]
Lagothrix lagothricha	6.875[1]
Ateles belzebuth	8.076[1]
Ateles belzebuth	8.076[1]
Ateles geoffroyi	8.168[1]
Ateles geoffroyi	8.168[1]
Paralouatta varonai	9.55–10.17[2]
Brachyteles arachnoides	13.500[1]
Caipora bambuiorum	20.5[3]
Protopithecus brasiliensis	24.85[4]

Sources: [1]Rosenberger (1992), [2]MacPhee and Meldrum (2006), [3]Cartelle and Hartwig (1996), [4]Hartwig and Cartelle (1996)

living forms means it is also possible that *Aotus* and *Callicebus* have had a long, extra-Amazonian history.

Segments of the NWM adaptive radiation, probably through episodes of vicariance and dispersal, became fixed in the Atlantic province, where they remain today. Another satellite evolved in the Caribbean, where they recently became extinct. These two groups each underwent modest adaptive radiations, probably arising from an ancestral community of primates in each region as opposed to a single-species last common ancestor seeding the two communities. Otherwise, multiple biogeographic events and multiple adaptive parallelisms must be postulated to explain the ecophylogenetic complimentarily of the Amazonian, Atlantic and Caribbean provinces.

The Patagonian offshoot has been extinct for nearly 15 Ma. This assemblage was species-poor by comparison with the Amazonian province and all forms seem to have been short lived in the region. Of the seven platyrrhine genera occurring in Argentina during the early middle Miocene over a five million year span, there may only be one genus that lived through any of the time horizons denoted by four geological formations. It is important that several are related to monkeys or lineages found later in the north, and they are generally more primitive than the La Venta forms. The apparently short duration and low diversity in the southern cone is probably a result of the progressive cooling and proliferation of grassland habitats that occurred after the EECO, when warm, moist environments were probably transient, short-term phenomena. In the nearby Falkland (Malvinas) Islands, sometime between the Middle Oligocene and Early Pliocene, there was a temperate broadleaf coniferous forest (Macphail and Cantrill 2006). Thus, the primate communities of Patagonia may have largely been consigned to pioneering taxa able to live in marginal habitats, more similar in character to a Mata Atlantica profile than to an Amazonian pattern. While only cebines and pitheciines are known to have been present, there are hints of alouattins as well (Tejedor 2002).

Sampling error must be partly responsible for this next point, but in all of the sites that have yielded any of these primates no more than two genera appear to co-occur. In the modern north, in the modest Mata Atlantica fauna, it is not unusual for four or five primate genera to live in sympatry. In the Caribbean, thus far we only have one genus per island, but this almost surely will change. Nevertheless, with respect to their limited diversity, the Patagonian localities do resemble islands. Were they subject to the same heightened extinction risks of insularity?

When the Patagonian primates became extinct is difficult to say. There is no record of them in formations younger than the Collon Cura, at about 16 Ma; they may not have survived into the middle Miocene. The "Mesopotamian" outcrops of eastern Argentina (Paraná Formation, *see* Aceñolaza 2000), for example, have been well prospected but without producing any primates, although it is believed to represent the same biogeogaphic province as the Acre Formation of the western Amazon. Acre has produced three primate genera thus far, at approximately 8 Ma (Kay and Cozzuol 2006). Younger primates may have been largely excluded from the southern cone by entrenched long-term trends toward increasing aridity, increasingly open, sparsely vegetated environments and geographical barriers that also limited dispersal (*see* below). Whatever were the immediate conditions in the early middle Miocene, the larger point here is that primates may have been almost predestined to extinction because NWM never did take root in Patagonia in the form of viable enduring communities resembling Amazonia or the Mata Atlantica forests.

There is evidence that during the late Middle Miocene and early Late Miocene, from about 11–9.5 Ma, the terrestrial faunas of the Patagonian region were separated from the mid-continent by an extensive marine incursion, the Paranaian Sea. Reconstructions of this embayment indicate it reached from the region of northern Argentina and Uruguay northwest to Bolivia, with its western edge running along the base of the rising Andes (e.g., Cozzuol 2006; Potter 1997). This would have been a barrier or strong filter against dispersal between Patagonia and Amazonia. Cozzuol (2006) argues that it coincided with tectonic, hydrogeologic and climatic events that divided the Patagonian and Amazonian biotas, and soon thereafter led to the extinction of northern, aquatic faunal elements that bridged between the western Amazonian and Paranaian basins. The recession of the Paranaian Sea may mark the beginnings of the terrestrial biota that evolved into the current configuration of the southern grasslands terrain.

In the far north, at the younger sites of Acre and La Venta, there is clear evidence of pitheciines, cebines and atelines, some quite advanced (*see* Hartwig and Meldrum 2002). From Acre, Kay and Cozzuol (2006) recently described a new, very large monkey closely related to *Cebus*, *Acrecebus fraileyi*, and a primitive atelin, *Solimoea acrensis*. The Acre fauna also includes a second species of *Stirtonia*, a genus known from two species at La Venta. There is a consensus that *Stirtonia* is very closely related to, and adaptively similar to, modern *Alouatta*, so much so that Delson and Rosenberger (1984) suggested modern howlers are living fossils, possibly descended directly from *Stirtonia*. Similar proposals have been made concerning *Neosaimiri* (or *Saimiri* (*Neoaimiri*)) *fieldsii* and *Aotus dindensis* being parts of the squirrel monkey and owl monkey generic lineages, respectively. Considering that the entire fauna is quite modern in the abundance of primate genera and their ecophylogenetic composition, we would infer that the role of atelines in that community was comparable to what it is today in Amazonia, even though at this point we have only one alouattin represented. Overall, it is likely that La Venta, and the lesser known Acre, form an ecological time-continuum with modern Amazonia. This suggests that the primates of the Amazonian province have been evolving as a regional community for at least 12 million years. It must be one of the reasons why a high proportion of La Venta genera are so closely related to modern forms. One would expect that latitudinal gradients, exacerbated by the Antarctic effect, would limit southern expansion of this biota, and there is evidence that the province was geographically contained. Well-sampled faunas of equivalent age, with abundant mammals but apparently no primates, are known from Quebrada Honda, Bolivia, but they are more similar in composition to high-latitude formations than to La Venta (Croft 2007).

Another critical facet of La Venta relates to callitrichines, a subject we have intentionally ignored since there is little question that a large part of the explanation for their rarity in the fossil record reflects our coarse collecting techniques. However, modern SAM platyrrhine communities are incomplete without them, so their presence would be an important indicator of ecological structure in a paleocommunity.

Four La Venta species have been proposed as possible callitrichines. No additional material has been assignable to the three teeth first allocated to *Micodon kiotensis* (Setoguchi and Rosenberger 1987), and this animal remains too poorly known to pursue much further. At the least, it tells us species of the right size range and morphology were present in this primate communinty. A second small-sized taxon is *Patasola magdalenae* (Kay and Meldrum 1997), based on a subadult lower jaw. Here, too, the evidence is compelling but not unquestionable. More controversial is a crushed skull, *Callicebus*-sized, that Kay (1994) described as a "giant tamarin," *Lagonimico conclucatus*, and a lower jaw, *Mohanamico hershkovitzi* (Lutcherhand et al. 1986). The latter has been vigorously debated (e.g., Rosenberger et al. 1990; Kay 1990). We believe that *Lagonimico* is a pithecine (Rosenberger 2002) and *Mohanamico* is probably a callimiconin callitrichine (Rosenberger et al. 1990). Thus La Venta, in keeping with its ecological modernity, did support callitrichines.

4.5.3 Subfossils of a Miocene Aspect

The Caribbean province and the Brazilian subfossils, while comprised of forms that are quite distinct, echo signals of the Miocene. *Protopithecus brasiliensis* and *Caipora bambuiorum* (Hartwig and Cartelle 1996; Cartelle and Hartwig 1996; *see* also MacPhee and Woods 1982) are a very large-sized alouattin and atelin, respectively. Their ecological role in the eastern Brazilian community is not easily discernable at this point; more functional morphology is required. However, preliminary work indicates that *Protopithecus* is considerably *less* folivorous than expected in an alouattin. It differs markedly from *Alouatta* and *Stirtonia*. Its dentition strongly emphasizes frugivory (Cooke et al. 2007). Other new specimens from the cave collections that produced these primates include a new species of *Alouatta* (Tejedor et al. 2008), a *Cebus*, and the original series collected at Lagoa Santa by Lund (1838) probably includes *Callicebus*. Thus the Quaternary of this region supported various forms of a modern ecological cast but, as noted above, the picture of their habitat is still fuzzy. Like Simpson's (1981; Fig. 4.2) stratified faunas of SAM, we think these giant subfossils reflect an earlier faunal layer of primates possibly derived from a greater Amazonia-Mata Atlantica continuum that existed during the Miocene. *Protopithecus* and *Caipora* were probably replaced by other forms such as *Brachyteles*, perhaps evolving *in situ*, when grasslands intervened and the Atlantic province began to take on its current physiographic shape. This involved the intrusion of the open country mega mammals that are associated with *Protopithecus* and *Caipora* (Cartelle 1993).

In the Greater Antilles, there are three known genera (*see* MacPhee and Woods 1982) and a fourth that is currently being described. There are strong disagreements about the origins and interrelationships of these forms (e.g., MacPhee and

Horovitz 2004) but we follow the arguments put forth by Rosenberger (2002) pending further study now underway. This suggests that atelines, pitheciines and cebines are present. *Paralouatta varonai* (and *P. marianae*) is, by cranial evidence (Rivero and Arredondo 1991), an alouattin, but its dentition, like *Protopithecus*, is adapted for frugivory (Cooke et al. 2007). MacPhee and Meldrum (2006) suggest it may have been semi-terrestrial. The two-molared *Xenothrix mcgregori*, which we consider a pitheciine, has very bunodont cheek teeth (Williams and Koopman 1952; Hershkovitz 1970; Rosenberger 1977, 1992) with large, broad premolars and thickly enameled cheek teeth superficially resembling *Cebus*, only relatively enlarged. The postcranial skeleton appears to be odd, but also quite arboreal (MacPhee and Fleagle 1991). *Xenothrix* may have had very broad incisor teeth (Rosenberger 2002). The dentition overall suggests a frugivorous diet, probably of hard-shelled fruits or grit-covered foods. Seed-eating cannot be disqualified, either, but leaf-eating as a staple seems unlikely. The enormously deepened mandible is a pitheciine feature. *Antillothrix bernensis* (MacPhee et al. 1995), which we regard as a cebid, is still difficult to interpret functionally, though a mandible attributed to it is relatively thick and it carries a molar that appears to be quite bunodont. This is a combination of features found in *Cebus*.

Thus the complexion of the Caribbean fauna, in spite of the fact that none of the three known genera have yet to be found on a single island, reflects the diversity of the NWM eco-clades but in a combination that fuses primitiveness with autapomorphy, and also new adaptive nuances within clades. *Xenothrix* is a stunning example of the latter. One of the implications of *Paralouatta* is that the dispersal of alouattins into the Caribbean was likely to have antedated La Venta times, or that it involved a form that existed elsewhere on the continent. This is consistent with the suggestion (Rosenberger 1978), now receiving confirmation from a radiometric date of 14.68 Ma, and a more likely stratigraphic date of 17.5–18.5 Ma, associated with *Paralouatta marianae* (MacPhee et al. 2003) – NWM have an ancient origin in the Caribbean. While MacPhee and colleagues (*see* MacPhee and Woods 1982) argue that these forms are the monophyletic issue of a single dispersal event, our view is that the ancestral population (assuming there was only one) was more likely a fauna, a splinter community of SAM primates comprised of at least three ecophylogenetic groups experiencing joint range extensions onto a newly accessible archipelago during the Miocene.

4.5.4 Origins and Evolution of the Amazonian Community – or, How Life in the Flooded Forest Weighed Against a Terrestrial Radiation

The Miocene date for *Paralouatta* thus becomes even more intriguing, as it is about as old as the Patagonian primates and several millions of years older than the La Venta primates (*see* also MacPhee 2005). With Caribbean platyrrhines having roots as deep as these, we need to entertain complex scenarios implied by NWM occupying continental space vastly larger than Amazonia, perhaps not all at once but surely over the expanse of geological time. Amazonia can no longer be considered the seat of platyrrhine evolution if Patagonian and Caribbean and the Quaternary platyrrhines of Brazil involve lineages more primitive than their ecophylogenetic counterparts living in the Amazonian province.

Thus, for Amazonia a key question is: How old is the basin? In a wide ranging synthesis, Cambpell et al. (2006) provide a detailed model and scenario of its evolution during the last 15 million years. They suggest that its beginnings were set in motion by middle Miocene tectonism in the Andes, as the eroding cratonic shields of the east filled the landscape with a massive amount of sediment to form a single, enormous flat plain. Rivers drained toward the northwest and into the Pacific, in the region of Ecuador, and in the north into the Caribbean through Venezuela. At about 9.5–9.0 Ma, Andean uplift began to block the northwest portal. The basin began to pool, forming a vast shallow lake, Lago Amazonas, or a system of coalescing lakes, swamps, and rivers. The whole environment was subject to tropical, seasonal monsoonal rains. The modern eastward drainage flow was initiated later, at about 2.5 Ma, when the mouth of the Amazon opened to funnel the immense river system into the Atlantic Ocean. But what type of environment was evolving within this gigantic inland sea? According to Campbell et al. (2006: 206):

> . . .it rarely preserved fossils, plant or animal. This is explained most readily by two factors. The first is that deposition occurred in a highly oxidizing, shallow water environment, which resulted in the rapid decomposing of most organic debris. Second, unlike Lago Pebas [an older paleo-lake complex to the south west], which had an abundant vegetation complex in and around it, Lago Amazonas probably had as its modern analog the llanos of eastern Bolivia. There, long-term, seasonal flooding effectively curtails much vegetation growth, including on the banks of numerous lakes that fill the region. Compared to the Amazonian forests that cover the modern landscape, lowland Amazonia of the late Neogene was probably a vast complex of shallow mega-lakes surrounded by swampy, grassland savanna that endured months-long periods of seasonal inundations. It is also reasonable to expect that climatic cycles leading to extended periods of exceptional precipitation could have produced years-long periods of inundation.

If Campbell et al. (2006) are correct – and there are other models which they review that posit the beginnings of the Rio Amazonas and its tributaries to 27 Ma – vast areas of the Amazonian province would at times have been a spongy, wet desert as far as monkeys are concerned. The llanos of Bolivia, which is the modern habitat analog of Campbell et al., does not support a primate fauna, and the Bolivian Chaco is depauperate. But in some areas of the basin, the long-term influence of inundation must have been an important source of selection for monkeys and their foods. For example, it may have been a crucial factor in selecting for viable dispersal mechanisms of falling fruits, enabling them to withstand being drenched for long periods. Ayres (1989) has argued cogently that a regime of pervasive forest floor inundation is behind the coevolution of water tight hard-shelled fruits and the husking and seed-eating adaptations in *Cacajao* and *Chiropotes*. Special locomotor adaptations in saki and uakaris, pedal hanging (Meldrum 1993), like-tail hanging in atelines, may also have been an advantageous way for larger-sized primates to hover safely beneath boughs and canopies in order to forage on a lower tier of vegetation arising from the flooded forest floor.

And, if our analysis of the Patagonian situation for primates is correct, that it was long ephemeral to the mainline evolutionary history of NWM that was taking place elsewhere on the continent, the Campbell et al. (2006) model of Amazonia's evolution exposes an old conundrum from a different perspective, i.e., why platyrrhines never produced a terrestrial radiation, or even heavily committed terrestrial sublineages or genera. The presence of NWM fossils woven into southern cone communities where other ground dwelling mammals were present accentuates this problem: terrestrial opportunities had definitely been broadly available there since the Miocene, even if they were less common later in the Amazonian province. While this matter will remain an open issue until someone finds a fossil that is an obligate terrestrialist, additional light can be shed on why the odds have been against NWM evolving terrestriality as an environmental adaptation. It relates to the basic structure of Amazonia and the profound depths to which NWM have adapted to it.

Unlike most Old World rain forest habitats, the Amazonian lowlands are flooded for six months of the year, when it grows to cover an area about three times the surface its encompasses during the dry season. This means there is simply no *terra firme* available to modern monkeys for much of the year (and half their lives). Following Campbell et al. (2006), this condition may have existed for millions of years, may even have been exaggerated for long stretches of geological time and would have been less cyclical before 2.7 Ma, when the newly opened eastern portal to the Atlantic began the massive drainage of accumulating inland rainwater. For lineages, this means eons of selection in a waterworld swamp. It also means that the monkeys who might be best able to find a transitional opportunity to shift activities from canopy to forest floor would be those that best exploit the subcanopy niches, the callitrichines. But on account of their small body size they are unlikely candidates for evolving terrestrial offshoots. The largest platyrrhines, on the other hand, are also adept at exploiting the subcanopy by using their feet and/or their tails in hanging toward it.

Thus, specialized positional behaviors that evolved within the Amazonian province would have included a large proportion of NWM adapting to efficiently use arboreal space above the seasonally flooded forest floor. This represents an environmental scenario and evolutionary trajectory that is the antithesis of adaptive scenarios toward terrestriality, for it reinforces arboreality. In other words, by maximizing subcanopy specializations in positional behavior, NWM would have minimized terrestriality as a selective option among callitrichines, atelines and pitheciines, barring the evolution of an additional layer of autapomorphies enabling an arbo-terrestrial transition. On the other hand, it would be interesting to consider the shortened tails of uakaris and the less acrobatic quadrupedalism of howlers and cebus monkeys (*see* also MacPhee and Meldrum 2006), all possibly secondary specializations within their larger clades, as potential solutions to enable some degree of ground use in these taxa. For uakaris, this would have evolved within the flooded forest regime, as Ayres (1989) anticipated. For *Alouatta* and *Cebus*, it would have evolved in marginal

Fig. 4.5 Rusconi's behavioral reconstruction of *Tremacebus harringtoni* (Rusconi 1935)

habitats, possibly outside the Amazon. Even so, the central point is that the *probability* of Amazonian platyrrhines evolving terrestrial taxa would have been very low across geological time.

A final point is that terrestriality within and among primate clades is a demonstrably rare phenomenon. Thus there is no cause to anticipate terrestriality evolving in platyrrhines even though it has been highly successful among catarrhines; our own catarrhine sensibilities may have falsely exaggerated the chances (Fig. 4.5). Among all the modern and subfossil strepsirhines, fossil adapiforms, living and extinct tarsiiforms, plesiadapiforms, all stem catarrhines – and platyrrhines so far as we know them – no more than a handful of genera appear to be obligate terrestrialists. Primates, as an order, are almost universally arboreal. As a radiation, platyrrhines are no different.

4.6 Summary

Platyrrhines have been evolving in parts of South America for more than 26 million years, adapting to many changes in the continent's structure, climate, flora and fauna. The Amazonian rain forest where New World monkeys are now most abundant may have begun to evolve its current configuration only 15 million years ago, long after platyrrhines arrived. Today's largest primate communities, in Amazonia and the Mata Atlantica, are different in character, and provide contrasting models for reconstructing past primate communities. The Patagonian province was inhabited by primates from the early to middle Miocene for at least 5 million years before becoming extinct. It was strongly influenced by proximity to Antarctica and the evolution of the Antarctic ice sheets, supporting a low-diversity primate community analogous to the Mata Atlantica but with a primate fauna that may have been dominated by middle-sized, non-seed eating, primitive pitheciines, able to tolerate highly seasonal resources and feeding predominantly on a mixture of tough fruits and leaves. The late middle Miocene La Venta region of Colombia was a very modern, diverse, tropical primate community, an extension of the Amazonian province.

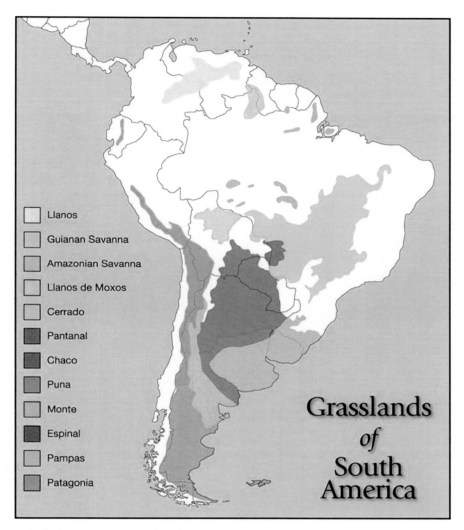

Llanos

Guianan Savanna

Amazonian Savanna

Llanos de Moxos

Cerrado

Pantanal

Chaco

Puna

Monte

Espinal

Pampas

Patagonia

Grasslands
of
South
America

Fig. 4.1a Grasslands of South America.

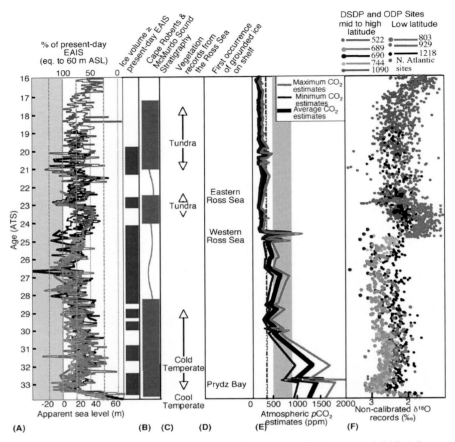

Fig. 4.3 Antarctica paleoenvironmental reconstruction from Early Oligocene to Middle Miocene (modified from Pekar and Christie-Blick, 2008). (**A**) Apparent sea-level (ASL) is global sea level plus the effects of water loading on the crust. The upper x-axis is the percent of the present-day East Antarctic Ice Sheet (EAIS) (equivalent to ~60 m ASL). The lower x-axis is ASL change: zero represents sea level resulting from ice volume lent to the present-day EAIS; increasing values represents sea-level rise; negative values represent ice volume greater than the present-day EAIS. Thick blue lines represent times when ice volume was ≥ than present-day EAIS. (**B**) Composite stratigraphy of cores drilled in the western Ross Sea. Thick brown lines represent times when sediment was preserved; red wavy lines represent hiatuses identified in cores. Note the excellent agreement when ice volume was ≥ than present-day EAIS and the timings of the hiatuses. (**C**) Reconstructed vegetation. (**D**) First occurrence of grounded ice based on core and seismic data around Antarctica. (**E**) Partial Pressure of Carbon Dioxide (pCO_2) estimates show decreasing values during the Oligocene reaching pre-industrial levels by the latest Oligocene and continuing into the early Miocene. Dashed line represents pre-industrial values (280 ppm); shaded box represents values predicted to occur this century. (**F**) Deep-sea oxygen isotope ($\delta^{18}O$) values. The abrupt decrease circa 24.5 Ma is due to a change in the source of data from high latitude to low latitude sites, with Southern Ocean sites below and mainly western equatorial Atlantic Site 929 above. (modified from Pekar and Christie-Blick, 2008.)

The poorly known Greater Antilles forms may have first entered in the Miocene as a small community involving at least three groups, pitheciines, atelines and cebines.

We propose that some platyrrhines that are now successful and very widely distributed in tropical forests across much of the South and Central American landscape, such as *Cebus* and *Alouatta*, are actually members of pioneering lineages that may have been more prone in the past to using the ground. However, we also argue that the modern platyrrhines more generally, as a consequence of exceptional and myriad adaptations to canopy and subcanopy locomotion and posture, have been unlikely candidates for evolving terrestrial lineages or communities. Apart from the earliest forms, most of the fossil record is consistent with the Long Lineage Hypothesis, that lengthy, enduring ecophylogenetic lineages have shaped the evolution of the platyrrhine adaptive radiation. Since some of these lineages, namely pitheciines and cebines, existed in Patagonia possibly before the modern wet lowlands in the north took on their contemporary aspect, it is an oversimplification to interpret the evolution of New World monkeys as a product of the Amazonian rain forest. The modern Amazonian primate community is an evolutionary mix of genera including: (1) ecologically flexible forms that may have arisen elsewhere and were first adapted to very different conditions, such as relatives of *Callicebus, Aotus* and other early pitheciines, and relatives of *Cebus* and *Alouatta*; (2) derived descendants of lineages who radiated and became increasingly specialized to unique habitat features of Amazonia, such as *Chiropotes* and *Cacajao*; and (3) a variety of monkeys that may have arisen *in situ*, including *Ateles*, *Lagothrix* and an assortment of callitrichines.

Acknowledgments As always, we are grateful to the many museums that enable us to study their collections, including the American Museum of Natural History, United States National Museum, Field Museum, Museum Nacional do Rio de Janeiro, Museo Argentino de Ciencias Naturales, Buenos Aires, and the Museo National de Historia Natural, in Havana, Cuba. Special thanks to Drs. Manuel Itturalde Vinent and Castor Cartelle for access to collections in their care. We gratefully acknowledge financial support from the Tow Travel Fellowship of Brooklyn College; Professional Staff Congress, CUNY; Wenner-Gren Foundation for Anthropological Research; New York Consortium in Evolutionary Primatology (NYCEP). We also thank Paul Garber and an anonymous reviewer for very helpful readings of the manuscript, as well as the co-editors of this volume for stimulating our thinking.

References

Aceñolaza, F.G. (2000) La Formación Paraná (Mioceno medio); estratigrafía, distribución regional y unidades equivalentes. In: Aceñolaza, F.G. and Herbst, R. (Eds.), *El Neógeno de Argentina* INSUGEO, Serie Correlación Geológica, 14, 9–27.

Alonso, R. (2000) El Terciario de la Puna en tiempos de la ingresión marina paranense. In: Aceñolaza, F.G. and Herbst, R. (Eds.), *El Neógeno de Argentina* INSUGEO, Serie Correlación Geológica, 14, 163–180.

Ameghino, F. (1906) Les formacions sédimentaires du Crétacé supérieur et du Tertiare de Patagonie avec un parálélle entre leur faunes mammalogiques et celles de l'ancien continent. Anales Museo de Historia Natural, Buenos Aires, ser. III 15, 1–568.

Anthony, M.R.L. and Kay, R.F. (1993) Tooth form and diet in ateline and alouattine primates: reflections on the comparative method. American Journal of Science, 283A, 356–382.

Ashworth, A.C., Harwood, D.M., Webb, P.-N. and Mabin, M.C.G. (1997) A weevil from the heart of Antarctica: Ion Studies in Quaternary entomology – an inordinate fondness for insects. Quaternary Proceedings, 5, 15–22.

Askin, R.A. (1997) Eocene-? earliest Olgicoene terrestrial palynology of Seymour Island, Antarctica, In: Ricci, C. (Ed.), the Antarctic Region: geological Evolution and Processes, Terra Antarctica Publications, Siena, pp. 993–996.

Ayres, J.M. (1989) Comparative, feeding ecology of the uakari and bearded saki, *Cacajao* and *Chiropotes*. Journal of Human Evolution, 18, 697–716.

Barreda, V. and Palamarczuck, S. (2000) Estudio palinoestratigráfico del Oligoceno tardío- Mioceno en secciones de la costa patagónica y plataforma continental argentina. In: Aceñolaza, F.G. and Herbst, R. (Eds.), *El Neógeno de Argentina* INSUGEO, Serie Correlación Geológica, 14, 103–138.

Barrett, P.J. (1989) Antarctic Cenozoic history from the CIROS-1 Drillhole, McMurdo Sound. *DSIR Bulletin*, 245, 1–254.

Beard, K.C. (2002) East of Eden at the Paleocene/Eocene boundary. Science, 295, 2028–2029.

Bigarella, J.J. and Ferreira, A.M.M. (1985) Amazonian geology and the Pleistocene and the Cenozoic environments and paleoclimates. In: Prance, G.T. and Lovejoy, T.E. (Eds.), *Amazonia. Key Environments Series*. Pergamon Press, New York, pp. 49–71.

Bond, M. (1986) Los ungulados fósiles de Argentina: evolución y paleoambientes. *Actas Congreso Argentino de Paleontología y Bioestratigrafía*, 2, 173–185.

Bown, T.M. and Larriestra, C.N. (1990) Sedimentary paleoenvironments of fossil platyrrhine localities, Miocene Pinturas Formation, Santa Cruz Province, Argentina. Journal of Human Evolution, 19, 87–119.

Browning, J.V., Miller, K.G. and Pak, D.K. (1996) Global implications of lower to middle Eocene sequence boundaries on the New Jersey coastal plain: The icehouse cometh. Geology, 24, 639–642.

Campbell, K.E., Frailey, C.D. and Romero-Pittman, L. (2006) The Pan-Amazonian Ucayali Peneplain, late Neogene sedimentation in Amazonia, and the birth of the modern Amazon River system. Palaeogeography, Palaeoclimatology, Palaeoecology, 239, 166–219.

Cartelle, C. (1993). Achado de Brachyteles do Pleistoceno Final. Neotropical Primates, 1, 8.

Cartelle, C. and Hartwig, W.C. (1996) A new extinct primate among the Pleistocene megafauna of Bahia, Brazil. Proceedings of the National Academy of Science, 93(13), 6405–6409.

Cartelle, C. (1994) *Tempo Passado. Mamíferos do Pleistoceno em Minas Gerais*. Editora Palco, ACESITA. Assessoria de Comunicação da Compania Aços Especiais Itabira, Belo Horizonte, 131pp.

Cerqueira, R. (1982) South American landscapes and their mammals. In: Mares, M.A. and Genonways, H.H. (Eds.), *Mammalian Biology in South America*, University of Pittsburgh, Linesville, pp. 53–75.

Cifelli, R.L. (1985) South American ungulate evolution and extinction. In: Stehli, F. and Webb, S.D. (Eds.), *The Great American Biotic Interchange*, Plenum Press, New York, pp. 249–266.

Cooke, S.B., Halenar, L., Rosenberger, A.L., Tejedor, M.F. and Hartwig, W.C. (2007) *Protopithecus, Paralouatta*, and *Alouatta:* The making of a platyrrhine folivore. American Journal of Physical Anthropology, 132(S44), 90.

Cooper, A.K., Barrett, P.J., Hinz, K., Traube, V., Leitchenkov, G. and Stagg, H.M.J. (1991) Cenozoic prograding sequences of the Antarctic continental margin: A record of glacio-eustatic and tectonic events. Marine Geology, 102, 175–213.

Cozzuol, M.A. (2006) The Acre vertebrate fauna: Age, diversity, and geography. Journal of South American Earth Sciences, 21, 185–203.

Croft, D.A. (2007) The middle Miocene (Laventan) Quebrada Honda fauna, southern Bolivia and a description of its notoungulates. Palaeontology, 50, 277–303.

Daválos, L.M. (2004) Phylogeny and biogeography of Caribbean mammals. Biological Journal of the Linnaean Society, 81, 373–394.

DeConto, R.M., Pollard, D. (2003) Rapid Cenozoic glaciation of Antarctica induced by declining atmospheric CO_2. Nature, 421, 245–249.

Delson, E. and Rosenberger, A.L. (1984) Are there any anthropoid primate living fossils? In: Eldredge, N. and Stanley, S.M. (Eds.), *Living Fossils*. Springer Verlag, New York pp. 50–61.

Fernandez-Duque, E. (2007) Aotinae: Social monogamy in the only nocturnal haplorhines. In: Campbell, C.J., Fuentes, A., MacKinnon, K.C., Panger, M. and Bearder, S.K. (Eds.), *Primates in Perspective*, Oxford, New York, pp. 139–154.

Fleagle J. and Rosenberger A.L. (1983) Cranial morphology of the earliest anthropoids. In: Sakka, M. (Ed.), *Morphologie, Evolutive, Morphogenese du Crane et Anthropogenese*. Paris: Centre National de la Recherche Scientifique, pp. 141–153.

Fleagle, J. and Tejedor, M.F. (2002) Early platyrrhines of southern South America. In: Hartwig, W.C. (Ed.), *The Primate Fossil Record*. Cambridge University Press, Cambridge, pp. 161–173.

Flynn, J.J. and Wyss, A.R. (1998) Recent advances in South American mammalian paleontology. TREE, 13(11), 449–454.

Flynn, J.J., Wyss, A.R., Charrier, R. and Swisher III, C.C. (1995) An Early Miocene anthropoid skull from the Chilean Andes. Nature, 373, 603–607.

Flynn, J.J., Wyss, A.R., Croft, D.A. and Charrier, R. (2003) The Tinguiririca Fauna,Chile: Biochronology, paleoecology, biogeography, and a new earliest OligoceneSouth American Land Mammal "Age". Palaeogeography, Palaeoclimatology, Palaeoecology, 195, 229–259.

Ford, S.M. (1990) Locomotor adaptations of fossil platyrrhines. Journal of Human Evolution, 19, 141–173.

Frailey, C.D. and Campbell, K.E. (2004) The rodents of the Santa Rosa local fauna. In: Campbell, K.E. (Ed.), *The Paleogene Mammalian Fauna of Santa Rosa, Peru*. Natural History Museum of Los Angeles County, Science Series, 40, 71–130.

Francis, J.E. and Hill, R.S. (1996) Fossil plants from the Pliocene Sirus Group, Transantarctic Mountains: Evidence for climate from growth rings and fossil leaves, Palaios, 11, 389–396.

Francis, J.E., Haywood, A.M., Ashworth, A.C. and Valdes, P.J. (2007) Tundra environments in the Neogene Sirius Group, Antarctica: evidence from the geological record and coupled atmosphere–vegetation models. Journal of the Geological Society, 164, 317–322.

Francis, J.E. and Poole, I. (2002) Cretaceous and early Tertiary climates of Antarctica: evidence from fossil wood: Palaeogeography, Palaeoclimatology, Palaeoecology, 182, 47–64.

Gebo, D.L. (2002) Adapiformes: Phylogeny and adaptation. In: Hartwig, W.C. (Ed.), *The Primate Fossil Record*. Cambridge University Press, Cambridge, pp. 21–43.

Gelfo J.N. (2006) *Los Didolodontidae (Mammalia: Ungulatomporpha) del Terciario Sudamericano. Sistemática, origen y evolución*. PhD thesis, Universidad Nacional de La Plata, Argentina.

Goin, F.J. (2003) Early marsupial radiations in South America. In: Jones, M., Dickman, C. and Archer, M. (Eds.), *Predators with Pouches: The Biology of Carnivorous Marsupials*. Commonwealth Scientific & Industrial Research Organization, Sydney, pp. 30–42.

Groves, C. (2001) *Primate Taxonomy*. Smithsonian Institution Press, Washinton, DC.

Gunnell, G.F. and Rose, K.D. (2002) Tarsiiformes: Evolutionary history and adaptation. In: Hartwig, W.C. (Ed.), *The Primate Fossil Record*. Cambridge University Press, Cambridge, pp. 45–82.

Hambrey, M.J., Ehrmann, W.U. and Larsen, B. (1991) Cenozoic glacial record of the Prydz Bay continental shelf, East Antarctica. In: Barron, J., et al. (Eds.), *Proceedings of the Ocean Drilling Program, Scientific Results 119*, Ocean Drilling Program, College Station, pp. 77–132.

Harrison, T. (2002) Late Oligocene to middle Miocene catarrhines from Afro-Arabia. In: Hartwig, W.C. (Ed.), *The Primate Fossil Record*. Cambridge University Press, Cambridge, pp. 311–338.

Hartwig, W. and Meldrum, D.J. (2002) Miocene platyrrhines of the northern Neotropics. In: Hartwig, W.C. (Ed.), *The Primate Fossil Record*. Cambridge University Press, Cambridge, pp. 175–187.

Hartwig, W.C. and Cartelle, C. (1996) A complete skeleton of the giant South American primate *Protopithecus*. Nature, 381(6580), 307–311.

Hass, C.A., Maxson, L.R. and Hedges, S.B. (2001) Relationships and Divergences Times of West Indian Amphibians and Reptiles: Insights from Albumin Immunology. In: Woods, C.A. and Sergile, F.E. (Eds.), *Biogeography of the West Indies: Patterns and Perspectives*. CRC Press, New York, pp. 157–174.

Hedges, S.B. (1996) The Origin of the West Indian Amphibians and Reptiles. In: Powell, R. and Henderson, R.W. (Eds.), *Contributions to West Indian Herpetology: a tribute to Albert Swartz.* Society for the Study of Amphibians and Reptiles, Ithaca, pp. 95–128.

Hedges, S.B. (2001) Biogeography of the West Indies: An Overview. In: Woods, C.A. and Sergile, F.E. (Eds.), *Biogeography of the West Indies: Patterns and Perspectives.* CRC Press, New York, pp. 15–33.

Hershkovitz, P. (1977) *Living New World Monkeys (Platyrrhini): with an introduction to Primates.* University of Chicago Press, Chicago.

Hess, L.L., Melack, J.M., Novo, E.M.L.M., Barbosa, C.C.F. and Gastil, M. (2003) Dual-season mapping of wetland inundation and vegetation for the central Amazon basin. Remote Sensing of Environment, 87, 404–428.

Heymann, E.W. (1998) Giant fossil New World primates: arboreal or terrestrial? Journal of Human Evolution, 34, 99–101.

Hinojosa, L.F. (2005) Cambios climáticos y vegetacionales inferidos a partir de paleofloras cenozoicas del sur de Sudamérica. Revista. Geologica Chile, 32, 95–115.

Hinojosa, L.F., (2005) Historia Terciaria de la Cordillera de la Costa: relaciones fisionómicas y fitogeográficas de la vegetación de la Cordillera de la Costa y las Paleofloras Terciarias del sur de Sudamérica. In: Smith, C., Armesto, J.J. and Valdovinos, C. (Eds.), *Biodiversidad y Ecología de los Bosques de la Cordillera de la Costa de Chile,* Editoral Universitaria, Santiago, pp. 90–104.

Hitz, R.B., Flynn, J.J. and Wyss, A.R. (2006) New Basal Interatheriidae (Typotheria, Notoungulata, Mammalia) from the Paleogene of Central Chile. America Museum Novitates, 3520, 1–32.

Horovitz, I. and MacPhee, R.D.E. (1996) New materials of *Paralouatta* and a new hypothesis for Antillean monkey systematics. Journal of Vertebrate Paleontology, 16, 42A.

Horovitz, I. and MacPhee, R.D.E. (1999) The quaternary Cuban platyrrhine *Paralouatta varonai* and the origin of the Antillean monkeys. Journal of Human Evolution, 36, 33–68.

Horovitz, I. (1997) *Platyrrhine Systematics and the Origin of the Greater Antillean Monkeys.* PhD dissertation. State University of New York, Stony Brook.

Horovitz, I. (1999) A Phylogenetic Study of the Living and Fossil Platyrrhines. American Museum Novitates, 3269, 1–40.

Iturralde-Vinent, M.A. and MacPhee, R.D.E. (1999) Paleogeography of the Caribbean Region: Implications for Cenozoic Biogeography. Bulletin of the American Museum of Natural History, 238, 1–95.

Jacobs, B.F., Kingston, J.D. and Jacobs, L.L. (1999) The Origin of Grass-Dominated Ecosystems. Annals of the Missouri Botanical Garden, 86(2), 590–643.

Jacobs, B.F., Kingston, J.D and Jacobs, L.L. (1999) The origins of grass-dominated ecosystems. Annals of the Missouri Botanical Gardens, 86, 590–643.

Jaramillo, C., Rueda, M.J. and Mora, G. (2006) Cenozoic Plant Diversity in the Neotropics. Science, 311, 1893–1896.

Kay, R.F. and Cozzuol, M.A. (2006) New platyrrhine monkeys from the Solimoes Formation (late Miocene, Acre State, Brazil). Journal of Human Evolution, 50, 673–686.

Kay, R.F. (1990) The phyletic relationships of extant and extinct fossil Pitheciinae (Platyrrhini, Anthropoidea). Journal of Human Evolution, 19, 175–208.

Kay, R.F. (1994) "Giant" Tamarin from the Miocene of Colombia. American Journal of Physical Anthropology, 95, 333–353.

Kay, R.F., Fleagle, J.G., Mitchell, T.R.T., Colbert, M., Bown, T. and Powers, D.W. (2008) The anatomy of *Dolichocebus gaimanensis,* a stem platyrrhine monkey from Argentina. Journal of Human Evolution, 54, 323–382.

Kay, R.F., Williams, B.A. and Anaya, F. (2001) The adaptations of *Branisella boliviana,* the earliest South American monkey, In: Plavcan, J.M., Kay, R.F., Jungers, W.L. and van Schaik, C. (Eds.), *Reconstructing Behavior in the Primate Fossil Record (Advances in Primatology),* Kluwer/Plenum, New York, pp. 339–370.

Kay, R.F., Campbell, V.M., Rossie, J.B., Colbert, M.W. and Rowe, T. (2004) The olfactory fossa of *Tremacebus harringtoni* (Platyrrhini, early Miocene, Sacanana, Argentina): Implications for activity pattern. Anatomical Record, 281A, 1157–1172 .

Kay, R.F., Madden, R.H., Cifelli, R.L. and Flynn, J.J. (1997) *Vertebrate paleontology in the neotropics : the Miocene fauna of La Venta, Colombia.* Smithsonian Institution Press, Washington, D.C.

Kay, R.F. and Meldrum, D.J. (1997) A new small platyrrhine from the Miocene of Colombia and the phyletic position of the callitrichines. In Kay, R.F., Madden, R.H., Cifelli, R.L. and Flynn, J. (eds.): *A History of Neotropical Fauna: Vertebrate Paleontology of the Miocene of Tropical South America.* Smithsonian Institution Press, Washington D.C., pp. 435–458.

Kinzey, W.G. (ed). (1997) *New World Primates: Ecology, Evolution, and Behavior.* Aldine de Gruyter, New York.

Kominz, M.A. and Pekar, S.F. (2001) Oligocene eustasy from two-dimensional sequence stratigraphic backstripping. Geological Society of America Bulletin, 113, 291–304.

Lawver, L.A. and Gahagan, L.M. (1998) The initiation of the Antarctic circumpolar current and its impact on Cenozoic climate. In: Crowley, T. and Burke, K. (Eds.), *Tectonic Boundary Conditions for Climate Model Simulations,* Oxford University Press, New York, pp. 213–226.

Lear, C.H., Rosenthal, Y., Coxall, H.K. and Wilson, P.A. (2004) Late Eocene to early Miocene ice sheet dynamics and the global carbon cycle. Paleoceanography, 19, PA4015.

Livermore, R., Eagles, G., Morris, P. and Maldonado, A. (2004) Shackleton Fracture Zone: No barrier to early circumpolar ocean circulation. Geology, 32, 797–800.

Lund, P.W. (1838) Blik paa Brasiliens dyreverden for sidste jordomvaeltning. *Det Kongelige Danske Videnskabernes Selskabs Naturvidenskabelige og Mathematiske Afhandlinger,* 8, 61–144.

Lund, P.W. (1840) View of the fauna of Brazil, previous to the last geological revolution. Charlesworth's Magazine of Natural History, 4, 1–8; 49–57; 105–112; 153–161; 207–213; 251–259; 3-7-317; 373–389.

Lutcherhand, K., Kay, R.F. and Madden, R.H. (1986) *Mohanimico herkovitzi,* gen. sp. nov. un primate du Miocene moyen d'Amerique du Sud. Comptes rendus de l'Academie de science de Paris, 303, 1753–1758.

MacDowell, S.B. (1958) The Greater Antillean insectivores. Bulletin of the American Museum of Natural History, 115, 117–214.

MacFadden, B.J. (1990) Chronology of Cenozoic primate localities in South America. Journal of Human Evolution, 19, 7–21.

MacFadden, B.J. (2000) Cenozoic Mammalian Herbivores from the Americas: Reconstructing Ancient Diets and Terrestrial Communities. Annual Review of Ecology and Systematics, 31, 33–59.

Macphail, M. and Cantrill, D. (2006) Age and implications of the Forest Bed, Falkland Islands, southwest Atlantic Ocean: Evidence from fossil pollen and spores. Palaeogeography, Palaeoclimatology, Palaeoecology, 240, 602–629.

MacPhee, R.D.E. (2005) "First" appearances in the Cenozoic land-mammal record of the Greater Antilles: significance and comparison with South American and Antarctic records. Journal of Biogeography, 32, 551–564.

MacPhee, R.D.E. and Iturralde-Vinent, M. (1994) First Tertiary Land Mammal from Greater Antilles: An Early Miocene Sloth (Xenarthra, Megalonychidae) From Cuba. American Museum Novitates, 3094, 1–13.

MacPhee, R.D.E. and Iturralde-Vinent, M. (1995) Origin of the Greater Antillean Land Mammal Fauna, 1: New Tertiary Fossils from Cuba and Puerto Rico. American Museum Novitates, 3141, 1–31.

MacPhee, R.D.E. and Fleagle, J.G. (1991) Postcranial Remains of *Xenothrix mcgregori* (Primates, Xenotrichidae) and Other Late Quaternary Mammals from Long Mile Cave, Jamaica. Bulletin of the American Museum of Natural History, 206, 287–321.

MacPhee, R.D.E. and Horovitz, I. (2004) New Craniodental Remains of the Quaternary Jamaican Monkey *Xenothrix mcgregori* (Xenotrichini, Callicebinae, Pitheciidae), with a Reconsideration of the *Aotus* Hypothesis. American Museum Novitates, 3434, 1–51.

MacPhee, R.D.E. and Meldrum, J. (X) Postcranial Remains of the Extinct Monkeys of the Greater Antilles (Platyrrhini, Callicebinae, Xenotrichini), with a Consideration of Semiterrestriality in *Paralouatta.* American Museum Novitates, 3516, 1–65.

MacPhee, R.D.E. and Horovitz, I. (2002) Extinct Quaternary platyrrhines of the Greater Antilles and Brazil. In: Hartwig, W.C. (Ed.), *The Primate Fossil Record*. Cambridge University Press, Cambridge pp. 189–200.

MacPhee, R.D.E. and Woods, C. (1982) A New Fossil Cebine from Hispaniola. American Journal of Physical Anthropology, 58, 419–436.

MacPhee, R.D.E., Horovitz, I., Arredondo, O. and Vasquez, O.J. (1995) A New Genus for the Extinct Hispaniolan Monkey *Saimiri bernensis* Rímoli, 1977, with Notes on Its Systematic Position. American Museum Novitates, 3134, 1–21.

MacPhee, R.D.E., Iturralde-Vinent, M. and Gaffney, E. (2003) Domo de Zaza, an Early Miocene Vertebrate Locality in South-Central Cuba, with Notes on the Tectonic Evolution of Puerto Rico and the Mona Passage. American Museum Novitates, 3394, 1–42.

Madden, R.H., Bellosi, E., Carlini, A.A., Heizler, M., Vilas, J.J., Re, G., Kay, R.F. and Vucetich, M.G. (2005) Geochronology of the Sarmiento Formation at Gran Barranca and elsewhere in Patagonia: Calibrating middle Cenozoic mammal evolution in South America. *Actas XVI Congreso Geológico Argentino*, La Plata, Tomo IV, 411–412.

Marchant, D.R., Denton, G.H., Swisher, C.C. and Potter, N. (1996) Late Cenozoic Antarctic paleoclimate reconstructed from volcanic ashes in the dry valleys region of southern Victoria Land. Geological Society of America Bulletin, 108, 181–194.

Marchant, D.R., Swisher, C.C., III, Lux, D.R., West, Jr. D. and Denton, G.H. (1993) Pliocene paleoclimate and East Antarctic ice sheet history from surficial ash deposits. Science, 260, 667–670.

Mares, M.A., Ernest, K.A. and Gettinger, D.D. (1986) Small Mammal Community Structure and Composition in the Cerrado Province of Central Brazil. Journal of Tropical Ecology, 2(4), 289–300.

Meldrum, D.J. (1993) Postcranial adaptations and positional behavior in fossil platyrrhines. In: Gebo, D.L. (Ed.), *Postcranial Adaptations in Nonhuman Primates*. Northern Illinois University, DeKalb, pp. 235–251.

Mikolajewicz, U., Maier-Reimer, E., Crowley, T.J. and Kim, K.-Y. (1993) Effect of Drake and Panamanian gateways on the circulation of an ocean model. Paleoceanography, 8(4), 409–426.

Miller, K.G., Wright, J.D. and Fairbanks, R.G. (1991) Unlocking the ice house: Oligocene-Miocene oxygen isotopes, eustasy, and margin erosion. Journal of Geophysical Research, 96, 6829–6848.

Miller, K.G., Fairbanks, R.G. and Mountain, G.S. (1987) Tertiary oxygen isotope synthesis, sea-level history, and continental margin erosion. Paleoceanography, 2, 1–19.

Miller, K.G., Kominz, M.A., Browning, J.V., Wright, J.D., Mountain, G.S., Katz, M.E., Sugarman, P.J., Cramer, B.J., Christie-Blick, N. and Pekar, S.F. (2005) The Phanerozoic Record of Global Sea-Level Change. Science, 310, 1293–1298.

Muizon, C. De and Cifelli, R. (2000) The "condylarths" (archaic Ungulata, Mammalia) from the early Paleocene of Tiupampa (Bolivia): implications on the origin of the South American ungulates. Geodiversitas, 22(1), 47–150.

Napier, P.H. (1976) *Catalogue of primates in the British Museum (Natural History) I. Families Callitrichidae and Cebidae*. British Museum Natural History, London.

Nong, G.T., Najjar, R.G., Seidov, D. and Peterson, W. (2000) Simulation of ocean temperature change due to the opening of Drake Passage. Geophysical Research Letters, 27, 2689–2692.

Norconk, M.A. (2007) Sakis, uakaris, and titi monkeys. In: Campbell, C.J., Fuentes, A., MacKinnon, K., Panger, M. and Bearder, S. (Eds.), *Primates in Perspective*. Oxford University Press, New York. pp. 123–138.

Opazo, J.C., Wildman, D.E., Prychitko, T., Johnson, R.M. and Goodman, M. (2006) Phylogenetic relationships and divergence times among New World monkeys (Platyrrhini, Primates). Molecular Phylogenetics and Evolution, 40, 274–280.

Pascual, R. (2006) Evolution and geography: The biogeographic history of South American land mammals. *Annals of the Missouri Botanical Garden*, 93, 209–230.

Pascual, R. and Ortíz Jaureguizar, E. (1990) Evolving climates and mammal faunas in cenozoic South America. Journal of Human Evolution, 19(1–2), 23–60.

Pascual, R., Ortíz Jaureguizar, E. and Prado, J.J. (1996) Land Mammals: Paradigm for Cenozoic South American geobiotic evolution. In: Arratia, G. (Ed.), *Contributions of Southern South America to Vertebrate Paleontology*. Münchner Geowissenschaftliche abhandlungen Verlag Dr. F. Pfeil (A) 30, 265–319.

Patterson, B. and Pascual, R. (1972) The fossil mammal fauna of South America. In: Keast, A., Erk, F. and Glass, B. (Eds.), *Evolution Mammals and Southern Continents*. State University of New York Press, Albany, pp. 247–309.

Pekar, S.F. and DeConto, R.M. (2006) High-resolution ice-volume estimates for the early Miocene: Evidence for a dynamic ice sheet in Antarctica. Palaeogeography, Palaeoclimatology, Palaeoecology, 231, 101–109.

Pekar, S.F., Hucks, A., Fuller, M. and Li, S. (2005) Glacioeustatic changes in the early and middle Eocene (51–42 Ma) greenhouse world based on shallow-water stratigraphy from ODP Leg 189 Site 1171 and oxygen isotope records. Geological Society of America Bulletin, 117(7), 1081–1093.

Pekar, S.F. and Christie-Blick, N. (2008) Resolving apparent conflicts between oceanographic and Antarctic climate records and evidence for a decrease in pCO_2 during the Oligocene through early Miocene (34–16 Ma). Palaeogeography, Palaeoclimatology, Palaeoecology, 260/1–2, 41–49.

Pekar, S.F., Christie-Blick, N., Kominz, M.A. and Miller, K.G. (2002) Calibrating eustasy to oxygen isotopes for the early icehouse world of the Oligocene. Geology, 30, 903–906.

Pekar, S.F., Harwood, D.M. and DeConto, R.M. (2006) Resolving a late Oligocene conundrum: deep-sea warming versus Antarctic glaciation. Palaeogeography, Palaeoclimatology, Palaeoecology, 231, 29–40.

Pirrie, D., Marshall, J.D., Crame, J.A. (1998) Marine high Mg calcite cements in Teredolites bored fossil wood: evidence for cool palaeoclimates in the Eocene La Meseta Formation, Seymour Island. Palaios, 13, 276–286.

Potter, P. E. (1997) The Mesozoic and Cenozoic paleodrainage of South America: a natural history. Journal of South American Earth Science, 10, 331–344.

Poux, C., Chevret, P., Huchon, D., DeJong, W.W. and Douzery, E.J.P (2006) Arrival and Diversification of Caviomorph Rodents and Platyrrhine Primates in South America. Systematic Biology, 55(2), 228–244.

Prebble, J.G., Raine, J.I., Barrett, P.J. and Hannah, M.J. (2006) Vegetation and climate from two Oligocene glacioeustatic sedimentary cycles (31 and 24 Ma) cored by the Cape Roberts Project, Victoria Land Basin, Antarctica. Palaeogeography, Palaeoclimatology, Palaeoecology, 231, 41–57.

Rivero, M. and Arredondo, O. (1991). *Paralouatta varonai*, a new Quaternary platyrrhine from Cuba. Journal of Human Evolution, 21, 1–11.

Rosenberger, A.L. and Strier, K.B. (1989) Adaptive radiation of the ateline primates. Journal of Human Evolution, 18, 717–750.

Rosenberger, A.L. (1977) *Xenothrix* and ceboid Phylogeny. Journal of Human Evolution, 6, 461–481.

Rosenberger, A.L. (1978) A new species of Hispaniolan primate: a comment. Anuario Cientifico Universidad Central Este 3, 248–251.

Rosenberger, A.L. (1979) Cranial anatomy and implications of *Dolichocebus*, a late Oligocene ceboid primate. Nature, 279, 416–418.

Rosenberger, A.L. (1980) Gradistic views and adaptive radiation of platyrrhine primates. Zeitschrift für Morphologie und Anthropologie, 71, 157–163.

Rosenberger, A.L. (1981) A mandible of *Branisella boliviana* (Platyrrhini, Primates) from the Oligocene of South America. International Journal of Primatology, 2, 1–7.

Rosenberger, A.L. (1992) Evolution of feeding niches in new world monkeys. American Journal of Physical Anthropology, 88(4), 525–562.

Rosenberger, A.L. (2002) Platyrrhine paleontology and systematics: The paradigm shifts. In: Hartwig, W. (Ed.), *The Primate Fossil Record*. Cambridge University Press, Cambridge, pp. 151–159.

Rosenberger, A.L. and Kinzey, W. (1976) Functional Patterns of Molar Occlusion in Platyrrhine Primates. American Journal of Physical Anthropology, 45, 281–298.

Rosenberger, A.L., Hartwig, W. and Wolf, R.G. (1991) *Szalatauus attricuspis*, an early platyrrhine primate from Salla, Bolivia. Folia Primatologica, 56, 225–233.

Rosenberger, A.L., Setoguchi, T. and Shigerhara. N. (1990) The fossil record of callitrichine primates. Journal of Human Evolution, 19, 209–236.

Rosenberger, A.L. and Tejedor, M.F. (in press) The misbegotten: long lineages, long branches and the interrelationhsips of Aotus, Callicebus and the saki-uakaris. In: Barnett, A., Viega, L., Ferrari, S., Norconk, M. Cambridge University Press, Cambridge. Evolutionary Biology and Conservation of Titis, Sakis and Uacaris. Cambridge University Press, Cambridge.

Rusconi, C. (1935) Los especies de primates del oligoceno de Patagonoa (gen. *Homunculus*). Revista argentina de paleonologia y anthropologia ameghinia, 1, 39–126.

Russo, A. and Flores, M.A. (1972) Patagonia Austral extra-Andina. In Leanza, A.P. (ed.) Geología Regional Argentina, Academia Nacional de Ciencias, 1, 707–725, Córdoba.

Scher, H.D. and Martin, E.E. (2004) Circulation in the Southern Ocean during the Paleogene inferred from neodymium isotopes. Earth and Planetary Science Letters, 228, 391–405.

Seiffert, E.R., Simons, E.L., Clyde, W.C., Rossie, J.B., Attia, Y., Bown, T.M., Chatrath, P. and Mathison, M.E. (2005) Basal anthropoids from Egypt and the antiquity of Africa's higher primate radiation. Science, 310, 300–304.

Setoguchi, T. and Rosenberger, A.L. (1987) A fossil owl monkey from La Venta, Colombia. Kyoto Univ. Overseas Res. Rep. New World Monkeys, 6, 1–6.

Sijp, W.P. and England, M.H. (2004) Effect of the Drake Passage in controlling the stability of the ocean's thermohaline circulation. Journal of Climate, 18, 1956–1966.

Simons, E.L. (1972) *Primate Evolution: An Introduction to Man's Place in Nature*. MacMillan Company, New York.

Simpson, G.G. (1981) *Splendid Isolation: The Curious History of South American Mammals*. Yale University Press, New Haven, p. 266.

Smith, T., Rose, K.D. and Gingerich, P.D. (2006) Rapid Asia–Europe–North America geographicdispersal of earliest Eocene primate Teilhardinaduring the Paleocene–Eocene Thermal Maximum. Proceedings of the National Academy of Sciences, 103(30), 11223–11227.

Stehli, F.G. and Webb, S.D. (1985) The Great American biotic interchange. Plenum Press, New York, p. 550.

Stirton, R.A. (1951) Ceboid monkeys form the Miocene of South America. Bulletin of the University of California Publications in Geology, 28, 315–356.

Sudgen, D.E., Marchant, D.R. and Denton, G.H. (1993) The case for a stable East Antarctic ice sheet: the background. Geografiska Annaler, 75A, 151–155.

Szalay, F. and Delson, E. (1979) *Evolutionary History of the Primates*. Academic Press, New York.

Tauber, A. (1991) *Homunculus patagonicus* Ameghino 1891 (Primates, Ceboidea), Mioceno temprano de la costa atlántica austral, Provincia de Santa Cruz, República Argentina. *Academia Nacional de Ciencias*, Córdoba, 82, 1–32.

Tauber, A. (1994) Estratigrafía y vertebrados fósiles de la Formación Santa Cruz (Mioceno inferior) en la costa atlántica entre las rías del Coyle y Río Gallegos, Provincia de Santa Cruz, República Argentina. Unpublished PhD Thesis, Facultad de Ciencias Exactas, Físicas y Naturales. Universidad Nacional de Córdoba, Argentina. p. 422.

Tauber, A.A. (1997a) Bioestratigrafía de la Formación Santa Cruz (Mioceno inferior) en el extremo sudeste de la Patagonia. Ameghiniana, 34(4), 413–426.

Tauber, A.A. (1997b) Paleoecología de la Formación Santa Cruz (Mioceno inferior) en el extremo sudeste de la Patagonia. Ameghiniana, 34(4), 517–520.

Tejedor, M.F. (2002) Primate canines from the early Miocene Pinturas Formation, Southern Argentina. Journal of Human Evolution, 43, 127–141.

Tejedor, M.F., Tauber, A.A., Rosenberger, A.L., Swisher, C.C. and Palacios, M.E. (2006) New primate genus from the Miocene of Argentina. Proceedings of the National Academy of Sciences, 103(14), 5437–5441.

Tejedor, M.F., Rosenberger, A.L. and Cartelle, C. (2008) Nueva Especie de *Alouatta* (Primates, Atelinae) del Pleistoceno Tardío de Bahía, Brasil. Ameghiniana, 45, 247–251.

Terborgh, J. (1983) *Five New World primates: a study in comparative ecology*. Princeton, New Jersey, p. 260

Terborgh, J. (1985) The ecology of Amazonian primates. In: Prance, G.T. and Lovejoy, T. (Eds.), *Key environments: Amazonia*. Pergamon, Oxford, pp. 284–304.

Toggweiler, J.R. and Bjornsson, H. (2000) Drake Passage and paleoclimate. Journal of Quaternary Science, 15, 319–328.

Vucetich, M.G. (1986) Historia de los roedores y primates en Argentina: su aporte al conocimiento de los cambios ambientales durante el Cenozoico. Actas Congreso Argentino de Paleontología y Bioestratigrafía, 2, 157–165.

Vucetich, M.G. (1994) La fauna de roedores de la Formación Cerro Boleadoras (Mioceno Inferior?) en la provincia de Santa Cruz (Argentina). Acta Geológica Leopoldensia, 39, 365–374.

Vucetich, M.G., Deschamps, C.M., Olivares, A.I. and Dozo, M.T. (2005) Capybaras, size, shape, and time: A model kit. Acta Palaeontologica Polonica, 50(2), 259–272.

Vucetich, M.G. and Verzi, D.H. (1994) Las homologías en los diseños oclusales de los roedores Caviomorpha: un modelo alternativo. Mastozoología Neotropical, 1, 61–72.

Webb, P.-N. and Harwood, D.M. (1993) Pliocene fossil *Nothofagus* (Southern beech) from Antarctica: phytogeography, dispersal strategies, and survival in high latitude glacial-deglacial environments. In: Alden, J., Mastrontonio, J.C and Ødum, S. (Eds.), *Forest Development in Cold Climates*. Plenum Press, New York, pp. 135–165.

Webb, P.-N. and Harwood, D.M. (1991) Late Cenozoic glacial history of the Ross Embayment, Antarctica. Quaternary Science Reviews, 10, 215–223.

Wesselingh, F.P. and Salo, J.A. (2006) A Miocene perspective on the evolution of the Amazonian biota. Scripta Geologica, 133, 439–458.

White, J.L. and MacPhee, R.D.E. (2001) The sloths of the West Indies: A systematic and Phylogenetic Review. In: Woods, C.A. and Sergile, F.E. (Eds.), *Biogeography of the West Indies: Patterns and Perspectives*. CRC Press, New York, pp. 35–54.

Wilf, P., Cúneo, N.R., Johnson, K.R., Hicks, J.F., Wing, S.L. and Obradovich, J.D. (2003) High Plant Diversity in Eocene South America: Evidence from Patagonia. Science, 300, 122–125.

Wilf, P., Labandeira, C.C., Johnson, K.R. and Cúneo, N.R. (2005) Richness of plant–insect associations in Eocene Patagonia: A legacy for South American biodiversity. Proceedings of the National Academy of Sciences, 102(25), 8944–8948.

Williams, E.E. and Koopman, K.F. (1952) West Indian Fossil Monkeys. American Museum Novitates, 1546: 16 p.

Wilson, G.S., Harwood, D.M., Askin, R.A. and Levy, R.H. (1998) Late Neogene Sirius Group strata in Reedy valley, Antarctica: a multiple-resolution record of climate, ice-sheet and sea-level events. Journal of Glaciology, 44, 437–447.

Wyss, A.R., Flynn, J.J., Norell, M.A., Swisher, C.C., Novacek, M.J., McKenna, M.C. and Charrier, R. (1994) Paleogene mammals from the Andes of central Chile: a preliminary taxonomic, biostratigraphic, and geochronologic assessment. American Museum Novitates, 3098, 1–31.

Zachos, J., Pagani, M., Sloan, L., Thomas, E. and Billups, K. (2001) Trends, Rhythms, and Aberrations in Global Climate 65 Ma to Present. Science, 292, 686–693.

Zamaloa, M.C. (1993) Hallazgos palinológicos en la Formación Pinturas, Sección Cerro de los Monos (Mioceno inferior), Provincia de Santa Cruz, Argentina. Ameghiniana, 30, 353.

Zamaloa, M.C., Gandolfo, M.A., Gonzalez, C.C., Romero, E.J., Cúneo, N.R. and Wilf, P. (2006) Casuarinaceae From The Eocene Of Patagonia, Argentina. International Journal of Plant Science, 167(6), 1279–1289.

Part III
Recent Theoretical Advances in Primate Behavior and Ecology

Chapter 5
Demographic and Morphological Perspectives on Life History Evolution and Conservation of New World Monkeys

Gregory E. Blomquist, Martin M. Kowalewski, and Steven R. Leigh

5.1 Introduction

As an order, primates are distinguished by several features of their life histories from other mammals. These include late achievement of sexual maturity, low female reproductive rates and potentially very long lives (Martin 1990). New World primates present a diverse array of life histories and social organizations (Ross 1991; Garber and Leigh 1997). In this chapter we explore primate life history variation relative to demography and development. We also note how an appreciation of life history and particularly the relations between life history, morphology, and demography can contribute solutions to vexing conservation problems.

Our interest in these questions comes at a critical moment in life history theory because much of this body of theory is undergoing important revisions. The traditional theoretical viewpoint arranges primate taxa along a continuum of "fast vs. slow" life histories. Primates with "fast" life histories bear young over short gestation periods. These young have brief infant and juvenile periods to begin reproducing at small sizes and young ages. Following these fleeting stages, animals with "fast" life history expect short adult lifespans. A "slow" life history species manifests an opposing set of attributes, with long pre- and postnatal developmental periods, large adult size, with few but protracted reproductive events. This paradigm or general theory of life history has been extremely productive in characterizing variation across the primate order. However, recent studies reveal serious deficiencies with this idea in understanding important variation in growth patterns and life history variables (Pereira and Leigh 2003). Furthermore, we maintain that a perspective of "fast" versus "slow" life histories is a heuristic that, unfortunately, inhibits the investigation and understanding of important variation in primate life histories and demography. This idea also hinders development of effective conservation programs.

G.E. Blomquist (✉)
Department of Anthropology, University of Missouri, Columbia, MO, USA
e-mail: blomquistg@missouri.edu

P.A. Garber et al. (eds.), *South American Primates,* Developments in Primatology: Progress and Prospects, DOI 10.1007/978-0-387-78705-3_5,

In this chapter, we review several areas in which a life history approach can enhance our perspectives on the biology of New World monkeys, with important implications for conservation efforts. Specifically, we briefly discuss demographic modeling in order to understand basic population processes in New World monkeys. These analyses show that even very limited data can provide critical insights into population dynamics. In addition, we discuss how these basic data can provide insights into understanding threats to populations. The data available for long-term, detailed demographic analyses are not yet available for Neotropical primates, except in a few important cases (Strier and Mendes this volume; for example *see* Crockett 1996; Rudran and Fernandez-Duque 2003 [*Alouatta seniculus*]; Strier et al. 2001 [*Brachyteles hypoxanthus*]). However, studies of primate life histories, even those that have analyzed rudimentary data, can provide information for basic demographic models. Our second objective is to consider how patterns of morphological development impact the course of life history. Traditionally, morphological analyses have been considered independent of both demographic analyses and conservation efforts. However, our morphological studies reveal considerable potential for morphology in understanding how life histories, and thus the basic demographic properties of populations, evolve (Garber and Leigh 1997; Pereira and Leigh 2003; Leigh 2004; Leigh and Blomquist 2007). Finally, attention to both modeling and morphology permits us to bring the concept of reaction norms to bear on conservation questions, along with ideas about genetics and conservation. In all, our contribution stresses the many, but often underappreciated, ways in which a life history perspective, by incorporating demographic models and ontogenetic data, can aid in understanding adaptation and solving conservation problems.

5.2 Demographic Models

5.2.1 Demography

The study of the size, structure, and distribution of populations and their dependence on vital rates of birth, death, and migration—is central to life history theory, providing ways in which ideas about life history can be linked directly to patterns of selection and to conservation concerns about the status of threatened primate populations. Most importantly, demography utilizes powerful quantitative tools to understand how population size and composition change. Demographic analyses can take many forms, ranging from broad approximations of population change to detailed analyses of the relations between individual reproductive performance and phenotypes (Kruuk et al. 2002). On the side of conservation concerns, one approach, *population viability analysis*, is an especially powerful tool. Such approaches are finding broad applications (Young and Isbell 1994; Strier 2000a), but we emphasize Caswell's (2001) thorough treatment of these techniques. He identifies central questions facing researchers and policy-makers developing conservation plans, including the following:

Is the population really declining? Is it declining more than that could be considered normal (i.e. outside the boundaries of what might happen given group structure and inter-annual patterns)? Defining the scope of demographic problems for a population is of obvious interest. The IUCN (2001) definitions of endangerment are based on population size, their degree of fragmentation, and estimated rates of decline.

What part(s) of the life cycle could be impacted to cause the population to decline? This may include factors such as increased emigration, mortality during specific phases of life, depressed fertility, and later maturation. Weighting their importance is key to targeting variables for management.

What are the immediate causes of these changes in the life cycle? Many factors could be involved in changing the lifecycle. For example, infant mortality may be higher because of increased numbers of displaced males raising levels of infanticide; females may have fewer live offspring because of nutritional stress, heavy parasite loads, or inbreeding depression.

What are the ultimate causes, and what can be done about them? Loss or fragmentation of required habitat, hunting, introduction of parasites or other novel environmental factors could be implicated.

A prominent example of how these questions can be answered with rudimentary species-average life history information is provided by conservation efforts on behalf of the loggerhead sea turtle (*Caretta caretta; see* Crouse et al. 1987; Crowder et al. 1994, 1995). Human disturbance through unintentional capture and death in fishing nets, the destruction of nesting sites on beaches, and mortality of eggs and hatchlings all impact loggerheads as well as other sea turtle species. Traditional conservation practices focused on protecting beach habitat, eggs, and hatchlings. However, demographic models demonstrated that neither doubling fertility nor increasing hatchling survival to 1.0 (no death) would induce population growth. Instead, modeling showed that increasing adult survival provided the most effective way to yield a healthy population (Crouse et al. 1987; Crowder et al. 1994, 1995).

At first glance, these questions and analyses seem only loosely related to the kinds of interspecific differences in socioecology and average life history frequently analyzed in the primate life history literature (e.g., Ross 1991; Lee and Kappeler 2003). Thus, such data seem to have limited significance for understanding conservation problems, mainly because such variables are seen as the outcome of long-term and species-specific evolutionary process, with little relevance to microevolutionary processes confronted by most field primatologists. However, a demographic model of the average life history has considerable power to identify variables with the greatest effects on population growth rates. For primates, these are typically survival variables (*see* below). Whether or not these variables can be easily manipulated in a favorable manner through changed land-management or other interventions is a separate, and often much more difficult, question. Such a question involves phenotypic and genetic variation in life history traits as well as covariation among such traits. Robust demographic data are rare for primates, mainly as a result of their long lifespans, and relatively few researchers have prioritized such data (e.g., Rudran and Fernandez-Duque 2003; Strier et al. 2006), at least in research predating the emergence of the major conservation threats facing primates. Primatologists play important roles in all of the steps of this process and must offer useful information for setting policy.

5.3 Simple Demographic Models for New World Monkeys

In an effort to provide some basic information on the connections among demography, conservation, and life history we explored a simple demographic model for a set of New World monkeys. This analysis provides an example of how demographic models can be used with primates even when data are very limited. One result of these models is a ranking of importance of the life history parameters for each species. Knowing which aspects of the life history impact population growth rate the most establishes baseline information for conservation programs to target these life history traits. The model relies on published data on average age of maturation, litter size, inter-birth/litter intervals, and maximum lifespan (Table 5.1). A mixture of captive and wild data was used, though wild data were prioritized when available. In the absence of survival data for these species we simulated average values for adult and juvenile rates. Because the comparative data available are thin, we utilized a very simple demographic model from Charlesworth (1980). The variables investigated can be measured relatively simply, and are commonly utilized in life history analyses. These are reported averages for species, so the data cannot address questions about variation within a species. Our model takes the form of an equation (Equation 1) exploring relations among key demographic parameters:

$$l_\alpha b \frac{\lambda^{-\alpha}}{1 - P\lambda^{-1}} = 1 \qquad (5.1)$$

Specifically, the terms in this model are: α, female age of first reproduction in years; l_α, percentage of newborns surviving to age α; b, birth rate (litter size/interlitter interval) $\times$ 0.5; P, annual survival rate of adult females; and λ, finite rate of increase–population growth rate. Data are input for l_α, b, P, and α and λ is solved for by iteration. In other words, values are substituted for λ until one that satisfies the equation (i.e., makes it equal to 1) is found. Survivorship terms (l_α and P) were simulated over a range of values (0.05 to 0.90 at intervals of 0.05 for l_α, and 0.5 to 0.995 at intervals of 0.005 for P). The resulting set of data includes non-functional life histories (e.g., extremely low survivorship) and parameter combinations that are very unlikely for these primate species. To correct this, we deleted simulated life histories based on three criteria. First, the combination of l_α and P had to yield survivorship to maximum recorded lifespan of between 10 and 1 percent. This ensures that the rates selected are reasonable for the species, given what little is known about their lifespans. Second, only combinations in which average sub-adult survival ($l_\alpha^{(1/\alpha)}$) was less than or equal to adult survival (P) were used. This is a typical pattern of most mammals and holds for those primates on which there is data. We only accepted sets of values that yielded λ less than 1.30. Higher values would indicate population growth rate higher than 30% per year, which is undoubtedly outside the capability of any primate species, for an extended period of time. Furthermore, the simulated values are intended to be averages. After culling the simulated data, between 74 and 304 life histories were available for each species shown in Table 5.1. Simulated life histories for all but one species included values that would result in both increasing

Table 5.1 Life history data used in study of elasticities of population growth rate to life history traits based on Equation (5.1). Data were selected to roughly capture the breadth of life history and taxonomic variation in New World monkeys. All variables are measured in years except litter size. Species are listed with phylogenetic relatives (Schneider 2000)

Species	Litter size	Inter-litter interval	Age of first reproduction	Maximum lifespan	Data type
Brachyteles hypoxanthus	1	3	9	35	wild[a]
Ateles paniscus	1	2.7	7.1	22	free-ranging[b]
Alouatta palliata	1	1.8	3.6	19	wild & free-ranging[b]
Pithecia pithecia	1	1	2.1	36	captive[c]
Cebuella pygmaea	2	0.6	1.9	18.6	captive[d]
Callithrix jacchus	2	0.4	1.3	20	captive[d]
Callimico goeldii	1	0.4[e]	1	22.5	captive[d]
Leontopithecus rosalia	2	0.9	1.5	31.6	wild & captive[d]
Saguinus oedipus	1	0.9	1.5	26.2	wild & captive[d]
Cebus apella	1	1.6	7	45.1	wild & captive[f]
Saimiri sciureus	1	1	2.5	30.2	wild & captive[f]
Aotus trivirgatus	1	1	5	21.2	wild & captive[g]

[a] Strier et al. 2006. Estaçao Biológica de Caratinga.
[b] Di Fiore and Campbell 2007. Barro Colorado, La Pacifica and Santa Rosa.
[c] Harvey and Clutton-Brock 1985. Maximum lifespan reported on captive animals in AnAge online database http://genomics.senescence.info/
[d] Digby et al. 2007 and AnAge database. Age of first reproduction data are for first ovulatory cycles are thus minima for first births. ILI for *Saguinus* and *Leontopithecus* are from wild populations.
[e] Harvey and Clutton-Brock 1985.
[f] Jack 2007 and maximum lifespan from AnAge database.
[g] Fernandez-Duque 2007 and maximum lifespan from AnAge database.

and decreasing populations ($\lambda > 1$ and $\lambda < 1$). The aberrant species (*Callithrix*) had values ranging from 1.015 to 1.297.

To compare the effect on population growth rate (λ) of a change in any of the model terms we used elasticities calculated through implicit differentiation of the model given in Equation 1. These formulae are provided in the Appendix. Elasticities record the proportional response of population growth rate (λ) to a proportional change in any of the other single model terms. For example, if we change adult survivorship (P), the percentage change recorded in λ is the elasticity of λ to P. A small percentage change in λ from altering P (low elasticity) indicates that adult survivorship has only a minimal effect on population growth rate. Elasticity calculations, especially with high quality data to parameterize a demographic model for a particular primate population with more detailed age-specific information on vital rates, can provide important information to assess the conservation consequences of changes in demographic parameters (Alberts and Altmann 2003). Elasticities are particularly useful for making comparisons among parameters that are measured on different scales (but *see* Caswell 2001 p. 243). In this model the elasticities of λ to l_α and b are equal (*see* Appendix).

Elasticities calculated for each of the New World primates in Table 5.1 show that changes in adult survival (P) consistently produce the largest demographic response. This is the highest ranking variable in our analyses of elasticities meaning that adult survival has the greatest impact on the demographic health of populations, and has been shown in more detailed population viability analyses of particular primate species (e.g., Strier 2000a). As a very general rule, then, conservation strategies and tactics should seek to identify factors that impact the survival of adults. The other variables in the model do not impact population growth rate nearly as much as adult survival (Fig. 5.1). The second ranking variables are survivorship to maturity (l_α) and birth rate (b). Their elasticities are equal in this model. Finally, the lowest ranking variable is age of maturation (α). No species deviated from this ranking pattern of elasticities. If extremely high values of λ are allowed, the small-bodied, twin-bearing genera (*Callithrix*, *Cebuella*, *Saguinus*, and *Leontopithecus*) have low frequencies of other ranking patterns. Most of the alternate ranking patterns involved l_α and b ranked above P.

Our analyses show that birth rate and age of maturation are far less important than adult survival in affecting population growth rate. Reproductive output measures are generally unimportant in these models. Minor variations in age of maturation, for example, though often reported, are unlikely to have much effect on population growth rate. We find this result to be an important theoretical contribution, because

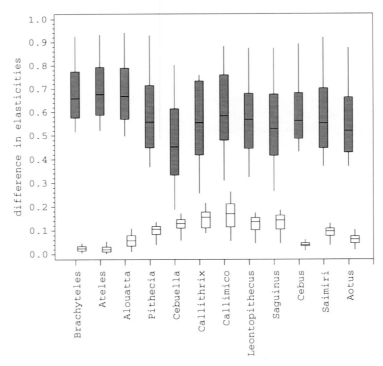

Fig. 5.1 Differences in elasticities of λ to P and l_α or b (*grey*), and between l_α or b and α (*white*). Species are ordered as in Table 5.1, which is roughly indicative of phylogenetic relationship

it suggests primate reproductive outputs can be viewed, in part, as simple consequences of negotiating the mortality risks of adulthood.

Changes in adult survival (P) clearly have a major impact on the demographic viability of primate populations. However, attempts to target P exclusively for management are misguided for several reasons. First, for all of the small-bodied species with high reproductive output the differences in elasticities of λ to P and l_α or b can be very small (<0.1) and are rarely larger than 0.6 (Fig. 5.1). This is not the case in larger species, where elasticity to P is typically between 0.5 and 0.9 units greater than that to l_α or b. In the small-bodied species, fairly modest changes in birth rates or survivorship to maturity can have a large effect on λ. Second, there may be no way of favorably influencing P due to economic costs or the population having reached a limiting value for P for internal physiological reasons (e.g., majority of deaths are due to senescent decay). Third, the elasticities for each parameter will change if any other parameter is changed, because λ will change. In this sense elasticities, or sensitivities, are "situational" (Stearns 1992 p. 34). This means that there are interactions among the model terms in determining the effects of each other on population growth rate. Fourth, the elasticity analysis will not identify responses in λ when there are correlated changes in the underlying life history traits. Elasticities are partial derivatives and identify the change in λ due to a single parameter when all other parameters are held constant. Documenting covariation among traits in response to habitat change is a more demanding, and interesting, empirical task for primatologists (Caswell 2000). Finally, because elasticities are derivatives, they only identify the effect of very small changes in a parameter on λ. Larger scale perturbations, such as a doubling of the inter-litter interval, can be made with a model explicitly targeting a certain population to explore how it might respond. Other limitations of elasticity analysis with limited data are discussed by Heppell et al. (2000).

5.4 Life History Modes and Ontogeny

At a broader theoretical level, the calculation of sensitivities or elasticities serves an important purpose aside from conservation applications. They are important because they identify the strength of selection on specific life phases for increases in fertility and survival thereby connecting life histories to evolutionary adaptations, particularly morphological and behavioral attributes. In this setting λ (normally a population growth rate) can be used to calibrate the *fitness* of the average life history, and sensitivities and elasticities document how much fitness changes with alterations in the life cycle.

The patterning of elasticities for primates shown here implies that adult survival will be particularly important in determining the overall pattern of selection. However, because primates take longer periods of time to reach adulthood than most other mammalian taxa, we expect to see diverse ways in which primates pass through development to reach adult sizes and shapes, and increased potential for decoupling of adult and juvenile morphological and behavioral attributes. Patterns of growth and development, if molded by selection, should be tailored to avoid mortality while growing into an adult state that will maximize survival and fertility. At the most general level, this means either growing *quickly* out of small size to minimize the exposure to dangerous periods (Williams 1966) or growing *slowly*, to reduce caloric needs, through phases in which nutritional stress would be particularly detrimental (Janson and van Schaik 1993). On a finer scale, there may be many ways in which different metabolically expensive tissues can be grown, that minimize their competition for limited energy, either by initiating or terminating their growth at different times, or by growing them at different rates.

Recent research shows that growth patterns are intimately related to adaptation, and in complex ways (Altmann and Alberts 2005; Schillaci and Stallmann 2005; Bolter and Zihlman 2003; Pereira and Leigh 2003; Badyaev et al. 2001; Starck and Ricklefs 1998; Leigh 1994). Specifically, primates, unlike many other mammalian species, seem to exhibit patterns of morphological dissociation during ontogeny. In other words, organs, organ systems, and functional units can develop on differing time scales within the same species. Primate postnatal development often lacks the tight coordination seen in species with shorter ontogenetic periods. This dissociation of developing structures is a core concept for understanding how ontogeny can be molded into adaptive patterns, and contrasts remarkably with traditional "fast vs. slow" models for mammalian life history evolution in which development is entirely absent or is the vacant space between neonatal and adult endpoints. A *life history mode* is a distinctive pattern or arrangement of ontogeny with respect to the rate and scheduling of growth for various organs, organ systems, or developmental modules.

Several examples substantiate the importance of life history modes for underpinning demographic and life history variation in New World monkeys. Across primate species, age of first reproduction and body size are modestly correlated ignoring phylogenetic nonindependence among species values (r = 0.881; Leigh and Blomquist 2007). This, accords well with mammalian life history models that see adult size as a function of the span of the growth period, and a constant growth rate for all species (Charnov 1993). However, a phylogenetically corrected correlation between adult mass and age of first reproduction is not significant (r = 0.059; Leigh and Blomquist 2007). Among New World monkeys, the phylogenetically corrected correlation between adult mass and age of first reproduction is also not significant. One reason for this lack of association is that there is substantial variation in body mass growth rates among platyrrhine species. Specifically, comparative analyses of small-bodied species (e.g., *Saimiri, Saguinus*) show that major differences in body size can be produced in the same amount of growth time (Fig. 5.2). Species differences in size may be produced entirely by growth rate differences, not time differences as is assumed by most classic life history models. This implies that these species support the energetic costs of growth in a variety of ways, including different parenting tactics and patterns of maternal investment (Garber and Leigh 1997).

Comparisons of brain growth patterns further substantiate the view that the developmental patterns of New World monkeys are highly variable, ultimately contributing to variation in demographic parameters (Leigh 2004). For example, brain

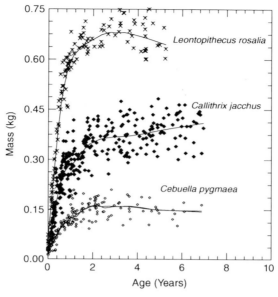

Fig. 5.2 Body mass growth trajectories in selected New World monkey species of differing body sizes (*Leontopithecus rosalia, Callithrix jacchus,* and *Cebuella pygmaea*). Different body size may be attained by over growth periods of similar duration. All data are from captive animals (*see* Garber and Leigh 1997)

growth curves in squirrel monkeys (*Saimiri sciureus*) and saddle-back tamarins (*Saguinus fuscicollis*) demonstrate one such difference (Fig. 5.3). Squirrel monkeys grow their brain quickly for a short period of time mostly during gestation, while tamarin brain growth occurs over a much longer interval at a much lower rate and extends through most of their post-partum somatic development. Goeldi's monkey (*Callimico goeldi*) may reveal a pattern much like squirrel monkeys. Despite being larger bodied, larger-brained and reaching reproductive maturity later than tamarins, squirrel monkeys (and possibly Goeldi's monkeys) have dissociated brain growth from body growth and sexual maturation. This means that patterns of brain growth do not necessarily determine the duration of life history stages in platyrrhines. Demographically, patterns of brain growth are important because they are correlated with a variety of patterns of maternal investment. For example, large brains that grow quickly are often associated with delayed female maturation (Leigh 2004), suggesting that the costs of brain growth have maturational, and thus demographic consequences.

In a larger sample of seven haplorhine species adult brain size and age of reproductive maturation are strongly correlated ($r = 0.93$) and age at brain growth cessation and age of reproductive maturation are moderately correlated ($r = 0.64$). Accounting for phylogenetic relatedness among the species using independent contrasts does not substantially diminish the adult brain size-age of reproductive

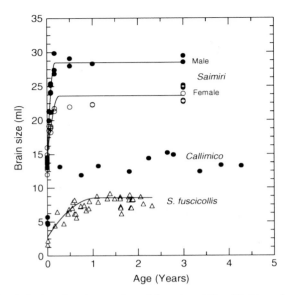

Fig. 5.3 Brain size (ml) plotted against age (years) for selected New World monkey species. Major differences in rates and timing of brain growth occur in these species. Piecewise regression lines are calculated for all species except *Callimico*, for which data are rare

maturation correlation ($r = 0.70$), but that between age of brain growth cessation and age of reproductive maturation evaporates ($r = 0.12$; Leigh and Blomquist 2007). Thus, age at brain growth cessation does not appear to determine directly and fully the age of reproductive maturation. Instead, once phylogenetic controls are applied, these variables are uncorrelated, suggesting that factors other than cessation of brain growth determine age at maturation.

While these patterns do suggest some association between adult brain size and life history, they do not fit models (e.g., Sacher and Staffeldt 1974) that identify the brain as a "pace-setter" of life histories. Instead, as we've shown, the length of the juvenile period and the time taken to grow the brain are unrelated. However, these relationships do emphasize the role of energetics and reducing mortality risk in primate life history evolution. The persistent correlation between adult brain size and age of reproductive maturation suggests indirect effects of brain ontogeny on both of these traits. Larger, faster-growing brains require larger, later-maturing mothers. Maternal energetics and mortality risk to both mother and offspring are crucial to understanding investment patterns (Martin 1983, 1996).

The contrast between *Saguinus* and other New World monkeys in terms of brain growth clearly relates to parental investment strategies in these species and the importance of social organization to these patterns. Mothers of the larger species in our comparison make heavy prenatal investments in their single offspring, while tamarin mothers invest little in their litter and deflect costs of growth to the offspring itself and other group members. These and previously mentioned patterns in these species suggests two distinctive life history modes. In squirrel monkeys maternal costs are high, body mass development is slow, but there is an extended period of mass growth yielding larger adults, and brain growth is rapid and early in development. Facing lower maternal costs, tamarins grow their small bodies quickly over a short period to small adult size, but extend brain growth over a much longer period than squirrel monkeys. These interesting results mean that the earliest periods of ontogeny are likely the most energetically costly for squirrel monkeys. Perturbations of resource bases that differentially affect foraging success of mothers and infants might be expected to have disproportionately large impacts on mortality or morbidity. Tamarins may be less impacted by such changes. Moreover, the costs of brain growth, and thus the susceptibility to environmental perturbation, are very high early in *Callimico* and *Saimiri*, but these costs are minimized after this brief period.

The demographic models explored previously add further insights to these interspecific comparisons. We expect patterns of elasticities for these species to be quite different, despite the stability of their rankings (Fig. 5.1). For *Saimiri* (and all other larger-bodied taxa examined) to offset a small decrease in adult survival, an extremely large increase in survivorship to maturation or birth rate, or decrease in age of maturation would be required. This is not the case in *Saguinus* and the smaller-bodied taxa, where relatively minor changes in the other variables can compensate for or exceed decreases in adult survival. Tamarins achieve an early sexual and somatic maturity because the potential mortality risk for the mother is offset by gains in reproduction. Such compensation is surely outside of the bounds of squirrel monkey life histories, such that maternal survival is more critical. The decision to mature is consequently delayed to when she has the requisite size to carry larger infants to term and not place herself at risk. In effect, morphological considerations, rarely the province of traditional conservation efforts, may play a central role in understanding population dynamics and challenges that species under resource stress face. Furthermore, ontogenetic perspectives on morphology greatly enhance our understanding of the diversity of ways primate life histories can be adaptations to environmental circumstances, balancing the many selective forces that impinge on mothers and offspring as they grow to independence.

5.4.1 Life History Responses to Habitat Change

Anthropogenic habitat change is a common threat to primate populations (Strier 2007). Tropical deforestation often results in the conversion of once continuous forest into patches or fragments of remaining forest in a matrix of non-forest vegetation and alterations of both structure and composition of the forests (Johns and Skorupa 1987; Plumptre and Reynolds 1994; Turner 1996; Marsh 2003; Norconk and Grafton 2003; Rivera and Calme 2006). Habitat fragmentation presents primate populations with changes that may occur over very short time spans. This rate of change means that the concept of a reaction norm (Schlichting and Pigliucci 1998) can be applied, in a general fashion, to contrast the ways in which primate species

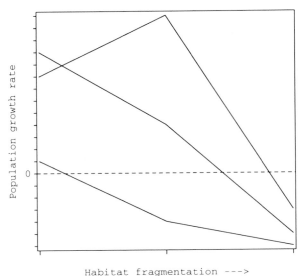

Fig. 5.4 Three imaginary reaction norms for species in three increasing levels of habitat fragmentation. Each solid line represents a population reaction norm. All populations are capable of positive growth in undisturbed habitat (*left*). One increases its growth rate in the mildly fragmented habitat (*middle*) while the growth rates of the others decline. In severely fragmented habitat (*right*) all have reduced growth rates such that they are experiencing population decline (growth rate below zero)

respond differently to changes in their habitats (Fig. 5.4). In a strict sense, a reaction norm is the set of phenotypes produced by a single genotype across a range of environments. In our loose application of the concept here, we will assume that primate populations experiencing rapid fragmentation are genetically unchanged, but simply expressing new phenotypes due to environmental changes. We hypothesize that the reaction norm view can be applied to suddenly fragmenting, essentially diversifying habitats, and has considerable power for assessing conservation risk. The reaction norm can interface directly with a demographic model. Specifically, greater accuracy and precision of models in projections of population change (growth or decline) can be obtained through models allowing for large-scale multi-trait perturbations. Such perturbations should only be performed with solid evidence on the variation and covariation in life history traits in response to environmental change. Unfortunately, little is known about life history responses to habitat fragmentation, and adult survival, the trait to which primate population growth rates are most responsive, is generally the most difficult to measure (Strier and Mendes this volume).

Feeding ecology and social organization often play important roles in predicting differing responses to disturbance. High quality habitats are likely to impact populations favorably both by increasing birth and survival rates, and reducing maturation ages. However, habitat quality is a subjective appraisal for species with differing habitat requirements. Species able to either meet or exceed their nutritional needs in a disturbed habitat are likely to persist (or even increase) in forest fragments, while those that cannot diminish as a result of nutritional stress (Chapman et al. 2006). Behavioral changes induced by fragmentation can also be important. Infanticide, a particular male reproductive strategy, was shown to increase in habitats under intense human alteration in *Alouatta caraya* (M. Kowalewski pers comm.) and *Semnopithecus entellus* (Sterck 1999).

Examples of primates encountering changing environmental conditions illustrate the importance of life history considerations in assessing conservation problems. Our first example emphasizes feeding ecology and changes in nutrition among hanuman langurs (*S. entellus*). Borries et al. (2001) compared the variation of life history traits in two populations followed for at least 7 years. One population was provisioned and raided crops while the other did neither, leading to energy intakes 5 to 10 times higher in the provisioned crop-raiding population than in the non-provisioned population. Borries et al. recorded remarkable life history trait differences between the two populations. For example, provisioned females gave birth year-round, had younger ages of first reproduction (means: 42.5 mo. vs. 80.4 mo.). They also had shorter interbirth intervals (16.7 mo. vs. 28.8 mo.), mainly as a result of a shorter lactation period (12.8 mo. vs. 24.9 mo.). It should be emphasized that these life history traits vary by about a factor of about two—a level of variation often far exceeding differences recorded among species. It would be important to know how much life history traits vary within a group at the individual level and how genetically differentiated the populations are to understand if the comparisons between groups are microevolutionary differences or can legitimately be interpreted as evidence of a reaction norm. However, these data are unavailable. Other examples of life history changes in response to food abundance or within-group life history variation due to dominance hierarchies that predict priority of access to resources are well documented in catarrhines (e.g., Watanabe et al. 1992; van Noordwijk and van Schaik 1999 [*Macaca*], Packer et al. 1995; Altmann and Alberts 2003; Wasser et al. 2004; Cheney et al. 2006 [*Papio*]; Pusey et al. 1997 [*Pan*]).

A second example emphasizes social pressures and male maturation patterns in black and gold howler monkeys (*A. caraya*). Adult males in this species have black fur and are much larger than females, which are either blond or beige. Both males and females are born blond, but males turn black upon maturity (Neville et al. 1988; Rumiz 1990; Bicca-Marques and Calegaro-Marques 1998). Explanations for dichromatism have been proposed (Zunino et al. 1986; Crockett and Eisenberg 1987; Bicca-Marques and Calegaro-Marques 1998; Dixson 1998). However, little research has been devoted to understanding how coloration relates to patterns of male maturation among groups. In particular, males in multi-male groups seem to maintain remnants of their golden pelage, especially on their upper torso and shoulders, and remain in their natal groups longer than males maturing in unimale groups. These males turn black rapidly and are quickly evicted by resident males (M. Kowalewski pers. obs.). The little evidence available on somatic growth patterns in male black and gold howlers indicates they have a sub-adult growth spurt (Leigh 1994). Both the pattern of color change and the presence of a sub-adult growth spurt suggest it is important for *A. caraya* males to make a rapid transition to adulthood. We expect that the pattern of somatic growth is generally species-typical, though nutritional changes will obviously impact the pace of mass gain. However, we hypothesize that the interesting differences in color change between males maturing in unimale and multi-male groups are decoupled from somatic growth and respond to social cues, possibly through some hormonal mechanism, perhaps similar to the hypothesized cause underlying bimaturism in male orangutans (Atmoko and van Hoof 2004). A more continuous habitat could increase the frequency of multi-male troops, causing delayed maturation for males. In a study of black and gold howlers in continuous forest, Kowalewski and Zunino (2004) reported only 3 of 27 groups (11%) were unimale. Zunino et al. (2007) present additional data on a population of black and gold howlers living in fragmented forests. Of these 34 groups, 27 (80%) groups were unimale. While males are "demographically disposable," meaning that their numbers do not limit population growth as much as females, changes in their maturation patterns could have important influences on patterns of inter-group gene flow and inbreeding depression that affect the viability of local populations.

If the general pattern is for multi-male groups in more continuous forests and unimale groups in more fragmented habitats, the low ability of maturing males to successfully migrate among groups—limiting intergroup gene flow—and high concentration of paternity within groups—reducing effective population size—will raise the risk for loss of genetic diversity and severity of inbreeding depression in unimale groups. This is particularly true if forest patches are becoming more isolated from the destruction of natural corridors among them. Male maturation patterns contribute to the problems of unimale groups. Were maturing males in unimale groups able to retain portions of their golden pelage, they might be able to gain further mass and experience prior to eliciting the evictionary violence of the group male. With greater mass and experience the young male has a better chance of becoming the group male himself, and even if the newly mature male loses he might have a better chance of successfully finding and entering a new group. Neville et al. (1988) also suggested that the retention of juvenile color in maturing male *Alouatta caraya* may allow them to become sexually active while they remain in their natal groups. However, in a captive study of *A. caraya* it was found that counts of abnormal sperm were higher in subadult individuals (n = 3) than in adult individuals (n = 3) (Moreland et al. 2001). Although these data are extremely limited, it could explain why adult males are tolerant of interactions between subadult males from neighboring groups with resident subadult and adult females during intergroup encounters (Kowalewski 2007). When, these subadult males are fully black, resident males actively stop any contact with resident females. Kingdon (1980) also suggested this juvenile color retention for *Cercopithecus neglectus* subadult males. For several primate species a relationship between the expression of secondary sexual traits and social suppression as a consequence of intermale competition (Gerald 2003) has been reported (Fontaine 1981 [*Cacajao calvus*]; van Noordwijk and van Schaik 1985 [*Macaca fascicularis*]; Dixson et al. 1993 [*Mandrillus sphinx*]; Kummer 1990 [*Papio hamadryas*]; Knott and Kahlenberg 2007 [*Pongo pygmaeus*]).

Under either outcome of successful replacement of the resident male or successful migration to a new group, a situation more amenable to population persistence would result. If the young male wins, we can imagine a long sequence of very short tenures of males in unimale groups. This will raise the effective population size. If he loses he may survive migration and increase inter-group gene flow. While both are results beneficial to the population such outcomes are usually of little importance for selection, which acts primarily on individuals and genes (Williams 1966). Instead males that mature rapidly under such situations must win eviction contests enough of the time that their mode of maturation is selectively favored. Males in multi-male groups must sire enough offspring while retaining their golden pelage, or they must survive at high enough rates to a fully adult state in which they perform well reproductively (Dixson 1998), that this mode of maturation is adaptive.

We would speculate that these differences are alternative tactics *A. caraya* males have evolved to deal with fringe habitats, where fragmentation may have been common in the past, though not on the present scale. These tactics are probably environmentally dependent—as opposed to microevolutionary differences among populations—and can be thought of as a norm of reaction. Males from either kind of group could follow either developmental mode but the social conditions, which may correlate very directly with habitat fragmentation, will induce which pattern is followed.

Some long-term studies have found relationships between human disturbance, habitat loss or fragmentation and a decrease in density of certain primate populations (e.g., *Gorilla gorilla* [Harcourt and Fossey 1981; Watts 1985]; *Cercopithecus mitis* [Lawes 1992]; *Cercocebus galeritus galeritus* [Medley 1993]). Despite inconsistencies in the responses of primates to different degrees of fragmentation across sites, some of this variation may relate to life history differences among species. Spider monkeys (*Ateles*) and muriquis (*Brachyteles*) have extremely long interbirth intervals compared to other platyrrhines and, aside from *Cebus*, much late ages of first reproduction. These traits normally imply that recovery from major environmental alterations will take many years (Strier 2000b). This may indicate a trade-off. Long interbirth intervals may afford these species a buffer during food shortfalls, if they forgo reproduction in such periods. However, such long interbirth intervals *certeris paribus* also depress population growth rates. How these responses relate to high levels of fragmentation is not yet known, though lowered adult survival rates would be expected to have a greater impact.

Primate populations respond in different ways to different environments (i.e., reaction norm differences) making generalization or extrapolation difficult (Marsh 2003). For example, howlers seem to cope easily with alteration of habitat, maintaining population numbers (Zunino et al. 2007) or increasing even when the fragments become smaller (Bicca-Marques 2003; Rodriguez-Luna et al. 2003). In contrast, spider monkeys simply cannot be found in certain fragments (Ferrari et al. 2003; Gilbert 2003; Marsh 2003). These responses vary across sites and populations, and seem to be related to both the characteristics of the fragments and the species under consideration. These differences are usually attributed to the variation in ecological specialization between these taxa. For example, howlers, being more folivorous than other atelines, are able persist in seasonal environments and higher latitudes because they can rely mostly on leaves when fruits and flowers decrease in availability. Howlers also have a shorter interbirth interval, shorter gestation length, and lower ages of weaning and first reproduction than other atelines (Di Fiore and Campbell 2007). These characteristics together possibly make howlers, a successful genus in habitats with lower diversity of plant species, and as such howlers are often considered colonizing species (Crockett and Eisenberg 1987). Morphological attributes, beyond those related solely to life history, clearly play roles in these kinds of contrasts.

5.5 Conclusions

A planned conservation effort should include concentrated study of life history schedules and their relationship with ontogeny, genetic patterns, and population dynamics (Hapke et al. 2001; Kappeler and Pereira 2003). Life-history strategies, which are not only constrained but interact with phylogeny, demography and ecology deeply affect the viability of endangered primate populations (Strier 2003).

We recommend abandoning "fast vs. slow" characterizations of life histories and focus on more general theoretical structures such as energetic trade-offs and the life history mode in addressing how life histories should be empirically researched and used in the planning of conservation efforts. Demographic modeling, even with limited data, can provide valuable information on which life history traits should be targeted to impact population growth rates most favorably from a strictly biological standpoint. Field study of the variation in and covariation among life history traits in response to habitat change further inform such efforts, and morphological investigations of development serve to highlight how primates negotiate demographically important mortality risks in the face of trade-offs.

We emphasize that the little available data on life history and genetic changes in response to habitat fragmentation indicate differing reaction norms among species. Some species, such as howlers, cope adequately with fragmentation while others simply cannot and some of these differences relate to feeding ecology and social organization (Kowalewski and Zunino 1999; Clarke et al. 2002; Bicca-Marques 2003; Estrada et al. 2006; Muñoz et al. 2006; van Belle and Estrada 2006; Pozo-Montuy and Serio-Silva 2006; Zunino et al. 2007). We recommend concentrating conservation studies on the understanding of the total life history strategy across species and the potentially differing responses of populations within species.

5.6 Summary

We explore the connections among demography, life histories, and growth and development in primate evolution assessing responses to habitat change, with an emphasis on New World monkeys. An appreciation of life history, and particularly the relations between life history, morphology, and demography can contribute solutions to vexing conservation problems and illuminate underappreciated adaptive diversity in New World monkey life histories. We briefly discuss demographic modeling, and relate how even very basic data can provide insights into understanding threats to populations. Second, we consider how patterns of morphological development impact the course of life history. Morphological analyses have traditionally played little role in either demographic analyses or conservation efforts, but our studies reveal considerable potential for morphology in understanding how life histories, and thus the basic demographic properties of populations, evolve. Finally, these dual foundations permit us to bring the concept of reaction norms to bear on conservation questions for primates in disturbed habitats. Throughout, we emphasize the limited value of a "'fast vs. slow" perspective on life histories for conservation planning and understanding adaptation.

Appendix

Partial derivatives of λ with respect to the life history parameters in Charlesworth's model (l_α, b, α, and P) can be obtained through implicit differentiation. The partial derivatives are the sensitivities of λ to each of the parameters, symbolized as $s(x)$ where x is the parameter in question. These are given below.

$$s(l_\alpha) = \frac{\partial \lambda}{\partial l_\alpha} = \frac{\dfrac{b\lambda^{-\alpha}}{1 - P\lambda^{-1}}}{\dfrac{l_\alpha b\lambda^{-\alpha}\alpha}{\lambda(1 - P\lambda^{-1})} - \dfrac{l_\alpha b\lambda^{-\alpha}P}{\lambda^2(1 - P\lambda^{-1})^2}} \qquad (5.2)$$

$$s(b) = \frac{\partial \lambda}{\partial b} = \frac{\dfrac{l_\alpha \lambda^{-\alpha}}{1 - P\lambda^{-1}}}{\dfrac{l_\alpha b\lambda^{-\alpha}\alpha}{\lambda(1 - P\lambda^{-1})} - \dfrac{l_\alpha b\lambda^{-\alpha}P}{\lambda^2(1 - P\lambda^{-1})^2}} \qquad (5.3)$$

$$s(\alpha) = \frac{\partial \lambda}{\partial \alpha} = \frac{\dfrac{l_\alpha b\lambda^{-\alpha}\log_e\lambda}{1 - P\lambda^{-1}}}{\dfrac{l_\alpha b\lambda^{-\alpha}\alpha}{\lambda(1 - P\lambda^{-1})} - \dfrac{l_\alpha b\lambda^{-\alpha}P}{\lambda^2(1 - P\lambda^{-1})^2}} \qquad (5.4)$$

$$s(P) = \frac{\partial \lambda}{\partial P} = \frac{\dfrac{l_\alpha b\lambda^{-\alpha}}{\lambda(1 - P\lambda^{-1})^2}}{\dfrac{l_\alpha b\lambda^{-\alpha}\alpha}{\lambda(1 - P\lambda^{-1})} - \dfrac{l_\alpha b\lambda^{-\alpha}P}{\lambda^2(1 - P\lambda^{-1})^2}} \qquad (5.5)$$

Elasticities can be obtained from the sensitivities by multiplying the sensitivity by the ratio of the parameter to λ.

$$e(x) = \frac{\partial \lambda}{\partial x}\frac{x}{\lambda} \qquad (5.6)$$

Note that this multiplication results in $e(l_\alpha) = e(b)$. An identical life cycle model and derivation of sensitivities can be found in Skalski et al. (2005).

References

Alberts, S. C. and Altmann, J. 2003. Matrix models for primate life history analysis. In P. M. Kappeler and M. E. Pereira (eds.), *Primate Life Histories and Socioecology* (pp. 66–102). Chicago: University of Chicago Press.

Altmann, J. and Alberts, S. C. 2003. Variability in reproductive success viewed from a life history perspective in baboons. *American Journal of Human Biology* 15: 401–409.

Altmann, J. and Alberts, S. C. 2005. Growth rates in a wild primate population: ecological influences and maternal effects. *Behavioral Ecology and Sociobiology* 57: 490–501.

Atmoko, S. U. and van Hoof, J. A. R. A. M. 2004. Alternative male reproductive strategies: male bimaturism in orangutans. In P. Kappeler and C. van Schaik (eds.), *Sexual Selection in Primates: New and Comparative Perspectives* (pp. 196–207). Cambridge: Cambridge University Press.

Badyaev, A. V., Whittingham, L. A., and Hill, G. E. 2001. The evolution of sexual size dimorphism in the house finch III. Developmental basis. *Evolution* 55: 176–189.

van Belle, S., and Estrada, A. 2006. Demographic features of *Alouatta pigra* populations in extensive and fragmented forests. In A. Estrada, P. A. Garber, M. S. M. Pavelka and L. Luecke (eds.), *New Perspectives in the Study of Mesoamerican Primates: Distribution, Ecology, Behavior, and Conservation* (pp. 121–142). New York: Springer Press.

Bicca-Marques, J. C. 2003. How do howler monkeys cope with habitat fragmentation? In L. K. Marsh (ed.), *Primates in Fragments: Ecology and Conservation* (pp. 283–303). New York: Kluwer Academic/Plenum.

Bicca-Marques, J. C. and Calegaro-Marques, C. 1998. Behavioral thermoregulation in a sexually and developmentally dichromatic neotropical primate, the black-and-gold howling monkey (*Alouatta caraya*). *American Journal of Physical Anthropology* 10: 533–546.

Bolter, D. R. and Zihlman, A. L. 2003. Morphometric analysis of growth and development in wild-collected vervet monkeys (Cercopithecus aethiops) with implications for growth patterns across Old World monkeys, apes, and humans. *Journal of Zoology* 260: 99–110.

Borries, C., Koenig, A. and Winkler, P. 2001. Variation of life history traits and mating patterns in female langur monkeys *Semnopithecus entellus*. *Behavioral Ecology and Sociobiology* 50: 391–402.

Caswell, H. 2000. Prospective and retrospective perturbation analyses: their roles in conservation biology. *Ecology* 81: 619–627.

Caswell, H. 2001. *Matrix Population Models: Construction, Analysis, and Interpretation*, 2nd edition. Sunderland: Sinauer Associates.

Chapman, C. A., Wasserman, M. D., Gillespie, T. R., Speirs, M. L., Lawes, M. J., and Ziegler, T. E. 2006. Do nutrition, parasitism, and stress have synergistic effects on red colobus populations living in forest fragments? *American Journal of Physical Anthropology* 131: 525–534.

Charlesworth, B. C. 1980. *Evolution in Age-Structured Populations*, 1st edition. Cambridge: Cambridge University Press.

Charnov, E. C. 1993. *Life History Invariants*. Oxford: Oxford University Press.

Cheney, D. L., Seyfarth, R. M., Fischer, J., Beehner, J. C., Bergman, T. J., Johnson, S. E., Kichen, D. M., Palombit, R. A., Rendall, D., and Silk, J. B. 2006. Reproduction, mortality, and female reproductive success in chacma baboons of the Okavango Delta, Botswana. In L. Swedell and S. R. Leigh (eds.), *Reproduction and Fitness in Baboons: Behavioral, Ecological, and Life History Perspectives* (pp. 147–176). New York: Springer.

Clarke, M. R., Collins, D. A., and Zucker, E. L. 2002. Responses to deforestation in a group of mantled howlers (*Alouatta palliata*) in Costa Rica. *International Journal of Primatology* 23: 365–381.

Crockett, C. M. 1996. The relation between red howler monkey (*Alouatta seniculus*) troop size and population growth in two habitats. In M. A. Norconk, A. L. Rosenberger, and P. A. Garber (eds.), *Adaptive Radiations of Neotropical Primates* (pp. 489–510 and 550–551). New York: Plenum Press.

Crockett, C. M. and Eisenberg, J. F. 1987. Howlers: variations in group size and demography. In B. B. Smuts, D. L. Cheney, R. M. Seyfarth, R. W. Wrangham and T. T. Struhsaker (eds.), *Primate Societies* (pp. 54–68). Chicago: University of Chicago Press.

Crouse, D. T., Crowder, L. B., and Caswell, H. 1987. A stage-based population model for the loggerhead sea turtle and implications for conservation. *Ecology* 68: 1412–1423.

Crowder, L. B., Crouse, D. T., Heppell, S. S., and Martin, T. H. 1994. Predicting the impact of turtle excluder devices on loggerhead sea turtle populations. *Ecological Applications* 4: 437–445.

Crowder, L. B., Hopkins-Murphy, S. R., and Royle, A. 1995. Estimated effect of Turtle-Excluder Devices (TEDs) on loggerhead sea turtle strandings with implications for conservation. *Copeia* 4: 773–779.

Di Fiore, A. and Campbell, C. J. 2007. The atelines: variation in ecology, behavior, and social organization. In C. Campbell, A. Fuentes, K. C. MacKinnon, M. Panger and S. Bearder (eds.), *Primates in Perspective* (pp. 155–185). Oxford: Oxford University Press.

Digby, L. J., Ferrari, S. F., and Saltzman, W. 2007. Callitrichines: the role of competition in cooperative breeding species. In C. Campbell, A. Fuentes, K. C. MacKinnon, M. Panger and S. Bearder (eds.), *Primates in Perspective* (pp. 85–105). Oxford: Oxford University Press.

Dixson, A. F. 1998.*Primate Sexuality: Comparative Studies of the Prosimians, Monkeys, Apes, and Human Beings*. Oxford: Oxford University Press.

Dixson, A. F., Bossi, T., and Wickings, E. J. 1993. Male dominance and genetically determined reproductive success in the mandrill (*Mandrillus sphinx*). *Primates* 34: 525–532.

Estrada, A., Saenz, J., Harvey, C., Naranjo, E., Munoz, D., and Rosales-Meda, M. 2006. Primates in agroecosystems: conservation value of some agricultural practices in Mesoamerican landscapes. In A. Estrada, P. A. Garber, M. S. M. Pavelka and L. Luecke (eds.),*New Perspectives in The Study of Mesoamerican Primates: Distribution, Ecology, Behavior, and Conservation* (pp. 437–470). New York: Springer.

Fernandez-Duque, E. 2007. Aotinae: social monogamy in the only nocturnal haplorhines. In C. Campbell, A. Fuentes, K. C. MacKinnon, M. Panger and S. Bearder (eds.), *Primates in Perspective* (pp. 139–154). Oxford: Oxford University Press.

Ferrari, S. F., Iwanaga, S., Ravetta, A. L., Freitas, F. C., Sousa, B. A. R., Souza, L. L., Costa, C. G., and Coutinho, P. E. G. 2003. Dynamics of primate communities along the Santarem-Cuiaba Highway in south-central Brazilian Amazonia. In L. K. Marsh (ed.), *Primates in Fragments: Ecology and Conservation* (pp. 123–144). New York: Kluwer Academic/Plenum.

Fontaine, R. 1981. The uakaris, genus *Cacajao*. In A. F. Coimbra-Filho and R. A. Mittermeier (eds.), *Ecology and Behavior of Neotropical Primates*, Vol. 1 (pp. 443–493). Rio de Janeiro: Academia Brasileira de Ciencias.

Garber, P. A. and Leigh, S. R. 1997. Ontogenetic variation in small-bodied New World primates: implications for patterns of reproduction and infant care. *International Journal of Primatology* 68: 1–22.

Gerald, M. S. 2003. How color may guide the primate world: Possible relationships between sexual selection and sexual dichromatism. In C. B. Jones (ed.), *Sexual Selection and Reproductive Competition in Primates: New Perspectives and Directions* (pp. 141–171). Norman, OK: American Society of Primatologists.

Gilbert, K. A. 2003. Primates and fragmentation of the Amazon forest. In L. K. Marsh (ed.), *Primates in Fragments: Ecology and Conservation* (pp. 145–157). New York: Kluwer Academic/Plenum.

Hapke, A., Zinner, D., and Zischler, H. 2001. Mitochondrial DNA variation in Eritrean hamadryas baboons (*Papio hamadryas hamadryas*): Life history influences population genetic structure. *Behavioral Ecology and Sociobiology* 50: 483–492.

Harcourt, A. H. and Fossey, D. 1981. The Virunga gorillas: Decline of an "island" population. *African Journal of Ecology* 19: 83–97.

Harvey, P. D. and Clutton-Brock, T. H. 1985. Life history variation in primates. *Evolution* 39: 559–581.

Heppell, S. S., Caswell, H., and Crowder, L. B. 2000. Life histories and elasticity patterns: perturbation analysis for species with minimal demographic data. *Ecology* 81: 654–665.

IUCN. 2001. *IUCN Red List Categories and Criteria. Version 3.1.* IUCN Species Survival Commission. Gland, Switzerland: IUCN.

Jack, K. M. 2007. The cebines: toward an explanation of variable social structure. In C. Campbell, A. Fuentes, K. C. MacKinnon, M. Panger and S. Bearder (eds.), *Primates in Perspective* (pp. 107–123). Oxford: Oxford University Press.

Janson, C. H. and van Schaik, C. P. 1993. Ecological risk aversion in juvenile primates: Slow and steady wins the race. In M. E. Perreira and L. A. Fairbanks (eds.), *Juvenile Primates* (pp. 57–74). Oxford: Oxford University Press.

Johns, A. D. and Skorupa, J. P. 1987. Responses of rain-forest primates to habitat disturbance: A review. *International Journal of Primatology* 8: 157–191.

Kappeler, P. M and Pereira, M. E. 2003. *Primate Life Histories and Socioecology*. Chicago: University of Chicago Press.

Kingdon, J. S. 1980. The role of visual signals and face patterns in African forest monkeys (guenons) of the genus *Cercopithecus*. *Transactions of the Zoological Society of London* 35: 431–475.

Knott, C. D. and Kahlenberg, S. M. 2007. Orangutans in perspective: forced copulations and female mating resistance. In C. Campbell, A. Fuentes, K. C. MacKinnon, M. Panger and S. Bearder (eds.), *Primates in Perspective* (pp. 290–305). Oxford: Oxford University Press.

Kowalewski, M. M. 2007. Patterns of affiliation and co-operation in howler monkeys: an alternative model to explain social organization in non-human primates. PhD dissertation, University of Illinois.

Kowalewski, M. M. and Zunino, G. E. 1999. Impact of deforestation on a population of *Alouatta caraya* in northern Argentina. *Folia Primatologica* 70: 163–166.

Kowalewski, M. M. and Zunino, G. E. 2004. Birth seasonality in *Alouatta caraya* in Northern Argentina. *International Journal of Primatology* 25: 383–400.

Kruuk, L. E. B., Slate, J., Pemberton, J. M., Brotherstone, S., Guinness, F. E., and Clutton-Brock, T. H. 2002. Antler size in red deer: heritability and selection but no evolution. *Evolution* 56: 1683–1695.

Kummer, H. 1990. The social system of hamadryas baboons and its presumable evolution. In M. T. de Mello, A. Whiten and R. W. Byrne (eds.), *Baboons: Behaviour and Ecology, Use And Care. Selected Proceedings Of The XIIth Congress of The International Primatological Society* (pp. 43–60). Brasilia, Brazil

Lawes, M. J. 1992. Estimates of population density and correlates of the status of the samango monkey *Cercopithecus mitis* in Natal, South Africa. *Biological Conservation* 60: 197–210

Lee, P. C. and Kappeler, P. M. 2003. Socioecological correlates of phenotypic plasiticity of primate life history. In P. M. Kappeler and M. E. Pereira (eds.), *Primate Life Histories and Socioecology* (pp. 41–65). Chicago: University of Chicago Press.

Leigh, S. R. 1994. Ontogenetic correlates of diet in anthropoid primates. *American Journal of Physical Anthropology* 94: 499–522.

Leigh, S. R. 2004. Brain growth, life history, and cognition in primate and human evolution. *American Journal of Primatology* 62: 139–164.

Leigh, S. R. and Blomquist, G. E. 2007 Life history. In C. Campbell, A. Fuentes, K. C. MacKinnon, M. Panger and S. Bearder (eds.), *Primates in Perspective* (pp. 396–407). Oxford: Oxford University Press.

Marsh, L. K. 2003. The nature of fragmentation. In L. K. Marsh (ed.), *Primates in Fragments: Ecology and Conservation* (pp. 1–10). New York: Kluwer Academic/Plenum.

Martin, R. D. 1983. *Human Brain Evolution in an Ecological Context. 52nd James Arthur Lecture on the Evolution of the Human Brain.* New York: American Museum of Natural History.

Martin, R. D. 1990. *Primate Origins and Evolution: A Phylogenetic Reconstruction*. Princeton: Princeton University Press.

Martin, R. D. 1996. Scaling of the mammalian brain: The maternal energy hypothesis. *News in the Physiological Sciences* 11: 149–156.

Medley, K. E. 1993. Primate conservation along the Tana River, Kenya: An examination of the forest habitat. *Conservation Biology* 7: 109–121.

Moreland, R. B., Richardson, M. E, Lamberski, N., and Long, J. A. 2001. Characterizing the reproductive physiology of the male southern black howler monkey, *Alouatta caraya*. *Journal of Andrology* 22: 395–403.

Muñoz, D., Estrada, A., Naranjo, E., and Ochoa, S. 2006. Foraging ecology of howler monkeys in a cacao (Theobroma cacao) plantation in Comalcalco, Mexico. *American Journal of Primatology* 68: 127–142.

Neville, M. K., Glander, K. E., Braza, F., and Rylands, A. B. 1988. The howling monkeys, genus *Alouatta*. In R. A. Mittermeier, A. B. Rylands, A. F. Coimbra-Filho and G. A. B. da Fonseca (eds.), *Ecology and Behavior of Neotropical Primates*, Vol. 2 (pp. 349–453). Washington, DC: World Wildlife Fund.

Norconk, M. A. and Grafton, B. W. 2003. Changes in forest composition and potential feeding tree availability on a small land-bridge island in Lago Guri, Venezuela. In L. K. Marsh (ed.), *Primates in Fragments: Ecology and Conservation* (pp. 211–227). New York: Kluwer Academic/Plenum.

van Noordwijk, M. A., and van Schaik, C. P. 1985. Male migration and rank acquisition in wild long-tailed macaques (*Macaca fascicularis*). *Animal Behaviour* 33: 849–861.

van Noordwijk, M. A. and van Schaik, C. P. 1999. The effects of dominance rank and group size on female lifetime reproductive success in wild long-tailed macaques, *Macaca fascicularis*. *Primates* 40: 105-130.

Packer, C., Collins, D. A., Sindimwo, A., Goodall, J. 1995. Reproductive constraints on aggressive competition in female baboons. *Nature* 377: 689–690.

Pozo-Montuy, G. and Serio-Silva, J. C. 2006. Movement and resource use by a group of *Alouatta pigra* in a forest fragment in Balancán, México. *Primates* 48: 102–107.

Plumptre, A. J. and Reynolds, V. 1994. The effect of selective logging on the primate populations in the Budongo Forest Reserve, Uganda. *Journal of Applied Ecology* 31: 631–641.

Pusey, A., Williams, J., Goodall, J. 1997. The influence of dominance rank on the reproductive success of female chimpanzees. *Science* 277: 828–831.

Rivera, A. and Calme, S. 2006. Forest fragmentation and its effects on the feeding ecology of black howlers (*Alouatta pigra*) from the Calakmul area is Mexico. In A. Estrada, P. A. Garber, M. S. M. Pavelka and L. Luecke (eds.), *New Perspectives in the Study of Mesoamerican Primates: Distribution, Ecology, Behavior, and Conservation* (pp. 189–213). New York: Springer.

Pereira, M. E. and Leigh, S. R. 2003. Modes of primate development. In P. M. Kappeler and M. E. Pereira (eds.), *Primate Life Histories and Socioecology* (pp. 149–176). Chicago: University of Chicago Press.

Rodriguez-Luna, E., Dominguez-Dominguez, L. E., Morales-Mavil, J. E., and Martinez-Morales, M. 2003. Foraging strategy changes in an *Alouatta palliata mexicana* troop released on an island. In L. K. Marsh (ed.), *Primates in Fragments: Ecology and Conservation* (pp. 229–250). New York: Kluwer Academic/Plenum.

Ross, C. 1991. Life history patterns of new world monkeys. *International Journal of Primatology* 12: 481–502.

Rudran, R. and Fernandez-Duque, E. 2003. Demographic changes over thirty years in a red howler population in Venezuela. *International Journal of Primatology* 24: 925-947.

Rumiz, D. I. 1990. *Alouatta caraya*: Population density and demography in northern Argentina. *American Journal of Primatology* 21: 279–294.

Sacher, G. A. and Staffeldt, E. F. 1974. The relation of gestation time to brain weight for placental mammals: Implications for the theory of vertebrate growth. *American Naturalist* 108: 593–615.

Schillaci, M. A. and Stallmann. R. R. 2005. Ontogeny and sexual dimorphism in booted macaques (Macaca ochreata). *Journal of Zoology* 267: 19–29.

Schneider, H. 2000. The current status of the New World monkey phylogeny. *Anais Da Academia Brasileira de Ciencias* 72: 165–172.

Schlichting, C. D. and Pigliucci, M. 1998. *Phenotypic evolution: A reaction norm perspective.* Sunderland, MA: Sinauer Associates.

Skalski, J. R., Ryding, K. E., and Millspaugh, J. 2005. *Wildlife Demography: Analysis of Sex, Age, and Count Data.* New York: Academic Press.

Starck, J. M. and Ricklefs, R. E. 1998. *Avian Growth and Development: Evolution within the Altricial-Precocial Spectrum.* Oxford: Oxford University Press.

Sterck, E. H. M. 1999. Variation in langur social organization in relation to the socioecological model, human habitat alteration, and phylogenetic constraints. *Primates* 40: 199–213.

Strier, K. B. 2000a. Population viabilities and conservation implications for muriquis (*Brachyteles arachnoides*) in Brazil's Atlantic forest. *Biotropica* 32: 903–913.

Strier, K. B. 2000b. *Primate Behavioral Ecology.* Boston: Allyn and Bacon.

Strier, K. B. 2003. Demography and the temporal scale of sexual selection. In C. B. Jones (ed.), *Sexual Selection and Reproductive Competition in Primates: New Perspectives and Directions* (pp. 45–63). Norman, OK: American Society of Primatologists.

Strier, K. B 2007. Conservation. In C. Campbell, A. Fuentes, K. C. MacKinnon, M. Panger and S. Bearder (eds.), *Primates in Perspective* (pp. 496–509). Oxford: Oxford University Press.

Strier, K. B., Boubli, J. P., Possamai, C. B., and Mendes, S. L. 2006. Population demography of northern muriquis (*Brachyteles hypoxanthus*) at the Estacao Biologica de Caratinga/Reserva Particular do Patrimonio Natural-Feliciano Miguel Abdala, Minas Gerais, Brazil. *American Journal of Physical Anthropology* 130: 227–237.

Strier, K. B., Mendes, S. L., and Santos R. R. 2001. Timing of births in sympatric brown howler monkeys (*Alouatta fusca clamitans*) and northern muriquis (*Brachyteles arachnoides hypoxanthus*). *American Journal of Primatology* 55: 87–100.

Turner, I. M. 1996. Species loss in fragments of tropical rain forest: A review of the evidence. *Journal of Animal Ecology* 33: 200–209.

Wasser, S. K., Norton, G. W., Kleindorfer, S., and Rhine, R. J. 2004. Population trend alters the effects of maternal dominance rank on lifetime reproductive success in yellow baboons (*Papio cynocephalus*). *Behavioral Ecology and Sociobiology* 56: 338–345.

Watanabe, K., Mori, A., and Kawai, M. 1992. Characteristic features of reproduction of Koshima monkeys, *Macaca fuscata fuscata*: A summary of thirty-four years of observation. *Primates* 33: 1–32.

Watts, D. P. 1985. Observations on the ontogeny of feeding behavior in mountain gorillas (*Gorilla gorilla beringei*). *American Journal of Primatology* 8: 1–10.

Williams, G. C. 1966. *Adaptation and Natural Selection: A Critique of Some Current Evolutionary Thought.* Princeton: Princeton University Press.

Young, T. C. and Isbell, L. A. 1994. Minimum group size and other conservation lessons exemplified by a declining primate population. *Biological Conservation* 68: 129–134.

Zunino, G. E., Chalukian, S. C., and Rumiz, D. I. 1986. Infanticidio y desaparición de infantes asociados al reemplazo de machos en grupos de *Alouatta caraya*. *Primatologia no Brasil* 2: 185–190.

Zunino, G. E., Kowaleski, M., Oklander, L., and Gonzalez, V. (2007). Habitat fragmentation and population size of the black and gold howler monkey (*Alouatta caraya*) in a semideciduous forest in northern Argentina. *American Journal of Primatology* 69: 1–10.

Chapter 6
Long-Term Field Studies of South American Primates

Karen B. Strier and Sérgio L. Mendes

6.1 Introduction

Field studies on South American primates have historically lagged behind those of Central American and Old World taxa, but research efforts on a variety of different taxa (e.g., *Alouatta, Aotus, Brachyteles, Callithrix, Cebus, Lagothrix, Leontopithecus, Saguinus*) have intensified over the last 25 years (Strier 1994; *see* also Estrada, this volume). Today, field studies on some populations of South American primates are approaching the longest-running Old World primate field studies in their multi-generational durations and in their empirical scopes (Strier 2003a). As a result, long-term studies of South American primates are contributing valuable comparative perspectives on the diversity of primate behavioral responses to ecological and demographic fluctuations.

In this chapter we review some of the key contributions that longitudinal studies on South American primates have made. We begin by considering the different kinds of data that different types of long-term studies can provide. Ecological and demographic changes can occur over a range of temporal scales, and they can have both immediate and persistent effects on behavior. Long-term studies can provide unique perspectives on the ways in which primates adjust their behavior in response to ecological and demographic fluctuations, and they are the only sources of information on life histories, which may mediate behavioral responses. Accumulating individual-based life history data require intensive, continuous research effort, and we therefore consider some of the trade-offs between these and other less intensive sampling protocols that can include a greater number of individuals and multiple groups.

We then review some of the factors that shape population dynamics of South American primates based on insights that have emerged from long-term field studies. We focus on variation in South American primate dispersal regimes and life history patterns, and how these affect population dynamics. The size, composition,

K.B. Strier (✉)
Department of Anthropology, University of Wisconsin-Madison, Madison, WI 53706, USA
e-mail: kbstrier@wisc.edu

P.A. Garber et al. (eds.), *South American Primates,* Developments in Primatology: Progress and Prospects, DOI 10.1007/978-0-387-78705-3_6,
© Springer Science+Business Media, LLC 2009

and density of primate groups are simultaneously affected by ecological variables, such as food availability, predators, and disease, and by social variables, such as the reproductive condition of females. Thus, long-term studies require maintaining perspectives on individual, group, and population level responses.

In the final section of this chapter, we evaluate what long-term studies contribute to our understanding of the viabilities of populations of endangered South American primates. Indeed, urgent conservation concerns on behalf of the endemic genera of Brazil's Atlantic forest were responsible for stimulating the ongoing long-term studies launched during the 1980s on critically endangered taxa such as the lion tamarins (Rylands et al. 2002) and muriquis (Mittermeier et al. 2005). Currently, some 30 of the 185 species and subspecies of primates endemic to South America are classified as endangered or critically endangered (Baillie et al. 2004; Rylands 1995; Rylands et al. 2000), and many are now the subjects of ongoing field studies as well.

6.2 Types of Long-Term Studies

There are no clear criteria for distinguishing long-term field studies from those of shorter durations because the distinction is relative and depends upon the questions being asked. Obviously, a 12-month study is longer than a two-month study, but even an ongoing field study initiated in the early 1980s, such as our own on northern muriquis, is short in comparison to some of the pioneering field studies on Japanese macaques begun in the 1950s (Fedigan and Asquith 1991; Kawai 1958; Kawamura 1958), or those on the Gombe chimpanzees (Goodall 1971, 1990) or the yellow baboons in Amboseli National Park , which were initiated soon after (Altmann and Altmann 1970, 2003).

One criterion for qualifying as a long-term study might be the number of primate generations it spans. Following the IUCN (2001) criterion, a generation is defined as the average age of parents when their offspring are born. A generation can therefore be greater than the minimum age at first reproduction, which also varies with the life history of each species. For South American primates, age at first reproduction ranges from about two years in wild marmosets and tamarins (Digby et al. 2007), to about eight years in wild spider monkeys, woolly monkeys, and muriquis (Di Fiore and Campbell 2007). Primate life spans are long enough for grandparents or even great-grandparents to overlap with their reproductively mature descendants, so at least three or four generations might also be considered. By these calculations, a six- to eight-year study of callitrichines might qualify as long-term, whereas atelins might require some 24–32 years of study.

Implicit in this example is an assumption that a long-term study will encompass at least one complete lifespan, and therefore span multiple generations, of the study subjects. This criterion explicitly emphasizes life histories, and the behavioral adjustments that individuals may make over the duration of their lifetimes. It thus captures behavioral changes associated with development and aging, and that occur in response to fluctuating ecological and demographic conditions encountered during the course of one's life.

Although our focus here is on long-term studies, we clearly recognize that study length is not relevant to certain types of questions, and that longitudinal field data are not the only sources of important discoveries. For example, brief censuses can provide valuable insights into the persistence of wild populations, and some problem-oriented studies, such as field experiments on primate decision-making rules (Bicca-Marques 2005; Bicca-Marques and Garber 2005; Di Bitetti 2005) can be conducted over relatively brief, highly focused periods of time. Even relatively brief studies can involve a substantial investment of time and effort to locate a suitable study site and population, habituate the animals, and learn enough about their behavior and ecology to develop and conduct the research. Moreover, although the probability of observing rare events, unusual behaviors, or extreme ecological or demographic transitions may increase over time and with sampling effort (Weatherhead 1986), such observations are no less informative when they are made fortuitously or while pursuing unrelated endeavors than when they are the focus of deliberately long-term studies.

6.2.1 Long-Term Ecological Studies

What constitutes a long-term ecologically focused field study is similarly affected by whether the goal is to describe seasonal variation or inter-annual variation in aspects of behavior such as diet, ranging, grouping, habitat use, or reproductive patterns. Clearly, a field study conducted during a few consecutive months within the same season will not be sufficient to evaluate patterns of behavioral variation across seasons, and therefore not comparable to a study that spans a full annual cycle and encompasses seasonal transitions. But even field studies that span a full annual cycle will miss potentially significant inter-annual variation in the availability of important food sources, and therefore be insufficient to evaluate how primates respond to unpredictable, catastrophic events or extended periods of surpluses in their food supplies. There is no guarantee that even long-term studies, as we have defined them here, can evaluate the effects of catastrophes. In fact, because long-term studies tend to encompass more variably ecological conditions, they may be less inclined to recognize unusual events as such than shorter-term studies, in which all deviations may seem more remarkable (Weatherhead 1986).

Catastrophic events, however rare, can exert a significant selective force on behavioral, as well as morphological, adaptations. Indeed, just as the "critical functions" of adaptations in molar morphology may not be apparent except during extreme periods of food scarcity (Rosenberger and Kinzey 1976), behavioral adaptations such as the ability of social groups to fission, may only become evident under extraordinary ecological or demographic conditions. The range of ecological conditions observed during a study will increase with the study's duration, and therefore long-term studies are more likely to capture the extreme ecological conditions that provide the evolutionary context for natural selection to operate than are short-term studies.

Fluctuations in the availability of preferred food resources can occur both predictably, such as during annual seasonal changes in rainfall, temperature, and day length, and unpredictably, such as during catastrophic events or extreme climatic regimes, or as a consequence of anthropogenic activities including fragmentation, selective logging, and ecological disturbances caused by livestock. Like other primates elsewhere, South American primates respond to predictable seasonal fluctuations by altering different aspects of their behavior. Changes in diet, activity budgets and ranging patterns often coincide in reinforcing ways. For example, a common response to seasonal scarcities in preferred foods is to minimize energy expenditure, usually through a dietary shift to include a greater reliance on lower quality foods, accompanied by correspondingly reduced activity levels and restricted day ranges. Many populations of South American howler monkeys, for example, shift their feeding behavior to include a higher proportion of mature leaves in their diets when preferred fruits and new leaves are seasonally scarce (Bicca-Marques and Calegaro-Marques 1994; Chiarello 1994; Mendes 1989). Mature leaves yield less readily available energy and are usually more evenly distributed than fruits and new leaves, simultaneously reducing both activity levels and the necessity of ranging widely. Alternatively, seasonal scarcities in the availability of preferred foods may result in increased activity and long distance ranging to maintain a high quality diet. Amazonian woolly monkeys, for example, may make long distance excursions even beyond their typically large home range when preferred fruit resources are seasonally scarce (Peres 1996).

Adjustments in grouping patterns can be brief and localized, such as when a subgroup of tufted capuchin monkeys in Parque Estadual Carlos Botelho, Brazil enters a small or medium sized fruit patch (Izar 2004), or longer lasting, such as when parties of white-bellied spider monkeys in Brazil (Nunes 1995) or Ecuador (Suarez 2006) fission and forage independently for days at a time. Adjusting the number of individuals foraging together reduces feeding competition and ranging costs at all times (Symington 1990), but may be particularly important when preferred foods are scarce. Flexible grouping patterns are thought to be a characteristic that distinguishes New World monkeys from Old World monkeys (Kinzey and Cunningham 1994), and could reflect any one or a combination of other distinctions from Old World primates, including their arboreal life styles, the trade-offs between risks of predation relative to feeding competition, or social bonds that are weaker or less dependent on proximity to maintain (Strier 1999). Yet, although many primates exhibit pronounced variation in the size and composition of their groups across populations and even in the same populations under different ecological conditions over time, some South American taxa, including *Callicebus*, *Aotus*, *Alouatta*, and most callitrichines, maintain consistently cohesive groups. For example, saddleback tamarins in Manu National Park, Peru maintained "stable, extended family groups" over a 13-year period (Goldizen et al. 1996), and the 11 groups of owl monkeys in the eastern Argentinean Chaco remained constant over a three-year period, with very small fluctuations (40–45 individuals) in total population size (Fernandez-Duque et al. 2001).

Responses to unpredictable climatic extremes are similar among South American primates to those reported among Central American and Old World primates. For example, the effects of an extreme drought led to 100% fetal loss among pregnant female cotton-top tamarins in Colombia (Savage et al. 1996), similar to the high infant mortality and reproductive failure described in ring-tailed lemurs during an extended drought in Madagascar (Gould et al. 2003). Over a 30-year period, from 1969 to 1999, the red howler monkey population at Hato Masaguaral in the Llanos of Venezuela experienced growth followed by a 74% decline in numbers, which could be speculatively attributed to disease or possibly drought (Rudran and Fernandez-Duque 2003). A similar population decline, along with the disintegration of existing social groups, occurred among black howler monkeys in Belize following the habitat devastation caused by Hurricane Iris (Pavelka and Behie 2005; Pavelka et al. 2003). By contrast, despite more than 20 years of prior systematic monitoring, we could not attribute the unusually high infant mortality that occurred among northern muriquis at the Reserva Particular do Patrimônio Natural Feliciano Miguel Abdala (RPPN/FMA) in Minas Gerais over a two-year period to any obvious ecological cause (Strier et al. 2006).

More positive demographic responses to ecological changes have been easier to document. For example, one unusually heavy, prolonged rainy season during 1996–1997 may have been responsible for extending the muriquis' mating and conception seasons, and therefore altered the distribution of births (Strier et al. 2003). Population growth has also been attributed to increased food availability in regenerating habitats for the Venezuelan red howler monkeys (Rudran and Fernandez-Duque 2003), and after regenerating growth permitted species such as brown capuchin monkeys and bearded sakis to move between 100 hectare forest plots (Gilbert 2003).

Populations of sympatric primate species in Madagascar (Lehman et al. 2006) and Africa (Chapman et al. 2005) are known to differ in how well they recover in regenerating habitats that have experienced past disturbances, and South American primate communities are no exception. Similar to Old World primates, populations of South American primates with more generalized, folivorous diets tend to fare better in forest fragments and regenerating habitats than those with more specialized, frugivorous diets (Bicca-Marques 2003; Gilbert 2003). Similarly, taxa that occur in a wide range of habitat types, such as capuchin monkeys and tamarin monkeys, tend to adjust to habitat disturbances better than those with narrower ecological niches, such as spider monkeys (Gilbert 2003). The flexibility of other South American primates, such as muriquis, marmosets and lion tamarins, may have evolved under expanding and contracting forest edges of the Atlantic Forest during the Pleistocene (Kinzey 1982). Their evolutionary histories, combined with current pronounced seasonality, may contribute to their ability to persist in disturbed habitats. For these primates, habitat fragmentation and its effects on the size and genetic structure of their populations may be a more serious obstacle to their survival than disturbances that require dietary and behavioral adjustments other than those that their evolutionary histories have prepared them to make.

6.2.2 Demography and Life Histories

The types and quality of demographic data that can be obtained from long-term studies of groups or populations of South American primates can be described by two axes, each of which represents a continuum along which the frequency of

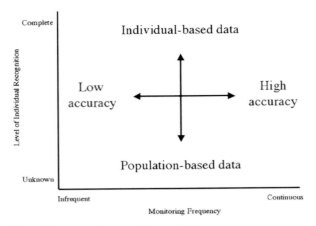

Fig. 6.1 Dimensions of long-term demographic monitoring

monitoring (X-axis) and the degree to which individuals can be identified (Y-axis) can vary (Fig. 6.1). Data accuracy will increase with monitoring frequency because more frequent monitoring will make it less likely to miss events, such as a birth followed by an infant's death. Similarly, the level of individual identification that is possible will affect whether vital rates can be calculated from individual-based life history data or population-level counts.

The ability to identify individuals is not essential for obtaining basic information about population vital rates. For example, systematic censuses can reveal the proportion of infants that survives from one year to the next, and thus provide a basis for calculating first year mortality rates. If periodic censuses are complete (i.e., total counts of all individuals), and ages can be accurately assigned based on visible developmental changes, a life table can be constructed from which vital rates among the distinguishable age-classes can be calculated.

Obtaining data on other life history landmarks, by contrast, requires the ability to identify and follow individuals. For example, determining age at dispersal or age at first reproduction requires following individuals of known age until they disperse or reproduce. A three-year field study on golden lion tamarins might be sufficient to document age at dispersal, but it would not yield a sufficient sample size to understand the sex differences in dispersal patterns that a 12-plus year study monitoring 18 groups has provided (Baker et al. 2002). In species with even slower rates of maturation, such as capuchin monkeys, howler monkeys, and the atelins, both dispersal and first reproduction ages can take much longer to document. Modal age of first birth in female tufted capuchin monkeys in Argentina, for example, was seven years (Di Bitetti and Janson 2001), and in atelins it is even higher.

The degree of continuity in demographic data defines the confidence intervals for determining other life history patterns involving reproductive rates and birth intervals. For example, daily observations of all adult females will yield birth dates to within ±1 day, and such continuous records are less likely to miss the birth of an infant that died than observations that occur at longer intervals. Unless reproduction occurs annually or is strongly seasonal (e.g., owl monkeys: Fernandez-Duque et al. 2002), birth intervals are usually significantly shorter if an infant dies before weaning. Births followed by deaths can be missed during the intervals between observations, contributing to possible error in calculations of birth intervals. For example, we monitored 10 troops of brown howler monkeys at two month intervals over a four year period at the RPPN-FMA (Strier et al. 2001). A total of 34 births were documented during this period, and 12 birth intervals involving ten females could be estimated to within the two-month sampling interval. Three females, however, were not seen with a second infant until 27–38 months after the birth of their previous infant. Because of the two-month intervals between observations, we have no way to confirm whether these birth intervals reflect actual variation, or whether the mothers had given birth, lost their infants, and reconceived again. Similar sources of error have been identified by other researchers with discontinuous demographic data on birth intervals in other South American primates, including white-faced saki monkeys (Norconk 2006) and red howler monkeys (Crockett and Rudran 1987a,b).

Both life history and demographic variables interact to influence the length of time required for a field study to accumulate sufficient data to understand mating patterns or assess whether reproduction is seasonal. Slow rates of maturation and long birth intervals may result in few observations of sexual behavior and small annual birth cohorts. Even in relatively large groups with multiple females, birth intervals of two or three years will result in only a corresponding fraction of the females being reproductively active in any particular year. It is difficult to draw rapid conclusions about sexual and reproductive patterns when the number of reproductive females and annual birth cohorts are small.

The effects of demographic conditions at the onset of a long-term study are even more difficult to anticipate than the influence of life histories. For example, it would have taken more than the six years it did to obtain the first data on female dispersal in northern muriquis if the cohort of six infants from the first year of the study had not happened to include any females. Similarly, adult sex ratios at the onset of a study may be unusually skewed due to past demographic events, yet the impact of this skew on competitive and cooperative behaviors may only become evident once the sex ratios change.

It is also difficult to interpret reproductive and other behavioral patterns in subjects whose groups are either growing or shrinking in size. We could predict, for example, that relationships among adult males might be more relaxed in a group whose adult sex ratio is strongly female-biased compared to another group whose sex ratio is male-biased, or even compared to the same group at another time in its history. Adult sex ratios reflect the cumulative effects of sex ratios at birth, sex-specific mortality, dispersal patterns, and the sex ratios of other groups, all of which can change over the course of an individual's lifetime due to adaptive or stochastic processes (Jones 2005; Strier 2003b). Consequently, it can take many years to be able to distinguish between typical and atypical demographic characteristics of a single study group.

Simultaneous monitoring of multiple groups in a population is one way to understand inter-group dynamics and population demography in less time and without the risks of biases due to individual group histories. Examples of multiple groups monitored over extended periods include those by Robinson (1988) on wedge-capped capuchin monkeys in Venezuela, as well as those by Goldizen et al. (1996) on saddle-back tamarins in Peru, and by Valladares-Padua et al. (1994) on black lion tamarins and Baker et al. (2002) on golden lion tamarins in Brazil (*see* also Kierulff et al. 2002). However, groups of the same species may vary in behavior as well as size and composition across populations. For example, differences in food availability are thought to be responsible for different reproductive patterns in two neighboring populations of *Alouatta caraya* in northern Argentina (Kowalewski and Zunino 2004).

6.2.3 Trade-Offs in Sampling and Research Effort

Like all other kinds of research, long-term field studies reflect compromises that are sometimes, but not always, deliberately made. One of the most influential of these trade-offs involves the intensity with which monitoring occurs, and which often varies inversely with the number of individuals or groups that can be monitored because of limited time, funds, and personnel. High intensity sampling will usually yield more detailed data on the behavior and ecology and life history patterns of a more limited number of individuals and groups. Low intensity sampling, by contrast, can yield more information about the range of variation across a larger number of individuals, groups or populations. Logistics and limitations in the resources available to researchers will often preclude the amount of time any individual or group can be followed, and may therefore necessitate the adoption of one type of protocol instead of another.

The choice between sample size and intensity is not a unique dilemma for long-term studies. As any student of statistics will know, the most appropriate solution will depend on whether or not a small subset of subjects is likely to be representative of the population, which will be influenced by how much variance the population exhibits in the trait of interest. Accumulating sufficient sample sizes to evaluate the range of variation in behavioral responses to ecological or demographic fluctuations, or in reproductive or life history events, is an especially challenging task for primate field researchers because of the slow rates at which these events typically occur.

One way to consider the trade-offs of sampling intensity relative to sample size is to consider the conditions in which one 25-year study of a single study group with 80 individuals in a population with 240 individuals might represent an $N = 1$ group, or an $N = 25$ years (or $N = 300$ months, or $N = 1,300$ weeks), or 30% of the total population size at this site. It might be legitimate to refer to the mean proportion of fruit in the diet across the 25 annual rainy seasons, or the mean sex ratio of each of the 25 annual birth cohorts, if we assume that each rainy season and each birth cohort is independent of the others, and that the 25 years will tell us something about the range of variation in rainy season fruit consumption (possibly related to rainfall that year or the previous year) and infant sex ratios (presumably approximate to 1:1 over time). But a demographic calculation, such as mean group size over the N = 25 year study period is of more dubious utility unless there are indications of demographic stability. In a growing or declining population, the mean group size will tell us little of biological value (Strier 2003a).

It is also important to consider the number of individuals upon which long-term data are based, and the distribution of these individuals among groups. One group with 20 adult females will yield less information about the relationship between group size and reproductive rates than four groups with four females each. However, more information about age at first reproduction may be gleaned from one group with 20 females than from 16 females from four groups.

Demographic and life history data vary in the rates at which they accumulate, not only due to the life histories themselves, but also because the number of individuals in each age-sex category at any time may be small. Understanding variation in these traits can require many years of monitoring unless annual birth cohorts are large, and even then, documenting inter-annual variation requires multiple cohorts. Ecological or demographic fluctuations can affect life histories, and the ability to distinguish responses to changing conditions from normal variation requires a respectable sample size.

6.3 Population Dynamics

Long-term field studies can provide unique perspectives on population dynamics even if they focus on high intensity sampling of a small number of discrete social groups whose individuals are known. The sensitivity of populations to natural demographic fluctuations varies directly with population size, and in small populations these effects can be long-lasting. For example, high infant mortality during an extended drought will affect the size and age structure of groups and populations long after typical rainfall patterns resume. Depending on the duration of the drought and the latency to recovery, entire cohorts may be lost or severely reduced in size. Population crashes can also destabilize existing social groups and reduce the size and density of groups, as has been documented in Belizean black howler monkeys (Pavelka and Behie 2005).

Similarly, successive years of female-biased infant sex ratios, for example, will result in female-biased adult sex ratios, and therefore influence the population's growth rate relative to what it would be if sex-ratios were balanced or male-biased. Infant sex ratios and mortality rates also affect the size and adult composition of groups, and therefore levels of reproductive competition within and between groups.

6.3.1 Dispersal Regimes

Whether dispersal regimes are female-biased, male-biased, or bi-sexual will influence whether expanding populations lead to increases in group size or increases in the number and density of groups (Strier 2000a). For example, bi-sexual dispersal in Venezuelan red howler monkeys may explain the increase in the number of groups instead of the sizes of groups during an extended period of population growth (Rudran and Fernandez-Duque 2003). When dispersal is sex-biased, by contrast, dispersal options are limited to joining extant groups instead of establishing new ones. One group of northern muriquis increased from 22 to 80 members over a 23-year period despite the greater number of female emigrants compared to immigrants. In the same population, however, another group fissioned on two separate occasions when a subset of the females established new groups and two subsets of males from that group ultimately joined them (Strier et al. 2006).

Changes in group size and the number of groups reflect individual dispersal options as well as population-wide conditions. When dispersal patterns are bi-sexual, male-biased sex ratios may limit the ability of dispersing males to establish new groups, and thus result in more multi-male than uni-male groups. When males are philopatric, or cannot secondarily disperse into groups with more favorable sex ratios, levels of male-male competition may rise both within groups and between groups. Female-biased sex ratios will contribute to population growth under any dispersal regime, but the costs of dispersal and both resource and reproductive competition may increase with the size and density of their groups (Strier 2000b).

Long-term studies have the potential to capture how dispersal patterns shift in response to population dynamics. For example, had our study of muriquis begun during a year when a subgroup of transient males was in flux, our obvious inference would have been that in this species it was males, instead of females, that dispersed. Yet, genetic analyses of wild primate populations can also provide insights into the flexibility of dispersal patterns in the absence of extensive demographic monitoring (*see* Di Fiore, this volume). For example, genetic analyses of woolly monkeys at Yasuni National Park, Ecuador revealed that overall relatedness was greater among males than among females within groups, as expected from their female-biased dispersal patterns. Nonetheless, relatedness among some males was lower than expected, suggesting that they were not all from the same natal group (Di Fiore and Fleischer 2005). Although it is not known whether these woolly monkey males fissioned with relatives from their natal group, as we know to have occurred twice among northern muriqui males, the genetic results suggest that dispersal by woolly monkey males may be similarly opportunistic under unusual demographic conditions.

6.3.2 Life History Patterns

As is true for other wild primates elsewhere, genetic data are necessary to confirm paternity, and therefore little is known about age at first reproduction in South American primate males beyond what can be inferred from observed copulations. By contrast, long-term field studies with individual monitoring have documented age at first reproduction among females in various species. Consistent patterns include an extended period of adolescent infertility or subfertility, and later ages at first reproduction among dispersing than among philopatric females.

Adolescent infertility has been documented in many Old World primates, and it is not an extraordinary finding to emerge from long-term field studies. South American primates do, however, provide an unusual diversity of dispersal patterns from which to evaluate the relationships with life histories. Moreover, long-term studies have revealed within population variation in female dispersal patterns, providing unique insights into the costs of dispersal on female age at first reproduction that few Old World primate field studies can match.

Variation in the dispersal patterns of female red howler monkeys at the Llanos of Venezuela is pronounced, with some females dispersing from their natal groups when they are as young as three years of age (Crockett and Pope 1993). When their natal groups are small, daughters may be permitted to remain and reproduce (Pope 2000). Male take-overs and replacements in this population typically occur often enough that risks of inbreeding between daughters and their fathers are low. Reproductive success among philopatric females is estimated to be higher than that among dispersing females, perhaps due to the earlier age at which daughters that remain in their established natal groups reproduced (Pope 2000).

Dispersal patterns among female northern muriquis at the RPPN are less variable than those among red howler monkeys. Northern muriqui females disperse from their natal groups at an average age of six years, and over a 24-year study period, all but three of the 38 females born in the main study group that survived to six years have dispersed into one of the other three muriqui groups in this population. The three females that remained and reproduced in their natal group gave birth to their first infants when they were 7.25–8.67 years old, or some two to three years earlier than the seven females whose first reproductions have been documented in their new groups (updated from Strier et al. 2006).

Dispersing female muriquis leave their natal groups prior to the onset of ovarian cycling (Strier and Ziegler 2000), and both dispersing and philopatric females are sexually active at least one mating season prior to the mating season in which they conceive (Martins and Strier 2004). Dispersing females reproduce later than philopatric females apparently because they begin to cycle later.

In contrast to red howler monkey and northern muriqui females, philopatric marmoset and tamarin females may have lower reproductive success than dispersing females. In captivity, ovarian cycling in sexually mature females is inhibited in their natal groups, but in wild populations, callitrichine females appear to cycle and pursue reproductive opportunities with resident or extra-group males (Arruda et al. 2005). Their lower reproductive rates compared to dominant females in their groups have been attributed to limited or interrupted mating opportunities, social stress resulting in spontaneous abortions, infanticidal attacks, and their inability to recruit other group members to help with infant care (Digby et al. 2007).

Philopatry among callitrichine females appears to be a consequence of limited dispersal opportunities. Females disperse when population density and habitat availability permit them to establish their own breeding groups. Otherwise they may be far better by waiting to inherit their natal territories than by dispersing (Baker et al. 2002).

6.4 Population Viabilities

The demographic and life history data that long-term field studies can provide are essential for evaluating the viability of small populations. Population viability analyses are only as accurate as the data that are used to define the demographic parameters, and are most powerful when the data include reliable estimates of the variance in demographic and life history variables. For example, age-specific mortality rates derived from years of high survival may be misleading unless the variance, which includes years of lower survival, can be included. Similarly, female-biased infant sex ratios will result in more optimistic assessments of population growth than those based on equal or male-biased sex ratios. The utility of long-term data sets with limited variance in these and other demographic parameters can be increased with simulations of a wider range of conditions than the favorable (or unfavorable) ones observed. For example, demographic conditions in one group of northern muriquis remained similarly favorable over a 20-year period, and in year 20 the group's size was similar to predictions based on demographic parameters calculated through year 10 (Strier 1993/1994; Strier et al. 2006). During years 21–25, however, demographic conditions dramatically shifted due to higher infant mortality rates and a strong male-bias among infants. The implications of these less favorable conditions to the group's future growth can be extrapolated from the simulations conducted previously.

Long-term field studies also have the potential to document the effects of catastrophes on populations. Sensitivity analyses, which measure the most vulnerable components of demography and life history, can be powerful analytical tools for evaluating the early warning signs that might signal a population's probability of crashing.

Perhaps the most critical contribution of long-term field studies is their unique ability to provide insights into the processes of population change. These processes include behavioral responses to ecological and demographic fluctuations, and both behavioral and demographic responses to environmental catastrophes. They can also provide informed perspectives on the rates at which small populations can recover, and thus, represent an important source of information for establishing conservation priorities.

We must be cautious, however, about relying on long-term demographic and life history data obtained from single groups or populations. Although useful for first approximations and estimates, comparative long-term data from multiple populations are critical for the development of informed conservation and management plans for endangered species.

6.5 Challenges for the Future

Despite advances in field research on South American primates, there are still many taxa, including some endangered and critically endangered species, for which long-term data are entirely lacking. In nearly all cases, these gaps can be attributed to some combination of logistical difficulties, political or bureaucratic obstacles, and limited funds and research personnel. Logistical difficulties can preclude the establishment of a long-term field study if the remoteness of an area inhibits regular access or the maintenance of supplies that even the most basic field studies and the most robust field researchers require. Some regions of the Amazon that support some of the most diverse primate assemblages in the world are also among the least accessible or hospitable to field researchers.

Political unrest can make field research dangerous, deterring both nationals and foreigners alike. Bureaucratic obstacles can similarly make obtaining and renewing research permissions unpredictable and time-consuming. We know of many dedicated field researchers who have been forced to change their plans to launch long-term studies in South America (and elsewhere in the world) due to insurmountable difficulties they have encountered with both local and national bureaucracies.

Long-term field research also requires a considerable ongoing financial commitment and sufficient interest among students and local people from which reliable personnel can be recruited. Conservation education and the engagement of local communities are important components of any long-term field study that can nonetheless entail an additional financial investment. Gaps in either funding or personnel can result in discontinuities in data collection, and therefore compromise the most unique contributions that long-term studies can make. Most of the funding agencies with which we are familiar tend to favor short-term, problem-oriented studies from which results are assured. Funding commitments to long-term projects, whose most valuable discoveries only emerge over time, are dauntingly difficult to obtain.

While funding is an equal concern throughout South America, access to interested personnel varies greatly. Some projects, such as our own and many we know of, have benefited greatly from long-term international collaborations among scientific colleagues, whether professors or students. Indeed, Brazil has a long tradition of both field primatology and international collaborations, which may be why primate research here has developed so rapidly (Coimbra-Filho 2004; Strier 2000c; Thiago de Mello 1984, 1995; Yamamoto and Alencar 2000).

Recognition of the importance of long-term field data for advancing our understanding of the behavioral ecology, life histories, and population dynamics of South American primates has grown in recent years. Growing awareness of the urgent need for protecting and managing endangered populations has also helped to focus attention on the importance of long-term field studies to conservation concerns. Sustaining the ongoing status of established long-term field studies, promoting new long-term initiatives on behalf of other populations and species of South American primates, and supporting the professional training and career development of new generations of field researchers in these countries should be high on the agendas of both scientists and conservationists alike.

6.6 Summary

Several populations of South American primates have been studied for a decade or more, approaching those on well-known Old World primates in their duration and scope. Long-term studies provide unique insights into the dynamics of wild populations, and are thus of great value in comparative analyses of how different species adjust their grouping patterns and behavior in response to ecological and demographic changes. They provide the demographic data necessary for assessing the viability of small populations of endangered and critically endangered species. South American primates exhibit a diverse range of responses to demographic fluc- tuations consistent with their diverse dispersal patterns and life histories. Changes in group and population sex ratios alter levels of intra-sexual competition and affect the potential for population growth. Results from long-term studies provide a basis for predictions about the ways in which ecology and demography interact to affect individual behavior and population persistence.

Acknowledgments We thank our many colleagues and students who have contributed to our knowledge of South American primates, and the funding agencies and government offices that have provided opportunities for us to conduct our field research in Brazil. We are also grateful to J.C. Bicca-Marques and an anonymous reviewer for their comments on an earlier version of this manuscript.

References

Altmann, S. A., and Altmann, J. 1970. Baboon Ecology. Chicago: University of Chicago Press.
Altmann, S. A., and Altmann, J. 2003. The transformation of behavior field studies. Anim. Behav. 65: 413–423.
Arruda, M. F., Araújo, A., and Sousa, M. B. C. 2005. Two-breeding females within free-living groups may not always indicate polygyny: alternative subordinate female strategies in common marmosets (*Callithrix jachus*). Folia Primatol. 76: 10–20.
Baillie, J. E. M., Hilton-Taylor, C., and Stuart, S. N. 2004. IUCN Red List of Threatened Species. A Global Species Assessment. IUCN, Gland, Switzerland.
Baker, A., Bales, K., and Dietz, J. 2002. Mating system and group dynamics in lion tamarins. In D. Kleiman and A. Rylands (eds.), Lion Tamarins: Biology and Conservation (pp. 188–212). Washington DC: Smithsonian Institution Press.
Bicca-Marques, J. C. 2003. How do howler monkeys cope with habitat fragmentation? In L. Marsh (ed.), Primates in Fragments: Ecology and Conservation (pp. 283–303). New York: Kluwer Academic/Plenum Publishers.
Bicca-Marques, J. C. 2005. The win-stay rule in foraging decisions by free-ranging titi monkeys (*Callicebus cupreus cupreus*) and tamarins (*Sagunius imperator imperator* and *Saguinus fusci- collis weddelii*). J. Comp. Psych. 119: 343–351.
Bicca-Marques, J. C., and Calegaro-Marques, C. 1994. Exotic plant species can serve as staple food sources for wild howler populations. Folia Primatol. 63: 209–211.
Bicca-Marques, J. C., and Garber, P. A. 2005. Use of social and ecological information in tamarin foraging decisions. Intl. J. Primatol. 26: 1321–1344.
Chapman, C. A., Struhsaker, T. T., and Lambert, J. E. 2005. Thirty years of research in Kibale National Park, Uganda, reveals a complex picture for conservation. Intl. J. Primatol. 26: 539–555.
Chiarello, A. 1994. Diet of the brown howler monkey *Alouatta fusca* in a semi-deciduous forest fragment of southeastern Brazil. Primates 35: 25–34.
Coimbra-Filho, A. F. 2004. Os primórdios da Primatologia no Brasil. In S. L. Mendes and A. G. Chiarello (eds.), A Primatologia no Brasil (Vol. 8, pp. 11–35). Vitória: IPEMA/SBPr.

Crockett, C., and Rudran, R. 1987a. Red howler monkey birth data I: Seasonal variation. Am. J. Primatol. 13: 347–368.

Crockett, C., and Rudran, R. 1987b. Red howler monkey birth data II: Interannual, habitat, and sex comparisons. Am. J. Primatol. 13: 369–384.

Crockett, C. M., and Pope, T. R. 1993. Consequences of sex differences in dispersal for juvenile red howler monkeys. In M. E. Pereira and L.A. Fairbanks (eds.), Juvenile Primates: Life History, Development, and Behavior (pp. 104–118). New York: Oxford University Press.

Di Bitetti, M. 2005. Food-associated calls and audience effects in tufted capuchin monkeys, *Cebus apella nigritus*. Anim. Behav. 69: 911–919.

Di Bitetti, M., and Janson, C. 2001. Reproductive socioecology of tufted capuchins (*Cebus apella nigritus*) in northeastern Argentina. Intl. J. Primatol. 22: 127–142.

Di Fiore, A., and Campbell, C. 2007. The Atelines. In C. Campbell, A. Fuentes, K. MacKinnon, M. Panger, S. Bearder (eds.), Primates in Perspective (pp. 155–185). New York: Oxford University Press.

Di Fiore, A., and Fleischer, R. 2005. Social behavior, reproductive strategies, and population genetic structure of *Lagothrix lagotricha poeppigii*. Intl. J. Primatol. 26: 1137–1173.

Digby, L., Ferrari, S., and Saltzman, W. 2007. Callithrichines. In C. Campbell, A. Fuentes, K. MacKinnon, M. Panger, S. Bearder (eds.), Primates in Perspective, (pp. 85–106). New York: Oxford University Press.

Fedigan, L., and Asquith, P. 1991. The Monkeys of Arashiyama: 35 Years of Research in Japan and the West. New York: SUNY Press, Albany.

Fernandez-Duque, E., Rotundo, M., and Sloan, C. 2001. Density and population structure of owl monkeys (*Aotus azarai*) in the Argentinean Chaco. Am. J. Primatol. 53: 99–108.

Fernandez-Duque, E., Rotundo, M., and Ramirez-Llorens, P. 2002. Environmental determinants of birth seasonality in night monkeys (*Aotus azarai*) of the Argentinean Chaco. Intl. J. Primatol. 23: 639–656.

Gilbert, K. 2003. Primates and fragmentation of the Am.azon forest. In L. Marsh (ed.), Primates in Fragments: Ecology and Conservation (pp. 145–157). New York: Kluwer Academic/Plenum Publishers.

Goldizen, A. W., Mendelson, J., van Vlaardingen, M., and Terborgh, J. 1996. Saddle-back tamarin (*Saguinus fuscicollis*) reproductive strategies: Evidence from a thirteen-year study of a marked population. Am. J. Primatol. 38: 57–83.

Goodall, J. 1971. In the Shadow of Man. London: Collins.

Goodall, J. 1990. Through a Window: Thirty Years with the Chimpanzees of Gombe. Boston: Houghton Mifflin.

Gould, L., Sussman, R. W., and Sauther, M. L. 2003. Demographic and life-history patterns in a population of ring-tailed lemurs (*Lemur catta*) at Beza Mahafaly Reserve, Madagascar: A 15-year perspective. Am. J. Phys. Anthropol. 120: 182–194.

IUCN. 2001. IUCN Red List Categories and Criteria: Version 3.1. IUCN Species Survival Commission. Gland, Switzerland and Cambridge, UK: IUCN.

Izar, P. 2004. Female social relationships of *Cebus apella nigritus* in a southeastern Atlantic Forest: An analysis through ecological models of primate social evolution. Behavior 141: 71–99.

Jones, C. B. 2005. Behavioral Flexibility in Primates: Causes and Consequences. New York: Springer Science+Business Media, Inc.

Kawai, M. 1958. On the system of social ranks in a natural group of Japanese monkeys. Primates 1: 11–48.

Kawamura, S. 1958. Matriarchal social order in the Minoo-B Group: A study on the rank system of Japanese macaques. Primates 1: 149–156.

Kierulff, M., Raboy, B., Procópio de Oliveira, P., Miller, K., Passos, F., and Prado, F. 2002. Behavioral ecology of lion tamarins. In D. Kleiman, D. and A. Rylands (eds.), Lion Tamarins: Biology and Conservation (pp. 157–187). Washington DC: Smithsonian Institution Press.

Kinzey, W. 1982. Distribution of Primates and forest refuges. In G. Prance (ed.), Biological Diversification in the Tropics (pp. 455–482). New York: Columbia University Press.

Kinzey, W. G., and Cunningham, E. P. 1994. Variability in platyrrhine social organization. Am. J. Primatol. 34: 185–198.

Kowalewski, M., and Zunino, G. 2004. Birth seasonality in *Alouatta caraya* in northern Argentina. Intl. J. Primatol. 25: 383–400.

Lehman, S., Rajaonson, A., and Day, S. 2006. Edge effects and their influence on lemur density and distribution in Southeast Madagascar. Am. J. Phys. Anthropol. 129: 232–241.

Martins, W. P., and Strier, K. B. 2004. Age at first reproduction in philopatric female muriquis (*Brachyteles arachnoides hypoxanthus*). Primates 45: 63–67.

Mendes, S. L. 1989. Estudo ecológico de *Alouatta fusca* (*Primates*: Cebidae) na Estação Biológica de Caratinga, MG. Rev. Nordestina Biol. 6: 71–104.

Mittermeier, R. A., da Fonseca, G. A. B., Rylands, A. B., and Brandon, K. 2005. A brief history of biodiversity conservation in Brazil. Conserv. Biol. 19: 601–607.

Norconk, M. 2006. Long-term study of group dynamics and female reproductive in Venezuelan *Pithecia pithecia*. Intl. J. Primatol. 27: 653–674.

Nunes, A. 1995. Foraging and ranging patterns in white-bellied spider monkeys. Folia Primatol. 65: 85–99.

Pavelka, M. S. M., and Behie, A. M. 2005. The effect of Hurricane Iris on the food supply of black howlers (*Alouatta pigra*) in southern Belize. Biotropica 37: 102–108.

Pavelka, M. S. M., Brusselers, O. T., Nowak, D., and Behie, A. M. 2003. Population reduction and social disorganization in *Alouatta pigra* following a hurricane. Intl. J. Primatol. 24: 1037–1055.

Peres, C. A. 1996. Use of space, spatial group structure, and foraging group size of gray woolly monkeys (*Lagothrix lagotricha cana*) at Urucu, Brazil. In M. A. Norconk, A. L. Rosenberger, and P. A. Garber (eds.), Adaptive Radiations of Neotropical Primates (pp. 467–488). New York: Plenum Press.

Pope, T. R. 2000. Reproductive success increases with degree of kinship in cooperative coalitions of female red howler monkeys (*Alouatta seniculus*). Behav. Ecol. Sociobiol. 48: 253–267.

Robinson, J. G. 1988. Demography and group structure in wedge-capped capuchin monkeys *Cebus olivaceus*. Behavior 104: 202–232.

Rosenberger, A. L., and Kinzey, W. G. 1976. Functional patterns of molar occlusion in platyrrhine primates. Am. J. Phys. Anthropol. 45: 281–298.

Rudran, R., and Fernandez-Duque, E. 2003. Demographic changes over thirty years in a red howler population in Venezuela. Intl. J. Primatol. 24: 925–947.

Rylands, A. B. 1995. A species list for the New World Primates (Platyrrhini): Distribution by country, endemism, and conservation status according to the Mace-Land system. Neotropical Primates (Supplement) 3: 113–160.

Rylands, A. B., Schneider, H., Langguth, A., Mittermeier, R. A., Groves , C. P., and Rodrigues-Luna, E. 2000. An assessment of the diversity of New World primates. Neotropical Primates 8(2): 61–93.

Rylands, A., Mallinson, J., Kleiman, D., Coimbra-Filho, A., Mittermeier, R., de Gusmão Câmara, I., Valladares-Padua, C., and Bampi, M. 2002. A history of lion tamarin research and conservation. In D. Kleiman and A. Rylands (eds.), Lion Tamarins: Biology and Conservation (pp. 3–41). Washington, DC: Smithsonian Institution Press.

Savage, A., Giraldo, L., Soto, L., and Snowdon, C. 1996. Demography, group composition, and dispersal in wild cotton-top tamarin (*Saguinus oedipus*) groups. Am. J. Primatol. 38: 85–100.

Strier, K. B. 1993/1994. Viability analyses of an isolated population of muriqui monkeys (*Brachyteles arachnoides*): implications for primate conservation and demography. Primate Conservation 14–15: 43–52.

Strier, K. B. 1994. Myth of the typical primate. Yrbk. Phys. Anthropol. 37: 233–271.

Strier, K. B. 1999. Why is female kin bonding so rare: Comparative sociality of New World Primates. In P. C. Lee (ed.), Primate Socioecology (pp. 300–319). Cambridge: Cambridge University Press.

Strier, K. B. 2000a. From binding brotherhoods to short-term sovereignty: the dilemma of male Cebidae. In P. M. Kappeler (ed.), Primate Males: Causes and Consequences of Variation in Group Composition (pp. 72–83). Cambridge: Cambridge University Press.

Strier, K. B. 2000b. Population viabilities and conservation implications for muriquis (*Brachyteles arachnoides*) in Brazil's Atlantic forest. Biotropica 32: 903–913.

Strier, K. 2000c. An American primatologist abroad in Brazil. In S. Strum and L. Fedigan, (eds.), Primate Encounters: Models of Science, Gender, and Society (pp. 194–207). Chicago: University of Chicago.

Strier, K. B. 2003a. Primatology comes of age: 2002 AAPA luncheon address. Yrbk. Phys. Anthropol. 122: 2–13.

Strier, K. B. 2003b. Demography and the temporal scale of sexual selection. In C. Jones (ed.), Sexual Selection and Reproductive Competition in Primates: New Perspectives and Directions (pp. 45–63). Norman, Oklahoma: American Society of Primatologists.

Strier, K. B., and Ziegler, T. E. 2000. Lack of pubertal influences on female dispersal in muriqui monkeys, *Brachyteles arachnoides*. Anim. Behav. 59: 849–860.

Strier, K. B., Mendes, S. L., and Santos, R. R. 2001. Timing in births in sympatric brown howler monkeys (*Alouatta fusca clamitans*) and northern muriquis (*Brachyteles arachnoides hypoxanthus*). Am. J. Primatol. 55: 87–100.

Strier, K. B., Lynch, J. W., and Ziegler, T. E. 2003. Hormonal changes during the mating and conception seasons of wild northern muriquis (*Brachyteles arachnoides hypoxanthus*). Am. J. Primatol. 61: 85–99.

Strier, K. B., Boubli, J. P., Possamai, C. B., and Mendes, S. L. 2006. Population demography of northern muriquis (*Brachyteles hypoxanthus*) at the Estação Biológica de Caratinga/Reserva Particular do Patrimônio Natural - Feliciano Miguel Abdala, Minas Gerais, Brazil. Am. J. Phys. Anthropol. 130: 227–237.

Suarez, S. 2006. Diet and travel costs for spider monkeys in a nonseasonal, hyperdiverse environment. Intl. J. Primatol. 27: 411–436.

Symington, M. M. 1990. Fission-fusion social organization in *Ateles* and *Pan*. Intl. J. Primatol. 11: 47–61.

Thiago de Mello, M. 1984. Treinamento de pessoal em primatologia. In M. Thiago de Mello (ed.), A Primatologia no Brasil, pp. 374–386. Brasília: Sociedade Brasileira de Primatologia.

Thiago de Mello, M. 1995. Treinamento em primatologia no Brasil. *Rev. Brasil. Ciência Veterinária* 2: 69–74.

Valladares-Padua, C., Pádua, S., and Cullen Jr, L. 1994. The conservation biology of the black lion tamarin (*Leontopithecus chrysopygus*: First ten years' report. Neotropical Primates (Supplement) 2: 36–39.

Weatherhead, P. J. 1986. How unusual are unusual events? Am. Nat. 128: 150–154.

Yamamoto, M., and Alencar, A. 2000. Some characteristics of scientific literature in Brazilian primatology. In S. Strum and L. Fedigan (eds.), Primate Encounters: Models of Science, Gender, and Society (pp. 184–193). Chicago: University of Chicago.

Chapter 7
Sexual Selection, Female Choice and Mating Systems

Patrícia Izar, Anita Stone, Sarah Carnegie, and Érica S. Nakai

7.1 General Introduction

In *The Descent of Man and Selection in Relation to Sex*, Darwin (1871) defined sexual selection as the process by which sexually dimorphic traits, which are advantageous to acquiring mates, evolved either by increasing competitive abilities against rivals (intrasexual selection) or by enhancing attractiveness to mates (intersexual selection). He further noted that in polygynous species the higher proportion of males to females would favor sexual selection, thus suggesting a relationship between mating systems and sexual selection – a point that received great attention from biologists with the flourishing of behavioral ecology (*see* Clutton-Brock 2004 for a historical view). The prevalence of competition among males for females, with consequent polygyny, has been attributed to differences between the sexes in parental investment, particularly in regard to the production of gametes, in that females are thought to provide a greater investment (Parker, Baker and Smith 1972; Trivers 1972).

Traditionally, sexual dimorphism in primates has been regarded as a product of intrasexual competition among males (Plavcan 2004), and in spite of more recent studies attempting to investigate the role of intersexual selection, evidence of direct female choice in primates remains scant (Kappeler and van Schaik 2004). However, a third component of sexual selection, "intersexual or sexual conflict", has recently been recognized as a strong selective force operating on both male and female reproductive strategies (Kappeler and van Schaik 2004). Sexual conflict can be defined as "the differences in the evolutionary interests between males and females" (Parker 1979). Although this concept was present in Triver's (1972) discussion of sexual selection, the first assessment of sexual conflict derived mainly from studies on insect reproduction while trying to identify the ultimate factors responsible for the phenomenon of male-induced harm to their mates (Parker 1979; see Rice 2000).

P. Izar (✉)
Department of Experimental Psychology, University of São Paulo, São Paulo Brazil
e-mail: patrizar@usp.br

P.A. Garber et al. (eds.), *South American Primates,* Developments in Primatology:
Progress and Prospects, DOI 10.1007/978-0-387-78705-3_7,

As reviewed by Chapman, Arnqvist, Bangham and Rowe (2003), sexual conflict differs from other forms of sexual selection because female choice represents *avoidance* strategies of male-imposed costs, rather than *preference* for indirect benefits like genetic quality, or for direct benefits like health, nuptial gifts, or parental care.

When the sexes take on different roles in reproduction, conflict will occur over any interaction between them including mating rate, fertilization efficiency and relative parental effort (Arnqvist and Rowe 2005). Sexual conflict theory suggests that males should evolve traits that function to manipulate female mate choice, and that females should evolve traits that counteract the manipulation by males (Clutton-Brock and Parker 1995; Kappeler and van Schaik 2004). Within this theoretical framework, infanticide by males has been interpreted as an adaptive reproductive strategy because it allows faster fertilization of the mother and reduces the possibility of parental investment in unrelated offspring (Hrdy 1979). While infanticidal males may benefit reproductively, this strategy is extremely costly for females. Therefore, infanticide has been argued to be one of the most important selective forces on many aspects of female primate behavior and social organization (Smuts and Smuts 1993; Sterck et al. 1997; Janson 2000; van Schaik 2000b; van Schaik et al. 2004).

If infanticide presents such a strong selective pressure on females, then natural selection is predicted to favor the evolution of female counterstrategies that influence mating behavior, reproductive physiology, and social affiliations including patterns of male–female association (van Schaik and Kappeler 1997; Ebensperger 1998; Nunn and van Schaik 2000; van Schaik 2000b; van Schaik et al. 2004). Within this context, female choice in primates might reflect avoidance of, or protection from, male sexual coercion rather than preference for other benefits (Kappeler and van Schaik 2004).

Female counterstrategies to reduce the risk of infanticide may involve tactics that attempt to confuse paternity among males. Females may do this by mating polyandrously during fertility periods and/or mating during periods of non-fertility. This reproductive strategy suggests that infanticide risk may lead to a multi-male promiscuous mating system (Hrdy 1979). However, counterstrategies may also function to concentrate paternity in one male. This may help explain why unrelated females form groups with one adult protector male leading to uni-male groups, which as is commonly seen in mountain gorillas (Sterck et al. 1997; Harcourt and Greenberg 2001).

The importance of sexual conflict in shaping reproductive strategies in primates has recently been suggested as the reason why observations of females choosing males for their intrinsic qualities are relatively rare. Therefore, obvious female preference for certain males would be expected in species where infanticide risk is low. Low infanticide risk is predicted to be present in highly seasonal breeders, sexually segregated groups, species with low sexual dimorphism, and when females are dominant over males (van Schaik 2000b; Kappeler and van Schaik 2004).

In this chapter, we investigate hypotheses about the relationship between mating system, female choice and male sexual coercion by examining data on female reproductive behavior, female mate choice, and male–female association patterns in three Neotropical primates; black-tufted capuchins, *Cebus nigritus*,[1] white-faced capuchins, *Cebus capucinus* and squirrel monkeys, *Saimiri sciureus*. Although closely related, these species differ in socioecological characteristics such as group size, group structure, patterns of male–female association and infanticide risk. We construct two sets of predictions, one comparing *Cebus* and *Saimiri* and one comparing *C. capucinus* and *C. nigritus*.

Both *Cebus* species live in multi-male/multi-female social groups that range in size from 3–30 individuals (Lynch and Rímoli 2000; Jack 2006). Males are typically dominant over females, but the alpha female usually ranks directly below the alpha male and therefore above the other adult males in the group (Fragaszy et al. 2004). Birth seasonality is variable within the genus. White-faced capuchins in Costa Rica, mate and give birth year round, but experience a slight birth peak during the dry season; however, they are not considered seasonal breeders (Fedigan 2003). Tufted capuchin monkeys present different degrees of birth seasonality in accordance with seasonality of food availability (Di Bitetti and Janson 2000; Bicca-Marques and Gomes 2005). Capuchins have a long life history due to their slow reproductive rates, high lactation to gestation ratio, slow postnatal growth (they are relatively altricial when born), and late age of puberty (van Schaik 2000; Fragaszy et al. 2004). In contrast to *Cebus*, squirrel monkeys have group sizes that are much larger, ranging from 25–50 animals (Mitchell 1990; Boinski 1999; Stone 2007). *Saimiri sciureus* groups are sexually segregated, with males generally remaining at the periphery of troops for most of the year and having weak associations with females. All species of squirrel monkeys are seasonal breeders (Boinski 1987; Mitchell 1990; Stone 2006), and are considered the most seasonal of all Neotropical primates (Di Bitetti and Janson 2000).

The specific traits of these species suggest that *Cebus* females are subjected to a higher risk of infanticide by males than *Saimiri* females. In fact, infanticide in white-faced capuchins is common after male takeovers (Fedigan 2003). Infanticide in tufted capuchin monkeys is less documented, but there is evidence of disappearance or death of infants after male rank takeovers (Rímoli pers. comm.; Izawa 1994). In contrast, infanticidal behavior has never been reported in *Saimiri sciureus* and therefore, infanticide risk is considered low or nonexistent in this species and other *Saimiri* populations (Stone 2004; Boinski et al. 2005). In accordance with our hypothesis that the type of female choice is determined by the risk of male sexual coercion, we predict that *Saimiri* females will display more behavioral patterns reflecting preference for male qualities, rather than patterns that reflect infanticide avoidance behaviors. We also predict that *Cebus* females will display

---

[1] *C. nigritus* is one of the three formerly considered subspecies of *Cebus apella* that were recently elevated to species status (for review see Rylands et al. 2005), and constitute the tufted capuchin group (*C. apella*, *C. nigritus*, *C. libidinosus* and *C. xanthosternos*). However, as noted by Fragaszy et al. (2004), the vast majority of the literature refers to tufted capuchin monkeys as one species; therefore, we will follow their solution and do the same here, except where it is possible to discuss specific features of *C. nigritus*.

more behavioral patterns reflecting strategies to reduce the risk of infanticide, rather than patterns that reflect a preference for male qualities.

While similar in many patterns of social organization, the two *Cebus* species differ in some important aspects. In *C. capucinus*, male dominance is relatively unstable over time due to the frequent dispersal of males (males may change groups more than two times during their lifetime) (Jack and Fedigan 2004), or through rank reversals within groups – although this is less common (Perry 1998a). Resident male eviction commonly occurs through the aggressive invasion by groups of two or three adult males (Fedigan and Jack 2004). In *C. nigritus*, males emigrate starting at around 5–9 years, but multiple dispersal events are not reported, nor collective invasion (Fragazsy et al. 2004), and male tenure is long (Lynch and Rímoli 2000; Di Bitetti and Janson 2001). Male infanticide in *Cebus* seems to occur most commonly after aggressive male takeovers, which are more frequent in *C. capucinus* than in *C. nigritus*. Therefore, if females of *C. nigritus* experience a lower risk of infanticide than *C. capucinus*, we predict that we will observe more behavioral patterns in *C. nigritus* that reflect a preference for male qualities, rather than infanticide avoidance behaviors. We also predict that in *C. capucinus* females, we will observe more behavioral patterns that reflect strategies to reduce the risk of infanticide, rather than preference for male qualities.

The remainder of this chapter is organized as follows: we first present case studies for each species independently, with some specific predictions and conclusions, and then we present a general discussion comparing the three species in terms of mating system, female choice and infanticide risk.

7.2 Case Studies: Black Tufted Capuchin Monkeys, (*Cebus nigritus*)

7.2.1 Introduction

Tufted capuchin monkeys live in multi-male/multi-female groups, but male reproductive success is strongly skewed toward the dominant male, characterizing a uni-male mating system (Janson 1986; Escobar-Páramo 1999). One male can monopolize several females leading to intersexual competition, as shown by active interruption of mating activity (Carosi et al. 2005), post-copulatory behavior consistent with sperm competition (Lynch-Alfaro 2005), and strong sexual dimorphism (Plavcan 2004). According to Plavcan (2004), dimorphism in body mass is exaggerated in species where female choice reinforces male–male competition. In fact, tufted capuchin monkeys are one of the few species with real evidence of direct female choice. Estrous females typically present proceptive behaviors toward the male (see details in Janson 1984; Carosi and Visalberghi 2002), and they usually solicit the dominant male more (Di Bitetti and Janson 2001; Lynch-Alfaro 2005).

But what does female choice in tufted capuchin monkeys mean? Preference for some good quality of dominant males that improves their offspring's chances of survival, such as access to better food sources controlled by the dominant male (Janson 1986, 1998), or avoidance against male coercion and male infanticide (Strier 1999)?

Here we present a study of female choice in a population of tufted capuchin monkeys (*Cebus nigritus*) at Carlos Botelho State Park, an area of Atlantic forest in São Paulo state, Brazil. Tufted capuchin monkeys in this area forage in a fission-fusion manner – social groups split into smaller subgroups of fluid size and composition during foraging for small dispersed foods, and coalesce to feed in large food sources (Izar and Nakai 2006; Nakai 2007). Usually, foraging subgroups contain half of the group total number of adult males (Nakai 2007). Moreover, in this population females may transfer between groups (Izar 2004). These social dynamics could potentially diminish the possibility of control by the dominant male over females, and in fact, experimental evidence indicates that females readily exploit the absence of the dominant male to copulate with other males (Visalberghi and Moltedo unpub.). Therefore, if female choice in tufted capuchin monkeys reflects avoidance of male coercion we expected a low reproductive skew toward the dominant male in the study population. On the other hand, if the dominant male can monopolize better food sources than other males, and female choice is a strategy to have access to these foods, then females should still prefer to copulate with the dominant male. Specifically, we analyze data from two study periods, one when groups were cohesive and one when the groups were foraging in a fission-fusion manner (see Izar 2004; Izar and Nakai 2006). We compared female choice, male–male competition and female–female competition between study periods and between mating and non-mating seasons.

7.2.2 Methods

7.2.2.1 Study Site

The Carlos Botelho State Park (PECB) is located in south-eastern Brazil (in São Paulo State), and along with three other frontier parks, forms a continuous protected area of more than $1200\,km^2$ within the Atlantic forest domain. The park comprises an area of about $380\,km^2$ of mainly undisturbed evergreen forest. Mean annual rainfall is 1683 mm, with high levels of precipitation between September and March (mean rainfall of 174 mm), and a mean rainfall of 78 mm in July, the driest month. The climatic diagram reveals the absence of hydric deficit (Dias et al. 1995). Population density of *C. nigritus* in the study area is low (Izar 2004).

7.2.2.2 Study Group

Data were collected by P. Izar and E.S. Nakai for two main groups: Orange, from November 2001 through December 2003 (totalling *ca* 1000 hours of direct observation), and Pimenta, from May–September 2003 through February 2004–September 2006 (totalling *ca* 750 hours of direct observation). Animals were fully habituated to human observers. Groups were followed daily from dawn to dusk, from 5 to 20 days per month. All adult members and some immatures were individually recognized based on color patterns, body sizes and cap shape. Of the seven adult females in Orange group, just one of them remained in the group during the whole study, three disappeared from the group in 2002, one returned in 2003 and three entered the group in 2002 (Izar 2004). Of the 11 adult females in Pimenta group, five of them were present during the whole study period, one disappeared in 2004, another one in 2005, and one entered the group in 2005. Orange group contained two adult males, the dominant one entered the group in 2002, and Pimenta group contained two adult and two sub-adult males.

7.2.2.3 Data Collection

While collecting systematic data during 5-minute interval scan samples for other purposes, we recorded all occurrences of proceptive behavior presented by females in the two social groups during the study. We consider proceptive behavior to include eyebrow raises with vocalizations, touching and running, following the male at short distances and soliciting, as described in the literature (Janson 1984; Carosi et al. 1999). We recorded the identity of individuals involved, the occurrence of copulations, and occurrence of aggressive interactions during the consortship. We also recorded all agonistic episodes, noting the individuals involved and the context.

Data on fission-fusion (subgroup size and membership) were collected recording the identity and number of individuals that were observed during the day with the initial daily party (Nakai 2007). These data allowed us to calculate an association index between pairs of adult males and females $S_{ij} = a/a + b + c$; where a is the number of times i and j were observed within the same party, b is the number of times i was observed but j was not, and c is the number of times j was observed but i was not.

In a previous paper (Izar 2004), fruits were shown to constitute the limiting food source for capuchin monkeys in this area. Therefore in the present study, to test the null hypothesis that fruit intake does not differ between parties containing different males, we compared fruit patch size and per capita energy intake of different parties. We followed the methodology adopted by Izar (2004), using the "feeding tree focal sample" method (FTFS) devised by Strier (1989) as a measure of fruit patch size, and per-capita individual minutes (PCIM) as a measure of per-capita food intake (Janson 1988).

7.2.3 Results

7.2.3.1 Female Proceptivity

We observed two types of proceptivity episodes (n episodes = 25; n females = 13) regarding the temporal extension of female behavioral changes. The first type (n = 16) involved the typical behavioral patterns described in the literature (Carosi et al. 1999) as associated with the periovulatory phase, such as eyebrow raises with vocalizations, touching and running and following the male at short distances from one to three entire days. The same female could present this type of behavior from three to five months in sequence before getting pregnant, as described for other populations (Di Bitetti and Janson 2001). We labeled these episodes as estrous proceptivity. The second type of proceptivity episodes were short-term, usually involving eyebrow raises and grins, but no vocalizations, and ending with copulation. This second type of proceptivity was presented mainly by females trying to enter a group (n = 7), or during agonistic encounters between groups (n = 1), or toward a newcomer male (n = 1), and in 70% of the cases, the female was carrying a dependent offspring. We considered this second type as non-estrous proceptivity.

We observed proceptivity in eight months out of 12 across 5 years of data collection (Fig. 7.1). Only one female presented proceptive behaviors at a given time throughout the course of the study. Estrous proceptivity occurred between January and August, with a peak between March and May, the end of the rainy season and the beginning of the dry season. Estrous proceptivity between June and August was presented only by a sub-adult female. In contrast, non-estrous proceptivity occurred between July and October, with one episode in January. Considering the mean gestation length of 5 months, most observed births (n = 26) occurred from August to December, with a peak in November, the beginning of the wet season (Fig. 7.1). These data indicate that, as described for *C. nigritus* in other study areas, most conceptions occurred during the dry season (Di Bitetti and Janson 2001; Lynch-Alfaro 2005).

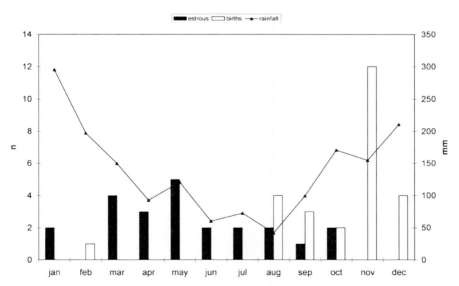

Fig. 7.1 Distribution of mean monthly rainfall from 2001 to 2005, of births, and of proceptive episodes of tufted capuchin monkeys (*Cebus nigritus*) in PECB

We could estimate the interbirth interval for 10 females, which was 30.22 ± 6.43 months, significantly longer (t-test $= 5.043$, p < 0.001) than the interbirth interval (19.4 mo) of *C. nigritus* females from Iguazú National Park, studied by Di Bitetti and Janson (2001).

7.2.3.2 Association Between Males and Females

There was no effect of estrous proceptivity in male–female associations. During the fission-fusion period, the most frequent party size was around half the total group size and the most frequent party composition contained half the total number of adult males (Nakai 2007). There was no difference (t-test $= -1.31$; df $= 24$; p $= 0.20$) in party size between the mating (mean $\pm$sd $= 7.7 \pm 2.0$ individuals) and the non-mating seasons (9.0 ± 2.4 individuals).

The association index of pairs of adult group members differed according to sex ($F = 3.93$, df $= 2$; p < 0.05). Male–female pairs were more frequently associated within the same party (mean $S_{ij} = 0.23 \pm 0.20$) than female–female pairs (mean $S_{ij} = 0.19 \pm 0.18$; t-test $= 2.79$, p < 0.01).

7.2.3.3 Female Choice

During the estrous proceptive episodes, the female would typically follow, solicit and copulate with only one male in the group. Females followed and copulated with the dominant male of the group in 75% (n $= 12$) of the estrous proceptive episodes. If we consider the two study periods, when groups were cohesive females always solicited the dominant male and when groups were forming subgroups females solicited another male in 33.3% of the episodes. There was a weak tendency for a higher preference for dominant males during the period of group cohesion in comparison to the period of fission-fusion (Binomial test; z $= 1.33$; p unilateral $= 0.09$). All four cases when a female solicited another male occurred during the fission-fusion period. In two cases, the female followed and copulated with the beta male, but they were in a subgroup without the dominant male. In only two cases the female followed a sub-adult male, that rejected her, and then she solicited the dominant male. In all four cases, the females had a lower association index with the dominant male than with the solicited male.

Females solicited the dominant male (or the largest one in the case of subgroups where the dominant male was not present and there was more than one male) of a group in seven out of nine non-estrous proceptive episodes, and in only three episodes the dominant male rejected the female. In these three episodes, a new female was apparently trying to enter the group and did not succeed (see Izar 2004 for details).

7.2.3.4 Male–Male Competition

During estrous episodes, as described in other studies, the chosen male was reluctant to accept the female's solicitations at the outset. Male behavior changed in the middle of the period, when the consortship pair would remain in close proximity and

Table 7.1 Frequency of agonistic conflicts between individuals of different age-sex classes of tufted capuchin monkeys (*Cebus nigritus*) in PECB. Period 1 = group cohesion; Period 2 = fission-fusion; m = adult male; f = adult female, j = juvenile

Participants	Period	Expected Count	Observed Count
AF-AF	1	10.4	13
	2	6.6	4
AF-AM	1	7.9	9
	2	5.1	4
AM-AM	1	17.1	14
	2	10.9	14
AF-Juvenile	1	21.3	32
	2	13.7	3
AM-AF	1	12.2	9
	2	7.8	11
AM-Juvenile	1	15.9	15
	2	10.1	11
Juvenile-Juvenile	1	23.8	17
	2	15.2	22

apart from the group for an entire day. Most of the copulations were observed in this situation (one to four per day between a given pair). The male would avoid the proximity of other group members, including immatures, by displaying aggressiveness. However, we never observed interruption of copulations by other males.

Comparing the two study periods, there was a significant drop in the frequency of observed agonistic episodes among all group members during the fission-fusion period in relation to the cohesion period ($\chi^2 = 91.94$; df = 3, p < 0.001). When we compared these rates, considering the age-sex class of aggressors and victims involved, we found that this drop was caused mainly by females (Table 7.1). There was no difference in frequency of male–male aggression between mating and non-mating seasons across the two study periods ($\chi^2 = 1.70$; df = 1, p = 0.68). In both periods, the major context of male–male agonism was exclusion of other males by the dominant male from his vicinity, either in the presence of an estrous female during the mating season, or from a sleeping or feeding tree.

The comparison of fruit acquisition between different parties revealed an effect of year so that significant differences were observed only for 2005: individuals in subgroups with the dominant male fed on fruit trees for more time than individuals in subgroups with subordinate males (t-test = 2.54; df = 42.6; p < 0.05), with a tendency to feed with less groups members in the same fruit tree (t-test = -2.020; df = 42.8; p = 0.05). There was no difference in subgroup and in *per capita* energy intake in subgroups with the dominant versus subordinate males.

7.2.4 Discussion

7.2.4.1 Mating Behavior

Tufted capuchin monkeys in PECB exhibit an exceptional social organization, considering that in most studied populations, females are philopatric and groups are cohesive (Izar 2004, see also Lynch-Alfaro 2007). The data presented here indicate that the reproductive strategies employed by female capuchins in PECB, albeit their exceptional social organization, are similar to those described for other wild and captive populations. Mating behavior occurred mainly during the dry season, which is considered the conception period of all wild populations and southern-captive populations of tufted capuchins studied (Di Bitetti and Janson 2001; Bicca-Marques and Gomes 2005). It is suggested that mating seasonality reflects birth seasonality, which is an adaptation to guarantee that infants are weaned during the period of higher food abundance (Di Bitetti and Janson 2000).

Reproductive rate was lower in PECB than in any other studied wild population. Female capuchin monkeys usually reproduce once every two years (Fragaszy et al. 2004), while in PECB they reproduced once every three years. Consequently, sexual activity was much less frequent here than in other studied areas. We observed 24 episodes of proceptive behavior involving 13 females in 5 years, while Lynch-Alfaro (2005) observed 31 episodes of 11 females in one year. This reproductive rate is probably related to the low food availability at PECB, and might explain the extremely low population density of capuchin monkeys found in this area compared to other sites, even in spite of the low predation risk that the monkeys face at PECB (Izar 2004).

The proceptive behavioral repertoire, which is considered indicative of active female choice for a male mate (Kappeler and van Schaik 2004) is also consistent with previous descriptions for the species (Carosi et al. 2005). Again, in accordance with most studied wild and captive populations, females more often chose the dominant male of the study groups in PECB. However, contrary to other populations where groups are cohesive and the dominant male can control females' choices, PECB females could have solicited other males while they were in separate subgroups, and at least two females did so. Although data are scarce, they suggest that female choice during estrous period reflects their bonds with particular males.

7.2.4.2 Female Choice

Most female capuchins at PECB chose to solicit the dominant male during their estrous period, and this preference tended to be higher when groups were cohesive than when groups were foraging in parties. This result might indicate that the choice for the dominant male by female tufted capuchins may represent preference for his ability to defend better food sources, avoidance of dominant male coercion or reduce the risk of infanticide. Our data do not support the first hypothesis, because when groups were foraging in subgroups, individuals that associated with the dominant male did not achieve a higher per capita food intake than individuals that associated with subordinate males.

Our data also do not support the hypothesis of avoiding male coercion because we have never observed evidence of male coercion such as harassment or interference during copulation with other males. Several studies show that dominant males rarely harass females during copulation with other males (Linn et al. 1995), but they can interrupt these copulations (Janson 1984; see Carosi et al. 2005 for a review; Lynch-Alfaro 2005). However, our result may be a consequence of lack of opportunity, because on only two occasions did females solicit another male in the presence of the dominant one, and in both cases were rejected. Therefore, males seem to behave in accordance with a formal dominance hierarchy so that subordinate males avoid aggression from the male (Carosi et al. 2005).

Why do females still choose the dominant male, more often, even when they could have "exploited the opportunity" of his absence, in foraging subgroups to solicit another male? The data presented here are most consistent with the hypothesis that female choice is a strategy against infanticide by males. We did not observe infant killing by males in our study, but the rarity of the phenomenon can indicate that females evolved counterstrategies that might have been actually effective (Sommer 2000). Some of our data reported here, such as the display of proceptive behavior outside the estrous period by females with dependent offspring when trying to enter a new group or toward a newcomer male, are indicative of a strategy against infanticide. This type of behavior has already been observed, usually toward newcomer males (Cooper et al. 2001; Carosi et al. 2005). It is possible that even if females are free to copulate with other males while in different subgroups, they might face the risk of infanticide by the dominant male when parties coalesce.

The suggestion that females are reducing the risk of infanticide by copulating only with the dominant male does not exclude other reasons for copulation including social benefits such as parental care or gaining his tolerance (Escobar-Páramo 1999), a better spatial position to reduce predation risk and increase access to food (Janson 1990a,b), a coalitionary partner for intragroup conflicts (Izar 2004), and a coalitionary partner for offspring protection (Ferreira et al. 2006).

7.3 White-Faced Capuchins, *Cebus capucinus*

7.3.1 Introduction

Cebus capucinus are one of four species referred to as the "un-tufted" group of capuchins (also includes *C. albifrons*, *C. olivaceus*, and *C. kaapori* (Fragaszy et al. 2004). They are the only *Cebus* species that ranges throughout Central America and can also be found on the north western edges of Columbia and Ecuador (Fragaszy et al. 2004). White-faced capuchins form multi-male/multi-female social groups that range in size from 10 to 30 individuals and consist usually of two or more matrilines (Jack 2006). Females remain philopatric and rarely (if ever) change groups, while males emigrate starting at a median age of four years and continue to change groups approximately every three to four years throughout their lives (Jack and Fedigan 2004a,b). *Cebus* females start to reproduce at six years of age and have singleton births every 2–2.5 years (Fragazy et al. 2004). White-faced capuchins are moderately sexually dimorphic, with males being 25–35% larger than females (Fedigan 1993).

C. capucinus are considered to be at a high risk of infanticide because adult males are not related, females can mate and give birth at any time of the year, and they have a slow life history. In this population of capuchins (Santa Rosa National Park, Costa Rica), group takeovers occur on average every three to four years (male tenure is relatively short), and are often associated with the deaths or disappearances of young infants (Fedigan 2003).

As suggested by van Schaik (2000a) for multi-male groups, female white-faced capuchins face a dilemma of having to concentrate paternity in the male most capable of protecting her offspring, yet confuse paternity among males to reduce the risk of infanticide. C. capucinus do not behaviorally display a conspicuous estrous period as is seen in some tufted capuchin species (e.g., C. nigritus). This "concealment" of fertility may have evolved to both concentrate and confuse paternity in their multi-male breeding system, and as such female mate choice is not conspicuous. These reproductive traits make C. capucinus an excellent species in which to study female mate choice strategies that have been shaped by a high degree of infanticidal pressure.

This study compares the occurrence of sexual behavior and female proceptive behaviors during the ovarian cycle and across reproductive states (i.e., cycling, lactating, pregnant) to investigate behavioral indicators of female mate choice and reproductive strategies in relation to hormonal patterns.

7.3.2 Methods

7.3.2.1 Study Site

This study took place between January and June, 2002, in Santa Rosa National Park (SNRP), Costa Rica; the original sector of the Area de Conservacion Guanacaste (ACG). SRNP is located in north western Costa Rica close to the Nicaraguan border and encompasses approximately 108 ha of dry deciduous forest fragments, including semi-evergreen, riparian and regenerating pasture land. The park is home to three primate species; Alouatta palliata, Cebus capucinus and Ateles geoffroyi, which have been the subjects of many behavioral ecology studies (Fedigan and Jack 2001). Santa Rosa experiences a dry season (December to May), when virtually no rain falls, and a wet season (June to November), with an average of 1470 mm of rainfall.

7.3.2.2 Study Subjects

The subjects of this study were 10 adult female white-faced capuchins who constituted all the adult female members of two habituated groups: CP and LV (five females per group). They ranged in age from six to 24 years and varied in rank within each group. Eight females were multiparous, and two were nulliparous at the start of the study. During this study, there were six adult males in the two groups; CP: one alpha male and one subordinate male; LV: one alpha male and three subordinate males. Individuals were easily identified by natural markings such as broken digits, scars, fur coloring, blotches on faces and the brow and peak shape. Both groups live in similar habitat types and are exposed to similar environmental pressures including water scarcity, predator pressure, food availability and level of human exposure.

7.3.2.3 Behavioral Data Collection

Continuous focal follows of 15 minutes were conducted on each subject female, during which time all behaviors and interactions were recorded (Altmann 1974) into a Psion Workabout computer (total of 443 hours of focal data, ranging between 38 and 41 hours/female). An exhaustive ethogram was used, which has been developed over the years by previous *C. capucinus* researchers to identify and code behaviors. When fast moving events occurred, such as fights or sexual displays, the behavioral sequences were dictated into a micro-cassette recorder (Sony M-430) and later transcribed and integrated into the behavioral data set for that day.

7.3.2.4 Fecal Collection and Field Extraction

Fecal samples were collected from each female on the same days that behavioral data was collected. A minimum of two to three samples per week per female was collected from each female, which is considered adequate to properly assess female ovarian patterns (Hodges and Heistermann 2003). Fecal samples were collected within 10 minutes of defecation, placed in plastic vials and stored in a cold pack until the end of the day, taking note of the individual and date. Methods followed Strier and Ziegler (1997) for the initial field extraction of the steroids. Samples were analyzed at the National Primate Research Center (NPRC) at the University of Wisconsin in Madison immediately after the field season. In total, 400 samples were analyzed; progesterone was measured using enzyme immunoassays (Ziegler et al. 1996), and estradiol using radio-immunoassays (Strier and Ziegler 1994; Carnegie 2004).

The results of the fecal assays were used to create hormone profiles for each adult female and from the profiles; each female was categorized as being in one of three reproductive states, cycling, lactating or pregnant. Of the 10 subject females, four were cycling, three were lactating, and three were pregnant. Among the cycling females, the three phases of the ovarian cycle (i.e., follicular, luteal and periovulatory) were difficult to interpret so we compared behavioral differences only between the periovulatory and non-ovulatory phases (see Carnegie et al. 2005 for further information on the interpretation of ovarian hormone profiles).

7.3.2.5 Behavioral Analysis

The behavioral variables analyzed for phase and reproductive state variation were copulations, courtship displays and behavioral indicators of proceptivity (i.e., indicators of female choice), and attractivity (i.e., indicators of male interest in the female). Behaviors considered indicative of proceptivity were approaches, follows and grooms *directed* by subject females to adult males. These same behaviors were used as indicators of attractive behaviors when they were *received* by adult females from adult males. These behavioral indicators were compared between ovarian cycle phases (i.e., periovulatory vs. non-ovulatory) and across each category of females (i.e., cycling, lactating, pregnant), by calculating a mean rate (frequency per hour) for each ovarian phase and for each reproductive state.

We used Hinde's Index (Hinde and Atkinson 1970) to decide which member of the male–female dyad was responsible for maintaining proximity. Hinde's Index is the proportion of all of the dyad's *approaches* directed by the subject female, minus the proportion of all of the dyad's *leaves* directed by the female. A negative index indicates that the male was responsible for maintaining proximity; a positive index suggests the female was responsible. *Approaches* to within three meters and *leaves* beyond three meters were used in this analysis.

To compare differences in female behavior between ovarian phases, we used a Randomization technique (Manly 1997), which creates a new distribution of the test statistic from the original data. For this analysis, we randomly scored the original data 10,000 times to create the new distribution that we used to compare our observed test statistic. This is a one-tailed test and significance is reached when probability is < 0.05. To compare behaviors across reproductive states, we used Kruskall Wallis one-way analysis of variance. All frequencies showing a significant difference were further analyzed using a multiple comparison test to determine where the difference existed among the three categories of females (Siegel and Castellan 1988).

7.3.3 Results

7.3.3.1 Sexual Behavior

Copulations ($n = 3$) occurred significantly more often in the periovulatory phases compared to the non-ovulatory phases in cycling females (mean paired difference (MPD) $= 0.247$; $p < 0.001$). Only alpha males copulated with periovulatory females and directed more courtship displays toward cycling females during the periovulatory phases compared to the non-ovulatory phases (MPD $= 0.475$; $p < 0.001$).

We found significant differences in the rates of copulations and courtship displays (copulations all females: $n = 22$; courtship displays all females: $n = 32$) across the three reproductive states (copulations: $\chi^2 = 7.52$; df $= 2$; $p = 0.023$; courtship displays: $\chi^2 = 7.891$; df $= 2$; $p = 0.019$). Pregnant females copulated and received more courtship displays from all adult males than did cycling females. However, 86% of copulations and 60% of courtship displays occurred between pregnant females and subordinate males. Lactating females were never observed to copulate with, or receive a courtship display from, any adult male.

7.3.3.2 Proceptivity

None of the proceptivity indicators showed significant phase-related changes between the periovulatory and non-ovulatory phases. Cycling females directed approaches to adult males at almost two times the rate during the periovulatory phases as compared to the non-ovulatory phases (periovulatory: 1.03/h; non-ovulatory: 0.59/h), but this difference was not significant. We never observed cycling females to either follow or direct a groom solicit to any adult male in any phase of their cycle.

Across reproductive states, the rate of grooming bouts directed to subordinate males by subject females varied significantly ($\chi^2 = 6.85$; df $= 2$; p $= 0.033$). Specifically, pregnant females directed grooming bouts to subordinate males at significantly higher rates (0.41/hr) than did cycling (0.08/hr) and lactating females (0.10/hr).

7.3.3.3 Attractivity

The rate of received grooming bouts from males and the rate of follows received from males occurred significantly more often during the periovulatory compared to the non-ovulatory phases for cycling females (grooming bouts: MPD $= 0.377$; p < 0.003; follows: MPD $= 1.512$; p < 0.001).

The rates of grooming bouts and groom solicits received from alpha males by subject females also varied significantly across reproductive states (grooming bouts; $\chi^2=7.63$; df $= 2$; p $= 0.02$; groom solicits: $\chi^2 = 6.82$; df $= 2$; p $= 0.032$). Cycling females received grooming bouts from alpha males at a significantly higher rate (0.13/h) than did pregnant (0.02/h) or lactating females (none). Cycling females also received groom solicits from alpha males at a higher rate (0.22/h) than did lactating females (0.01/h), but the rate was comparable to pregnant females (0.15/h).

7.3.3.4 Proximity Patterns

Among cycling females only, alpha males were responsible for maintaining proximity in all alpha male/female dyads during the periovulatory phase (4 dyads), whereas they were responsible for maintaining proximity with only two of the dyads in the non-ovulatory phase. Females in the periovulatory phase maintained proximity in all of the dyads (n=2) involving subordinate males.

Among pregnant females, alpha males maintained proximity in 67% of the dyads, and subordinate males maintained proximity in 80% of the dyads. In contrast, all of the dyads involving alpha males and lactating females were positive, and all dyads involving subordinate males were negative.

7.3.4 Discussion

7.3.4.1 Female Choice in White-Faced Capuchins

The results of this study suggest that female mate choice among periovulatory or conceptive females is not conspicuous in this species. In contrast to female tufted capuchins that are known to actively solicit the alpha male as a mate (e.g., *C. nigritus* this chapter; Janson 1984; Lynch-Alfaro 2005), white-faced capuchin females do not display any conspicuous proceptive behaviors toward males during their periovulatory periods. Instead, alpha males in this species intensely followed, directed more grooming bouts and were solely responsible for maintaining proximity to females during their periovulatory phase of their ovarian cycle, suggesting that alpha males may be "guarding" conceptive females. Additionally, subordinate males directed very few sexual solicitations to any of the females during their periovulatory phases, suggesting that alpha males may also have exclusive mating rights to conceptive females and that male reproductive skew is high.

Jack and Fedigan (2006) found that in this same population of capuchins, alpha males sire about 80% of the offspring, proving that females are able to successfully reproduce with subordinate males. In support of this, we found that periovulatory females were entirely responsible for maintaining proximity with subordinate males, suggesting that they may be attempting to prevent alpha male monopolization. Only recently (current study, Carnegie pers. obs.), have cycling females been observed to mate secretly with subordinate males. These events occurred at the periphery of the group, lower in the canopy and were extremely quiet, which contrasts to copulations with alpha males that involve elaborate displays within the center of the group and accompanied by loud vocalizations (Carnegie unpub.; see also Lynch-Alfaro 2005 for *C. a. nigritus*). Therefore, if female choice is present in this species it is inconspicuous, and it is possible that alpha male mate guarding is functioning to prevent females from soliciting other males, rather than to prevent other males from soliciting conceptive females.

7.3.4.2 Counterstrategies to Infanticide in White-Faced Capuchins

Infanticide occurs at a high frequency in this population of capuchins when aggressive male takeovers or reversals in rank among males occur (Fedigan 2003). Hrdy (1979) suggested, if females are moderately receptive throughout their cycle and mate polyandrously, they may reduce the risk of infanticide by confusing paternity among group males. Furthermore, van Schaik, Pradhan and van Noordwijk (2004) suggest that if female sexual behavior can raise the estimated paternity probability for a male enough so that he is near-ignorant to the true paternity of the offspring, then he should refrain from committing infanticide as he would lose more reproductively if he killed his own offspring than he would, if he allowed another male's offspring to survive.

The most significant finding from the analysis of behavioral differences among reproductive states was that pregnant females mated more often than any of the other females and they did so almost always with subordinate males. These females were also significantly more proceptive to subordinate males compared to cycling or lactating females. Additionally, we found that subordinate males were responsible for maintaining proximity to pregnant females more than to any other category of female.

It is likely that pregnant females have uncoupled their sexual receptivity from their ovarian activity as a strategy to reduce the risk of infanticide. They may choose to mate with subordinate males to confuse the male as to the paternity of the soon-to-be-born infant. Since alpha males appear to mate guard periovulatory females and are known to sire the majority of the infants, one could assume that pregnant females have already mated and conceived infants with the alpha male. Because lower ranked and immigrating males may pose a high risk to a new infant in terms of infanticide, being receptive to the male's solicitations during gestation and even within days of parturition, a pregnant female may be confusing paternity and forming a positive bond with a male, thereby reducing the likelihood of aggressive encounters with him once the infant is born. However for this strategy to be successful, males must not be able to associate the timing of mating with the timing of births (Zinner and Deschner 2000), and whether or not this is the case is still unknown.

7.4 Squirrel Monkeys, *Saimiri sciureus*

7.4.1 Introduction

Compared to closely related neotropical primates, squirrel monkeys have group sizes that are much larger, ranging from of 25–50 animals (Mitchell 1990; Boinski 1999; Stone 2007). Marked sexual segregation has been observed in at least two populations (*S. boliviensis* in Peru: Mitchell 1990; *S. sciureus* in Brazil: Stone 2004), as well as in captivity (*S. sciureus*: Leger et al. 1981) and in semi-free ranging settings (*S. boliviensis*: Baldwin 1969), with females and juveniles forming the core of social groups. In addition, as a genus, squirrel monkeys are characterized by the presence of several breeding females, seasonal breeding (Janson and Di Bitteti 2000; Stone 2004), and synchronized births (Boinski 1987; Stone 2004; Stone 2006), decreasing the chances of female monopolization by males. Therefore, it is expected that opportunities for female choice occur in squirrel monkeys, as males have fewer opportunities to coerce females. In addition, infanticide risk is low in *Saimiri* (Boinski et al. 2005). Infanticidal behavior has never been reported and the seasonality of births makes infanticide an unlikely male strategy (van Schaik 2000c).

Finally, all squirrel monkey species are sexually dimorphic, with males 30–35% larger than females (Mitchell 1990; Boinski 1999). In addition, the unusual reproductive physiology of male squirrel monkeys is highly suggestive of strong sexual selection. Males show seasonal weight gain during the brief (usually 8 week) mating period (85–222 g; DuMond and Hutchison 1967) due to fat deposition and water retention, which produces a "fatted" appearance in the upper torso, arms and shoulders (Mendoza et al. 1978; Mitchell 1990; Boinski 1992; Stone 2004).

Data for this study were collected by A. Stone over 14 months in Eastern Amazonia (including two mating seasons) on *S. sciureus*, in order to examine the relationship between social organization, infanticide risk/male coercion and possible female choice in the species. Data are presented on patterns of male–female association during mating and non-mating periods and on mating behaviors of females in order to test the hypothesis that females are more likely to engage in mate choice than in avoidance of male coercion. These findings are also compared to those of other species of *Saimiri* to examine the likelihood of female choice in the genus.

7.4.2 Methods

7.4.2.1 Study Area

This study was conducted from March 2002 to March 2003, following a 2-month preliminary study (June-July 2000) in the village of Ananim (VA), 150 km east of Belém, Brazil. The 800-hectare site includes *terra firme* and inundated primary forest and adjacent secondary forests. Rainfall is highly seasonal, with a wet season from January to June and a dry season from July to December (Stone 2007). Fruit availability, including that of the most common fruit in the monkeys' diet (*Attalea maripa*), is highest during the wet season. Mating in this population occurs during an 8-week period from mid-July to mid-September and births in this population occur in January and February of each year (Stone 2006). Thus, the rainy season corresponded to birth and lactation (infants are weaned at 8 months of age; Stone 2006), while the dry season corresponded to mating and gestation.

7.4.2.2 Study Troops

Data were collected during 1,190 observation hours on two groups of squirrel monkeys: troop A (44 individuals) and troop B (50 individuals). Both groups were well habituated and could be differentiated based on the presence of identifiable individuals. The troops were followed at least 10 days per month (17 ± 4.5 days) from 0600 until approximately 1600 h, although on 16 occasions observations extended until the monkeys settled into a sleeping area (around 1830 h). Most of the data presented here are from troop A (858 contact hours, including *ad libitum* observations), though detailed observations were made in both troops.

7.4.2.3 Determination of Age–Sex Classes

Initial group composition for troop A was as follows: 9 adult males, 16 adult females, 11 juveniles and 8 infants. Fifteen infants were born in January and February 2003, 12 of which were still alive at the end of the study in March 2003. Subjects were classified as: adult male (AM; $\geq$ 5 years), adult female (AF; $\geq$ 3 years), juveniles (8 months–3 years) or infants (0 to 8 months). The eight infants present at the start of the study period entered the juvenile classification in September (dry season), when nursing observations became rare. In the course of the study, 17 individuals were reliably recognized in troop A (6 males, 7 females and 4 juveniles), either by physical characteristics or by having been marked with Nyanzol fur dye (Albanyl Dyestuffs International).

7.4.2.4 Data Collection

Focal animal samples (Altmann 1974) lasting 5–10 minutes were conducted, with instantaneous observations taken at 1-minute intervals. Because not all troop members were recognized, behavioral samples were grouped by sex rather than by individual subjects. Although non-identification of individuals is a potential limitation of the study, several steps were taken to ensure adequate sampling within age-sex classes. First, the order of observations of age-sex classes was not completely random in order to avoid oversampling certain individuals. For example, if the first sample of the day was an adult male (determined randomly), often a second male was sampled immediately after the first in order to minimize the risk of resampling the same individual. Second, since the group was often spread over 50–150 m, sequential samples on individuals that were distantly located were conducted whenever possible.

A total of 139 focal animal samples on adults were collected over 12 months in order to quantify association patterns of males and females. At each 1-min interval, the following variables were recorded: party size (number of individuals within a 5 m radius); distance to nearest neighbor up to 5 m; party score (50% females as neighbors, 50% males as neighbors, equal, 50% juveniles as neighbors, alone.). Aggressive behaviors and the context in which they occurred (e.g., feeding, mating) were noted opportunistically each month. During each instance, the age-sex class of the initiator, recipient and winner was noted. Data on consortships and copulations were collected on an *ad libitum* basis during the 2-month mating period (both in 2000 and 2002).

7.4.2.5 Data Analyses

Instantaneous observations within each focal sample are not statistically independent. Therefore, each sample, rather than each individual observation within a sample, was treated as a data point for analyses. Party score was converted to a quantitative variable as the proportion of intervals an animal was engaged in that activity, and numerical variables, such as party size, were averaged within a sample (Boinski 1988; Di Fiore 2003). Samples were pooled by sex class (Leger and Didrichsons 1994; Treves 1999). T-tests were performed to examine the effects of sex on selected behavioral variables within each season. Variables that did not conform to a normal distribution were transformed prior to statistical tests. Proportional data (party score) were arcsine-transformed. Square-root transformations

Table 7.2 Occurrence of agonism among age-sex classes of *S. sciureus* (AF = adult female; AM = adult male). Infants are not included

Participants	Expected (%)	Observed Mating season (n = 52) (%)	Observed Non-mating season (n = 162) (%)
AF-AF	13	6	11
AF-AM	15	17	20
AM-AM	4	31	7
AF-Juvenile	31	6	20
AM-Juvenile	18	17	13
Juvenile-Juvenile	19	23	30

were performed on count data (party size). Statistical significance was set at 0.05 for all tests, and data are reported as the mean ± 1 SE of untransformed data. All tests are two-tailed. Based on average group composition (9 males, 16 females and 19 juveniles), the probability of agonistic interactions occurring between age-sex classes was calculated (Table 7.2). This allowed us to compare expected values to observed values using a Chi-squared analysis.

7.4.3 Results

7.4.3.1 Associations Between Adult Males and Females

Females associated predominantly with other females and with juveniles, while males spent more time with other males or alone. Due to the large size and dispersion of troops (up to 140 m), an individual's position in the troop during focal samples (e.g., center vs. periphery) could not be definitively classified. However, qualitative observations strongly suggest that adult females and juveniles formed the core of the troop. Adult males maintained proximity to females during the 8-week mating season (*see* below), then gradually drifted toward the periphery of the troop once matings subsided. Indeed, locating adult males for collection of focal samples proved particularly challenging during the birthing and early nursing periods (January to May).

When foraging for *A. maripa* fruits, age-sex classes separated their foraging in time and space despite the large number of animals that fed in a tree. Adult females (particularly those carrying infants) and juveniles arrived first at a tree on 82% of visits (n = 80 visits). Adult males rarely foraged in the initial party, usually appearing in a second "wave". Adult males also often moved temporarily off the infructescence into the foliage if juveniles returned to the tree.

7.4.3.2 Spatial Associations During Mating Season

In the mating season, there was no significant difference in the party sizes to which adult males and adult females belonged (t-test $_{(33)}$ = 0.57; p = 0.57), with the average party size at 1.7 individuals. Adult males also did not differ from females in their interindividual distances (t-test $_{(33)}$ = −0.99; p = 0.33; mean: 3.3 m). As indicated in Fig. 7.2a, party scores also did not vary significantly across sexes. No sex differences existed in time spent alone (t-test $_{(33)}$ = −0.084; p = 0.93), in proximity to females (t-test $_{(33)}$ = 1.07; p = 0.29) or in proximity to males (t-test $_{(33)}$ = 0.26; p = 0.69).

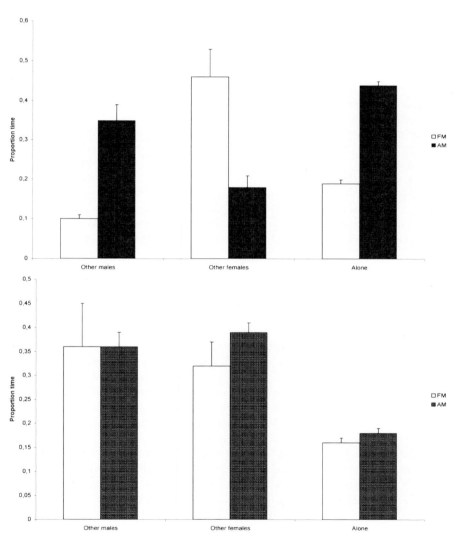

Fig. 7.2 Spatial association between males and females by season: (**a**) mating and (**b**) non-mating. Proportions do not add up to 100% due to time spent with juveniles (not shown)

7.4.3.3 Spatial Associations During Gestation and Lactation

During the remainder of the year, adult males had an average of two fewer neighbors than did females (t-test $_{(102)}$ = 2.82; p = 0.06). Adult males also had higher interindividual distances (t-test $_{(102)}$ = -2.95; p = 0.004; 3.9 m males vs. 3.0 m females). Party scores are shown in Fig. 7.2b. Males spent more time alone (t-test $_{(102)}$ = -4.43; p < 0.001) or in proximity to males (t-test $_{(102)}$ = -1.97; p = 0.04) and less time in proximity to females (t-test $_{(102)}$ = 2.18; p = 0.03).

7.4.3.4 Agonism

Dominance Patterns

Although individual dominance ranks were not determined in this study, relative rank ordering by age and sex could be established based on the outcome of agonistic interactions. Observations of both troops indicated that adult males were subordinate to adult females and, possibly, to juveniles. Females won 74% of all conflicts against adult males (n = 33 male–female conflicts). Adult females were equally likely to initiate agonism toward adult males as toward other females. Parties of females and juveniles often chased adult males from a tree when resting or foraging (n = 34 observations). The strong avoidance of juveniles by adult males, particularly evident when foraging on palm fruit, suggests that adult males may be subordinate to juveniles. Infants as young as three months were observed to chase males with threatening vocalizations and genital displays if the male attempted to approach the infant or its mother. Aggression of males toward infants was never observed.

Context of Agonistic Interactions and Participants

In the course of 700 hours of observation, 277 agonistic interactions were recorded (0.008 events/hour/individual). Levels of agonism were higher in the mating season (0.015 events/hour/individual), than at other times (0.008 events/hour/individual). The 2-month mating season accounted for 62% of all male–male agonistic interactions observed over the year (n = 47 observations). The frequencies of agonistic interactions between age-classes were significantly different from those expected (Table 7.2), with male–male agonism higher than expected and female–female agonism lower than expected in the mating season (χ^2 = 17.8; df = 5; p = 0.003) and juvenile-juvenile agonism higher than expected in the non-mating season (χ^2 = 11.7; df = 5; p = 0.04).

Males competed aggressively for females during the mating season. Injuries (cuts and gashes) on adult males were extremely common during this period and fights between males often resulted in one or both individuals falling to the ground. For example, in the 2000 mating season, an adult male lost a portion of his upper lip during a conflict with another male. Although troops (troops A and B and other troops observed in the area) had overlapping ranges and sometimes came in contact at feeding trees during non-mating periods, intergroup encounters seldom occurred during the breeding season. Small bands of males from troop A (2–4 males) were often seen chasing unfamiliar males (which presumably were from another troop) during the mating season. Following such chases, these males often positioned themselves on high portions of the canopy, where they continued to "bark" and remain vigilant (*S. boliviensis*: cf. Mitchell 1990)

Mating Behaviors

As has been observed in other *Saimiri* populations, both in the wild and in captivity (DuMond and Hutchison 1967), adult males experienced seasonal enlargement ("fatting") during the breeding season, and returned to their normal state after mid-September. As indicated above, the mating season was characterized by males moving from the periphery to the center of the group, where they remained in proximity to females (though males continued to patrol the periphery for unfamiliar males). Observed "consortship associations" between males and females lasted anywhere from 17 minutes to over 3 hours (n =12 observations), but likely lasted longer because the same pair was not necessarily followed for the entire day. Consortships often started by a male approaching a female (and sometimes her newly weaned infant) and remaining in close proximity (1–3 m) to her during foraging and travel activities. Females did not aggressively reject males when approached, although females sometimes moved rapidly away (although the male often followed). Females were never seen presenting or soliciting to males. In the course of the consortship, the male–female pair often remained at the back of the group, although they never completely left the group. On three occasions, two males were seen following a female together, though copulations were not observed. On six occasions, male follows to the female were interrupted by the male chasing off a second approaching male.

Mating behaviors such as mounts and intromissions often were interspersed with activities such as insect foraging. Copulations (including those with ejaculations) were not accompanied by vocalizations by either males or females. Following a series of intromissions, ejaculation usually occurred. Males would then quietly dismount females, both individuals would engage in autogrooming while facing away from each other, and then the male would move away. No evidence of "mate guarding" following ejaculation was observed.

7.4.4 Discussion

The data for *S. sciureus* in Brazil indicate that males and females live in large, sexually segregated groups, where mating is confined to a short period. No evidence of infanticide was seen. Thus, females are expected to have more opportunities for choosing males rather than simply avoiding male coercion. However, behaviors associated with female choice such as females presenting to males, following males or calling to males (Boinski 1992) were not observed during two mating seasons (2000 and 2002). Furthermore, the high levels of male–male aggression during the mating season over access to females suggest that intrasexual selection may be a stronger force than female choice for *S. sciureus*. Female–female agonism (which includes chases, threats and displacements) was also lower than expected during the mating weeks, suggesting that females were not competing for first access to males. Similarly, female choice has not been observed in *S. boliviensis* studied in Peru, where females are dominant to males, but males compete aggressively for access to females (Mitchell 1990; Mitchell 1994). In Mitchell's population, fatting also appeared to protect males from wounds caused by the high levels of intrasexual aggression. The largest male in this population (established visually) was responsible for 90% (n = 10) of long consortships (> 6 h) observed (Mitchell 1990).

In contrast to South American squirrel monkeys, there is evidence that female *S. oerstedii* in Costa Rica solicit copulations, and preferentially solicit from the most fatted males in the troop (established visually). Boinski (1987; 1992) reports that during one breeding season, the most fatted male obtained 70% of copulations (n = 35), while the smaller males mated only when the large males rejected female solicitations. Boinski suggests that females in Costa Rica prefer to mate with these enlarged males because fatting indicates that males have attained a stable social position and will remain in the group, providing more vigilance against predators. Older males appear more vigilant against raptors, a major cause of neonate mortality (Boinski 1987). Therefore, in at least one species of squirrel monkeys, there is evidence for female choice. Why the variability in the genus? One possibility is that predation risk is higher in certain populations, making it advantageous for females to select males on the basis of vigilance and protection. Boinski (1986) reports a 50% infant mortality rate for *S. oerstedii* in the first year of life, mostly due to predation by raptors. Predation on infants has not been reported to such an extent for other squirrel monkey populations (Boinski et al. 2003). These qualitative observations suggest that, for male *S. sciureus* and *S. boliviensis*, a higher degree of fatting may confer an advantage during agonistic encounters with other males, rather than serving as a mechanism for female choice.

Finally, evidence from several mammals indicates that species with high levels of male–male competition achieve sexual dimorphism via bimaturism, i.e., males mature later and grow for a longer period of time compared to females (Jarman 1983; Leigh 1995). For example, in primates where female choice is strong (e.g., lemurs), neither sexual dimorphism nor bimaturism is observed (Leigh and Terranova 1998). Conversely, squirrel monkeys show both bimaturism and sexual dimorphism, supporting the hypothesis that male–male competition is occurring. Thus, the ontogenetic pattern of squirrel monkeys is more consistent with a system of male–male competition than of female choice.

It is important to note, however, that female choice in the genus *Saimiri* remains to be examined systematically. For example, no study has quantified whether a higher degree of fatting (seen in all *Saimiri* populations studied to date) is correlated with higher offspring production. Furthermore, while fatting may serve a male-competition function, this does not exclude the possibility that females are selecting males based on other (yet unstudied) factors. For example, older males are usually more aggressive and dominant and copulate more frequently (Coe et al. 1985; Boinski 1987; Mitchell 1990), such that females might be selecting males on the basis of age or dominance for good genes. This remains to be examined in careful studies where any female mating preferences are documented, along with the paternity of their offspring.

7.5 General Discussion

In this chapter we compared data on three Neotropical primate species in order to investigate hypotheses about the relationship between mating system, female choice and male sexual coercion. Our first prediction, that female choice would reflect preference for male qualities in *Saimiri* and avoidance of male coercion in *Cebus*, was supported by the data from capuchin monkeys only, but not by the squirrel monkey data. Although females have more freedom to choose among males that are subordinate to them, the currently available data suggest that *S. sciureus* females do not actively solicit particular males and that male copulatory success seems to be the result of their competitive ability. If there is little female choice occurring in South American squirrel monkeys, then it is interesting to consider why this is so, given that females have the opportunity to be choosy. We suggest that female choice in *Saimiri sciureus* might reflect the fact that males do not provide many benefits to females. Given the pattern of female dominance, the lack of paternal care and the lack of territory defense, males do not provide benefits to females in the form of protection from male aggression, caring for infants or securing territories. The same appears to be true in *S. boliviensis*, where females are dominant to males and intersexual selection is lacking (Mitchell 1990).

Female *Saimiri* are subject to the high energetic costs of supporting expensive infants without significant extra-maternal help. Gestation length in *Saimiri* is long (5 months; Mitchell 1990; Stone 2004), similar to those of the larger *Cebus* (Hartwig 1996). In addition, despite the production of a single infant, prenatal growth rates in *Saimiri* are high (Ross 1991; Hartwig 1996; Garber and Leigh 1997). Not surprisingly, *Saimiri* neonates are well developed, representing 16–20% of maternal weight, (Long and Cooper 1968; Elias 1977; Kaack et al. 1979). During the first three months of life, infants exhibit a nearly constant growth rate, resulting in maternal costs that are equivalent to the cost of callitrichine twins (Garber and Leigh 1997). The brains of *S. sciureus* neonates also are about 61% complete at birth (Elias 1977) and post-natal brain growth in *Saimiri* infants occurs extremely rapidly (Manocha 1979). Despite the high allocation of maternal resources through lactation and transport, females do not receive paternal assistance with infant care (Mitchell 1990; Stone 2004). Since reproduction is expensive for females and the males do not participate in infant care, females may not have an incentive to select males on that basis, as seems to occur in callitrichines (Price 1990; Garber 1997). Furthermore, since females are dominant to males, mating with more dominant and aggressive males may not necessarily provide females with social benefits such as improved access to food resources to better support offspring. In addition, the absence of infanticidal pressure makes it unlikely that females are selecting males based on their ability to protect them from infanticidal males. Thus, infanticide has little effect on female mating strategies and sexual behavior (e.g., confusing paternity). However, females still benefit from permanent association with males if males provide benefits such as intergroup vigilance and predator detection. For example, because *S. sciureus* males are often positioned in the group's periphery, high in the canopy, they are often the first to detect and give alarms to aerial predators. In addition, it is possible that males support females in aggressive interactions with other females, although this has not been examined.

Our second prediction, that female choice would reflect preference for male qualities in *Cebus nigritus* and strategies to reduce the risk of infanticide in *C. capucinus*, was not supported by data present here. Despite the apparent lower infanticide pressure in *C. nigritus* compared to *C. capucinus*, female reproductive strategies in both species seem to reflect avoidance of male sexual coercion, apparent paternity concentration in the first case and apparent paternity confusion in the second. Fragaszy et al. (2004) suggest that the mating system of tufted capuchin monkeys can be characterized as uni-male "since it is thought that only the alpha male breed and father young", whereas white-faced capuchins are truly multi-male, since all adult males mate with estrous females. They suggest that it is the difference in the dominance relationships between the males of each species that influence the different mating strategies seen in capuchin monkeys (i.e., despotic relationships among tufted male capuchins and tolerant relationships among male white-faced capuchins).

The data presented here on *C. capucinus* show that during the conceptive phase, females copulated exclusively with the dominant male, and only during non-conceptive periods (including post-conception) did they copulate with subordinate males. In accordance, DNA analyses on both species show that reproductive success is skewed toward the dominant male (Escobar-Páramo 1999; Jack and Fedigan 2006). This is similar to observations of female tufted capuchin monkeys that mate exclusively with the dominant male when most likely to conceive but might solicit subordinate males when less likely to conceive (Janson 1984; Lynch-Alfaro 2005). Thus, the mating systems of both species are quite similar in that they display dominant male reproductive skew; however, the mechanisms used to accomplish this, which have been shaped by female reproductive strategies, are different. Although it can be argued that in both capuchin species females practice both strategies of paternity concentration and paternity confusion, *C. nigritus* estrous females actively solicit the dominant male, whereas in *C. capucinus*, dominant males are responsible to maintaining proximity to cycling females.

To avoid male sexual coercion, *C. nigritus* females behave in a way that facilitates paternity concentration and *C. capucinus* females behave in a way that facilitates paternity confusion, but the reasons why these two species choose different strategies still remain to be clarified by further investigations. However, we propose that the difference in the stability of male group membership and of the dominance relationships between males of each species may be shaping the different female strategies. In white-faced capuchins, females can experience replacements of some or all of their adult male group members relatively often and these replacements coincide with higher infant mortality (Fedigan 2003). In this scenario, it would not be advantageous for females to conspicuously concentrate paternity in one male, since the probability is low that he would remain in the group long enough to protect his offspring from extra-group males. By confusing paternity among group males, any remaining subordinate males after a takeover may help to protect offspring since their knowledge of paternity is confused. In tufted capuchin monkeys, the dominant male presents the highest potential of killing infants that were not sired by him (Izar et al. 2007) and the length of the dominant male tenure may be more than ten years (Lynch and Rímoli 2000; Di Bitetti and Janson 2001). When male tenure is long, it would be more advantageous for females to clearly indicate their choice for the strongest male (through a "flamboyant" courtship, in the words of Carosi et al. 2005), thus reducing the risk of male sexual coercion (i.e., infanticide). This strategy would have evolved if the display of proceptive behaviors conspicuously signaling mate choice is an inheritable feature that indeed promotes higher chances of survival of offspring carrying these features.

If our hypothesis is correct, than we might expect that in groups or populations of tufted capuchin monkeys where the dominant male is less powerful (for whatever ecological, social or demographic causes), females will be less strict in their choice of sexual partner. In support of this, Janson (1998) reports that female tufted capuchins are more promiscuous in unstable wild social groups. In addition, there is evidence showing that male competition for estrous females in tufted capuchin monkeys seems to follow their formal dominance hierarchy; when dominance hierarchies are stable, subordinate males refuse female solicitations but they are more willing to attempt copulations when hierarchies are unstable (Carosi et al. 2005).

In conclusion, our data on *Cebus* species corroborate the importance of sexual conflict in shaping reproductive strategies in primates and support the hypothesis that in species which females that are at risk of male infanticide, female choice reflects avoidance of male coercion, adopting strategies of paternity concentration or paternity confusion. In contrast, our data on *Saimiri* do not support the hypothesis that females in species free from infanticide risk would present female choice for male qualities. We argue that this would be the case only when males offer direct benefits to females, such as parental care.

7.6 Summary

Contemporary evolutionary biologists have recognized that intersexual conflict is an important third component to Darwin's theory of sexual selection. The importance of sexual conflict in shaping reproductive strategies in primates may be one reason why direct observations of female choice of male intrinsic qualities are relatively rare. We investigated the relationships between mating system, female choice and male sexual coercion by examining data on female reproductive behavior, female mate choice and male–female association patterns in three Neotropical primates; black-tufted capuchins, *Cebus nigritus*, white-faced capuchins, *Cebus capucinus* and squirrel monkeys, *Saimiri sciureus*. Although these species are closely related, they differ in their socioecological characteristics such as patterns of male–female association and infanticide risk and therefore, we hypothesized that the type of female choice is determined by the risk of male sexual coercion. We predicted that *Saimiri* would display more behavioral patterns indicative of female choice for male qualities rather than infanticide avoidance, and *Cebus* would display behavioral patterns reflecting strategies to reduce infanticide risk rather than behaviors indicative of female choice for male qualities. Our predictions for capuchin monkeys were supported by the data, but our predictions for squirrel monkeys were not. We suggested that female choice in *Saimiri sciureus* might reflect the fact that males do not provide many benefits to females (e.g., protection from male aggression, caring for infants or securing territories). We found that female reproductive strategies in *C. nigritus* and *C. capucinus* seem to reflect avoidance of male sexual coercion; however, the mechanisms differ in that *C. nigritus* attempts to concentrate paternity and *C. capucinus* attempts to confuse paternity. Our *Cebus* data illustrates the importance of sexual conflict in shaping reproductive strategies and supports the hypothesis that female choice reflects avoidance of male coercion. However, our data on *Saimiri* does not support the hypothesis that females chose males based on intrinsic qualities when infanticide risk is low.

Acknowledgments Patrícia Izar and Érica S. Nakai thank Eraldo Vieira, Juliana T. Taira, Andréa Presotto, Mariana D. Fogaça, and Lucas Peternelli for help with field data collection. They thank the Instituto Florestal de São Paulo, specially the PECB manager, José Carlos Maia, for permission to conduct this research, which was funded by grants from the São Paulo State Research Foundation (FAPESP) and the Brazilian Research Council (CNPq). A. Stone wishes to thank Edmilson Viana da Silva, for field assistance in Brazil, and Dr. Paul Garber, for help in all phases of this project. Discussions with Dr. Pat Weatherhead also provided valuable insights into sexual selection in squirrel monkeys. SD. Carnegie would like to graciously thank R. Seguro-Blanco and Parque Nacional Santa Rosa, Dr. LM Fedigan for her supervision and funding support, Dr. TE Ziegler for laboratory guidance, G. McCabe for field assistance, and J. Addicott for statistical advice. This research was funded by operating grant from the Natural Sciences and Engineering Research Council of Canada (LMF), the American Society of Primatologists, Sigma-Xi, and FGSR/Dept of Anthropology, University of Alberta Research Fund (SDC), and by the National Primate Research Center in Madison, Wisconsin (TEZ). They would also like to thank Karen B. Strier and Jessica Lynch-Alfaro who contributed helpful comments on an early version of the manuscript.

References

Altmann, J. 1974. Observational study of behavior: sampling methods. Behavior 49:227–265.
Arnqvist, G., and Rowe, L. 2005. Sexual conflict. Princeton: Princeton University Press.
Baldwin, J.D. 1969. The ontogeny of social behavior of squirrel monkeys (*Saimiri sciureus*) in a seminatural environment. Folia Primatol. 18:161–184.
Bicca-Marques, J.C., and Gomes, D.F. 2005. Birth seasonality of *Cebus apella* (Platyrrhini, Cebidae) in Brazilian zoos along a latitudinal gradient. Am. J. Primatol. 65:141–147.
Boinski S., Kauffman L., Westoli A., Stickler, C.M., Cropp S., and Ehmke E. 2003. Are vigilance, risk from avian predators and group size consequences of habitat structure? A comparison of three species of squirrel monkey (*Saimiri oerstedii*, *S. boliviensis* and *S. sciureus*). Behavior 140:1421–1467

Boinski, S. 1986. The ecology of squirrel monkeys in Costa Rica. Ph.D. dissertation, University of Texas, Austin, Texas.

Boinski, S. 1987. Birth synchrony in squirrel monkeys (*Saimiri oerstedii*). Behav. Ecol. Sociobiol. 21:393–400.

Boinski, S. 1988. Sex differences in the foraging behavior of squirrel monkeys in a seasonal habitat. Behav. Ecol. Sociobiol. 23:177–186.

Boinski, S. 1992. Monkeys, with inflated sex appeal. Nat. Hist. 101:42–49.

Boinski, S. 1999. The social organization of squirrel monkeys: implications for ecological models of social evolution. Evol. Anthropol. 8:101–112.

Boinski, S., Kauffman, L., Ehmke, E., Schet, S., and Vreedzaam. A. 2005. A comparison of three species of squirrel monkey (*Saimiri oerstedii, S. boliviensis and S. sciureus*). Behavior140:1421–1467

Carnegie, S.D. 2004. Reproductive behavior and hormonal patterns in white-faced capuchins, *Cebus capucinus*. MA thesis, University of Alberta, Edmonton, Alberta Canada.

Carnegie, S.D., Fedigan, L.M., and Ziegler, T.E. 2005. Behavioral indicators of ovarian phase in white-faced capuchins, *Cebus capucinus*. Special Issue: Advances in Field Endocrinology. Am. J. Primatol. 67:51–68.

Carosi, M., and Visalberghi, E. 2002. Analysis of tufted capuchin (*Cebus apella*) courtship and sexual behavior repertoire: Changes throughout the female cycle and female interindividual differences. Am. J. Phys. Anthropol. 118:11–24.

Carosi, M., Heistermann, M., and Visalberghi, E. 1999. The display of proceptive behaviors in relation to urinary and fecal progestin levels over the ovarian cycle in female tufted capuchin monkeys. Horm. Behav. 36:252–265.

Carosi, M., Linn, G.S., and Visalberghi, E. 2005. The sexual behavior and breeding system of tufted capuchin monkeys (*Cebus apella*). Adv. Stud. Behav. 35:105–149.

Chapman, T., Arnqvist, G., Bangham, J., and Rowe, L. 2003. Sexual conflict. Trends in Ecol. Evol. 18:41–47.

Clutton-Brock, T.H., and Parker, G. 1995. Sexual coercion in animal societies. Anim. Behav. 49:1345–1365.

Clutton-Brock, T.H. 2004. What is sexual selection? In P.M. Kappeler and C.P. van Schaik (eds.), *Sexual Selection in Primates: New and Comparative Perspectives* (pp. 24–36). Cambridge: Cambridge University Press

Coe, C.L., Smith, E.R., and Levine, S. 1985. The endocrine system of the squirrel monkey. In L.A. Rosenblum and C.L. Coe (eds.), *Handbook of Squirrel Monkey Research* pp. 191–218. New York: Plenum Press.

Cooper, M.A., Bernstein, I.S., Fragaszy, D.M., and de Waal, F.B.M. 2001. The immigration of new males into four social groups of tufted capuchin monkeys (*Cebus apella*). Int. J. Primatol. 22:663–683.

Darwin, C. 1871. The Descent of Man and Selection in Relation to Sex. London: John Murray.

Di Bitetti, M. S., and Janson, C. H. 2000. When will the stork arrive? Patterns of birth seasonality in neotropical primates. Am. J. Primatol. 50:109–130.

Di Bitetti, M.S., and Janson, C.H. 2001. Reproductive socioecology of tufted capuchins (*Cebus apella nigritus*) in northeastern Argentina. Int. J. Primatol. 22:127–142.

Di Fiore, A. 2003. Ranging behavior and foraging ecology of lowland monkeys (*Lagothrix lagotricha poeppiggi*) in Yasuni National Park, Ecuador. Am. J. Primatol. 59:47–66.

Dias, A.C., Custodio Filho, A., Franco, G.A.D.C., and Couto, H.T.Z. 1995. Estrutura do componente arbóreo em um trecho de floresta pluvial atlântica secundária – Parque Estadual Carlos Botelho. Rev. do Inst. Florestal 7:125–155.

DuMond, F. A., and Hutchison, T. C. 1967. Squirrel monkey reproduction: the "fatted" male phenomenon and seasonal spermatogenesis. Science 58:1067–1070.

Ebensperger, L. 1998. Strategies and counterstrategies to infanticide in mammals. Biol. Rev. 73:321–346.

Elias, M.F. 1977. Relative maturity of *Cebus* and squirrel monkeys at birth and during infancy. Develop. Psychobiol. 10:519–528.

Escobar-Páramo, P. 1999. Inbreeding avoidance and the evolution of male mating strategies. Ph.D. dissertation, State University of New York at Stony Brook, New York

Fedigan, L.M. 1993. Sex differences and intersexual relations in adult white-faced capuchins (*Cebus capucinus*). Int J Primatol 14:853–877.

Fedigan, L.M. 2003. Impact of male take-overs on infant deaths, births and conceptions in *Cebus capucinus* at Santa Rosa, Costa Rica. Int. J. Primatol. 24:723–741.

Fedigan, L.M. and Jack, K. 2001. Neotropical primates in a regenerating Costa Rican dry forest: a comparison of howler and capuchin population patterns. Int. J. Primatol. 22:689–713.

Fedigan, L.M., and Jack, K.M. 2004. The demographic and reproductive context of male replacements in *Cebus capucinus*. Behavior 141:755–775.

Ferreira, R.G., Izar, P., and Lee, P.C. 2006. Exchange, affiliation and protective interventions in semi-free ranging brown capuchin monkeys (*Cebus apella*). Am. J. Primatol. 68:765–776.

Fragaszy, D.M., Visalberghi, E.M., and Fedigan. L.M. 2004. The Complete Capuchin: The Biology of the Genus Cebus. Cambridge: Cambridge University Press.

Garber P.A., and Leigh S.R. 1997. Ontogenetic variation in small-bodied new world primates: implications for patterns of reproduction and infant care. Folia Primatol. 68:1–22.

Garber, P.A. 1997. One for all and breeding for one: cooperation and competition as a tamarin reproductive strategy. Evol. Anthropol. 5:187–199.

Harcourt, A.H., and Greenberg, J. 2001. Do gorilla females join males to avoid infanticide? A quantitative model. Anim. Behav. 62:905–915.

Hartwig, W.C. 1996. Perinatal life history traits in New World monkeys. Am. J. Primatol. 40: 99–123.

Hinde, R.A., and Atkinson, S. 1970. Assessing the roles of social partners in maintaining mutual proximity as exemplified by mother infant relations in rhesus monkeys. Anim. Behav. 18: 169–176.

Hodges, J.K., and Heistermann, M. 2003. Field endocrinology: monitoring hormonal changes in free-ranging primates. In J.M. Setchell and D.J. Curtis (eds.), *Field and Laboratory Methods in Primatology: A Practical Guide* (pp. 282–294). Cambridge: Cambridge University Press.

Hrdy, S.B. 1979. Infanticide among animals: a review, classification, and examination of the implications for the reproductive strategies of females. Ethol. and Sociobiol. 1:13–40.

Izar, P. 2004. Female social relationships of *Cebus apella nigritus* in southeastern Atlantic Forest: an analysis through ecological models of primate social evolution. Behavior 141:71–99.

Izar, P., and Nakai, E.S. 2006. Fission-fusion in tufted capuchin monkeys (*Cebus apella nigritus*) in Brazilian Atlantic Forest. Int. J. Primatol. 27: 226.

Izar, P., Ramos-da-Silva, E.D.A, Resende, B.D., and Ottoni, E.B. 2007. A case of infant killing in tufted capuchin monkeys (*Cebus apella*). Mastozoologia Neotropical 14:73–76.

Izawa, K. 1994. Group division of wild black-capped capuchins. Field Studies of New World Monkeys, La Macarena, Colombia 9:5–14.

Jack, K. M. 2006. The Cebines: toward an explanation of variable social structure. In B. Campbell, A. Fuentes, K. MacKinnon, M. Pangerand S.K. Bearder (eds.), *Primates in Perspective* (pp. 107–123). Oxford: Oxford University Press.

Jack, K.M., and Fedigan, L.M. 2004a. Male dispersal patterns in white-faced capuchins, *Cebus capucinus* Part 1: patterns and causes of natal emigration. Anim. Behav. 67:761–769.

Jack, K.M., and Fedigan, L.M. 2004b. Male dispersal patterns in white-faced capuchins, *Cebus capucinus* Part 2: patterns and causes of secondary dispersal. Anim. Behav. 67:761–769.

Jack, K.M., and Fedigan, L.M. 2006. Why be alpha male? Dominance and reproductive success in wild, white-faced capuchins (*Cebus capucinus*). In A. Estrada, P. Garber, M.S.M. Pavelka and L. Luecka (eds.), *New Perspectives in the Study of Mesoamerican Primates: Distribution, Ecology, Behavior and Conservation* (pp. 367–386). New York: Springer.

Janson, C.H. 1984. Female choice and mating system of the brown capuchin monkey *Cebus apella* (Primates: Cebidae). Z. Tierpsychol. 65:177–200.

Janson, C.H. 1986. The mating system as a determinant of social evolution in capuchin monkeys (*Cebus*). In J. Else and P.C. Lee (eds.), *Primate Ecology and Conservation* Vol. II (pp. 169–180). Cambridge: Cambridge University Press.

Janson, C.H. 1988. Food competition in brown capuchin monkeys (*Cebus apella*): Quantitative effects of group size and tree productivity. Behavior 105:53–76.

Janson, C.H. 1990a. Social correlates of individual spatial choice in foraging groups of brown capuchin monkeys, *Cebus apella*. Anim. Behav. 40:910–921.

Janson, C.H. 1990b. Ecological consequences of individual spatial choice in foraging groups of brown capuchin monkeys, *Cebus apella*. Anim. Behav. 40:922–934.

Janson, C.H. 1998. Capuchin counterpoint. In R.L. Ciochon and R.A. Nisbett (eds.), The *Primate Anthology* (pp. 153–160). New Jersey: Prentice Hall

Janson, C.H. 2000. Primate socioecology: the end of a golden age. Evol. Anthropol. 9:73–86

Jarman, P. 1983. Mating system and sexual size dimorphism in large, terrestrial, mammalian herbivores. Biol. Rev. 58:485–520.

Kaack, B., Walker, L., and Brizzee, K.R. 1979. The growth and development of the squirrel monkey, *Saimiri sciureus*. Growth 43:116–135.

Kappeler, P.M., and van Schaik, C.P. 2004. Sexual selection in primates: review and selective preview. In P. Kappeler and C.P. van Schaik (eds.), *Sexual Selection in Primates, New and Comparative Perspectives* (pp. 3–23). Cambridge: Cambridge University Press.

Leger D.W., and Didrichsons, I.A. 1994. An assessment of data pooling and some alternatives. Anim. Behav. 48:823–832.

Leger, D.W., Mason, W.A., and Fragaszy, D.M., 1981. Sexual segregation, cliques, and social power in squirrel monkey (*Saimiri*) groups. Behavior 76:163–181.

Leigh, S.R. 1995. Socioecology and the ontogeny of sexual size dimorphism in anthropoid primates. Am. J. Phys. Anthropol. 97:339–356.

Leigh, S.R., and Terranova, C.J. 1998. Comparative perspectives on bimaturims, ontogeny and dimorphism in lemurid primates. Int. J. Primatol. 19:723–749.

Linn, G., Mase, D., LaFrancois, D., O'Keeffe, R., and Lifshitz, K. 1995. Social and menstrual cycle phase influences on the behavior of group housed *Cebus apella*. Am. J. Primatol. 35:41–57.

Long, J.O., and Cooper, R.W. 1968. Physical growth and dental eruption in captive-bred squirrel monkeys, *Saimiri sciureus* (Leticia, Colombia). In L.A. Rosenblum and C.W. Cooper (eds.), *The Squirrel Monkey* (pp. 193–205). New York: Academic Press.

Lynch, J.W., and Rímoli, J. 2000. Demography and social structure of group of *Cebus apella nigritus* (Goldfuss, 1809, Primates/Cebidae) at Estação Biológica de Caratinga, Minas Gerais. Neotr. Prim. 8:44–49.

Lynch-Alfaro, J.W. 2005. Male mating strategies and reproductive constraints in a group of wild tufted capuchin monkeys (*Cebus apella nigritus*). Am. J. Primatol. 67:313–328.

Lynch-Alfaro, J.W. 2007. Subgrouping patterns in a group of wild *Cebus apella nigritus*. Int. J. Primatol. 28:271–289.

Manly, B.F.J. 1997. Randomization, Bootstrap and Monte Carlo Methods in Biology, 2nd edn. London: Chapman and Hall.

Manocha, S.L. 1979. Physical growth and brain development of captive-bred male and female squirrel monkeys, *Saimiri sciureus*. Experientia 35:96–98.

Mendoza, S.P., Lowe, E.L., Davidson, J.M., and Levine, S. 1978. Annual cyclicity in the squirrel monkey (*Saimiri sciureus*): the relationship between testosterone, fatting, and sexual behavior. Horm. Behav. 11:295–303.

Mitchell, C.L. 1990. The ecological basis of female dominance: a behavioral study of the squirrel monkey (*Saimiri sciureus*) in the wild. Ph.D. dissertation, Princeton University.

Mitchell, C.L. 1994. Migration alliances and coalitions among adult male South American squirrel monkeys (*Saimiri sciureus*). Behaviour. 130:169–190.

Nakai, E.S. (2007). Fissão-fusão em *Cebus apella nigritus*: flexibilidade social como estratégia de ocupação de ambientes limitantes. Master Thesis, University of São Paulo, Brazil.

Nunn, C.L., and van Schaik, C.P. 2000. Social evolution in primates: the relative roles of ecology and intersexual conflict. In C.P. van Schaik and C.H. Janson (eds.), *Infanticide by Males and its Implications* (pp. 388–412). Cambridge: Cambridge University Press

Parker, G.A. 1979. Sexual selection and sexual conflict. In M.S. Blum and N.A. Blum (eds.), *Sexual Selection and Reproductive Conflict in Insects* (pp. 123–166). New York: Academic Press

Parker, G.A., Baker, R.R., and Smith, V.G.F. 1972. The origin and evolution of gametic dimorphism and the male–female phenomenon. J. Theor. Biol. 36:529–553.

Perry, S. 1998a. A case report of a male rank reversal in a group of wild white-faced capuchins. Primates 39:51–70.

Plavcan, J.M. 2004. Sexual selection, measures of sexual selection, and sexual dimorphism in primates. In P.M. Kappeler and C.P. van Schaik (eds.), *Sexual Selection in Primates, New and comparative perspectives* (pp. 230–252). Cambridge: Cambridge University Press

Price, E.C. 1990. Infant carrying as a courtship strategy of breeding male cotton-top tamarins. Anim. Behav. 40:784–786.

Rice, W.R. 2000. Dangerous liaisons. Proc. Nat. Acad. Scien. 97:12953–12955.

Ross C. 1991. Life history patterns on New World monkeys. Int. J. Primatol. 12:481–502.

Rylands, A.B., Kierulff, M.C.M., and Mittermeier, R.A. 2005. Notes on the taxonomy and distributions of the tufted capuchin monkeys (*Cebus*, Cebidae) of South America. Lundiana 6:97–110.

Siegel, S., and Castellan N.J., Jr. 1988. Non Parametric Statistics for the Behavioral Sciences. 2nd edn. Mexico: McGraw-Hill Inc.

Smuts, B., and Smuts, R. 1993. Male aggression and sexual coercion in females in nonhuman primates and other mammals: evidence and theoretical implications. Adv. Stud. Behav. 22: 1–63.

Sommer, V. 2000. The holy wars about infanticide: which side are you on and why? In C.P. van Schaik and C.H. Janson (eds.), *Infanticide by Males and its Implications* (pp. 9–26). Cambridge : Cambridge University Press.

Sterck, E., Watts, D.P., and van Schaik, C. 1997. The evolution of female social relationships in nonhuman primates. Behav. Ecol. Sociobiol. 31:291–309.

Stone, A. 2004. Juvenile feeding ecology and life history in a neotropical primate, the squirrel monkey (*Saimiri sciureus*). Ph.D. dissertation, University of Illinois at Urbana-Champaign.

Stone, A.I. 2006. Foraging ontogeny is not linked to delayed maturation in squirrel monkeys. Ethology 112:105–115.

Stone, A.I. 2007. Responses of squirrel monkeys to seasonal changes in food availability in an Eastern Amazonian rainforest. Am. J. Primatol. 69:142–157.

Strier, K.B. 1989. Effects of patch size on feeding associations in muriquis (*Brachyteles arachnoides*). Folia Primatol. 52:70–77.

Strier, K.B. 1999. Why is female bonding so rare? Comparative sociality of Neotropical primates. In P.C. Lee (ed.), Comparative Primate Socioecology (pp. 300–319). Cambridge: Cambridge University Press.

Strier, K.B., and Ziegler, T.E. 1994. Insights into ovarian function in wild muriqui monkeys (*Brachyteles arachnoids*) Am. J. Primatol. 32:31–40.

Strier, K.B. and Ziegler, T.E. 1997. Behavioral and endocrine characteristics of the reproductive cycle in wild muriqui monkeys, *Brachyteles arachnoids*. Am. J. Primatol. 42:299–310.

Treves, A. 1999. Within-group vigilance in red colobus and redtail monkeys. Am. J. Primatol. 48:113–126.

Trivers, R.L. 1972. Parental investment and sexual selection. In B.C. Campbell (ed.), *Sexual Selection and the Descent of Man* (pp. 136–179). Chicago: Aldine-Atherton

van Schaik, C.P. 2000a. Paternity confusion and the ovarian cycle of female primates. In C.P. van Schaik and C.H. Janson (eds.), *Infanticide by Males and its Implications* (pp. 361–387). Cambridge: Cambridge University Press.

van Schaik, C.P. 2000b. Social counterstrategies against infanticide by males in primates and other mammals. In: P.M. Kappeler (ed.), *Primate Males: Causes and Consequences of Variation in Group Composition* (pp. 34–52). Cambridge: Cambridge University Press.

van Schaik, C.P. 2000c. Vulnerability to infanticide by males: patterns among mammals. In C.P. van Schaik and C.H. Janson (eds.), *Infanticide by Males and its Implications* (pp. 61–71). Cambridge: Cambridge University Press.

van Schaik, C.P., and Kappeler, P.M. 1997. Infanticide risk and the evolution of male–female association in primates. Proc. Royal. Soc. London, B 264:1687–1694.

van Schaik, C.P., Pradhan, G.R., and van Noordwijk, M.A. 2004. Mating conflict in primates: infanticide, sexual harassment and female sexuality. In P.M. Kappeler and C.P. van Schaik (eds.), *Sexual Selection in Primates, New and comparative perspectives* (pp. 131–150). Cambridge: Cambridge University Press

Visalberghi, E., and Moltedo, G. (unpub). Social influences on the sexual behavior of tufted capuchin monkeys (*Cebus apella*): an experimental approach.

Ziegler, T.E., Scheffler, G., Wittwer, D.J., Schultz-Darken, N., Snowdon, C.T., and Abbott, D.H. 1996. Metabolism of reproductive steroids during the ovarian cycle in two species of Callitrichids, *Saguinus oedipus* and *Callithrix jacchus*, and estimation of the ovulatory period from fecal steroids. Biol. Reprod. 54:91–99.

Zinner, D., and Deschner, T. 2000. Sexual swellings in female hamadryas baboons after male take-overs: "deceptive" swellings as a possible female counter strategy against infanticide. Am. J. Primatol. 52:157–168.

Chapter 8
The Reproductive Ecology of South American Primates: Ecological Adaptations in Ovulation and Conception

Toni E. Ziegler, Karen B. Strier, and Sarie Van Belle

8.1 Introduction

The process of ovulation does not necessarily end in conception and production of offspring. Although some South American primates, such as common marmosets (*Callithrix jacchus*), regularly conceive following ovulation, other species, such as northern muriquis (*Brachyteles hypoxanthus*), are more likely to experience multiple ovarian cycles prior to conception. The development of non-invasive fecal collection and analyses techniques permit us to investigate the factors that affect the variation in conception processes of wild primates.

Insights into the reproductive endocrinology of South American primates were previously limited to specialized studies of captive species such as squirrel monkeys, marmosets and tamarins, capuchin monkeys and owl monkeys (Hearn 1983). More recently, researchers studying wild populations have provided new information on their reproductive biology, which allows us to understand their reproductive strategies from a more ecological perspective. Factors such as food quality, group composition, rainfall, and other seasonal environmental conditions can influence the fertility of males or females within a species. In addition, many aspects of the ovarian cycles and endocrine metabolism of platyrrhines (New World monkeys) differ from those of catarrhines (Old World monkeys, apes and humans). This chapter will examine the variables affecting female fertility and the timing of ovulation and conception in South American primates.

8.1.1 The Primate Ovulatory Cycle

Unlike other mammals, anthropoid primates have ovarian cycles that are not generally considered as estrous cycles. Mating activity is not restricted to the

T.E. Ziegler (✉)
Wisconsin National Primate Research Center, Department of Psychology, University
of Wisconsin-Madison, Madison, WI 53715, USA
e-mail: Ziegler@primate.wisc.edu

P.A. Garber et al. (eds.), *South American Primates,* Developments in Primatology:
Progress and Prospects, DOI 10.1007/978-0-387-78705-3_8,
© Springer Science+Business Media, LLC 2009

periovulatory period as occurs in other mammals (Dixson 1998). Among anthropoid primates, a rigid control of the female receptivity by ovarian hormones does not occur. However, as with other mammals, the process of ovulation and ovarian cycling is reflected by circulating steroid hormones. The ovarian cycle begins with the follicular phase when the follicles grow to maturity, and culminates in ovulation. This phase is characterized by increased estrogen production by the granulosa cells of the follicle or follicles and low circulating concentrations of progesterone (Yen and Jaffe 1978). As the process of ovulation occurs, the levels of progesterone begin to increase due to luteinization of the granulosa cells within the follicle. Estrogen levels peak as the follicle matures, and the peak levels feed back to the hypothalamus and to the pituitary to increase the frequency and concentration of gonadotropin releasing hormone (GnRH), also known as luteinizing hormone releasing hormone (LHRH; Weiner et al. 1988). GnRH then stimulates the release of luteinizing hormone (LH) and follicle stimulating hormone (FSH) from the pituitary. The LH surge reaches the ovaries via the circulatory system and causes the follicle to rupture due to a local inflammatory process (Bonello et al. 2004). LH causes a rapid secretion of follicular steroid hormones and induces the steroid conversion from cholesterol within the thecal cells to secreting progesterone that occurs shortly before ovulation (Stocco et al. 2007). A corpus hemorrhagicum forms immediately following ovulation and then develops into the corpus luteum upon full differentiation of the luteinized thecal layers of the follicle (Bonello et al. 2004). The luteal phase of the ovarian cycle consists of elevated progesterone and a lesser peak of estrogens. It is during the early luteal phase of the cycle that fertilization can take place. Fertilization can only occur during the first 24 hours after ovulation due to the short life span of the oocyte. If the ovum (or ova) has been fertilized, then the corpus luteum remains intact and continues to secrete progesterone until the placenta is developed. Steroids and gonadotropins secreted by the embryo and the placental unit will sustain the pregnancy independent of the luteal secretions (Hodgen and Itskovitz 1988). The timing of the luteoplacental shift may differ among species and is related to the secretion of chorionic gonadotropin from the embryonic trophoblast, a signal unique to primates (Hodgen and Itskovitz 1988). Platyrrhine species, such as the common marmoset and the cottontop tamarin (*Saguinus oedipus*) have a delayed implantation relative to humans and other catarrhines (Ziegler et al. 1987; Hearn 2001).

The profiles of steroid secretion typical during the ovarian cycle are shown in Fig. 8.1. The ovarian-derived steroids produced prior to and post-ovulation readily define the phases of the ovarian cycle. All steroids are synthesized from cholesterol via specific steroid pathways (Yen and Jaffe 1978). The thecal cells of the follicle synthesize androgens, such as androstenedione and testosterone. These androgens are transferred to the granulosa cells where they are aromatized into estrogens. A greater proportion of androgens are converted to estrogens as the follicle matures toward the preovulatory size. At this time, the granulosa cells switch to producing primarily progesterone and the follicle is converted to the corpus luteum, with both androgens and estrogens continuing to be produced in smaller amounts (Stocco et al. 2007).

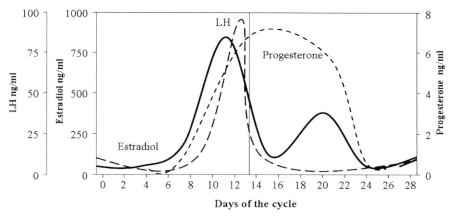

Fig. 8.1 A typical nonconceptive ovulatory cycle of circulating estradiol, progesterone and LH in a human ovarian cycle is shown across the days of the cycle. The vertical line represents the timing of ovulation. The follicular phase occurs prior to ovulation, during the elevation in estradiol. The luteal phase occurs after ovulation

8.1.2 Steroid Metabolism and Excretion in Platyrrhines

The development of techniques of measuring steroid metabolites excreted in urine or feces has allowed primatologists to collect noninvasive samples from wild and captive primates to monitor their ovarian cycles (Hodges and Heistermann 2003; Ziegler and Wittwer 2005). Steroids produced in the adrenals and the gonads circulate in the blood and then are filtered by the liver and kidneys. Circulating steroids are excreted into both urine and feces. Unique to the platyrrhines, the urinary and fecal steroid profiles of ovarian cycles do not match the profiles of the corresponding circulating steroid hormones (Ziegler et al. 1987, 1989, 1996, 1993). In contrast to catarrhines, urinary and fecal estrogen profiles do not exhibit a follicular surge prior to ovulation in any of the platyrrhines examined to date (e.g., cotton-top tamarin [Ziegler et al. 1987; Heistermann et al. 1993; Savage et al. 1997], saddle back tamarin, *Saguinus fuscicollis* [Heistermann et al. 1993], golden lion tamarin, *Leontopithecus rosalia* [French et al. 1992], common marmoset [Eastman et al. 1984; Heistermann et al. 1993; Ziegler et al. 1996], Goeldi's monkey, *Callimico goeldi* [Ziegler et al. 1989; Pryce et al. 1994], northern muriqui [Strier and Ziegler 1994, 1997; Ziegler et al. 1997], black-handed spider monkey, *Ateles geoffroyi* [Campbell et al. 2001]; white-faced capuchin monkey, *Cebus capucinus* [Carnegie et al. 2005]). Instead, platyrrhine estrogen metabolites increase similarly to progesterone metabolites with a sustained elevation throughout the luteal phase of the ovulatory cycle. While this unique pattern makes it more difficult to discern the timing of ovulation from urine or fecal samples, several studies have indicated that the onset of the sustained increase in urinary and fecal progesterone concentration occurs shortly after the serum LH peak, while the delay in the excretion of estrogens is more variable (Ziegler et al. 1993, 1996, 1997). Figure 8.2 demonstrates the delayed excretion of

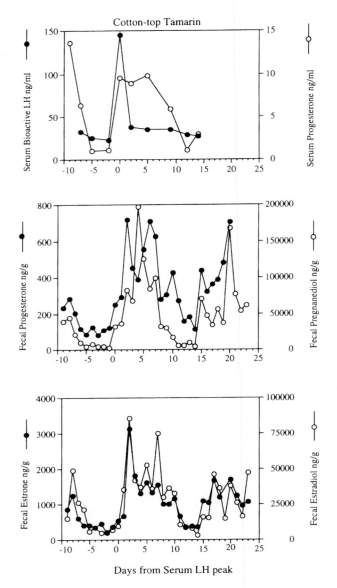

Fig. 8.2 Comparisons of hormones from serum and fecal values in the cotton-top tamarin are shown across the ovarian cycle. The upper panel shows the serum LH peak associated with ovulation in an ovarian cycle from a cotton-top tamarin and concurrent changes in serum progesterone. The day of the LH peak is recorded as day zero on the X-axis for all graphs. Patterns of fecal progesterone and the progesterone metabolite, pregnanediol, are shown on the middle graph for the same ovarian cycle. The fecal estrogens, estradiol and estrone, are shown in the bottom graph for the ovarian cycle. Originally published in Ziegler et al. 1996, Biol. Reprod. 54:91–99

fecal estrogens relative to the LH peak and progesterone metabolites for the cotton-top tamarin.

There are several important considerations concerning steroid metabolism and excretion in platyrrhines. First, both estradiol and estrone, the major ovarian estrogens, are conjugated to water-soluble conjugates, such as glucuronides and sulfates in the urine and feces (Ziegler and Wittwer 2005). The amount of conjugation that occurs in the liver is much higher in platyrrhines than in other anthropoid primates. Only small amounts of estrogens are secreted in the free form in platyrrhines, whereas the majority of estrogens are excreted in the free form in catarrhines. This delays the excretion of the conjugated steroids into the urine and feces. Second, both estrone and estradiol are known to interconvert as a consequence of peripheral metabolism. Interconversion of estradiol and estrone also occurs in the intestine due to 17ß-dehydrogenase activity from bacterial flora (Ziegler et al. 1996). Because estrone is generally secreted in higher amounts during the luteal phase compared to the follicular phase, the interconversion will lead to the highest levels of estrogens found concurrent with or after the sustained rise in progesterone or its metabolites. Third, the high level of conjugations seen for excreted steroids from South American primates is most likely due to their typically high level of steroid production. All Platyrrhini show high levels of circulating and excreted steroids compared to Catarrhini (see Chen et al. 1997 for review). The high level of conjugation and interconversion of steroids into less bioactive metabolites that occurs in platyrrhines means that the majority of the steroids excreted into the urine and feces are not biologically active. For instance, the majority of estradiol is either converted to the less active steroid, estrone, or it is multiply conjugated to a form that is difficult to convert back to free estradiol.

The common marmoset shows a typical platyrrhine pattern of delayed excretion of fecal estrogens relative to fecal progesterone metabolites (Fig 8.3; Ziegler et al. 1996). Figure 8.3 demonstrates the excretion of fecal steroids relative to the preovulatory LH peak in serum. Relative to the timing of the preovulatory LH surge measured in serum, the onset of excretion of fecal progesterone is delayed by three days and the onset of fecal estrogen metabolites excretion is delayed by five days (Fig. 8.3). Similarly, in the black-handed spider monkey, the onset of a sustained estrogen increase occurs a day following progestins increase in fecal extracts (Campbell et al. 2001). In the white-faced capuchin monkeys, fecal estradiol elevation is excreted at the same time or shortly after the progesterone excretion (Carnegie et al. 2005). Comparing the profiles of these species indicates great variability in the length of the delay of estrogen metabolites excretion relative to progesterone metabolites excretion seen in platyrrhines. Furthermore, steroid metabolites measured in fecal material represents the accumulation of circulatory levels over several hours to days (Whitten et al. 1998). Ideally, fecal steroids should be calibrated to serum levels to determine the lag time of steroid excretion into the feces for each species (Ziegler et al. 1997; Campbell et al. 2001). Because collecting near daily blood samples of many primate species is impossible, fecal steroids can be calibrated to urinary steroid levels, in particular the urinary LH surge. This alternative method is based on the strong relationship between serum and urinary steroid levels

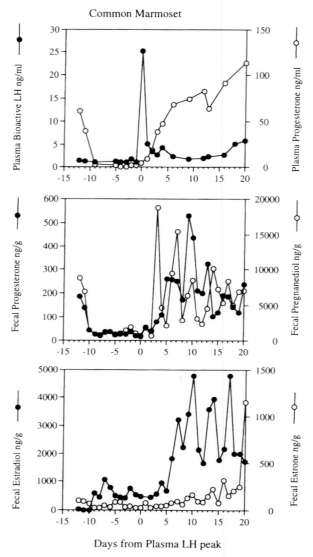

Fig. 8.3 Comparisons of hormones from serum and fecal values for the common marmoset are shown across the ovarian cycle. The upper panel shows the LH peak associated with ovulation in an ovarian cycle from a common marmoset and concurrent changes in serum progesterone. The day of the LH peak is recorded as day zero on the X-axis for all graphs. Patterns of fecal progesterone and the progesterone metabolite, pregnanediol, are shown on the middle graph for the same ovarian cycle. The fecal estrogens, estradiol and estrone are shown in the bottom graph for the ovarian cycle. Originally published in Ziegler et al. 1996, Biol. Reprod. 54:91–99

in both catarrhines and platyrrhines (Wasser et al. 1988; Heistermann et al. 1993; Pryce et al. 1994; Shideler et al. 1994; Ziegler et al. 1996). Consistent patterns have been found, and fecal progesterone levels tend to lag 0–2 days while fecal estrogen levels increase 1–5 days following corresponding serum/urine levels (e.g., Shideler et al. 1994; Ziegler et al. 1997; Campbell et al. 2001).

8.1.3 Ovarian Cycle Length

The length of the ovarian cycle is calculated as the period between two consecutive days of ovulation. Platyrrhine cycle lengths range from 9 days in the squirrel monkey (*Saimiri sciureus*) to 33 days in the pygmy marmoset. Table 8.1 lists the average cycle lengths reported for Platyrrhini. Most cycle length estimates are derived from captive studies, but wild and captive cycle lengths are similar within the species for which both have been examined (Ziegler et al. 1997; Carosi et al. 1999; Campbell et al. 2001; Carnegie et al. 2005). Except for three species (the common marmoset, pygmy marmoset and red howler monkey, *Alouatta seniculus*) the average ovarian cycle length of Platyrrhini is much shorter than the approximately 30-day cycle duration (24–38 days) reported for Catarrhini (Dixson 1998).

8.1.4 Conceptive and Nonconceptive Cycles

Many female primates experience several ovarian cycles before conceiving. Conceptive and nonconceptive cycles are identical in steroid patterns and concentrations of estradiol and progesterone during the period when ovulation takes place and during early and mid luteal phase. During the late luteal phase of nonconceptive cycles, however, concentrations of these steroids drop. The late luteal rise in estradiol has been found to be a sensitive indicator of pregnancy in women (Laufer et al. 1982). We have also found this to be the case for northern muriqui monkeys, although the sustained elevation in estradiol following conception occurs about 5 days later than the sustained rise in progesterone (Strier and Ziegler 1997). During the northern muriqui's conceptive cycle, the levels of both progesterone and estradiol remain elevated, as occurs in other anthropoid primates. Figure 8.4 demonstrates the levels and patterns of ovarian cycling in a wild female muriqui through the use of fecal steroid analyses. This female had successive ovarian cycles prior to conception. Conception was indicated by the failure of fecal estradiol to return to the baseline levels and the further rise of elevated progesterone.

Unlike many platyrrhines, marmoset females, such as the common marmoset, tend to conceive on the first postpartum ovulation and do not undergo multiple non-conceptive cycles, both in captivity and in the wild (Albuquerque et al. 2001; Ziegler and Sousa 2002). In the wild, female marmosets have been reported to have two births per year (Stevenson and Rylands 1988; Vilela and Faria 2004). This reflects their 5-month gestational length with a postpartum conception within two weeks

Table 8.1 Ovarian cycle length for New World primates

Species	Days (mean ± SD)	Range	# of cycles	# of females	Captive/ Wild	References
Callimico goeldii	24.1 ± 0.7	(21–27)	9	6	C	Carroll et al. 1989
	27.8 ± 5.2	(20–35)	9	2	C	Christen et al. 1989
	23.9 ± 0.4	(23–26)	18	9	C[1]	Pryce et al. 1993; Dettling 2002
Callithrix jacchus	28.6 ± 1.0	–	19	–	C[2]	Hearn 1983
	26.2 ± 2.4	(24–29)	8	5	C	Kholkute 1984
Callithrix kuhlii	24.9 ± 0.6	–	4	5	C	French et al. 1996
Cebuella pygmea	33.3 ± 5.5	(26–37)	5	3	C[3]	Ziegler et al. 1990; Converse et al. 1995
Leontopithecus rosalia	19.6 ± 1.4	(14–31)	14	4	C[4]	French and Stribley 1985, 1987
Leontopithecus chrysomelas	18.3 ± 2.1	(16–20)	3	2	C	Chaoui and Hasler-Gallusser 1999
	21.5 ± 2.5	(18–25)	11	6	C	De Vleeschouwer et al. 2000
Saguinus fuscicollis	17.4 ± 3.1	–	34	6	C	Epple and Katz 1983
	25.5 ± 1.0	(19–31)	11	4	C	Heistermann and Hodges 1995
Saguinus oedipus	23.2 ± 1.4	(18–31)	32	9	C[5]	Brand 1981; French et al. 1983
Aotus trivirgatus	15.9 ± 0.3	(13–19)	11	4	C[6]	Bonney et al. 1979, 1980
Callicebus moloch	17.7 ± 1.2	–	12	6	C	Valeggia et al. 1999
Pithecia pithecia	16.9 ± 1.6	(14–19)	20	3	C	Savage et al. 1995
	–	(14–16)	6	3	C	Shideler et al. 1994
Cebus albifrons	–	(18–19)	2	1	C	Hodges et al. 1981
Cebus apella	21.0 ± 1.1	(18–24)	10	10	C	Nagle et al. 1979
	21.1	–	–	–	C	Wright and Bush 1977
Cebus capucinus	20.0 ± 6.1	(14–26)	5	4	W	Carnegie et al. 2005
	20.6 ± 1.6	(17–24)	20	5	C	Carosi et al. 1999
Saimiri sciureus	9.0 ± 1.2	(8–9)	–	14	C	Ghosh et al. 1982
Alouatta seniculus	29.5 ± 1.5	(29–31)	2	2	W	Herrick et al. 2000
Alouatta palliate	16.1 ± 4.3	–	37	24	W	Jones 1985
Ateles fusciceps	–	(20–22)	2	1	C	Hodges et al. 1981
Ateles geoffroyi	22.7 ± 6.9	(13–34)	10	5	W	Campbell et al. 2001
	25.3 ± 3.0	(22–28)	2	2	C	Hernández-López et al. 1998
Brachyteles hypoxanthus	21.3 ± 5.2	(16–38)	34	10	W[7]	Strier and Ziegler 1997; Strier et al. 2003

[1] Only data from Pryce et al. 1993 shown.
[2] These were selected from many captive studies.
[3] Both studies combined.
[4] Only data from French and Stribley 1985 shown.
[5] Both studies combined.
[6] Only data from Bonney et al. 1979 shown.
[7] Both studies combined.

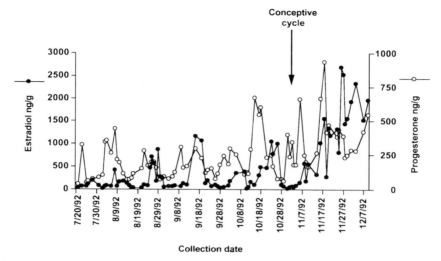

Fig. 8.4 Ovarian cycling and the onset of a conceptive cycle and pregnancy in the muriqui monkey as indicated by fecal estradiol and progesterone profiles. Note that for the conceptive cycle, the estradiol values do not return to baseline indicating pregnancy. Adapted from data presented in Strier and Ziegler 1997, Am. J. Primatol. 42:229–310)

twice a year. There is no seasonal effect on the onset of ovarian cycling or conception in the common marmoset, but the alignment of marmoset births at the end of the dry season and the end of the rainy season may help maximize food availability for mothers who experience high energetic costs while they are simultaneously nursing twin infants and are pregnant (Di Bitetti and Janson 2000). Additionally, marmosets are able to exist on gum from trees during dry seasons when fruit availability is limited (Ferrari 1988). This may explain why marmosets can maintain continuous pregnancies, while the twin-bearing tamarins, who do not rely on gum exudates, tend to have a longer delay in their reproductive events.

Tamarin species from the genera *Saguinus* and *Leontopithecus* show differences in reproductive potential between captivity and the wild. Cotton-top tamarins can conceive over 80% of the time on their first postpartum ovulations under optimal conditions in captivity (Ziegler et al. 1987), but they have been reported to conceive only once a year in the wild (Savage et al. 1997). Food availability may impose energetic limits on postpartum conceptions, and therefore delay the resumption of postpartum ovarian activity in the wild (Savage et al. 1997). This holds true for golden lion tamarins as well. Female golden lion tamarins are seasonal in pregnancies, and it appears that their ovaries may be quiescent for several months out of the year in their native environment (French et al. 2003). Although there is limited seasonal variation in the feeding patterns of two tamarin species, the moustached tamarin (*Saguinus mystax*) and the saddle back tamarin (Garber 1993), data from moustached tamarins suggest periods of anovulation after parturition for the breeding female. Measuring fecal steroids, Löttker et al. (2004) found that the period of postpartum ovarian inactivity in wild moustached tamarins lasts between 50 and 80 days, and conception is delayed following the onset of postpartum ovarian activity. In this species, unlike other callitrichids, breeding females experience periods of inactive ovaries as well as a delay in conception following the onset of ovarian cycling. Tamarin fertility appears to be more seasonal, but the data on anovulation during the non-breeding season is limited in most studies and the frequency of sampling in reported studies is often is too low to reliably evaluate reproductive function. While the callitrichids can potentially ovulate and conceive while nursing young infants, this has not been reported for any other primates except humans (McNeilly et al. 1988; Ziegler et al. 1990). In captivity, frequent nursing and alternating twin infants can induce delays in ovulation in both marmosets and tamarins (Ziegler et al. 1990), but in the wild only marmosets appear to conceive regularly on their first postpartum ovulation.

Seasonal effects on breeding may be limited in species such as the common marmoset, where ovulation usually ends in conception, or there may be a few ovulations prior to conception, as is seen in captive squirrel monkeys (Schiml et al. 1999). The observation that females experience several ovarian cycles before conceiving is rather the norm for most platyrrhines. There exists great variability in the number of nonconceptive cycles prior to conception between species, between females, and within individual females over different periods of ovarian cycling. Captive squirrel monkeys have just a few nonconceptive cycles (Schiml et al. 1999), while wild northern muriquis and wild black-handed spider monkeys, have on average 2–6 ovulations prior to conception (Campbell et al. 2001; Strier et al. 2003). Unfortunately, there are few long-term field studies on other platyrrhines that have monitored reproductive steroids in sufficient detail over multiple years to examine the effects of seasonality on ovarian cycles and conception. Consistent and long-term data are needed to determine the differences in ovulatory and conception cycles.

Northern muriqui females at the RPPN-Feliciano Miguel Abdala (previously, Estação Biológica de Caratinga) in Minas Gerais, Brazil, show seasonal mating patterns that resume concurrently with ovulatory cycles (Strier and Ziegler 1994). The onset of ovarian cycling varies within the breeding season, even between females whose last parturitions were only a few days apart (Strier and Ziegler 2005). Additionally, there is considerable variability in when females conceive, which usually occurs later in the breeding season (Strier and Ziegler 1997, 2005; Strier et al. 2003). The reasons for this variation in ovarian cycle onset and conception are currently unknown, but may be related to food and energetics. We have found that captive muriqui females in outdoor enclosures at the Centro de Primatologia in Rio de Janeiro have an earlier onset of ovarian cycling and conception than the wild muriquis (Ziegler et al. 1997). In fact, two captive females studied were nursing their 6-month-old infants while ovulating and mating whereas wild muriqui mothers do not usually resume mating for at least two years after the birth of a surviving infant. The captive females received high protein diets and were restricted in their activity compared to the wild muriquis, which might explain their early resumption of ovulation while lactating.

8.1.5 *Reproductive Seasonality*

As with cercopithocines, there is a continuum of reproductive seasonality in platyrrhines. Seasonal reproduction is generally indicated by the occurrence of the majority of births at a certain time of year within a particular population (Lindburg 1987). Birth peaks may be concentrated during the wet season or dry season, and when preferred foods are most abundant or least abundant. The degree of seasonality may be attributed to seasonal supplies of food that can be directly related to climate. Seasonality in births indicates seasonality in breeding. However, neither seasonal breeding nor the corresponding seasonal birth peaks always occur at precisely the same time of the year, even in sympatric species existing in the same environmental conditions (Strier 2007).

Differences in diet, adult body size, and life history traits between sympatric primates may explain their different responses to fluctuating environmental conditions (Di Bitetti and Janson 2000; Strier 2007). For instance, in the Brazilian Atlantic forest where four species of primates share the same habitat, the northern muriquis show seasonal breeding that coincides with the peak rainy season months (Strier 1996) while the tufted capuchin monkeys (*Cebus nigritus*) have their primary breeding season in the dry season months (Lynch et al. 2002). The buffy-headed marmoset (*Callithrix flaviceps*) breeds twice a year (Ferrari and Mendes 1991) while the brown howler monkey (*Alouatta guariba clamitans*) breeds year round (Strier et al. 2001). The different seasonality patterns of these sympatric primates living in the Atlantic forest diverge from those observed for sympatric white faced capuchin monkeys (*Cebus capucinus*), mantled howler monkeys (*Alouatta palliata*), and black handed spider monkeys living in Costa Rica, where births of all three species are concentrated during the dry season (Fedigan and Rose 1995). Even populations of the same species living 40 km apart but in different forest types can experience different reproductive seasonality patterns. A black-and-gold howler monkey population (*Alouatta caraya*) living in a flooded forest with a consistent supply of young and mature leaves did not show seasonality in births, while a population living in a forest on the mainland with seasonal fluctuation in availability of young and mature leaves did show birth seasonality (Kowalewski and Zunino 2004).

The squirrel monkey has been extensively studied for birth and breeding seasonality in captivity and in the wild. Squirrel monkeys are seasonal even in enclosed, controlled captive conditions with consistent supplies of food. In the wild and in captivity, female squirrel monkeys give birth every other year with distinct seasons of mating and births (DuMond 1968; Rosenblum 1968; Harrison and Dukelow 1973; Taub et al. 1978; Logdberg 1993; Schiml et al. 1996; Stone 2007). Males begin the breeding season with an onset of fatting and weight gain, and females follow with the onset of ovarian cycling (Coe and Rosenblum 1978). Environmental cues appear to be responsible for the onset of the breeding season. In a captive population, Trevino (2007) found that photoperiod has a significant positive influence on the number of births per month, while temperature has a significant effect on both the number of births and matings per month. Additional proximate mechanisms influence seasonal breeding in both male and female squirrel monkeys. The presence of multiple females in a group has a pronounced effect on seasonal changes in male reproductive hormones, and dominant males inhibit gonadal hormone production in subordinates (Schiml et al. 1996). Interestingly, female squirrel monkeys are more likely to undergo ovarian cyclicity when in the presence of other females than when living with a single male. Apparently, a consortium of females is required for both male and female optimal reproduction.

Less seasonal species may not be as tied to environmental cues as the squirrel monkey. In species such as the white-faced capuchin monkey, limited access to high quality food is likely to be the ultimate cause of infertility or nonconceptive cycling and the lack of ovulatory cycles at certain times of the year. However, there may be more proximate causes for the delays in conception after the onset of ovulation. For example, wild female white-faced capuchin monkeys living in Santa Rosa National Park in Costa Rica have been shown to exhibit cyclical variation in fecal progesterone and estradiol levels (Carnegie et al. 2005). While four females were displaying ovarian steroid cycling, three other females were at the end of pregnancy and subsequently gave birth. During the same 6-month study period, three other females failed to display any ovarian cycling. The four cycling females had one to three ovarian cycles before their reproductive hormones declined to basal levels. The cessation of ovarian cycling in these four females coincided with the parturitions of the pregnant females. These females most probably started ovulating again later that year because they gave birth the following year. While there may be several energetic explanations for the cessation in cycling (Carnegie et al. 2005), the birth of infants might have caused a suppression of reproduction in the cycling females similar to the effects observed on captive cotton-top tamarin daughters when their reproductive mothers gave birth. The tamarin daughters were not cycling but had high levels of urinary estrogens and LH, which abruptly declined to basal levels when their mothers gave birth (Snowdon et al. 1993). Social influences, mediated through chemical communication, may therefore inhibit the hormonal release in daughters.

Fecal steroid monitoring of wild pregnant and cycling female black-handed spider monkeys at Barro Colorado Island, Panama have also revealed considerable variation. Campbell et al. (2001) detected conceptions in four of eight females from the sustained elevations of their estrogen and progestin levels. Two of the pregnant females showed three to six cycles prior to conception, and one of the pregnant females had steroid patterns that reflected early fetal loss followed by another conception two months later. Only one of the eight adult females never showed ovarian cycling during the 11-month study period. Two females that were considered to be subadult also showed cyclical steroid activity. These data suggest that spider monkeys, at least during this study period at this site, might not have a discrete conception season and may instead reproduce year-round. However, in a captive population of black-handed spider monkeys, fecal estradiol and progesterone levels of five females were higher in their ovarian cycles during the fall season compared to those during other times of the year (Cerda-Molina et al. 2006). The lower fecal levels of estradiol and progesterone in winter, spring and summer might have resulted in anovulatory cycles. The ovulatory cycles combined with the higher sperm quality in males during the same season could lead to higher probability for conception during the fall, even though conception might occur year round (Hernández-López et al. 2002; Cerda-Molina et al. 2006).

Understanding the effects of diet and energetics on reproductive patterns in South American primates requires long-term behavioral and ecological data. This is particularly true when rainfall and its effects on food availability, are known to fluctuate inter-annually, as well as seasonally. For example, in one group of wild northern muriquis, extended birth seasons, indicative of extended mating and conception seasons, occur in years of prolonged and heavy rainfall (Strier 1996).

8.2 Discussion

Primate ovaries are sensitive to energy balance, and changes in energy levels affect female conception probabilities. Conception is most likely to occur when: (1) sufficient energy has been stored in the body and can be mobilized for reproductive events; (2) energy intake exceeds energy expenditure; and (3) the absolute level of energy turnover is independent of energy balance (Ellison 2003). For some seasonally breeding primates, intermediate levels of energy that can be quickly mobilized may allow for ovulatory cycles, but may still be insufficient for conception. Thus, interspecific and inter-individual variation in the postpartum resumption of cycling, and in the number of cycles prior to conception may reflect variation in ecology and in individual condition (e.g., age, rank, health) among South American primates.

Individual differences in energy reserves have been shown to correlate with the time to conception in strepsirrhines (Lewis and Kappeler 2005) and cercopithecoids (Bercovitch 1987). For instance, heavier wild female sifaka (*Propithecus verreauxi*) are more likely to reproduce than lighter females (Richard et al. 2000), and heavier female olive baboons (*Papio anubis*) conceive earlier than lighter females (Bercovitch 1987). Hanuman langurs (*Semnopithecus entellus*) that look to be in better physical condition conceive sooner than females in visibly poorer conditions (Koenig et al. 1997). Verreaux's sifaka live in highly seasonal environments and females adapt their reproductive cycles to the fluctuation in food availability (Lewis and Kappeler 2005). In these, as in other primates, limited food resources limit the energy necessary for reproduction.

The callitrichid monkeys also have the ability to increase their reproductive output by increasing litter size. Callitrichids have one of the highest reproductive potentials among primates (Tardif et al. 1993). The common marmoset has been shown to ovulate a variable number of follicles with each cycle. The number of follicles that ovulate can be between two to four and can occur on both or either ovary (Ziegler and Stott 1986; Tardif et al. 1993). The follicles are not necessarily synchronized so fertilization of the oocytes may not be at the same time. Their reproductive rate is influenced by their nutritional status and body weights, with heavier females producing larger litters than lighter females (Tardif and Jaquish 1997). This is also true for the golden lion tamarins (Bales et al. 2001). Additionally, there are developmental effects on reproduction. The energy conditions of the environment where a female develops can influence her lifetime reproduction. For instance, wild caught, but captive, golden-headed lion tamarins (*Leontopithecus chrysomelas*) have lower body weights than captive born golden-headed tamarins even with similar diets (De Vleeschouwer et al. 2003). Wild golden-headed tamarins have similar litter sizes to captive wild-born golden-headed tamarins but these are smaller than litter sizes found in captive-born females (De Vleeschouwer et al. 2003).

In humans, variation in ovarian steroid levels during cycling has been demonstrated to influence the probability of conception (Lipson and Ellison 1996). Our analyses of female northern muriquis also suggest that conception in this species may require reaching a threshold in peak estradiol levels during cycling (Strier and Ziegler 2005).

Once conception has occurred, a sufficient level of energy intake is required to sustain the pregnancy. Studies on energy restriction during pregnancy in the common marmoset have shown that females without adequate energy intake will incur fetal loss (Tardif et al. 2004, 2005). When females lose weight, there is a reduction in estradiol, cortisol and chorionic gonadotropin levels, which can be measured in the mother's urine (Tardif et al. 2005). These markers indicate that there is impaired placental or fetal function that leads to fetal and pregnancy loss. Pregnancy loss has not been reported for wild common marmosets (Albuquerque et al. 2001) but this may be difficult to detect without extensive sampling. Additionally, the social effects on ovarian activity may confound the interpretation of the data. For instance, subordinate daughters in groups of monogamous common marmosets may show a rapid reduction in ovarian hormones during the postpartum period of their mothers (Albuquerque et al. 2001).

Understanding the proximate mechanisms underlying variation in female ovarian cycles remains an important area for future investigations. Physiological adjustments help to insure that energetic conditions are sufficient to sustain the costs of pregnancy and subsequent lactation in humans (Ellison 2003). Similar selective pressures undoubtedly affect the evolutionary physiology of nonhuman primates and may therefore account for the diversity of cycling and conception patterns between platyrrhines and catarrhines, and among the South American primates that have been studied to date.

8.3 Summary

South American primates are similar to other primates in having well-defined ovarian cycles without having mating restricted to the periovulatory period. However, platyrrhines are different from catarrhines in having much higher levels of circulating and excreted steroids. The use of fecal steroid metabolites for monitoring ovarian cycles, conception and pregnancies requires an understanding of the pattern of delayed excretion of fecal estrogens in each species. In the platyrrhines for which information is available, estrogen patterns are highly variable in relation to progesterone excretion and to the timing of ovulation. Therefore, patterns of reproductive steroids need to be assessed on a species-by-species basis to determine the relationship of the estrogens to the progestins and the timing of ovulation and conception.

South American primates also exhibit considerable variation in the degree of seasonality in births and mating. Much of this variability can be attributed to seasonality in food resources and the influence of energetics on reproductive function. Data from captive primates have provided us with a good understanding of reproduction in some species of South American primates, but only systematic studies on these and other wild primates will allow us to understand the factors that regulate the onset of ovarian cycling and conception, and environmental influences on reproductive patterns. Advances in the development of non-invasive methods for measuring reproductive steroids have extended our knowledge of primate reproduction, but indepth behavioral, ecological, and hormonal studies on many species are still needed to fully characterize the variation in the reproductive ecology of South American primates.

Acknowledgments   The authors wish to thank E. W. Heymann and two anonymous reviewers for their helpful feedback on an earlier version of this manuscript. They also thank all the investigators whose work was referenced in this manuscript and their dedication to understanding the reproductive physiology of South American primates is well noted. This work was supported by the NIH base grant to the Wisconsin National Primate Research Center, RR000167 and other grants to individual investigators for the various species.

References

Albuquerque, A. C. S. R., Sousa, M. B. C., Santos, H. M., and Ziegler, T.E. 2001. Behavioral and hormonal analysis of social relationships between oldest (reproductive and non-reproductive) females in a wild monogamous group of common marmosets (*Callithrix jacchus*). Int. J. Primatol. 22:631–645.

Bales, K., O'Herron, M., Baker, A. J., and Dietz, J. M. 2001. Sources of variability in numbers of live births in wild golden lion tamarins (*Leontopithecus rosalia*). Am. J. Primatol. 54:211–221.

Bercovitch, F. B. 1987. Female weight and reproductive condition in a population of olive baboons (*Papio anubis*). Am. J. Primatol. 12:189–195.

Bonello, N., Jasper, M. J., and Norman, R. J. 2004. Periovulatory expression of intercellular adhesion molecule-1 in the rat ovary. Biol. Reprod. 71:1384–1390.

Bonney, R. C., Dixon, A. F., and Fleming, D. 1979. Cyclic changes in the circulating and urinary levels of ovarian steroids in the adult female owl monkey (*Aotus trivirgatus*). J. Reprod. Fertil. 56:271–280.

Bonney, R. C., Dixon, A. F., and Fleming, D. 1980. Plasma concentrations of oestradiol-17β, oestrone, progesterone and testosterone during the ovarian cycle of the owl monkey (*Aotus trivirgatus*). J. Reprod. Fertil. 60:101–107.

Brand, H. M. 1981. Urinary oestrogen excretion in the female cotton-topped tamarin (*Saguinus oedipus oedipus*). J. Reprod. Fertil. 62:467–473.

Campbell, C. J., Shideler, S.E., Todd, H. E., and Lasley, B. L. 2001. Fecal analysis of ovarian cycles in female black-handed spider monkeys (*Ateles geoffroyi*). Am. J. Primatol. 54:79–89.

Carnegie, S. D., Fedigan, L. M., and Ziegler, T. E. 2005. Behavioral indicators of ovarian phase in white-faced capuchins (*Cebus capucinus*). Am. J. Primatol. 67:51–68.

Carosi, M., Heistermann, M., and Visalberghi, E. 1999. Display of proceptive behaviors in relation to urinary and fecal progestin levels over the ovarian cycle in female tufted capuchin monkeys. Horm. Behav. 36:252–265.

Carroll, J. B., Abbott, D. H., George, L. M., Martin, R. D. 1989. Aspects of urinary oestrogen excretion during the ovarian cycle and pregnancy in Goeldi's monkeys, *Callimico goeldii*. Folia Primatol. 52:201–205.

Cerda-Molina, A. L., Hernández-López, L., Páez-Ponce, D. L., Rojas-Maya, S., and Mondragón-Ceballos, R. 2006. Seasonal variation of fecal progesterone and 17β-estradiol in captive female black-handed spider monkeys (*Ateles geoffroyi*). Theriogenology 66:1985–1993.

Chaoui, N. J., and Hasler-Gallusser, S. 1999. Incomplete sexual suppression in *Leontopithecus chrysomelas*: a behavioral and hormonal study in a semi-natural environment. Folia Primatol. 70:47–54.

Chen, H., Arbelle, J. E., Gacad, M. A., Allegretto, E. A., and Adams, J. S. 1997. Vitamin D and gonadal steroid-resistant New World primate cells express an intracellular protein which competes with the estrogen receptor for binding to the estrogen response element. J. Clin. Invest. 99:669–675.

Christen, A., Dobeli, M., Kempken, B., Zachmann, M., and Martin, R. D. 1989. Urinary excretion of oestradiol-17ß in the female cycle of Goeldi's monkeys (*Callimico goeldii*). Folia Primatol. 52:191–200.

Coe, C. L., and Rosenblum, L. A. 1978. Annual reproductive strategy of the squirrel monkey (*Saimiri sciureus*). Folia Primatol. 29:19–42.

Converse, L. J., Carlson, A. A., Ziegler, T. E., and Snowdon, C. T. 1995. Communication of ovulatory state to mates by female pygmy marmosets, *Cebuella pygmaea*. Anim. Behav. 49:615–621.

Dettling, A. C. 2002. Reproduction and development in Goeldi's monkey (*Callimico goeldii*). Evol. Anthropol. Suppl. 1:207–210.

De Vleeschouwer, K., Heistermann, M., and Van Elsacker, L. 2000. Signaling of reproductive status in captive female golden-headed lion tamarins (*Leontopithecus chrysomelas*). Int. J. Primatol. 21:445–465.

De Vleeschouwer, K., Leus, K., and Van Elsacker, L. 2003. Characteristics of reproductive biology and proximate factors regulating seasonal breeding in captive golden-headed lion tamarins (*Leontopithecus chrysomelas*). Am. J. Primatol. 60:123–137.

Di Bitetti, M. S., and Janson, C. H. 2000. When will the stork arrive? Patterns of birth seasonality in neotropical primates. Am. J. Primatol. 50:109–130.

Dixson, A. F. 1998. Primate Sexuality: Comparative Studies of the Prosimians, Monkeys, Apes, and Human Beings. Oxford: Oxford University Press.

DuMond, F. V. 1968. The squirrel monkey in a seminatural environment. In L. A. Rosenblum (ed.), *The Squirrel Monkey* (pp. 87–145). New York: Academic Press.

Eastman, S. A. K., Makawiti, D. W., Collins, W. P., and Hodges, J. K. 1984. Pattern of excretion of urinary steroid metabolites during the ovarian cycle and pregnancy in the marmoset monkey. J. Endocrinol. 120:19–26.

Ellison, P. T. 2003. Energetics and reproductive effort. Am. J. Human Biol. 15:342–351.

Epple, G., and Katz, Y. 1983. The saddle back tamarin and other tamarins. In J. P. Hearn (ed.), *Reproduction in New World Primates: New Models in Medical Science* (pp. 115–148). Boston: MTP Press.

Fedigan, L. M., and Rose, L. M. 1995. Interbirth interval variation in three sympatric species of neotroptical monkeys. Am. J. Primatol. 37:9–24.

Ferrari, S. F. 1988. The behavior and ecology of the buffy-headed marmoset, *Callithrix flaviceps*. Ph.D. thesis, University College, London.

Ferrari, S. F., and Mendes, S. L. 1991. Buffy-headed marmosets 10 years on. Oryx 25:105.

French, J. A., and Stribley, J. A. 1985. Patterns of urinary oestrogen excretion in female golden lion tamarins (*Leontopithecus rosalia*). J. Reprod. Fertil. 75:537–546.

French, J. A., and Stribley, J. A. 1987. Synchronization of ovarian cycles within and between social groups in golden lion tamarins (*Leontopithecus rosalia*). Am. J. Primatol. 12:469–478.

French, J. A., Abbott, D. H., Scheffer, G., Robinson, J. A., and Goy, R. W. 1983. Cyclic excretion of urinary oestrogen in female tamarins (*Saguinus oedipus*). J. Reprod. Fertil. 68:177–184.

French, J. A., Bales, K. L., Baker, A. J., and Dietz, J. M. 2003. Endocrine monitoring of wild dominant and subordinate female *Leontopithecus rosalia*. Int. J. Primatol. 24:1281–1300.

French, J. A., Brewer, K. J., Schaffner, C. M., Schalley, J., Hightower-Merritt, D., Smith, T. E., and Bell, M. 1996. Urinary steroid and gonadotropin excretion across the reproductive cycle in female Wied's black tufted-ear marmosets (*Callithrix kuhlii*). Am. J. Primatol. 40:231–245.

French, J. A., De Graw, W. A., Hendricks, S. E., Wegner, F., and Bridson, W. E. 1992. Urinary and plasma gonadotropin concentrations in golden lion tamarins (*Leontopithecus r. rosalia*). Am. J. Primatol. 26:53–59.

Garber, P. 1993. Seasonal patterns of diet and ranging in two species of tamarin monkeys: Stability versus variability. Int. J. Primatol. 14:146–166.

Ghosh, M., Hutz, R. J., and Dukelow, W. R. 1982. Serum estradiol 17ß, progesterone, and relative luteinizing hormone levels in *Saimiri sciureus*: cyclic variations and the effect of laparoscopy and follicular aspiration. J. Med. Primatol. 11:312–318.

Harrison, R. M., and Dukelow, W. R. 1973. Seasonal adaptation of laboratory maintained squirrel monkey (*Saimiri sciureus*). J. Med. Primatol. 2:277–283.

Hearn, J. P. 1983. Reproduction in New World Primates: New Models in Medical Science. Lancaster, England: MTP Press.

Hearn, J. P. 2001. Embryo implantation and embryonic stem cell development in primates. Reprod. Fertil. Dev. 13:517–522.

Heistermann, M., and Hodges, J. K. 1995. Endocrine monitoring of the ovarian cycle and pregnancy in the saddle-back tamarin (*Saguinus fuscicollis*) by measurement of steroid conjugates in urine. Am. J. Primatol. 35:117–127.

Heistermann, M., Tari, S., and Hodges, J. K. 1993. Measurement of faecal steroids for monitoring ovarian function in New World primates, Callitrichidae. J. Reprod. Fertil. 99:243–251.

Hernández-López, L., Mayagoita, L., Esquivel-Lacroix, C., Rojas-Maya, S., and Mondragón-Ceballos, R. 1998. The menstrual cycle of the spider monkey (*Ateles geoffroyi*). Am. J. Primatol. 44:183–195.

Hernández-López, L., Cerezo-Parra, G., Cerda-Molina, A. L., Pérez-Bolaños, S. C., Díaz-Sánchez. V., and Mondragón-Ceballos, R. 2002. Sperm quality differences between the rainy and dry season in captive black-handed spider monkeys (*Ateles geoffroyi*). Am. J. Primatol. 57:35–41.

Herrick, J. R., Agoramoorthy, G., Rudran, R., and Harder, J. D. 2000. Urinary progesterone in free-ranging red howler monkeys (*Alouatta seniculus*): preliminary observations of the estrous cycle and gestation. Am. J. Primatol. 51:257–263.

Hodgen, G. D., and Itskovitz, J. 1988. Recognition and maintenance of pregnancy. In E. Knobil and J. Neill (eds.), *The Physiology of Reproduction*, New York: Raven Press, Ltd.

Hodges, J. K., and Heistermann, M. 2003. Field endocrinology: monitoring hormonal changes in free-ranging primates. In J. M. Setchell and D. J. Curtis (eds.), *Field and Laboratory Methods in Primatology: A Practical Guide*, Cambridge: Cambridge University Press.

Hodges, J. K., Gulick, B. A., Czekala, N. M., and Lasley, B. L. 1981. Comparison of urinary oestrogen excretion in South American primates. J. Reprod. Fertil. 61:83–90.

Jones, C. B. 1985. Reproductive patterns in mantled howler monkeys: Estrus, mate choice and copulation. Primates 26:130–142.

Kholkute, S. D. 1984. Plasma progesterone levels throughout the ovarian cycle of the common marmoset (*Callithrix jacchus*). Primates 25:123–126.

Koenig, A., Borries, C., Chalise, M. K., and Winkler, P. 1997. Ecology, nutrition, and the timing of reproductive events in an Asian primate, the Hanumann langur (*Presbytis entellus*). J. Zool. Lond. 24:215–235.

Kowalewski, M., and Zunino, G. E. 2004. Birth seasonality in *Alouatta caraya* in Northern Argentina. Int. J. Primatol. 25:383–400.

Laufer, N., Navot, D., and Schenker, J. G. 1982. The pattern of luteal phase plasma progesterone and estradiol in fertile cycles. Am. J. Obstet. Gynecol. 143:808–813.

Lewis, R. J., and Kappeler. P. M. 2005. Seasonality, body condition, and timing of reproduction in *Propithecus verreauxi verreauxi* in the Kirindy Forest. Am. J. Primatol. 67:347–364.

Lindburg, D. G. 1987. Seasonality of reproduction in primates. In G. Mitchell and J. Erwin (eds.), *Comparative primate biology, Vol 2A: Behavior, cognition, and motivation* (pp. 167–218). New York: Alan r. Liss, Inc.

Lipson, S. F., and Ellison, P. T. 1996. Comparison of salivary steroid profiles in naturally occurring conception and nonconception cycles. Hum. Reprod. 11:2090–2096.

Löttker, P., Huck, M., Heymann, E. W., and Heistermann, M. 2004. Endocrine correlates of reproductive status in breeding and non-breeding wild female moustached tamarins, *Saguinus mystax*. Intl. J. Primatol. 25:919–937.

Logdberg, B. 1993. Methods for timing of pregnancy and monitoring of fetal body and brain growth in squirrel monkeys. J. Med. Primatol. 22:374–379.

Lynch, J. W., Ziegler, T. E., and Strier, K. B. 2002. Individual and seasonal variation in fecal testosterone and cortisol levels of wild male tufted capuchin monkeys, *Cebus apella nigritus*. Horm. Behav. 41:275–287.

McNeilly, A. S., Howie, P. W., and Glasier, A. 1988. Lactation and the return of ovulation. In P. Diggory, M. Potts and S. Teper (eds.), *Natural Human Fertility: Social and Biological Determinants* (pp. 102–117). London: MacMillan Press.

Nagle, C. A., Denari, J. H., Quiroga, S., Riarte, A., Merlo, A., Germino, N. I., Gomez-Argana, F., and Rosner, J. M. 1979. The plasma pattern of ovarian steroids during the menstrual cycle in capuchin monkeys (*Cebus apella*). Biol. Reprod. 21:979–983.

Pryce, C. R., Jurke, M., Shaw, H. J., Sandmeier, I. G., and Doebeli. M. 1993. Determination of ovarian cycle in Goeldi's monkey (*Callimico goeldii*) via the measurement of steroids and peptides in plasma and urine. J. Reprod. Fertil. 99:427–435.

Pryce, C. R., Schwarzenberger, F., and Dobeli, M. 1994. Monitoring fecal samples for estrogen excretion across the ovarian cycle in Goeldi's monkey (*Callimico goeldii*). Zoo. Biol. 13: 219–230.

Richard, A. F., Dewar, R. E., Schwartz, M., and Ratsirarson, J. 2000. Mass change, environmental variability and female fertility in wild *Propithecus verreauxi*. J. Hum. Evol. 39:381–391.

Rosenblum, L. A. 1968. Some aspects of female reproductive physiology in the squirrel monkey. In L. A. Rosenblum and R. W. Cooper (eds.), *The Squirrel Monkey* (pp. 147–169). New York: Academic Press.

Savage, A., Lasley, B. L., Vecchio, A. J., Miller, A. E., and Shideler, S. E. 1995. Selected aspects of female White-faced saki (*Pithecia pithecia*) reproductive biology in captivity. Zoo. Biol. 14:441–452.

Savage, A., Shideler, S. E., Soto, L. H., Causado, J., Giraldo, L. H., Lasley, B. L., and Snowdon, C. T. 1997. Reproductive events of wild cotton-top tamarins (*Saguinus oedipus*) in Colombia. Am. J. Primatol. 43:329–337.

Schiml, P., Mendoza, S., Saltzman, W., Lyons, D., and Mason, W. 1996. Seasonality in squirrel monkeys (*Saimiri sciureus*): social facilitation by females. Physiol. Behav. 60:1105–1113.

Schiml, P. A., Mendoza, S. P., Saltzman, W., Lyons, D. M., and Mason, W. A. 1999. Annual physiological changes in individually housed squirrel monkeys (*Saimiri sciureus*). Am. J. Primatol. 47:93–103.

Shideler, S. E., Savage, A., Ortuna, A. M., Moorman, E. A., and Lasley, B. L. 1994. Monitoring female reproductive function by measurement of fecal estrogen and progesterone metabolites in the white-faced saki. (*Pithecia pithecia*). Am. J. Primatol. 32:95–108.

Snowdon, C. T., Ziegler, T. E., and Widowski, T. M. 1993. Further hormonal suppression of eldest daughter cotton-top tamarins following the birth of infants. Am. J. Primatol. 31:11–21.

Stevenson, M. F., and Rylands, A. B. 1988. The marmosets, genus *Callithrix*. In R. A. Mittermeier, A. B. Rylands, A. F. Coimbra-Filho and G. A. B. da Fonseca (eds.), *Ecology and Behavior of Neotropical Primates*, Vol. 2 (pp. 131–222). Washington DC: World Wildlife Fund.

Stocco, C., Telleria, C., and Gibori, G. 2007. The molecular control of corpus luteum formation, function, and regression. Endocr. Rev. 28:117–149.

Stone, A. 2007. Responses of squirrel monkeys to seasonal changes in food availability in an eastern Amazonian forest. Am. J. Primatol. 69:142–157.

Strier, K. B. 1996. Reproductive ecology of female muriquis. In M. A. Norconk, A. L. Rosenberger and P. A. Garber (eds.), Adaptive Radiations of Neotropical Primates, (pp. 511–532). New York: Plenum Press.

Strier, K. B. 2007. Primate Behavioral Ecology, Third Edition. Massachusetts: Allyn & Bacon.

Strier, K. B., and Ziegler, T. E. 1994. Insights into ovarian function in wild muriqui monkeys (*Brachyteles arachnoides*). Am. J. Primatol. 32:31–40.

Strier, K. B., and Ziegler, T. E. 1997. Behavioral and endocrine characteristics of the reproductive cycle in wild muriqui monkeys, *Brachyteles arachnoides*. Am. J. Primatol. 42:299–310.

Strier, K. B., and Ziegler, T.E. 2005. Variation in the resumption of cycling and conception by fecal androgen and estradiol levels in female northern muriquis (*Brachyteles hypoxanthus*). Am. J. Primatol. 67:69–81.

Strier, K. B., Lynch, J. W., and Ziegler, T. E. 2003. Hormonal changes during the mating and conception seasons of wild northern muriquis (*Brachyteles arachnoides hypoxanthus*). Am. J. Primatol. 61:85–99.

Strier, K. B., Mendes, S. L., and Santos, R. R. 2001. Timing of births in sympatric brown howler monkeys (*Alouatta fusca clamitans*) and northern muriquis (*Brachyteles arachnoides hypoxanthus*). Am. J. Primatol. 55:87–100.

Tardif, S. D., and Jaquish, C.E. 1997. Number of ovulations in the marmoset monkey (*Callithrix jacchus*): relation to body weight, age and repeatability. Am. J. Primatol. 42:323–329.

Tardif, S. D., Harrison, M. L., and Simek, M. A. 1993. Communal infant care in marmosets and tamarins: relation to energetics, ecology and social organization. In A. B. Rylands (ed.), Systematics, Ecology and Behavior (pp. 220–234). Oxford: Oxford University Press.

Tardif, S. D., Lacker, H. M., and Feuer, M. 1993. Follicular development and ovulation in the marmoset monkey as determined by repeated laparoscopic examination. Biol. Reprod. 48: 1113–1119.

Tardif, S., Power, M., Lane, D., Smucny, D., and Ziegler, T. 2004. Energy restriction initiated at different gestational ages has varying effects on maternal weight gain and pregnancy outcome in common marmoset monkeys (*Callithrix jacchus*). Brit. J. Nutrit. 92:841–849.

Tardif, S. D., Ziegler, T. E., Power, M., and Layne, D. G. 2005. Endocrine changes in full-term pregnancies and pregnancy loss due to energy restriction in the common marmoset (*Callithrix jacchus*). J. Clin. Endocrin. Metab. 90:335–339.

Taub, D. M., Adams, M. R., and Auerback, K. G. 1978. Reproductive performance in a breeding colony of Brazilian squirrel monkeys (*Saimiri sciureus*). Lab Anim. Sci. 28:562–566.

Trevino, H. S. 2007. Seasonality of reproduction in captive squirrel monkeys (*Saimiri sciureus*). Am. J. Primatol. 69:1–12.

Valeggia, C. R., Mendoza, S. P., Fernandez-Duque, E., Mason, W. A., and Lasley, B. 1999. Reproductive biology of female titi monkeys (*Callicebus moloch*) in captivity. Am. J. Primatol. 47:183–195.

Vilela, S. L., and Faria, D. S. de. 2004. Seasonality of the activity pattern of *Callithrix penicillata* (Primates, Callitrichidae) in the cerrado (scrub savanna vegetation). Braz. J. Biol. 64:363–370.

Wasser, S. K., Risler, L., and Steiner, R. A. 1988. Excreted steroids in primate feces over the menstrual cycle and pregnancy. Biol. Reprod. 39:862–872

Weiner, R. I., Findell, P. R., and Kordon. C. 1988. Role of classic and peptide neuromediators in the neuroendocrine regulation of LH and prolactin. In E. Knobil and J. Neill (eds.), The Physiology of Reproduction (pp. 1235–1282). New York: Raven Press, Ltd.

Whitten, P. L., Brockman, D. K., and Stavisky, R. C. 1998. Recent advances in non-invasive techniques to monitor hormone-behavior interactions. Yrbk. Phys. Anthropol. 41:1–23.

Wright, E. M. Jr., and Bush, D. E. 1977. The reproductive cycle of the capuchin (*Cebus apella*). Laboratory Animal Science 27:651–654.

Yen, S. C., and Jaffe, R. B. 1978. Reproductive Endocrinology: Physiology, Pathophysiology and Clinical Management. Philadelphia: Saunders Company.

Ziegler, T., and Stott, G. 1986. Determination of estrogen concentrations and ovulation detection in the common marmoset (*Callithrix jacchus*) by an enzymatic technique. In D. M. Taub and F. E. King (eds.), Current Perspectives in Primate Biology (pp. 42–58). New York: Van Nostrand Reinhold.

Ziegler, T. E., and Sousa, M. B. C. 2002. Parent-daughter relationships and social controls on fertility in female common mamrosets, *Callithrix jacchus*. Horm. Behav. 42: 356–367.

Ziegler, T. E., and Wittwer, D. J. 2005. Fecal steroid research in the field and laboratory: Improved methods for storage, transport, processing and analysis. Am. J. Primatol. 67:159–174.

Ziegler, T. E., Bridson, W. E., Snowdon, C. T., and Eman, S. 1987. Urinary gonadotropin and estrogen excretion during the postpartum estrus, conception and pregnancy in the cotton-top tamarin (*Saguinus oedipus oedipus*). Am. J. Primatol. 12:127–140.

Ziegler, T. E., Sholl, S. A., Scheffler, G., Haggerty, M. A., and Lasley, B. L. 1989. Excretion of estrone, estradiol and progesterone in the urine and feces of the female cotton-top tamarin (*Saguinus oedipus oedipus*). Am. J. Primatol. 17:185–195.

Ziegler, T. E., Snowdon, C. T., and Bridson, W. E. 1990. Reproductive performance and excretion of urinary estrogens and gonadotropins in the female pygmy marmoset (*Cebuella pygmaea*). Am. J. Primatol. 22:191–203.

Ziegler, T. E., Snowdon, C. T., Warneke, M., and Bridson, W. E. 1990. Urinary excretion of oestrone conjugates and gonadotropins during pregnancy in the Goeldi's monkey, *Callimico goeldii*. J. Reprod. Fertil. 89:163–168.

Ziegler, T. E., Wittwer, D. J., and Snowdon, C. T. 1993. Circulating and excreted hormones during the ovarian cycle in the cotton-top tamarin, *Saguinus oedipus*. Am. J. Primatol. 31:55–65.

Ziegler, T. E., Scheffler, G., Wittwer, D. J., Schultz-Darken, N. J., Snowdon, C.T., and Abbott, D. H. 1996. Metabolism of reproductive steroids during the ovarian cycle in two species of callitrichids, *Saguinus oedipus* and *Callithrix jacchus*, and estimation of the ovulatory period from fecal steroids. Biol. Reprod. 54: 91–99.

Ziegler, T. E., Santos, C. V., Pissinatti, A., and Strier, K. B. 1997. Steroid excretion during the ovarian cycle in captive and wild muriquis, *Brachyteles arachnoides*. Am. J. Primatol. 42: 311–321.

Chapter 9
Genetic Approaches to the Study of Dispersal and Kinship in New World Primates

Anthony Di Fiore

9.1 Introduction

Among social animals such as primates, "kinship" or genetic relatedness is commonly invoked as a key factor underlying and organizing the expression of within-group social behavior (Alexander 1974; Wilson 1975; Gouzoules 1984; Bernstein 1991; Silk 2001, 2002). Indeed, kin-correlated behavior – particularly kin-directed beneficent behavior or "nepotism" – is often considered a hallmark feature of the social lives of group-living primates (Gouzoules 1984; Gouzoules and Gouzoules 1987).

Within primate social groups, the patterns of genetic relatedness among group members are influenced principally by the dispersal and mating behaviors of those individuals. Dispersal directly shuffles genes across the physical and social landscapes, reassorting how the genetic variation present in a population is partitioned geographically and both within and among the various demographic units (e.g., social groups) into which the population is divided. Individuals' social and reproductive behaviors (e.g., mating frequency, choice of partners) likewise can influence the structuring of genetic variation within and among social groups across time. For example, high reproductive skew among males within a social group can lead to cohorts of similarly aged individuals being more closely related to one other through common paternity than are animals of different ages. Similarly, extra-group mating by either males or females can act to reduce the extent of genetic differentiation between groups.

These two key behavioral factors influencing the kinship structure of primate groups – the dispersal and reproductive tactics of individual animals – are some of the most intractable features of primate social systems for researchers to study in the field. For long-lived species such as primates, dispersal events tend to be rare – i.e., individual animals typically disperse only once or a small number of times during their lives. Even in the most detailed, long-term field studies, it is often difficult

A. Di Fiore (✉)
Center for the Study of Human Origins, Department of Anthropology, New York University, 25 Waverly Place, New York, NY 10003, USA
e-mail: anthony.difiore@nyu.edu

P.A. Garber et al. (eds.), *South American Primates,* Developments in Primatology: Progress and Prospects, DOI 10.1007/978-0-387-78705-3_9,
© Springer Science+Business Media, LLC 2009

to accurately discern the fates of individuals who disappear from the social groups under investigation. For many primate species, sexual behavior is also not always easily observed. Moreover, sexual behavior in primates can serve many different social functions, and it is often largely decoupled from reproduction. Thus, even for those taxa in which matings are relatively conspicuous, the actual pattern of reproductive success may not be well predicted from behavioral observations.

Fortunately, molecular genetic data (e.g., multilocus genotypes, DNA sequence data, various kinds of DNA "fingerprints") provide a powerful, indirect means of studying the dispersal and reproductive behavior of individuals and for examining patterns of relatedness among animals both within and among primate social groups (Di Fiore 2003). With the development of high-throughput hardware for DNA sequencing and genotyping and with the optimization of methods for storing and extracting DNA suitable for molecular analysis from noninvasively collected samples such as feces or hair (Morin et al. 2001; Nsubuga et al. 2004; Roeder et al. 2004), such data are becoming ever easier and more cost-effective to collect. Interestingly, however, to date relatively few studies of wild nonhuman primates – and fewer still of New World monkeys – have used molecular data either to investigate dispersal patterns or to examine genetic relatedness among the animals within social groups.

In this chapter, I outline some of the ways in which genetic data can be applied to the study of dispersal patterns and kinship, and I review a range of case studies drawn from South American primates to illustrate some of these methods. The New World monkeys are a particularly interesting taxonomic group within which to consider these issues because dispersal patterns and social systems within this clade are so varied. For example, based on observational studies, some taxa of New World monkeys are characterized by predominantly female dispersal (e.g., *Ateles*: Symington 1987; *Brachyteles*: Strier 1990, 1994a, b; *Lagothrix*: Nishimura 2003; Central American squirrel monkeys, *Saimiri oerstedii*: Mitchell et al. 1991), some by predominantly male dispersal (e.g., *Cebus*: Jack and Fedigan 2004a, 2004b; western Amazonian squirrel monkeys, *Saimiri boliviensis*: Mitchell et al. 1991), and some by routine dispersal of individuals of both sexes (e.g., callitrichines: Baker and Dietz 1996; Savage et al. 1996; *Aotus*: Fernandez-Duque and Huntington 2002; *Callicebus*: Kinzey 1981; *Alouatta*: Clarke and Glander 1984; Rumiz 1990; Glander 1992; Crockett and Pope 1993). In still other genera (e.g., *Cacajao*, *Chiropotes*), dispersal patterns are either completely unknown or poorly understood. Genetic data have been collected on only a handful of these taxa, but in some cases reveal interesting dispersal patterns not anticipated from observational work.

The few studies of New World monkeys that have looked at within-group kinship have provided insight into the mating systems of several platyrrhine species, and these studies hint at an impressive and underappreciated diversity in reproductive patterns and behavior within the platyrrhine clade. But much additional work remains to be done. Given the widespread acceptance of kinship as a key explanatory principle underlying and structuring much of primate social lives, it is imperative that future primate studies pay more attention to exploring the link between relatedness and individual behavior using molecular data.

9.2 A Brief Review of Theory and Methods

A range of analytical methods have been developed for **evaluating dispersal patterns** using molecular data (e.g., Paetkau et al. 1995; Rannala and Mountain 1997; Goudet et al. 2002), for **estimating the degree of relatedness** between pairs of individuals using various kinds of marker data (e.g., Queller and Goodnight 1989; Lynch 1990; Li et al. 1993; Lynch and Ritland 1999; Wang 2002), and for **inferring the likely pedigree relationship** among those individuals (e.g., Goodnight and Queller 1999; Epstein et al. 2000; Milligan 2003). In recent years, a number of excellent review papers have been published that discuss in detail many of these methods and their important assumptions and limitations (e.g., van de Casteele et al. 2001; Prugnolle and de Meeus 2002; Blouin 2003; Piry et al. 2004; Manel et al. 2005; Csilléry et al. 2006; Weir et al. 2006; Lawson Handley and Perrin 2007), and thus only a brief introduction to some of these methods is given here.

9.2.1 Evaluating Gene Flow

At its most fundamental level, dispersal reflects a process by which genes are shuffled among populations or social groups that exist in a spatial landscape. If the dispersal rate is high enough, then the genetic variation present is effectively homogenized. By contrast, if the dispersal rate is low or if significant physical or social barriers to dispersal exist – i.e., in a geographically or socially subdivided population – then, as random mutation and genetic drift alter local gene frequencies, different portions of the landscape should come to be characterized by different local gene pools. One way, then, to evaluate the extent of gene flow among different local gene pools is to infer backwards from some measure of population subdivision.

Traditionally, population geneticists take a hierarchical view of population subdivision based on Wright's (1931) simple island model, which imagines that local populations are semi-isolated demes connected to one another via the movement of dispersers (Fig. 9.1). In the general n-island model, dispersers can move between any subpopulation, although other migration models (e.g., "stepping stone" island models or isolation by distance models) may better approximate animals' true dispersal options. Under the basic island model, Wright's (1965) fixation index F_{ST}, which summarizes the proportion of the total genetic variation found in a population that is explained by subpopulation or group membership, is inversely related to subpopulation size (N_e) and the proportion of individuals that migrate among subpopulations per generation (m). Thus, a crude estimate of the total number of migrants per generation ($N_e m$) among subpopulations can be estimated from empirical measures of F_{ST}, which can be derived from a variety of molecular marker data. Obviously, a number of crucial assumptions made under the island model are unlikely to be met in natural populations (e.g., that subpopulations are of equal and constant size, that there is a symmetric rate of migration among subpopulations), which makes the interpretation of estimated rates of gene flow among subpopulations based on F_{ST} problematic. Nonetheless, estimates of relative $N_e m$ for species with similar

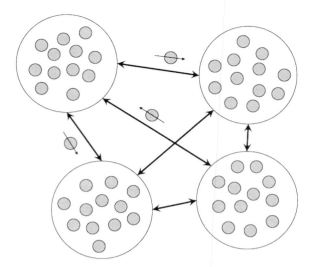

Fig. 9.1 Sewall Wright's (1931) island model of population structure and migration. Local populations are denoted by *large circles* and individuals within these populations are denoted by smaller, *filled circles*. Under the model, mating is random within each of several equal-sized local populations. Local populations are connected to one another via the movement of dispersers, with dispersal possible between any pair of populations

grouping patterns or for different demographic subgroups within populations of the same species (e.g., males versus females) can be very informative.

9.2.2 *Identifying Dispersing Individuals*

When an individual changes social groups or changes the area in space it normally occupies, it often becomes a member of a different deme from the one in which it was born. Because dispersing individuals carry with them genetic material characteristic of their natal group and local population, genetic data provide an indirect way to identify dispersing animals – one that does not rely on observations of animals immigrating into or transferring between social groups. Specifically, if the genetic variation characterizing the population or social group that an animal joins is sufficiently distinct from that of its deme of origin, then it should be possible to identify which individuals in a population are immigrants and, potentially, the source populations from which those immigrants came.

A variety of analytical methods have been proposed in recent years that use genotype data (e.g., multilocus microsatellite or SNP genotypes) for identifying immigrants and for assigning individuals to a likely population of origin (Paetkau et al. 1995; Favre et al. 1997; Rannala and Mountain 1997; Cornuet et al. 1999; Banks and Eichert 2000; Pritchard et al. 2000; Piry et al. 2004; Manel et al. 2005). Many of these methods are based on calculating, for each individual of interest, a so-called **assignment index** (AI), which is a measure of the likelihood that their genotype originated in the population in which they were sampled versus other sampled populations for which genotype data are available. When assignment indices are standardized such that the mean index within a population equals 0 – i.e., by subtracting the mean assignment index (mAI) for the sampled population from each individual's AI (Favre et al. 1997) – animals with positive corrected assignment index (AIc) values are those more likely than average to have been born in the sampled population, while those with negative AIc values are more likely to be immigrants.

9.2.3 Sex-Biases in Dispersal Behavior

As in most vertebrates (Greenwood 1980; Waser and Jones 1983; Johnson and Gaines 1990), dispersal in most primate species tends to be heavily sex-biased (Melnick and Pearl 1987; Pusey and Packer 1987) – i.e., individuals of predominantly one sex leave their natal range and social group and join another prior to beginning reproduction. Sex-biased dispersal generates clear predictions for sex differences in structuring of genetic variation within and between social groups in a population (Melnick and Hoelzer 1992; Avise 1994; Melnick and Hoelzer 1996; Di Fiore 2003; Avise 2004; Hoelzer et al. 2004; Lawson Handley and Perrin 2007), and a number of analytical approaches can be used to evaluate sex-biases in dispersal patterns using molecular data.

One set of approaches looks for genetic signatures of **sex-biased gene flow** over the population's past history (Prugnolle and de Meeus 2002; Lawson Handley and Perrin 2007), either by examining the diversity, phylogeny, and geographic distribution of non-recombining, uniparentally inherited markers (e.g., Y chromosome microsatellite haplotypes, mitochondrial DNA sequence haplotypes) or by taking a classical population genetics approach (Wright 1943, 1965) and examining, for the set of post-dispersal age individuals, how genetic variation is partitioned hierarchically among and within various demographic units from the sampled population. Briefly, if sex-biased gene flow has characterized a population's past demographic history, we would expect to see differences between post-dispersal males and post-dispersal females in how genetic variation is partitioned, with the more philopatric sex showing greater evidence of genetic substructuring because of its more restricted gene flow.

A second set of approaches focuses on **sex differences in individual or instantaneous dispersal** rather than on population-level assessments of historical gene flow. These approaches take advantage of the assignment techniques discussed above (Prugnolle and de Meeus 2002; Lawson Handley and Perrin 2007), evaluating for each post-dispersal age individual the likelihood of its genotype having originated in the population in which it was sampled. In this case, under sex-biased dispersal, we would expect aspects of the assignment indices of post-dispersal age males and post-dispersal age females to differ in predictable ways. Below are summarized a number of the key predictions of various molecular tests of sex- biased dispersal.

9.2.3.1 Genetic Relatedness Among Nonjuvenile Animals

Where dispersal is predominantly by individuals of one sex, nonjuvenile, post-dispersal aged group members of the more philopatric sex are expected, on average, to be more closely related to one another than are group members of the dispersing sex (Morin et al. 1994b; Goudet et al. 2002; Di Fiore 2003; Hammond et al. 2006). Thus, if dispersal is predominantly by males and females are the philopatric sex – as is common in most cercopithecine and many cercopithecoid primates – then measures of average relatedness among dispersal-aged females within groups are predicted to be greater than measures of average relatedness among males. The opposite pattern is expected for taxa in which males are philopatric and where dispersal is argued to be primarily by females, such as chimpanzees, bonobos, red colobus monkeys, and some atelin primates of the New World (Morin et al. 1994b; Di Fiore 2003; Hammond et al. 2006). Additionally, if individuals of both sexes disperse but members of one sex travel farther, on average, than those of the opposite sex, then we would expect to see higher average relatedness among same-sexed members of different social groups within the same local population for the more philopatric sex (Di Fiore and Fleischer 2005). Some recent theoretical work suggests that the expected pattern of greater average relatedness among individuals of the philopatric sex should hold true mainly in small social groups and in groups where the reproductive skew among males is high (Lukas et al. 2005).

9.2.3.2 Diversity and Structuring of Genetic Variation Among Nonjuveniles

For portions of the genome that are transmitted to offspring through only one parent (e.g., mitochondrial DNA from the mother, Y chromosomal DNA from the father) the structuring of genetic variation is also expected to covary with sex differences in dispersal behavior. Thus, within social groups of species where females are philopatric, much lower diversity is expected in the mitochondrial DNA of post-dispersal age females compared to males, because of the dual processes of restricted female-mediated gene flow (as females are recruited into the adult, breeding population primarily from within their natal social group) and stochastic lineage sorting (Melnick and Hoelzer 1996; Wallman et al. 1996). Under female philopatry, too, greater genetic substructuring is expected for the mitochondrial versus the nuclear genome among nonjuvenile females. This is because mitochondrial genes would not be shuffled among social groups to the extent that nuclear genes are by the process of male dispersal (Avise 1995; Melnick and Hoelzer 1996; Avise 2000; Di Fiore 2003). For evaluating these predictions, the extent of population substructuring – i.e., the amount of genetic differentiation seen between different subpopulations or social groups – is typically characterized using F_{ST}, one of Wright's (1965) fixation indices or F-statistics, or an analogous summary statistic (e.g., R_{ST}, G_{ST}, δ_{ST}, θ_{ST}: Weir and Cockerham 1984; Nei 1987; Michalakis and Excoffier 1996; Goodman 1997), which summarizes the proportion of the total genetic variation in the population that is explained by subpopulation or group membership. Fixation indices are commonly calculated within the general framework of analysis of molecular variance (AMOVA: Excoffier et al. 1992, 2005).

For male philopatric taxa, by contrast, there is no expectation of much lower mitochondrial DNA diversity within groups for post-dispersal age males versus females, because dispersing females carry their mitochondrial haplotypes with them when they move. Similarly, little or no difference in the extent of population sub-structuring is expected for the mitochondrial versus nuclear genomes of either males or females (Hapke et al. 2001; Di Fiore 2003). In male philopatric taxa, however, Y chromosomal diversity among males is expected to be low (and lower than mito-chondrial DNA diversity among the same males), while F_{ST} values between groups for Y chromosomal markers should be high – the opposite pattern to that expected for mitochondrial DNA in female-philopatric taxa (Eriksson et al. 2006). In fact, comparison of the degree of structuring seen in maternally inherited mitochondrial DNA versus paternally inherited Y chromosomes for the same individuals from the same populations can also provide strong insight into the direction and degree of sex-biased dispersal (Hammond et al. 2006). For female-philopatric taxa, then, the ratio of $F_{ST \bullet mtDNA}$ to $F_{ST \bullet Y}$ found in the population is expected to be much greater than one – that is, mitochondrial DNA is expected to be more divergent among groups than is Y chromosomal DNA (Laporte and Charlesworth 2002). Note that this assumes roughly equivalent effective population sizes (N_e) for males and females, given that F_{ST} is inversely proportional to the product of N_e and the mutation rate (Wright 1943, 1965). For male-philopatric taxa, by contrast, Y chromosomal DNA should be far more divergent between groups than mitochon-drial DNA, and $F_{ST \bullet Y}$ is expected to be much greater than $F_{ST \bullet mtDNA}$ (Laporte and Charlesworth 2002; Eriksson et al. 2006; Hammond et al. 2006).

Another of Wright's (1965) F-statistics – the inbreeding coefficient F_{IS} – can also be used to evaluate sex biased gene flow. F_{IS} measures the extent of excess homozygosity in a sample. Because within any social group, nonjuvenile members of the dispersing sex will consist of a mix of immigrants coming from different social groups plus some nondispersing residents, homozygosity among these indi-viduals (and thus F_{IS}) is expected to be increased (i.e., be less negative or more positive) relative to that found within nonjuvenile members of the more philopatric sex (Goudet et al. 2002).

9.2.3.3 Intraspecific Phylogeny of Uniparentally Inherited Markers Among Nonjuveniles

An additional way to investigate sex-biased dispersal is to apply phylogenetic methods haplotype data derived from non-recombining portions of the genome (e.g., mitochondrial and Y chromosomal DNA) to infer the evolutionary relation-ships among the various haplotypes segregating in the population. For example, where females are philopatric and female-mediated gene flow is thus restricted, we would expect to see a strong association between the inferred intraspecific phylogeny for female mitochondrial DNA and the geographic location or group of origin from which the sample was collected. That is, under female philopatry, females from the same social group are expected have closely related or identical mitochondrial DNA haplotypes, and female mitochondrial DNA haplotypes are expected to show evidence of isolation by distance, i.e., greater divergence between geographically distant samples than among less-separated samples (Wright 1943). No such strong geographic clustering of closely related mitochondrial DNA types or evidence of isolation by distance is expected for males (Melnick 1987; Melnick and Hoelzer 1992, 1996; Di Fiore 2003). By contrast, when males are philopatric and females regularly disperse from their social groups, we do not expect to see a strong geographic structuring to mitochondrial DNA variation among females. Rather, females from the same social group are likely to possess widely divergent mitochondrial DNA types (Morin et al. 1994b; Di Fiore 2003) and female mito-chondrial DNA types should not show evidence of significant isolation by distance (Hapke et al. 2001; Douadi et al. 2007).

9.2.3.4 Mean and Variance in Assignment Indices Among Nonjuveniles

As noted above, individuals with negative corrected assignment indices possess genotypes that are more likely than average to have originated outside of the social group in which they were sampled and are likely to have been immigrants. Ani-mals with positive AIc values, by contrast, are more likely than average to be natal, philopatric individuals. Thus, if dispersal is sex-biased, then the mean corrected assignment index is expected to differ significantly between the sexes: under female philopatry, the mean AIc of females is expected to be greater than that of males, while the reverse should be true if males are the more philopatric sex (Goudet et al. 2002; Lawson Handley and Perrin 2007).

If dispersal is sex-biased, then the variance in assignment index scores of males and females is also expected to differ significantly. Members of the dispersing sex are expected to show greater variance in assignment index scores (vAIc) because any set of sampled group members will theoretically contain a mix of resident and immigrant individuals, the latter of whom will presumably have originated in multiple other social groups. By contrast, the variance in assignment index scores is expected to be less for the more philopatric sex. Simulation experiments have demonstrated that these various molecular tests for sex-biased dispersal are sensitive to not only the extent of the bias in dispersal seen between males and females but also on the sampling strategy – e.g., the number of populations and individuals per population sampled as well as the variation present at the loci used to genotype sampled animals (Goudet et al. 2002).

9.2.3.5 Comparing Pre-dispersal and Post-dispersal Age Individuals

Yet another approach to evaluating sex differences in dispersal patterns using molecular data is to compare aspects of classical population genetic structure and assignment indices for pre-dispersal and post-dispersal age individuals. The expec-tation is that there will be no difference between the sexes in average related-ness (mean r), F_{ST} (for either mitochondrial, Y chromosomal, or autosomal loci), F_{IS}, mAIc, vAIc, or evidence of isolation by distance in uniparentally inherited markers among same-sexed individuals at the pre-dispersal (i.e., juvenile) stage. Amongst post-dispersal individuals, however, the various patterns outlined above with respect to these parameters should obtain. More importantly, comparison of some of these parameter values between pre-dispersal and post-dispersal individuals can yield quantitative estimates of sex-specific dispersal rates (Vitalis 2002; Fontanillas et al. 2004).

9.2.4 Within-Group Kinship and Behavior

The association patterns and social behaviors of group-living primates have long been argued to be shaped heavily by patterns of kinship among group mates (Wrangham 1980; Gouzoules 1984; Gouzoules and Gouzoules 1987; van Schaik 1989; Silk 2001, 2002, 2006). The theoretical case for the importance of kinship's influence on behavior was articulated formally by W.D. Hamilton (1964a, 1964b) in his discussion of the concept of "inclusive fitness". Hamilton (1964a, 1964b) noted that because animals share genes with relatives, natural selection should favor those individuals who behave in ways that maximize their "inclusive fitness" – defined as personal or direct fitness augmented or decremented by the effects of their behavior on the reproductive success of relatives. Even behaviors that are costly to an animal's personal fitness can nonetheless be favored by natural selection, provided that the benefits (**b**) to the recipient, discounted by the degree of genetic relatedness (**r**) between the recipient and the actor, exceed the net cost (**c**) to the actor of performing the behavior (Hamilton 1964a), i.e., if $br - c > 0$. This condition, often referred to as "Hamilton's Rule", forms the basis of "kin selection theory" in behavioral ecology, which generates simple predictions about how actors should interact socially with conspecifics of different degrees of relatedness. All else being equal, given a choice of behaviors and related recipients toward whom those behaviors might be directed, actors are expected to choose that combination which maximizes the value of $br - c$.

Molecular marker data can be used to generate quantitative estimates of **r** among pairs of individuals, and a variety of relatedness estimators (and softwares for calculating these estimators) have been proposed for both codominant (e.g., microsatellites, SNPs) and dominant markers (e.g., AFLP, RAPD, and minisatellite DNA "fingerprints") (Queller and Goodnight 1989; Li et al. 1993; Goodnight and Queller 1999; Lynch and Ritland 1999; Wang 2002, 2004; Ritland 2005; Kalinowski et al. 2006; *see also* reviews by Milligan 2003; van de Casteele et al. 2001; Blouin 2003; Weir et al. 2006). If kinship is an important predictor of affiliative and agonistic social behavior in primates, as primatologists have long assumed, then patterns of within-group relatedness revealed by genetic data are expected to correlate positively with patterns of spatial and affiliative social associations (e.g., grooming, support in coalitions) and negatively with agonistic social interactions. Additionally, given the clear significance that social dominance has within groups of many primate taxa, genetic estimates of reproductive success are expected to correlate positively with dominance status, although until recently it has proved difficult to test this assumption. Finally, assuming that there is a risk of inbreeding depression associated with mating with close relatives, then animals of both sexes are expected to avoid mating and breeding with close kin.

9.3 Genetic Studies of Dispersal and Kinship in Primates

9.3.1 A Brief Review of Results for Non-Platyrrhines

To date, very few primate studies – and fewer still of platyrrhines – have actually used genetic data to examine dispersal patterns or to investigate directly the relationship between genetic relatedness and social behavior. With respect to predictions based on sex-biased dispersal, Melnick (1987, 1988) and Melnick and Hoelzer (1992, 1996) have documented greater mitochondrial diversity in males versus females within groups of several species of macaques, as would be expected for female philopatric taxa. More recently, Altmann et al. (1996) and de Ruiter and Geffen (1998) examined average female and average male pairwise relatedness in groups of baboons (*Papio hamadryas cynocephalus*) and long-tailed macaques (*Macaca fascicularis*), respectively, using genotype data. As expected, given observations of predominantly male exogamy and female philopatry in these taxa, they found that females within social groups were more closely related to one another, on average, than were males. Similar results have been reported for several group-living strepsirrhines (e.g., Verreaux's sifaka: Lawler et al. 1995; Alaotran gentle lemurs: Nievergelt et al. 2002; red fronted lemurs: Wimmer and Kappeler 2002). For chimpanzees and bonobos – where observational studies suggest that female exogamy and male philopatry are the rule – males do not generally appear to be more closely related to one another than females, contrary to expectation (Gerloff et al. 1999; Vigilant et al. 2001; Lukas et al. 2005). Nonetheless, among bonobos paternally inherited Y chromosomal markers do show much greater geographic differentiation than maternally inherited mtDNA (Eriksson et al. 2006), which is a strong signature that dispersal is female-biased, and a similar pattern has been found in Arabian hamadryas baboons (Hammond et al. 2006). Among Eritrean hamadryas baboons, too, the lack of population structure in mitochondrial DNA variation seen in a country-wide sample likewise strongly implicates female-biased dispersal (Hapke et al. 2001).

With respect to associations between genetic relatedness and affiliative within-group social behavior, results from Old World primates have been mixed. For example, two seminal studies of chimpanzees that examined the association between within-group social behavior and (matrilineal) kinship found that, contrary to expectation, males who were matrilineal kin (e.g., shared a mtDNA haplotype) were not more cooperative or affiliative with one another than males from different matrilines (Goldberg and Wrangham 1997; Mitani et al. 2000). More recently, Langergraber et al. (2007) used a large suite of autosomal, X, and Y chromosomal microsatellite markers to identify pairs of maternal and paternal half-siblings in one large community of chimpanzees and indeed found evidence that males preferentially affiliated and cooperated with their maternal half-brothers, though not their paternal ones. Nonetheless, most pairs of males that were highly affiliative and cooperative were not closely related, suggesting that male chimpanzees' decisions about their social interactions have more to do with direct rather than indirect fitness benefits. Thus, it is unclear the extent to which affiliative social behavior among male chimpanzees is, in fact, generally kin-based, although this has long been assumed (Goodall 1986). By contrast, several recent studies of female rhesus macaques and baboons have found that both maternally and paternally related half-siblings do behave more affiliatively with one another than nonkin, as is predicted by kin selection theory, although maternal half-siblings tend to be more affiliative than paternal ones (Widdig et al. 2001, 2002; Smith et al. 2003).

With respect to reproductive behavior, molecular data provide the only tractable means for assessing parentage in wild populations where it is often impossible to observe matings. In several studies of wild and free-ranging cercopithecines, researchers have found support for a positive relationship between male dominance rank and paternity success (Melnick 1987; de Ruiter et al. 1992; de Ruiter and Inoue 1993; Paul et al. 1993; Altmann et al. 1996; Alberts et al. 2003; Widdig et al. 2004). However, for other species, including male-philopatric chimpanzees (Morin et al. 1994a; Constable et al. 2001) and bonobos (Gerloff et al. 1999), reproductive skew toward dominant males is less pronounced, and multiple males sire offspring within the same groups.

None of these aspects of primate behavioral ecology have been well explored in very many platyrrhine taxa using molecular data. In the remainder of this chapter, I review those few studies that have used molecular data to address the issues of dispersal, within-group kinship, and social behavior among platyrrhines. I then conclude with a discussion of the implications of these studies for our appreciation of platyrrhine social organization and dispersal patterns.

9.3.2 Cebids: Callitrichines

Two studies of callitrichines living in multimale groups – one of common marmosets (*Callithrix jacchus*: Nievergelt et al. 2000) and one of moustached tamarins (*Saguinus mystax*: Huck et al. 2005) – have used molecular data to investigate issues of kinship, dispersal, and mating patterns. Using genotype data from nine variable microsatellite marker loci, Nievergelt et al. (2000) evaluated patterns of genetic relatedness among a set of 40 individual marmosets that comprised most of the individuals in three wild social groups plus a portion of the animals resident in two adjacent groups. Within two of the well-sampled groups, adult females were closely related to one another – on the order of mother–daughter or full sibling pairs – and in all groups, all reproductively inactive adults and immatures were closely related to either the dominant female or a second, breeding female. Resident adult males were not closely related either to the adult females in their social groups or to one another, though several showed a high average relatedness to the members of another well-sampled group. A single, solitary male was also more closely related to the members of a different social group from the one he was commonly seen following. The limited paternity analyses afforded by the dataset found that the dominant males within each study group were the likeliest sires of most infants born during the study, although for more than half of offspring one or more extragroup males also could not be excluded as possible sires. Together, these genetic results are consistent with the idea that common marmosets live in polygynous, extended family groups. The results are also suggestive of male transfer among groups (Nievergelt et al. 2000), although observational data confirm that females also emigrate from their natal groups (Digby and Barreto 1993).

In a similar study, Huck et al. (2005) evaluated patterns of kinship and paternity among 62 moustached tamarins from eight social groups using a set of 12 microsatellite marker loci. As in common marmosets, a single male in each group was identified as the most likely sire for almost all of the offspring born within each group, indicative of strong reproductive skew among males, although behavioral data indicated that more than one resident male typically mated with a group's sole breeding female. Interestingly, the study also found one case where a breeding female's twin offspring were apparently sired by different males. Among the tamarins, the average degree of relatedness among individuals from the same group was much greater than the average relatedness of members of different groups (mean R = 0.31 versus −0.03), and mating partners tended to be unrelated to one another (mean R = −0.06), two patterns also seen in *Callithrix jacchus* (Nievergelt et al. 2000). However, where the mean relatedness among resident adult males in common marmoset groups was low, among moustached tamarins the mean relatedness among males (mean R = 0.34) was not significantly less than that among females (mean R = 0.38). This might suggest that dispersal patterns among the tamarins are less biased toward males or may simply reflect the fact that animals of both sexes more commonly remain in their natal groups past the juvenile stage. Finally, based on an allele-sharing analysis among adult group members, Huck et al. (2005) concluded that while some closely related adults were likely to be parent-offspring pairs, others are more likely to be siblings or half siblings. This observation suggests that pairs of animals may sometimes join a group or inherit a territory together, a phenomenon that has been seen in behavioral studies of other populations of moustached tamarins (Garber et al. 1993).

9.3.3 Cebids: Cebines

Among the cebines, few studies have used molecular data to examine patterns of kinship or dispersal or social behavior within social groups, which is interesting given the variation in dispersal patterns seen among species of squirrel monkeys (*Saimiri*) (Mitchell et al. 1991; Boinksi 1999; Boinski et al. 2002; Boinski et al. 2005a, 2005b) and the strong same-sex affiliative behavior that has been reported among males (Robinson 1988b; Perry 1998; Jack 2003) and among females (O'Brien and Robinson 1991; O'Brien 1993) in various species of capuchins (*Cebus*). Two notable exceptions include studies by Valderrama Aramayo (2002) examining male reproductive success and population genetic structure in wedge-capped capuchins (*Cebus olivaceus*) in Venezuela and by Muniz et al. (2006) examining patterns of paternity in white-faced capuchins (*Cebus capucinus*) in Costa Rica.

Valderrama Aramayo (2002) evaluated paternity for 22 offspring born in a multi-male group of wedge-capped capuchins over a ten-year period that encompassed the tenure of two closely related alpha males. She found that paternities were skewed in favor of the two alpha males, who together sired just over half of all the offspring born during this period, a proportion that conformed well with predictions based on a priority-of-access model of male breeding success (Altmann 1962). The remaining offspring were sired by a combination of eight other resident or extragroup males. Nonetheless, alpha males sired a smaller proportion of the offspring of females from high ranking matrilines than low ranking matrilines, possibly implicating some degree of female behavioral preference for particular non-alpha males (Valderrama Aramayo 2002).

Valderrama Aramayo (2002) also examined patterns of relatedness among the sampled individuals in the capuchin population. She found that among the set of immigrant males in her main study group, there existed many pairs who were esti-mated to be related to one another at the level of either half or full siblings, highlight-ing a potential role of kinship for influencing male dispersal decisions and possibly dispersal success. Mean female relatedness was higher within groups than between groups, and within groups the mean relatedness among females was somewhat greater than that seen among males, a pattern consistent with observation-based reports of male-biased dispersal (Robinson 1988a; Valderrama Aramayo 2002)

For white-faced capuchins, *Cebus capucinus*, a recent paternity analysis of 41 infants born over a 14-year period in three social groups revealed that the long-term alpha males resident in these groups sired from 38% to 80% of the offspring born in their respective social groups, indicative of a strong association between male rank and reproductive success (Muniz et al. 2006). More interestingly, however, is the fact that while long-term alpha males sired 79% of offspring born to adult females other than their daughters, only 1 of 17 infants (6%) born to daughters' resulted from father-daughter inbreeding. After discounting other explanations for these results, Muniz et al. (2006) conclude that behavioral avoidance of father-daughter mating is the most likely. If correct, this would be one of the very few documented cases among platyrrhines of individuals biasing their social interactions with conspecifics on the basis of relatedness.

9.3.4 Atelids: Howler Monkeys

The earliest molecular studies of kinship, dispersal patterns, and social behavior in New World primates focused on red howler monkeys (*Alouatta seniculus*) liv-ing in central Venezuelan llanos (Pope 1990, 1992, 1996, 1998, 2000). Howler monkeys belong to the platyrrhine family Atelidae, a monophyletic grouping that also includes woolly monkeys, spider monkeys, and muriquis. In Venezuela, red howler monkeys live in both single-male and age-graded multimale groups typi-cally containing two to three adult males (Crockett and Eisenberg 1987; Rudran and Fernandez-Duque 2003). Prior observational data concluded that both male and female red howler monkeys often disperse from their natal social groups prior to breeding, although some individuals of either sex may remain as adults in their natal groups (Rudran 1979; Crockett 1984; Pope 1989; Crockett and Pope 1993). Dispersal was observed to be female-biased, in the sense that females appeared to disperse greater distances from their natal groups, on average, than males (six versus one home range diameters: Pope 1989).

Pope (1992) examined the genetic structure of the llanos population using geno-type data for a suite of 9 variable allozyme markers scored in 137 animals from 18 social groups that were split into two local populations separated by over four kilometers of open savanna. Based on F_{ST} values, Pope (1992) found that groups within the same local population were highly differentiated from one another, a result of both the strong monopolization of reproduction within groups by a single male and the fact that groups often contained matrilines of closely related females.

With respect to the possible influence of within-group relatedness on social behavior, molecular studies of red howler monkeys provide perhaps the strongest evidence from any primate in support of positive fitness consequences of kin-directed nepotism. For example, in an early study, Pope (1990) found that multimale groups held together by coalitions of related males lasted longer and enjoyed greater overall fitness than did multimale groups where the resident males were unrelated. Later, Pope (1992) demonstrated that long-established groups of red howler mon-keys also tend to be characterized by a greater average degree of relatedness among the resident females, who cooperate with one another to prevent unrelated females from joining the group. Importantly, females in these groups also enjoyed greater per capita reproductive output than females in groups with lower mean relatedness among the resident females. Finally, with respect to mating patterns, Pope (1990) also examined paternity in five single-male and four multimale red howler groups. In no case were offspring found to have been sired by extragroup males. Additionally, within each of the multimale groups, paternities appeared to be limited to solely the dominant male, highlighting a clear link between dominance rank and male fitness.

In contrast to red howler monkeys, mantled howler monkeys (*Alouatta palli-ata*) typically live in much larger social groups containing up to six adult males. Ellsworth (2000) investigated patterns of population genetic structure among Costa Rican mantled howler monkeys using genotype data from suite of eight microsatel-lite marker loci for 65 individual animals from nine social groups. In contrast to the results from red howler monkeys, groups within the same local population were not highly genetically differentiated from one another. The global F_{ST} calculated for these nine groups, while significantly differing from zero indicating some genetic structure to the population, was nonetheless very low (0.02), suggesting that the population is essentially panmictic.

This dramatic difference from red howler monkeys is likely due in part to dif-ferences in the dispersal patterns seen in the two species. While dispersal by both males and females is common in both species, in red howler monkeys, dispersing females are seldom able to integrate themselves into established social groups, and instead they must form new social groups with other dispersing animals before they begin breeding (Crockett 1984; Crockett and Pope 1988, 1993; Pope 2000).

The situation is somewhat different in mantled howler monkeys where female dispersers sometimes succeed in directly moving into established groups (Jones 1980; Glander 1992). Differences in mating system also probably play a role in explaining differences in population structure between red and mantled howler monkeys. While a single male tends to monopolize both mating opportunities and paternity in red howler monkeys, within mantled howler monkey groups multiple males may mate with females and, presumably, sire offspring. Indeed, in a very limited study of paternity, Ellsworth (2000) found that out of five cases where DNA samples were available for a mother, her offspring, and the alpha male resident at the time of conception, the alpha male could be excluded as a potential sire in three cases.

Ellsworth (2000) also examined patterns of relatedness within groups and found that, on average, males were slightly more closely related to one another than were females, though this pattern was not significant. More importantly, the mean relatedness among group members was low relative to that seen in red howler monkeys, most likely reflective of the fact that groups of mantled howler contain greater numbers of immigrant individuals of both sexes.

9.3.5 Atelids: Woolly and Spider Monkeys

Within the family Atelidae, woolly monkeys (genus *Lagothrix*), spider monkeys (genus *Ateles*), and muriquis (genus *Brachyteles*) form a monophyletic clade – the atelins – that is a sister group to the howler monkeys. Among atelins, dispersal has long been thought to be strongly female-biased (Di Fiore and Campbell 2007). Observational studies of spider monkeys (Symington 1987, 1988; Ahumada 1989) and muriquis (Strier 1987, 1990, 1991) suggest that dispersal is solely or largely by females. For woolly monkeys, too, females have been observed to transfer between groups, sometimes multiple times during their lifetimes (Nishimura 1990; Stevenson et al. 1994; Stevenson 2002; Nishimura 2003). However, at two different sites in Yasuní National Park, Ecuador, solitary adult and subadult males have been seen, as well as small bachelor groups of ~5 individuals of various ages (Di Fiore 2002, unpublished data). In La Macarena, Colombia, too, Nishimura (1990) has reported that animals of both sexes occasionally disappear from their social groups and even join other groups temporarily. Thus, whether strongly female-biased dispersal characterizes all atelins remains an important question.

As noted above, the patterns of relatedness seen among individuals within a population are expected to reflect dispersal patterns. In turn, kin selection theory predicts that the quality of social interactions among individuals should be sensitive to their genetic relatedness. However, apart from one published study on patterns of within-group relatedness in several social groups of woolly monkeys (*Lagothrix poeppigii*) in lowland Ecuador (Di Fiore and Fleischer 2005), molecular data have not been used to investigate the link between dispersal patterns, kinship, and social behavior in any atelin taxon. Here, I revisit the results of that study and present new, preliminary data on these subjects for both woolly and spider monkeys.

9.4 Methods

Between 1998 and 2007, tissue and fecal samples were collected from animals in multiple social groups of woolly and spider monkeys at two different sites in lowland Ecuador – the Proyecto Primates Research Area (PPRA) and the region around the Tiputini Biodiversity Station (TBS) – as well as opportunistically from several other sites within Yasuní National Park (Fig. 9.2). Sampling was done intensively at both sites in 1998 and at the TBS site between 2005 and 2007. Intermittent samples from the intervening years were also collected at one or the other site. The sampling procedures used follow those reported elsewhere (Di Fiore and Fleischer 2005). DNA was extracted from each sample using commercially available nucleic acid isolation kits for tissue (Qiagen DNeasy Blood and Tissue Kit) or feces (QIAmp DNA Stool Mini Kit). These samples were then genotyped via PCR for a suite of polymorphic microsatellite (SSR) markers – 8 and 16 loci, respectively, for woolly and spider monkeys (Di Fiore and Fleischer 2004) (*see also* Table 1A,B) – using a modification of the multiple tubes approached followed in other studies (Taberlet et al. 1996; Alberts et al. 2006). Finally, most samples were also sex-typed using PCR-based sexing assays (Wilson and Erlandsson 1998; Di Fiore 2005a, 2005b) to confirm or correct field assignments of sex.

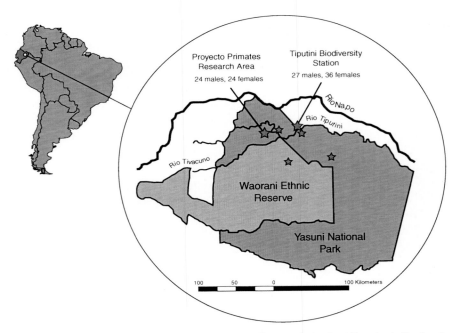

Fig. 9.2 Map of Yasuní National Park and the surrounding area in lowland Ecuador indicating the locations where samples were collected and the number of individuals of each sex sampled at the PPRA and TBS sites. Small stars indicate several other locales from which a handful of samples were collected

During the periods of intensive sampling many samples were collected without individual identification – particularly for woolly monkeys, which are difficult to distinguish individually – hence, two or more samples from the same species at the same site often yielded identical multilocus genotypes. In all such cases, the assigned genetic sexes also always matched. These samples were therefore assumed to be replicates from the same individual and were analyzed as such, which is a reasonable assumption given the low probability of identity (PI) afforded by both the woolly and spider monkey genotyping panels: the chance of even two full siblings sharing the same multilocus genotype was less than one in 2000 for woolly monkeys and less than one in 1,000,000 for spider monkeys.

Genotypes were derived for a total of 35 individual woolly monkeys from at least four social groups at the PPRA site and 16 individuals from at least three social groups at the TBS site sampled in 1998. An additional nine individuals were sampled at the PPRA site between 2000 and 2002, and four more were sampled in 2006. At the TBS site, an additional 47 individuals from at least three social groups were sampled between 2005 and 2007. Finally, ten individuals from various other sites in the region were sampled at various times between 1998 and 2005. For spider monkeys, 24 individuals were sampled from one social group at the PPRA

Table 1A Locus characteristics for woolly monkeys by population and overall

Population	Locus	N	Na	Ho	He	
PPRA	**LL1-1#10**	48	11	0.85	0.85	ns
	LL1-1#18	48	14	0.92	0.88	ns
	Locus 5	48	8	0.73	0.78	ns
	LL1-1#15	48	11	0.83	0.80	ns
	LL1-1#3	48	20	0.92	0.85	ns
	LL1-5#7	47	6	0.79	0.73	ns
	Leon21	48	7	0.77	0.71	ns
	LL3-1#2	48	4	0.77	0.60	*
Average			**10.1**	**0.82**	**0.78**	
TBS	**LL1-1#10**	63	11	0.81	0.83	*
	LL1-1#18	63	11	0.86	0.88	ns
	Locus 5	62	8	0.74	0.70	ns
	LL1-1#15	63	10	0.84	0.81	ns
	LL1-1#3	62	19	0.92	0.92	ns
	LL1-5#7	63	5	0.75	0.68	ns
	Leon21	63	7	0.67	0.73	ns
	LL3-1#2	63	4	0.57	0.54	ns
Average			**9.4**	**0.77**	**0.76**	
Overall	**LL1-1#10**	111	13	0.83	0.85	ns
	LL1-1#18	111	15	0.88	0.89	ns
	Locus 5	110	10	0.74	0.74	ns
	LL1-1#15	111	13	0.84	0.83	ns
	LL1-1#3	110	24	0.92	0.91	ns
	LL1-5#7	110	6	0.76	0.71	ns
	Leon21	111	8	0.71	0.73	ns
	LL3-1#2	111	4	0.66	0.57	ns
Average			**11.6**	**0.79**	**0.78**	

Table 1B Locus characteristics for spider monkeys by population and overall

Population	Locus	N	Na	Ho	He	
PPRA	D17S804	24	5	0.71	0.72	ns
	D5S111	24	7	0.75	0.67	ns
	D8S165	24	7	0.83	0.74	ns
	D8S260	24	13	0.88	0.89	ns
	Leon 15	24	3	0.46	0.61	ns
	Leon 2	24	8	0.88	0.81	ns
	Leon 21	24	8	0.83	0.78	ns
	LL 1-1#10	24	10	0.83	0.87	ns
	LL 1-1#18	24	12	0.92	0.88	ns
	LL 1-5#7	24	9	0.83	0.87	ns
	LL 3-1#1	24	2	0.04	0.04	ns
	LL 3-1#2	21	5	0.76	0.71	ns
	Locus 5	24	8	0.83	0.80	ns
	SB 19	24	3	0.67	0.57	ns
	SB 30	24	3	0.63	0.63	ns
	SB 38	24	8	0.83	0.84	ns
Average			**6.9**	**0.73**	**0.71**	
TBS	D17S804	25	4	0.64	0.57	ns
	D5S111	25	8	0.64	0.67	ns
	D8S165	25	6	0.64	0.70	ns
	D8S260	25	10	0.84	0.83	ns
	Leon 15	25	4	0.48	0.61	ns
	Leon 2	25	7	0.92	0.82	ns
	Leon 21	25	9	0.92	0.84	ns
	LL 1-1#10	25	12	0.88	0.83	ns
	LL 1-1#18	25	8	0.84	0.83	ns
	LL 1-5#7	25	11	0.92	0.86	ns
	LL 3-1#1	25	2	0.08	0.08	ns
	LL 3-1#2	25	5	0.84	0.76	ns
	Locus 5	25	5	0.72	0.71	ns
	SB 19	25	3	0.56	0.61	ns
	SB 30	25	4	0.40	0.54	*
	SB 38	25	6	0.48	0.57	
Average			**6.5**	**0.68**	**0.68**	
Overall	D17S804	49	5	0.67	0.67	ns
	D5S111	49	8	0.69	0.71	ns
	D8S165	49	7	0.73	0.78	ns
	D8S260	49	13	0.86	0.87	ns
	Leon 15	49	4	0.47	0.61	ns
	Leon 2	49	8	0.90	0.85	ns
	Leon 21	49	10	0.88	0.82	ns
	LL 1-1#10	49	13	0.86	0.87	ns
	LL 1-1#18	49	13	0.88	0.87	ns
	LL 1-5#7	49	12	0.88	0.88	ns
	LL 3-1#1	49	2	0.06	0.06	ns
	LL 3-1#2	46	5	0.80	0.74	ns
	Locus 5	49	8	0.78	0.76	ns
	SB 19	49	3	0.61	0.60	ns
	SB 30	49	4	0.51	0.62	ns
	SB 38	49	8	0.65	0.75	ns
Average			**7.7**	**0.70**	**0.72**	

site in 1998 and subsequently 25 were sampled from one social group at the TBS site between 2005 and 2007. The total dataset then, comprised genotypes from 121 woolly monkeys and 49 spider monkeys.

For both woolly and spider monkeys, genotype data for each locus in each of the two best sampled populations (PPRA and TBS) were checked for deviation from Hardy-Weinberg expectations and for the likely presence of null alleles using the softwares GenAlEx version 6 (Peakall and Smouse 2006) and ML-Relate (Kalinowski et al. 2006). For this analysis, genotype data were combined across all sampling years, which is appropriate given the long lifespans of individual animals and the demographic continuity of populations at each site. Indeed, some of the same animals sampled in 1998 or 1999 were resampled in 2006 or 2007. For woolly monkeys, in each population genotype data for one locus deviated significantly from Hardy-Weinberg expectations (Table 1A). At the TBS site, this was due to the presence of a particular private allele seen only in this population and only one individual. In the PPRA population, an excess in the frequency of one particular heterozygous genotype at locus LL 3-1#2 is not easily explained, thus, this locus was both included and excluded from the calculation of F statistics and from the estimation of pairwise relatedness among individual animals in the sample. The results are qualitatively unchanged whether the locus is included or excluded, and the analyses presented below exclude data from this locus.

Looking at each population of spider monkeys separately, genotype frequencies showed significant deviation from Hardy-Weinberg expectations in only one population, the TBS site, and at only one of 16 loci, SB30 (Table 1B). The slight excess homozygosity at this locus in the PPRA population could be explained by the presence of a null allele. Looking at the complete set of spider monkey genotypes from both populations, at one locus, Leon 15, the observed heterozygosity was both low (less than 0.50 in both populations and overall) and substantially lower than the expected heterozygosity (0.61). Analyses in ML-Relate (Kalinowski et al. 2006) suggested the possible presence of a null allele at this locus as well.

Genotype data for each taxon were used to derive two estimators of pairwise relatedness – the regression-based estimator of Queller and Goodnight (1989) and the maximum-likelihood-based estimator of Kalinowski et al. (2006) – for all individuals sampled in the PPRA and TBS populations, again using GenAlEx (Peakall and Smouse 2006) and ML-Relate (Kalinowski et al. 2006). Results are qualitatively unchanged when these different relatedness estimators are used, thus I present results on average relatedness based on the estimator of Queller and Goodnight (1989). Sex-biased dispersal was evaluated using assignment tests as implemented in the softwares FSTAT version 2.9.3 (Goudet 2001) and GenAlEx version 6 (Peakall and Smouse 2006).

Finally, for woolly monkeys sampled in 1998 at both sites, I sequenced up to 528 base pairs of hypervariable region I (HV1) of the mitochondrial control region. This region was amplified from tissue or fecal sample derived DNA extracts using ateline specific primers, either as one large fragment or as three smaller, overlapping fragments. Unincorporated bases and excess primers were removed from the amplified PCR products either via QiaQuick cleanup procedures or by subjecting the products to ExoSap treatment. Cycle-sequencing of each strand of the amplified products was performed using either ABI Dye-Terminator or Big-Dye sequencing chemistries, and then fragments were separated and visualized on ABI 373XL and ABI 3730 automated DNA analyzers. Sequences were aligned using the software Sequencher (GeneCodes) and checked by eye. MEGA 3.1 (Kumar et al. 2004) was used to estimate haplotype sequence divergence. The phylogenetic relationships among haplotypes were inferred using Bayesian maximum-likelihood methods, as implemented in Mr. Bayes 3.1.2 (Huelsenbeck and Ronquist 2001). Additional control region sequences extracted from GenBank for the other atelids (*Alouatta*, *Brachyteles*, *Ateles*, and *Lagothrix*) were included in the phylogeny inference, with *Alouatta seniculus* specified as an outgroup. A reticulating network of relationships among the Yasuní woolly monkey control region haplotypes was also inferred using the softwares Network 4.2.0.1 (Fluxus Technology Ltd 2007) and TCS 1.2.1 (Clement et al. 2000).

With respect to these sequence data, it is important to note that no special procedures were undertaken to overcome the potential for preferentially amplifying nuclear copies of the mitochondrial genome (i.e., "numts": Lopez et al. 1994; Zhang and Hewitt 1996). However, several lines of evidence suggest that this was not a problem. First, for a subset of three individuals I subsequently amplified a much larger section of mtDNA (~3000 base pairs) using long-range PCR and then sequenced the same HV1 fragment from within this larger amplicon, and for each sample the resultant HV1 sequences were identical, suggesting the same target was amplified using the ateline HV1 primer sequences as was recovered via long-range PCR. Second, even if numts were present among the set of recovered woolly monkey HV1 sequences, it should only serve to diminish any signal of sex-biased dispersal seen in the patterning of mtDNA variation among males and females. The fact that a definite pattern is still seen (*see below*) suggests that the pattern is robust. Finally, looking within each study site, the members of each of four likely mother-offspring pairs identified by ML-Relate shared the same mitochondrial DNA haplotype, again strongly suggesting that these were true maternally inherited mitochondrial sequences.

9.5 Results

9.5.1 Average Relatedness Within and Gene Flow Among Populations

For woolly monkeys, the mean relatedness, r_{xy}, among all individuals sampled at the PPRA site was 0.008 and at the TBS site was 0.012. For both sites, mean r_{xy} within the population is significantly greater than the average relatedness among all individuals sampled, regardless of population of origin (overall mean $r_{xy} = -0.009$, $p \leq 0.05$ for both populations, tested by permutation). F_{IS} for both populations was close to, and did not differ significantly from zero, (PPRA $F_{IS} = -0.026$, TBS $F_{IS} = 0.004$), suggesting that the overall level of inbreeding within each population was minimal.

Woolly monkeys at the PPRA and TBS sites were slightly but significantly differentiated genetically from one another, regardless of whether the complete dataset, including juveniles, is used (overall $F_{ST} = 0.019$, 99% C.I. $= 0.009$ to 0.028, p ≤ 0.001), or whether the data is limited to only adult individuals – i.e., those past the expected age of natal dispersal ($F_{ST} = 0.014$, 99% C.I. $= 0.005$ to 0.025, p ≤ 0.001). The significant F_{ST} value suggests some restriction to gene flow between the two sites. Focusing on adults only, the degree of genetic differentiation seen between the sites was somewhat higher for males than for adult females ($F_{ST} = 0.023$ versus 0.014, p ≤ 0.005 for both sexes), consistent with the idea that gene flow between the sites for males has been somewhat more restricted than it has for females.

For spider monkeys, the mean relatedness among animals sampled at the PPRA site was -0.005, while at the TBS site the mean relatedness was 0.049. The latter value is significantly greater than the average relatedness among all the individuals sampled across sites (mean $r_{xy} = -0.021$, p ≤ 0.001 tested by permutation). The inbreeding coefficient for spider monkeys at the TBS site was higher than that for the PPRA population, indicating more local inbreeding, although neither inbreeding coefficient differed significantly from zero (PPRA: $F_{IS} = -0.002$, TBS: $F_{IS} = 0.023$).

As for woolly monkeys, the PPRA and TBS populations of spider monkey were significantly differentiated from one another genetically, both when juveniles are included in the dataset ($F_{ST} = 0.042$, 99% C.I. $= 0.015$ to 0.073, p ≤ 0.001) and when only adults are considered ($F_{ST} = 0.031$, 99% C.I. $= 0.006$ to 0.060, p ≤ 0.001). Among adults, the degree of genetic differentiation seen between the sites was much higher for males ($F_{ST} = 0.091$, p ≤ 0.005) than for females, for whom differentiation between the sites was not significant at an alpha level of 0.05 ($F_{ST} = 0.017$, p $= 0.094$). This suggests that, historically, gene flow between the sites has been much more restricted for males than for females.

9.5.2 Genetic Relatedness Among Nonjuvenile Animals in Each Site

Woolly monkeys were sampled from a minimum of five different social groups at the PPRA site, although only two were sampled thoroughly. When genotype data for all of these groups are considered together, the average relatedness among 11 sampled adult males was 0.008 while that among 15 adult females was 0.024. While the mean relatedness among adult females at the PPRA site was slightly higher than that among males, a permutation test revealed that the difference was not significant (p $= 0.45$). At the TBS site, samples were also collected from woolly monkeys in a minimum of five social groups, each sampled only sparsely. The average relatedness among 19 sampled adult males was 0.017, while that among 23 adult females was 0.005. Here, the mean relatedness among adult males was slightly greater than that among adult females, but again difference was not significant (p $= 0.42$), suggesting comparable levels of dispersal by both males and females.

Di Fiore and Fleischer (2005) previously examined the mean pairwise relatedness among males versus females within groups in two well-sampled social groups at the PPRA site, but they included subadults in their male and female datasets (presuming these to be post-dispersal individuals) and used a slightly different set of loci. Repeating their analysis with the current, more conservative dataset focusing only on adults, the mean pairwise relatedness among the three adult males in PPRA Group 4 was 0.347 and among the six adult females was 0.078. For this group, adult males, on average, were more closely related than adult females, but this result is driven by the fact that two of the three resident males were more closely related to one another than is expected even for full siblings ($r = 0.834$ versus $r = 0.50$). For PPRA Group 5, the mean relatedness among five adult males in PPRA Group 5 was 0.058, while the mean relatedness among seven adult females was only slightly less at 0.032, a difference that is not significant ($p = 0.49$). Note that this result differs from the earlier report of greater mean r_{xy} among males than among females for this same group when subadults were included in the dataset (Di Fiore and Fleischer 2005).

For spider monkeys, the average relatedness among six adult males from one completely sampled social group at the TBS site was 0.220, corresponding roughly to the degree of relatedness expected among half siblings ($r = 0.25$), which was significantly higher than the average relatedness among nine adult females sampled from the same group (mean $r_{xy} = 0.004$, $p < 0.01$). At the PPRA site, by contrast, the average relatedness among five resident adult males in one completely sampled social group was not significantly different from that among the 14 adult females sampled in the group (mean $r_{xy\ male} = -0.027$, mean $r_{xy\ female} = -0.012$, $p = 0.40$).

9.5.3 Presence of Close Kin Within Groups

The software ML-Relate (Kalinowski et al. 2006) was used to evaluate the likely kinship relationships among all pairs of woolly and spider monkeys within both well-sampled woolly monkey groups at the PPRA site and in each of the completely sampled groups of spider monkeys at the PPRA and TBS sites. The software estimates the likelihood that a particular pair of individuals falls into one of the four kinship classes "unrelated", "half siblings" (HS), "full siblings" (FS), or "parent-offspring" (PO), taking into account the allele frequencies found in the population and the possible presence of null alleles. In each of these groups, most adult individuals of both sexes had at least one other adult individual resident in their social group who was a likely close relative (i.e., an individual whose most likely category of relatedness was HS, FS, or PO rather than unrelated), and often more than one.

In one woolly monkey group at the PPRA site, Group 4, two of the three males were very closely related (HS or closer), as noted above, and five of six females had at least one same-sexed adult kin resident in the group (Fig. 9.3A). For these females, an average of 20%, of same-sexed group members were likely to be close kin. For Group 5 at the PPRA site, all five adult males had at least one same-sex close kin in the group, as did five of seven adult females (Fig. 9.3B). An average of

31% of same sexed adults were identified as likely close relatives for males, while for females an average of only 14% of same-sexed adults were likely to be close kin.

For the spider monkey group at the TBS site, all but one adult male could be linked to one another in a patriline comprising likely PO, FS, and HS relationships. Only three of nine females in this group showed likely close-kin relationships with same-sexed adults, and all of these were more likely to be HS than FS or PO relationships (Fig. 9.3C). For males, an average of 40% of the same-sexed individuals in this group were close kin, while an average of only 8% of same-sexed adults were close kin for females. The situation was somewhat different in the PPRA spider monkey group, where three of five males were linked to other males by likely HS relationships and nine of 13 females could be linked to other females by at least one likely HS relationship (Fig. 9.3D). Still, 20% of same-sexed adults were close kin for males versus only 13% for females. Thus, in all groups, a greater proportion of same-sexed pairs residing in the same group were likely to be HS, FS, or PO dyads for males than for females.

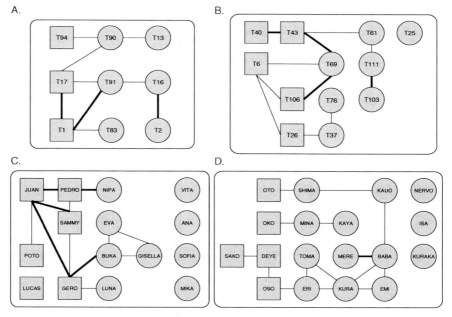

Fig. 9.3 Likely close-kin relationships among adults in two groups of woolly monkeys and two groups of spider monkeys, identified using ML-Relate (Kalinowski et al. 2006). Darker lines indicate a likely PO or FS relationship and thinner lines indicate a likely HS relationship. Males are squares, females are circles. (**A**) Woolly monkey Group 4 in the PPRA site. (**B**) Woolly monkey Group 5 in the PPRA site. (**C**) Spider monkey Group MQ-1 at the PPRA site. (**D**) Spider monkey group MQ-1 at the TBS site

Table 2 Mean and variance in assignment indices for males and females

		Male mAIc	Female mAIc	Male vAIc	Female vAIc
Spider	**PPRA**	0.074	−0.027	1.653	1.283
	TBS	1.438	−0.959	4.955	1.315
Woolly	**PPRA**	−0.035	0.026	0.754	1.666
	TBS	0.000	0.000	1.088	1.409

9.5.4 Assignment Tests for Nonjuvenile Animals

Assignment indices were calculated for all sampled adult woolly and spider monkeys at both the PPRA and TBS sites (Table 2). For woolly monkeys at both sites, the mean corrected assignment index (mAIc) of both males and females was close to zero and the distribution of assignment indices for the two sexes was similar, suggesting little bias in dispersal among the sexes (Fig. 9.4). For spider monkeys at the PPRA site, the mean corrected assignment index (mAIc) for males was slightly positive while that for females was slightly negative, and the variance in the corrected assignment index (vAIc) for males was greater than that for females. Overall, in the PPRA site assignment indices for male versus female spider monkeys were not significantly different (Mann-Whitney Test: $U_{(5, 14)} = 33$, p $= 0.891$). By contrast, at the TBS site, assignment indices for males were significantly higher than those of females (Mann-Whitney Test: $U_{(6, 9)} = 8$, p < 0.05), clearly implicating female-biased dispersal. Unexpectedly, the variance in assignment index scores was much higher for males, but this was due to the very low assignment index of one adult male, who was thus a likely immigrant into the community.

9.5.5 Mitochondrial DNA Diversity and Intraspecific Phylogeny

Mitochondrial sequence data was only collected for woolly monkeys. A total of 23 mitochondrial haplotypes were found among individuals sampled in the PPRA population and 13 were found among individuals in the TBS population. Three additional unique haplotypes were found in other populations within the Yasuní region that were sampled much less intensively. Of the 36 total unique mitochondrial DNA haplotypes recovered, only five were shared by individuals at more than one sampling site. Assuming a Tamura-Nei model of nucleotide substitution, the mean haplotype divergence across the whole set of unique haplotypes was 3.4 ± 1.5%. Within the set of females sampled, the mean haplotype divergence was 2.9 ± 1.6%, while among males, the mean divergence was 3.2 ± 1.7%.

At both the PPRA and TBS sites, the haplotype diversity seen among both females and males was high, suggesting high levels of transfer by females. Thirteen different haplotypes were seen among the 14 adult females sampled at the PPRA site, and a total of nine haplotypes were found among the 11 adult females sampled at the TBS site. For males, seven haplotypes were found in the set of eleven adult

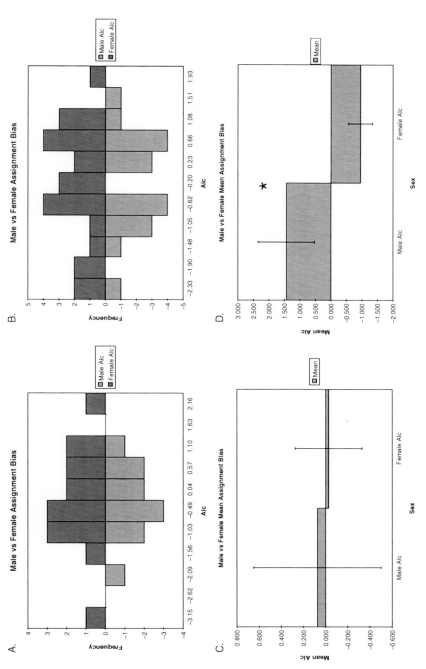

Fig. 9.4 (**A**) Distribution of assignment scores for male and female woolly monkey at the PPRA site. (**B**) Distribution of assignment scores for male and female woolly monkey at the TBS sites. (**C**) Average assignment indices for males versus females in spider monkey Group MQ-1 at the PPRA site. (**D**) Average assignment indices for males versus females in spider monkey Group MQ-1 at the TBS site. Assignment indices were calculated using GenAlEx (Peakall and Smouse 2006)

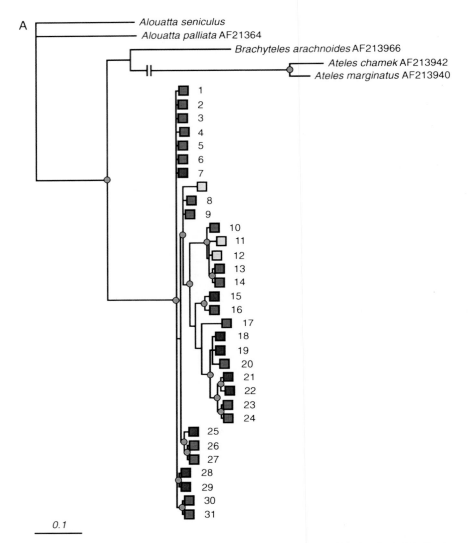

Fig. 9.5 (**A** and **B**) Phylogenies of male and female woolly monkey mitochondrial DNA haplotypes sampled at the PPRA and TBS sites and other locales within Yasuní National Park, as inferred by Bayesian maximum likelihood methods implemented in the software Mr. Bayes 3.1.2 (Huelsenbeck and Ronquist 2001). For the analysis, control region sequence from *Alouatta seniculus* (sampled in Yasuní) was used as an outgroup, and additional control region sequences extracted from GenBank for *Alouatta palliata*, *Brachyteles arachnoides*, *Ateles chamek*, and *Ateles marginatus* were also included (accession numbers indicated on figure). The analysis was run for 1×10^6 generations, with each of two simultaneous runs sampling four Markov chains every 100 generations under a GTR+I evolutionary model and using the software's default priors. The male and female trees shown are 50% majority rule consensus phylogenies derived from all alterantive topologies stored after a burnin period of 25000 generations. Darkest shaded samples come from the TBS site (in A: individuals 7, 15, 18-19, 21-22, 25, and 28-29; in B: individuals 5-6, 12, 17, 22, 24, 28, 31,

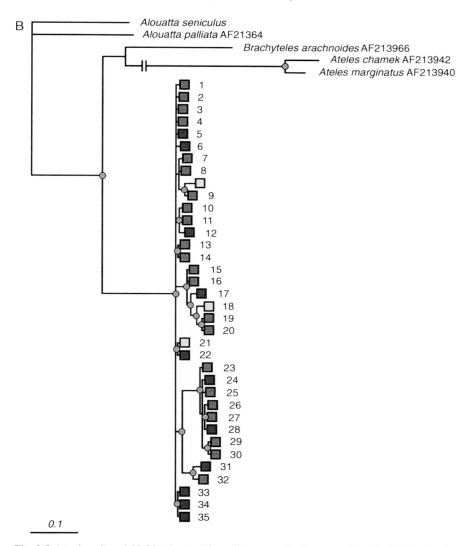

Fig. 9.5 (continued) and 33-35), those with medium gray shading come from the PPRA site (in A: individuals 1-6, 8-10, 13-14, 16-17, 20, 23-24, 26-27, and 30-31; in B: individuals 1-4, 7-8, 9-11, 13-16, 19-20, 23, 25-27, 29-30, and 32), and those with the lightest shading come from other locales. Nodes demarcated by filled circles reflect clades with >70% posterior probability on the consensus tree. (**C** and **D**) Networks of relationships among mitochondrial DNA haplotypes of males and females inferred using the Median-Joining algorithm (Bandelt et al. 1999) implemented in the software Network 4.2 (Fluxus Technology Ltd 2007). For the analysis, the value of epsilon was set at 0 for males and 10 for females, and all variable sites were weighted equally. Very similar networks are also recovered using TCS 1.2.1 (Clement et al. 2000) (data not shown). In each figure, on the terminal nodes the darkest shading refers to haplotypes sampled at the TBS site, medium gray shading refers to haplotypes sampled at the PPRA site, and the lightest shading to haplotypes sampled at one of the other locales indicated in Figure 9.2. Terminal node size reflects the relative

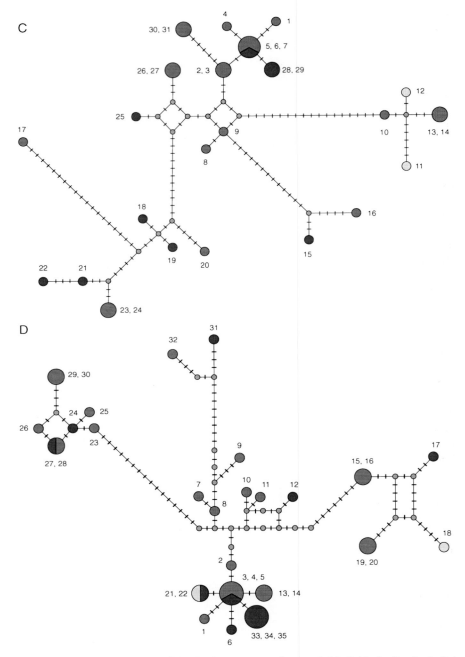

Fig. 9.5 (continued) frequency of that haplotype among the sampled individuals. Small, shaded internal nodes indicate reconstructed (unsampled) median haplotypes and black tick marks indicate the number of nucleotide differences between nodes. Numbers in each figure refer to unique individuals: males for Fig. 9.5A,C and females for Fig. 9.5B,D. *See* Color Insert.

individuals sampled at the PPRA site, and five of the six adult males sampled at the TBS site had different haplotypes.

The phylogenetic relationships and haplotype networks inferred for males and females are shown in Fig. 9.5A–D, along with an indication of the sampling locale (i.e., "PPRA", "TBS", or "other") where each haplotype was found. Importantly, for both sexes, closely related haplotypes (i.e., members of the same clade or haplotypes located close to one another in the network) were sampled at different sampling sites, while some haplotypes found at the same site were only distantly related (i.e., occurred in the same clade or portion of the network), suggesting a high level of female-mediated gene flow across the sampling sites.

Finally, pairs of adult males tended to share the same mitochondrial xDNA haplotype with one another more often than pairs of females. Four pairs of adult males – three pairs within one social group and one pair in a second group – shared mitochondrial DNA haplotypes at the PPRA site, and one pair of adult males, both in the same social group, shared haplotypes at the TBS site, implying they were close matrilineal kin. Among adult females, at the PPRA site only one pair (whose members resided in different groups) shared a haplotype, and at the TBS site one trio of females, two from the same group and one from a different group, shared the same haplotype.

9.6 Discussion

The studies reviewed above exemplify some of the utility of using molecular data to inform our understanding of primate dispersal patterns, even in the absence of long-term observational data and without directly observing dispersal events. These studies likewise highlight the variation in dispersal patterns seen among New World primates and reiterate the fact that, outside of the cercopithecoids, dispersal by females is fairly common among primates, a fact that several researchers have noted previously (Di Fiore and Rendall 1994; Strier 1994b). Additionally, the molecular data reviewed here provide several new insights into the dispersal behavior of woolly and spider monkeys and into patterns of kinship among group members.

First, for woolly monkeys, genetic analyses reveal that dispersal by females is indeed common. However, contrary to the traditional classification of *Lagothrix* as a male-philopatric taxon, a significant degree of dispersal by males appears to be occurring as well. Adult males, in general, are not more closely related to one another than are adult females within most woolly monkey groups or within two local populations, and assignment tests do not detect evidence of a strong female-bias to dispersal patterns. Some males do reside as adults in groups with close same-sexed kin but some adult females do as well, and members of both sexes often live in groups with close opposite-sexed kin as well. Still, based on mitochondrial DNA haplotype sharing, it may be more common for male matrilineal kin to co-reside in social groups than female matrilineal kin. Overall, these patterns might indicate that some individuals of both sexes remain philopatric while others disperse, or they could reflect a tendency for animals of both sexes to disperse but settle in groups where close kin are already resident. Obviously, distinguishing among these possibilities will require significant longitudinal data on dispersal by known individuals, something that, at present, is lacking.

Second, with respect to spider monkeys, molecular data for one well-sampled social group at the TBS site conform to what has long been suspected about the social structure of *Ateles* – i.e., that females disperse while males remain philopatric, thus the adult members of groups comprise primarily close male relatives and unrelated females. Interestingly, however, this pattern was not seen in the well-studied group of spider monkeys at the PPRA site, where many adult females seemed to reside with likely close kin and where the mean degree of relatedness among both adult males and females was close to zero. It is worthwhile noting, however, that we suspect that several animals from this study group were lost to local hunters during the mid-1990s, prior to genetic sampling. It is thus possible that males related to the current residents were lost from the study population or that hunting opened up opportunities for unrelated males to immigrate into the community. Since 2003, however, the only animals known to have moved into this group have been females (Link and Di Fiore, unpublished data; Shimooka et al. in press), reinforcing the idea that dispersal in spider monkeys is, in general, strongly female-biased. The fact that one adult male in the TBS group was unrelated to the remaining adult males, however, suggests that male immigration into spider monkey groups can sometimes occur. Indeed, several cases of male immigration have been documented for *Ateles geoffroyi* in Costa Rica (Filippo Aureli, personal communication).

Differences between woolly monkeys and at least some groups of spider monkeys in the mean level of relatedness among males relative to females may help to explain some of the clear differences in social behavior among same-sexed individuals in these two species. In spider monkeys, males are generally more affiliative with one another than females and cooperate with each other in territory defense (Fedigan and Baxter 1984; Di Fiore and Campbell 2007). In the TBS study population, for example, male spider monkeys often travel together in the same subgroups, and coalitions of males also jointly aggress against females, cooperate in intergroup encounters with males from adjacent groups, and join together to conduct raids or patrols into other groups' territories (Link et al. in review; Di Fiore and Link, unpublished data), as has been reported for chimpanzees (Watts and Mitani 2001) and for spider monkeys at other sites (Aureli et al. 2006).

In woolly monkeys, by contrast, male-male cooperation is rare, although it occasionally occurs in the context of intergroup interactions, and adult males are tolerant but not overly affiliative with one another (Di Fiore 1997; Di Fiore and Fleischer 2005; Di Fiore and Campbell 2007). Interestingly, high intensity aggression among adult male woolly monkeys has been observed on several occasions. In one case, an adult male resident disappeared from one group under observation after he was seen participating in aggression with another male over a period of several days. In a second case, several adult males from a second study group were seen aggressively interacting with an unfamiliar, non-resident adult male over one to two days, after which the unfamiliar animal was not seen again. The fact that adult

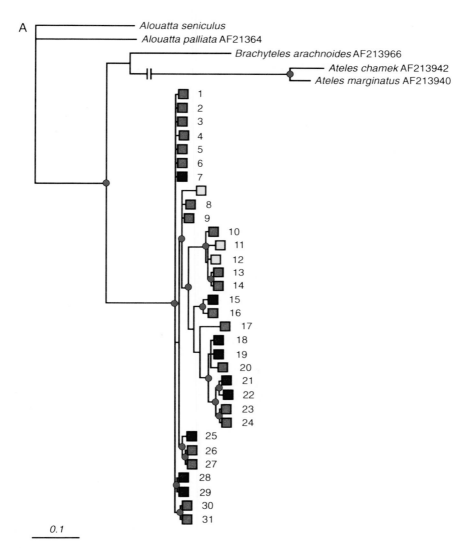

Fig. 9.5 (**A** and **B**) Phylogenies of male and female woolly monkey mitochondrial DNA haplotypes sampled at the PPRA and TBS sites and other locales within Yasuní National Park, as inferred by Bayesian maximum likelihood methods implemented in the software Mr. Bayes 3.1.2 (Huelsenbeck and Ronquist 2001). For the analysis, control region sequence from *Alouatta seniculus* (sampled in Yasuní) was used as an outgroup, and additional control region sequences extracted from GenBank for *Alouatta palliata*, *Brachyteles arachnoides*, *Ateles chamek*, and *Ateles marginatus* were also included (accession numbers indicated on figure). The analysis was run for 1×10^6 generations, with each of two simultaneous runs sampling four Markov chains every 100 generations under a GTR+I evolutionary model and using the software's default priors. The male and female trees shown are 50% majority rule consensus phylogenies derived from all alterantive topologies stored after a burnin period of 25000 generations. Darkest shaded samples come from the TBS site (in A: individuals 7, 15, 18-19, 21-22, 25, and 28-29; in B: individuals 5-6, 12, 17, 22, 24, 28, 31,

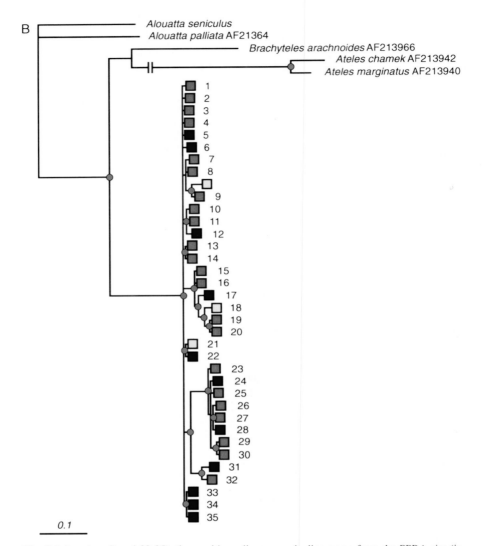

Fig. 9.5 (continued) and 33-35), those with medium gray shading come from the PPRA site (in A: individuals 1-6, 8-10, 13-14, 16-17, 20, 23-24, 26-27, and 30-31; in B: individuals 1-4, 7-8, 9-11, 13-16, 19-20, 23, 25-27, 29-30, and 32), and those with the lightest shading come from other locales. Nodes demarcated by filled circles reflect clades with >70% posterior probability on the consensus tree. (**C** and **D**) Networks of relationships among mitochondrial DNA haplotypes of males and females inferred using the Median-Joining algorithm (Bandelt et al. 1999) implemented in the software Network 4.2 (Fluxus Technology Ltd 2007). For the analysis, the value of epsilon was set at 0 for males and 10 for females, and all variable sites were weighted equally. Very similar networks are also recovered using TCS 1.2.1 (Clement et al. 2000) (data not shown). In each figure, on the terminal nodes the darkest shading refers to haplotypes sampled at the TBS site, medium gray shading refers to haplotypes sampled at the PPRA site, and the lightest shading to haplotypes sampled at one of the other locales indicated in Figure 9.2. Terminal node size reflects the relative

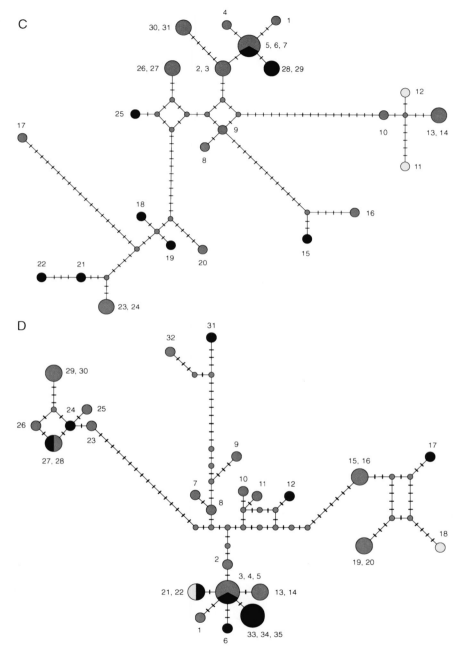

Fig. 9.5 (continued) frequency of that haplotype among the sampled individuals. Small, shaded internal nodes indicate reconstructed (unsampled) median haplotypes and black tick marks indicate the number of nucleotide differences between nodes. Numbers in each figure refer to unique individuals: males for Fig. 9.5A,C and females for Fig. 9.5B,D.

males occasionally bear broken canines and/or broken digits on the hands while such injuries are less common among adult females may be due to competition among males for residence in a group, although it is important to stress that overt aggression among males is almost never seen. Nonetheless, the fact that only some of the adult males within a group seem to be close relatives sets up a situation whereby we might expect to see some males be the targets of high intensity aggression and attempted repulsion by coalitions of other males. Whether patterns of affiliative within-group social behavior in woolly monkeys correlate with genetic relatedness among individuals is currently under investigation.

9.7 Conclusions

The molecular genetic techniques discussed above clearly hold great promise for the study of dispersal, kinship, and social organization, but as yet have not been widely applied in primates. Over the past 10 years, the costs of DNA sequencing and multilocus genotyping have dropped dramatically, while laboratory techniques have been refined sufficiently such that it is possible to reliably collect genetic data from many animals using relatively low quality or degraded samples collected noninvasively. At present, the major impediment to broader implementation of the techniques discussed here is the time and money it takes to identify a sufficient number of suitably variable loci within a taxon of interest for unambiguous identification of individuality and for estimating the relatedness between individuals using multilocus genotyping. However, with the growing availability of primate genomic data and the development of new bioinformatic tools for rapidly searching genomes comparatively for points of homology, even the marker identification phase of molecular studies will become much faster. Thus, the use of genetic data as a complement to observational data in field studies is expected to become more common and should be encouraged. Field primatologists must familiarize themselves with the utility of these methods and be encouraged to collect valuable samples (e.g., hair, feces, tissue) from their subjects whenever possible.

9.8 Summary

Among social animals such as primates, "kinship" (i.e., genetic relatedness) has commonly been invoked as a key factor underlying and organizing the expression of social behavior. Patterns of genetic relatedness within groups in turn are linked to individuals' behavior by the effect that behavior has on the distribution of genetic variation. Dispersal and reproductive behavior, in particular, act to shuffle genes across the social and geographic landscapes and to a large extent determine how genetic variation within a population is partitioned within and between social groups. In recent years, a number of analytical techniques have been developed that allow researchers to use molecular genetic data from a variety of markers (e.g., mitochondrial DNA sequences, multilocus microsatellite genotypes, AFLP fingerprints, etc.) to characterize the kinship relationships among the individuals in a sample, to investigate mating systems, and to make inferences about dispersal patterns. Somewhat surprisingly, however, remarkably few studies of wild primates have taken advantage of these techniques, and fewer still have taken the further critical step of examining whether the genetic relatedness among individuals is in fact a reliable predictor of the social behaviors it has often been invoked to explain. In this chapter, I outline a number of the key theoretical links that can be drawn among between dispersal patterns, genetic relatedness, and population genetic structure and describe some of the analytical methods that can be used to explore these links. I then review the few published studies of New World monkeys that have used genetic data to investigate dispersal and patterns of within-group relatedness. Finally, I supplement this review with results from my own research group's long-term work on two species of sympatric platyrrhines, lowland woolly monkeys (*Lagothrix poeppigii*) and white-bellied spider monkeys (*Ateles belzebuth*).

References

Ahumada, J. A. 1989. Behavior and social structure of free ranging spider monkeys (*Ateles belzebuth*) in La Macarena. Field Studies of New World Monkeys, La Macarena, Colombia 2: 7–31.

Alberts, S. C., Buchan, J. C. and Altmann, J. 2006. Sexual selection in wild baboons: From mating opportunities to paternity success. Animal Behavior 72:1177–1196.

Alberts, S. C., Watts, H. E. and Altmann, J. 2003. Queuing and queue-jumping: Long-term patterns of reproductive skew in male savannah baboons, *Papio cynocephalus*. Animal Behavior 65:821–840.

Alexander, R. D. 1974. The evolution of social behavior. In R. F. Johnston, P. W. Frank and C. D. Michener (eds.), *Annual Review of Ecology and Systematics* (pp. 325–383). Palo Alto: Annual Reviews, Inc.

Altmann, J., Alberts, S. C., Haines, S. A., Bubach, J., Muruthi, P., Coote, T., Geffen, E., Cheesman, D. J., Mututa, R. S., Saiyalel, S. N., Wayne, R. K., Lacy, R. C. and Bruford, M. W. 1996. Behavior predicts genetic structure in a wild primate group. Proceedings of the National Academy of Sciences 93:5797–5801.

Altmann, S. A. 1962. A field study of the sociobiology of the rhesus monkey, Macaca mulatta. Annals of the New York Academy of Sciences 102:338–435.

Aureli, F., Schaffner, C. M., Verpooten, J., Slater, K. and Ramos-Fernandez, G. 2006. Raiding parties of male spider monkeys: Insights into human warfare? American Journal of Physical Anthropology 131:486–497.

Avise, J. C. 1994. Molecular Markers, Natural History and Evolution. New York: Chapman and Hall.

Avise, J. C. 1995. Mitochondrial DNA polymorphism and a connection between genetics and demography of relevance to conservation. Conservation Biology 9:686–690.

Avise, J. C. 2000. Phylogeography: The History and Formation of Species. Cambridge: Harvard University Press.

Avise, J. C. 2004. Molecular Markers, Natural History, and Evolution, 2nd Edition. Sunderland, MA: Sinauer Associates, Inc.

Baker, A. and Dietz, J. 1996. Immigration in wild groups of golden lion tamarins (*Leontopithecus rosalia*). American Journal of Primatology 28:47–56.

Bandelt, H.-J., Forster, P. and Röhl, A. 1999. Median-joining networks for inferring intraspecific phylogenies. Molecular Biology and Evolution 16:37–48.

Banks, M. A. and Eichert, W. 2000. WHICHRUN (Version 3.2) a computer program for population assignment of individuals based on multilocus genotype data. Journal of Heredity 91:87–89.

Bernstein, I. S. 1991. The correlation between kinship and behavior in nonhuman primates. In P. G. Hepper (ed.), *Kin Recognition* (pp. 6–29). Cambridge: Cambridge University Press.

Blouin, M. S. 2003. DNA-based methods for pedigree reconstruction and kinship analysis in natural populations. Trends in Ecology and Evolution 18:503–511.

Boinksi, S. 1999. The social organizations of squirrel monkeys: Implications for ecological models of social evolution. Evolutionary Anthropology 8:101–112.

Boinski, S., Ehmke, E., Kauffman, L., Schet, S. and Vreedzaam, A. 2005a. Dispersal patterns among three species of squirrel monkeys (*Saimiri oerstedii, S. boliviensis* and *S. sciureus*): II. Within-species and local variation. Behavior 142:633–677.

Boinski, S., Kauffman, L., Ehmke, E., Schet, S. and Vreedzaam, A. 2005b. Dispersal patterns among three species of squirrel monkeys (*Saimiri oerstedii, S. boliviensis* and *S. sciureus*): I. Divergent costs and benefits. Behavior 142:625–532.

Boinski, S., Sughrue, K., Selvaggi, L., Quatrone, R., Henry, M. and Cropp, S. 2002. An expanded test of the ecological model of primate social evolution: Competitive regimes and female bonding in three species of squirrel monkeys (*Saimiri oerstedii, S. boliviensis,* and *S. sciureus*). Behavior 139:227–261.

Clarke, M. R. and Glander, K. E. 1984. Female reproductive success in a group of free-ranging howling monkeys (*Alouatta palliata*) in Costa Rica. In M. F. Small (ed.), *Female Primates: Studies by Women Primatologists* (pp. 111–126). New York: Alan R. Liss.

Clement, M., Posada, D. and Crandall, K. A. 2000. TCS: A computer program to estimate gene genealogies. Molecular Ecology 9:1657–1660.

Constable, J. L., Ashley, M. V., Goodall, J. and Pusey, A. E. 2001. Noninvasive paternity assignment in Gombe chimpanzees. Molecular Ecology 10:1279–1300.

Cornuet, J.-M., Piry, S., Luikart, G., Estoup, A. and Solignac, M. 1999. New methods employing multilocus genotypes to select or exclude populations as origins of individuals. Genetics 153:1989–2000.

Crockett, C. M. 1984. Emigration by female red howler monkeys and the case for female competition. In M. F. Small (ed.), *Female Primates: Studies by Women Primatologists* (pp. 159–173). New York: Alan R. Liss.

Crockett, C. M. and Eisenberg, J. F. 1987. Howlers: Variations in group size and demography. In B. B. Smuts, D. L. Cheney, R. M. Seyfarth, R. W. Wrangham and T. T. Struhsaker (eds.), *Primate Societies* (pp. 54–68). Chicago: University of Chicago Press.

Crockett, C. M. and Pope, T. R. 1988. Inferring patterns of aggression from red howler monkey injuries. American Journal of Primatology 14:1–21.

Crockett, C. M. and Pope, T. R. 1993. Consequences of sex differences in dispersal for juvenile red howler monkeys. In M. E. Pereira and L. A. Fairbanks (eds), *Juvenile Primates: Life History, Development, and Behavior* (pp. 104–118). New York: Oxford University Press.

Csilléry, K., Johnson, T., Beraldi, D., Clutton-Brock, T., Coltman, D., Hansson, B., Spong, G. and Pemberton, J. M. 2006. Performance of marker-based relatedness estimators in natural populations of outbred vertebrates. Genetics 173:2091–2101.

de Ruiter, J. and Inoue, M. 1993. Paternity, male social rank, and sexual behavior. Primates 34: 553–555.

de Ruiter, J. R. and Geffen, E. 1998. Relatedness of matrilines, dispersing males and social groups in long-tailed macaques (*Macaca fascicularis*). Proceedings of the Royal Society of London, B 265:79–87.

de Ruiter, J. R., Scheffrhan, W., Trommelen, G. J. J. M., Uitterlinden, A. G. and Martin, R. D. 1992. Male social rank and reproductive success in wild long-tailed macaques. In R. D. Martin, A. F. Dixson and E. J. Wickings (eds.), *Paternity in Primates: Genetic Tests and Theories* (pp. 175–191). Basel: Karger.

Di Fiore, A. 1997. Ecology and Behavior of Lowland Woolly Monkeys (*Lagothrix lagotricha poeppigii*, Atelinae) in Eastern Ecuador. Unpublished Ph.D. thesis, Davis, CA.

Di Fiore, A. 2002. Molecular perspectives on dispersal in lowland woolly monkeys (*Lagothrix lagotricha poeppigii*). American Journal of Physical Anthropology S34:63.

Di Fiore, A. 2003. Molecular genetic approaches to the study of primate behavior, social organization, and reproduction. Yearbook of Physical Anthropology 46:62–99.

Di Fiore, A. 2005a. A rapid genetic method for sex-typing primate DNA. American Journal of Physical Anthropology Supplement 40:95.

Di Fiore, A. 2005b. A rapid genetic method for sex assignment in nonhuman primates. Conservation Genetics 6:1053–1058.

Di Fiore, A. and Campbell, C. J. 2007. The atelines: Variation in ecology, behavior, and social organization. In C. J. Campbell, A. Fuentes, K. C. MacKinnon, M. Panger and S. K. Beader (eds.), *Primates in Perspective* (pp. 155–185). New York: Oxford University Press.

Di Fiore, A. and Fleischer, R. C. 2004. Microsatellite markers for woolly monkeys (*Lagothrix lagotricha*) and their amplification in other New World primates (Primates: Platyrrhini). Molecular Ecology Notes 4:246–249.

Di Fiore, A. and Fleischer, R. C. 2005. Social behavior, reproductive strategies, and population genetic structure of *Lagothrix poeppigii*. International Journal of Primatology 26:1137–1173.

Di Fiore, A. and Rendall, D. 1994. Evolution of social organization: A reappraisal for primates by using phylogenetic methods. Proceedings of the National Academy of Sciences, USA 91: 9941–9945.

Digby, L. J. and Barreto, C. E. 1993. Social organization in a wild population of *Callithrix jacchus*: I. Group composition and dynamics. Folia Primatologica 61:123–134.

Douadi, M. I., Gatti, S., Levero, F., Duhamel, G., Bermejo, M., Vallet, D., Menard, N. and Petit, E. 2007. Sex-biased dispersal in western lowland gorillas (*Gorilla gorilla gorilla*). Molecular Ecology 16:2247–2259.

Ellsworth, J. A. 2000. Molecular Evolution, Social Structure, and Phylogeography of the Mantled Howler Monkey (*Alouatta palliata*). Unpublished Ph.D. thesis, Reno, NV.

Epstein, M. P., Duren, W. L. and Boehnke, M. 2000. Improved inference of relationship for pairs of individuals. American Journal of Human Genetics 67:1219–1231.

Eriksson, J., Siedel, H., Lukas, D., Kayser, M., Erler, A., Hashimoto, C., Hohmann, G., Boesch, C. and Vigilant, L. 2006. Y-chromosome analysis confirms highly sex-biased dispersal and suggests a low effective male population size in bonobos (*Pan paniscus*). Molecular Ecology 15:939–949.

Excoffier, L., Laval, G. and Schneider, S. 2005. Arlequin ver. 3.0: An integrated software package for population genetics data analysis. Evolutionary Bioinformatics Online 1:47–50.

Excoffier, L., Smouse, P. and Quattro, J. 1992. Analysis of molecular variance inferred from metric distances among DNA haplotypes: Application to human mitochondrial DNA restriction data. Genetics 131:479–491.

Favre, L., Balloux, F., Goudet, J. and Perrin, N. 1997. Female-biased dispersal in the monogamous mammal *Crocidura russula*: Evidence from field data and microsatellite patterns. Proceedings of the Royal Society of London, B 264:127–132.

Fedigan, L. M. and Baxter, M. J. 1984. Sex differences and social organization in free-ranging spider monkeys (*Ateles geoffroyi*). Primates 25:279–294.

Fernandez-Duque, E. and Huntington, C. 2002. Disappearances of individuals from social groups have implications for understanding natal dispersal in monogamous owl monkeys (*Aotus azarai*). American Journal of Primatology 57:219–225.

Fluxus Technology Ltd. 2007. NETWORK (version 4.2.0.1). Suffolk, England: Distributed by fluxus-engineering.com.

Fontanillas, P., Petit, E. and Perrin, N. 2004. Estimating sex-specific dispersal rates with autosomal markers in hierarchically structured populations. Evolution 58:886–894.

Garber, P. A., Encarnación, F., Moya, L. and Pruetz, J. D. 1993. Demographic and reproductive patterns in moustached tamarin monkeys (*Saguinus mystax*): Implications for reconstructing Platyrrhine mating systems. American Journal of Primatology 29:235–254.

Gerloff, U., Hartung, B., Fruth, B., Hohmann, G. and Tautz, D. 1999. Intracommunity relationships, dispersal pattern, and paternity success in a wild living community of bonobos (*Pan paniscus*) determined from DNA analysis of faecal samples. Proceedings of the Royal Society of London, B 266:1189–1195.

Glander, K. E. 1992. Dispersal patterns in Costa Rican mantled howling monkeys. International Journal of Primatology 13:415–436.

Goldberg, T. L. and Wrangham, R. W. 1997. Genetic correlates of social behavior in wild chimpanzees: Evidence from mitochondrial DNA. Animal Behavior 54:559–570.

Goodall, J. 1986. The Chimpanzees of Gombe: Patterns of Behavior. Cambridge, MA: Belknap Press.

Goodman, S. J. 1997. RSTCALC: A collection of computer programs for calculating unbiased estimates of genetic differentiation and gene flow from microsatellite data and determining their significance. Molecular Ecology 6:881–886.

Goodnight, K. F. and Queller, D. C. 1999. Computer software for performing likelihood tests of pedigree relationship using genetic markers. Molecular Ecology 8:1231–1234.

Goudet, J. 2001. FSTAT, a program to estimate gene diversity and fixation indices (version 2.9.3). Institute of Ecology, Laboratory for Zoology, University of Laussane.

Goudet, J., Perrin, N. and Waser, P. 2002. Tests for sex-biased dispersal using bi-parentally inherited genetic markers. Molecular Ecology 11:1103–1114.

Gouzoules, H. and Gouzoules, S. 1987. Kinship. In B. B. Smuts, D. L. Cheney, R. M. Seyfarth, R. W. Wrangham and T. T. Struthsaker (eds.), Primate Societies (pp. 299–305). Chicago: University of Chicago Press.

Gouzoules, S. 1984. Primate mating systems, kin associations, and cooperative behavior: Evidence for kin recognition? Yearbook of Physical Anthropology 27:99–134.

Greenwood, P. J. 1980. Mating systems, philopatry and dispersal in birds and mammals. Animal Behavior 28:1140–1162.

Hamilton, W. D. 1964a. The genetical evolution of social behavior I. Journal of Theoretical Biology 7:1–16.

Hamilton, W. D. 1964b. The genetical evolution of social behavior II. Journal of Theoretical Biology 7:17–52.

Hammond, R. L., Lawson Handley, L. J., Winney, B. J., Bruford, M. W. and Perrin, N. 2006. Genetic evidence for female-biased dispersal and gene flow in a polygynous primate. Proceedings of the Royal Society of London, B: Biological Sciences 273:479–484.

Hapke, A., Zinner, D. and Zischler, H. 2001. Mitochondrial DNA variation in Eritrean hamadryas baboons (Papio hamadryas hamadryas): Life history influences population genetic structure. Behavioral Ecology and Sociobiology 50:483–492.

Hoelzer, G. A., Morales, J. C. and Melnick, D. J. 2004. Dispersal and the population genetics of primate species. In B. Chapais and C. M. Berman (eds.), Kinship and Behavior in Primates (pp. 109–131). Oxford: Oxford University Press.

Huck, M., Löttker, P., Böhle, U.-R. and Heymann, E. W. 2005. Paternity and kinship patterns in polyandrous moustached tamarins (Saguinus mystax). American Journal of Physical Anthropology 127:449–464.

Huelsenbeck, J. P. and Ronquist, F. 2001. MRBAYES: Bayesian inference of phylogeny. Bioinformatics 17:754–755.

Jack, K. and Fedigan, L. 2004a. Male dispersal patterns in white-faced capuchins, Cebus capucinus. Part 1: Patterns and causes of natal emigration. Animal Behavior 67:761–769.

Jack, K. and Fedigan, L. 2004b. Male dispersal patterns in white-faced capuchins, Cebus capucinus. Part 2: Patterns and causes of secondary dispersal. Animal Behavior 67:771–782.

Jack, K. M. 2003. Explaining variation in affiliative relationships among male white-faced capuchins (Cebus capucinus). Folia Primatologica 74:1–16.

Johnson, M. L. and Gaines, M. S. 1990. Evolution of dispersal: Theoretical models and empirical tests using birds and mammals. Annual Review of Ecology and Systematics 21:449–480.

Jones, C. B. 1980. Seasonal parturition, mortality, and dispersal in the mantled howler monkey, Alouatta palliata Gray. Brenesia 17:1–10.

Kalinowski, S. T., Wagner, A. P. and Taper, M. L. 2006. ML-Relate: A computer program for maximum likelihood estimation of relatedness and relationship. Molecular Ecology Notes 6:576–579.

Kinzey, W. G. 1981. The titi monkeys, Genus *Callicebus*. In A. F. Coimbra-Filho and R. A. Mittermeier (eds.), *Ecology and Behavior of Neotropical Primates* (pp. 241–296). Rio de Janeiro, Brasil: Academia Brasileira de Ciências.

Kumar, S., Tamura, K. and Nei, M. 2004. MEGA3: Integrated software for Molecular Evolutionary Genetics Analysis and sequence alignment. Briefings in Bioinformatics 5:150–163.

Langergraber, K. E., Mitani, J. C. and Vigilant, L. 2007. The limited impact of kinship on cooperation in wild chimpanzees. Proceedings of the National Academy of Sciences, USA 104: 7786–7790.

Laporte, V. and Charlesworth, B. 2002. Effective population size and population subdivision in demographically structured populations. Genetics 162:501–519.

Lawler, S. H., Sussman, R. W. and Taylor, L. L. 1995. Mitochondrial DNA of the Mauritian macaques (*Macaca fascicularis*): An example of the founder effect. American Journal of Physical Anthropology 96:133–141.

Lawson Handley, L. J. and Perrin, N. 2007. Advances in our understanding of mammalian sex-biased dispersal. Molecular Ecology 16:1559–1578.

Li, C. C., Weeks, D. E. and Chakravarti, A. 1993. Similarity of DNA fingerprints due to chance and relatedness. Human Heredity 43:45–52.

Lopez, J. V., Yuhki, N., Masuda, R., Modi, W. and O'Brien, S. J. 1994. Numt, a recent transfer and tandem amplification of mitochondrial DNA to the nuclear genome of the domestic cat. Journal of Molecular Evolution 39:174–191.

Lukas, D., Reynolds, V., Boesch, C. and Vigilant, L. 2005. To what extent does living in a group mean living with kin? Molecular Ecology 14:2181–2196.

Lynch, M. 1990. The similarity index and DNA fingerprinting. Molecular Biology and Evolution 7:478–484.

Lynch, M. and Ritland, K. 1999. Estimation of pairwise relatedness with molecular markers. Genetics 152:1753–1766.

Manel, S., Gaggioti, O. E. and Waples, R. S. 2005. Assignment methods: Matching biological questions with appropriate techniques. Trends in Ecology and Evolution 20:136–142.

Melnick, D. J. 1987. The genetic consequences of primate social organization: A review of macaques, baboons, and vervet monkeys. Genetica 73:117–135.

Melnick, D. J. 1988. Genetic structure of a primate species: Rhesus macaques and other cercopithecine monkeys. International Journal of Primatology 9:195–231.

Melnick, D. J. and Hoelzer, G. A. 1992. Differences in male and female macaque dispersal lead to contrasting distributions of nuclear and mitochondrial DNA variation. International Journal of Primatology 13:379–393.

Melnick, D. J. and Hoelzer, G. A. 1996. The population genetic consequences of macaque social organization and behavior. In J. E. Fa and D. G. Lindburg (eds.), *Evolution and Ecology of Macaque Societies* (pp. 413–443). Cambridge: Cambridge University Press.

Melnick, D. J. and Pearl, M. C. 1987. Cercopithecines in multi-male groups: Genetic diversity and population structure. In B. B. Smuts, D. L. Cheney, R. M. Seyfarth, R. W. Wrangham and T. T. Struhsaker (eds.), *Primate Societies* (pp. 121–134). Chicago: University of Chicago Press.

Michalakis, Y. and Excoffier, L. 1996. A generic estimation of population subdivision using distances between alleles with a special reference for microsatellite loci. Genetics 142: 1061–1064.

Milligan, B. 2003. Maximum-likelihood estimation of relatedness. Genetics 163:1153–1167.

Mitani, J. C., Merriwether, A. and Zhang, C. 2000. Male affiliation, cooperation and kinship in wild chimpanzees. Animal Behavior 59:885–893.

Mitchell, C. L., Boinski, S. and van Schaik, C. P. 1991. Competitive regimes and female bonding in two species of squirrel monkeys (*Saimiri oerstedi* and *S sciureus*). Behavioral Ecology and Sociobiology 28:55–60.

Morin, P. A., Chambers, K. E., Boesch, C. and Vigilant, L. 2001. Quantitative polymerase chain reaction analysis of DNA from noninvasive samples for accurate microsatellite genotyping of wild chimpanzees (*Pan troglodytes*). Molecular Ecology 10:1835–1844.

Morin, P. A., Moore, J. J., Chakraborty, R., Jin, L., Goodall, J. and Woodruff, D. S. 1994a. Kin selection, social structure, gene flow, and the evolution of chimpanzees. Science 265: 1193–1201.

Morin, P. A., Wallis, J., Moore, J. J. and Woodruff, D. S. 1994b. Paternity exclusion in a community of wild chimpanzees using hypervariable simple sequence repeats. Molecular Ecology 3: 469–478.

Muniz, L., Perry, S., Manson, J. H., Gilkenson, H., Gros-Louis, J. and Vigilant, L. 2006. Father–daughter inbreeding avoidance in a wild primate population. Current Biology 16:R156–R157.

Nei, M. 1987. Molecular Evolutionary Genetics. New York: Columbia University Press.

Nievergelt, C. M., Digby, L. J., Ramakrishnan, U. and Woodruff, D. S. 2000. Genetic analysis of group composition and breeding system in a wild common marmoset (*Callithrix jacchus*) population. International Journal of Primatology 21:1–20.

Nievergelt, C. M., Mutschler, T., Feistner, A. T. C. and Woodruff, D. S. 2002. Social system of the Alaotran gentle lemur (*Hapalemur griseus alaotrensis*): Genetic characterization of group composition and mating system. American Journal of Primatology 57:157–176.

Nishimura, A. 1990. A sociological and behavioral study of woolly monkeys, *Lagothrix lagotricha*, in the Upper Amazon. The Science and Engineering Review of Doshisha University 31:87–121.

Nishimura, A. 2003. Reproductive parameters of wild female *Lagothrix lagotricha*. International Journal of Primatology 24:707–722.

Nsubuga, A. M., Robbins, M. M., Roeder, A. D., Morin, P. A., Boesch, C. and Vigilant, L. 2004. Factors affecting the amount of genomic DNA extracted from ape feces and the identification of an improved sample storage method. Molecular Ecology 13:2089–2094.

O'Brien, T. G. 1993. Allogrooming behavior among adult female wedge-capped capuchin monkeys. Animal Behavior 46:499–510.

O'Brien, T. G. and Robinson, J. G. 1991. Allomaternal care by female wedge-capped capuchin monkeys: Effects of age, rank and relatedness. Behavior 119:30–50.

Paetkau, D., Calvert, W., Stirling, I. and Strobeck, C. 1995. Microsatellite analysis of population structure in Canadian polar bears. Molecular Ecology 4:347–354.

Paul, A., Kuester, J., Timme, A. and Arnemann, J. 1993. The association between rank, mating effort, and reproductive success in male Barbary macaques (*Macaca sylvanus*). Primates 34:491–502.

Peakall, R. and Smouse, P. E. 2006. GENALEX 6: Genetic analysis in Excel. Population genetic software for teaching and research. Molecular Ecology Notes 6:288–295.

Perry, S. 1998. Male-male social relations in wild white-faced capuchins, *Cebus capucinus*. Behavior 135:139–142.

Piry, S., Alapetite, A., Cornuet, J.-M., Paetkau, D., Baudouin, L. and Estoup, A. 2004. GeneClass2: A software for genetic assignment and first-generation migrant detection. Journal of Heredity 95:536–539.

Pope, T. R. 1989. The Influence of Mating System and Dispersal Patterns on the Genetic Structure of Red Howler Monkey Populations. Unpublished Ph.D. thesis, Gainesville.

Pope, T. R. 1990. The reproductive consequences of male cooperation in the red howler monkey: Paternity exclusion in multi-male and single-male troops using genetic markers. Behavioral Ecology and Sociobiology 27:439–446.

Pope, T. R. 1992. The influence of dispersal patterns and mating systems on genetic differentiation within and between populations of the red howler monkey (*Alouatta seniculis*). Evolution 46:1112–1128.

Pope, T. R. 1996. Socioecology, population fragmentation, and patterns of genetic loss in endangered primates. In J. C. Avise and J. L. Hamrick (eds.), *Conservation Genetics: Case Studies from Nature* (pp. 119–159). New York: Chapman & Hall.

Pope, T. R. 1998. Effects of demographic change on group kin structure and gene dynamics of populations of red howling monkeys. Journal of Mammalogy 79:692–712.

Pope, T. R. 2000. Reproductive success increases with degree of kinship in cooperative coalitions of female red howler monkeys (*Alouatta seniculus*). Behavioral Ecology and Sociobiology 48:253–267.

Pritchard, J. K., Stephens, M. and Donnelly, P. 2000. Inference of population structure using multilocus genotype data. Genetics 155:945–959.

Prugnolle, F. and de Meeus, T. 2002. Inferring sex-biased dispersal from population genetic tools: A review. Heredity 88:161–165.

Pusey, A. E. and Packer, C. 1987. Dispersal and philopatry. In B. B. Smuts, D. L. Cheney, R. M. Seyfarth, R. W. Wrangham and T. T. Struhsaker (eds.), *Primate Societies* (pp. 250–266). Chicago: University of Chicago Press.

Queller, D. C. and Goodnight, K. F. 1989. Estimating relatedness using genetic markers. Evolution 43:258–275.

Rannala, B. and Mountain, J. L. 1997. Detecting immigration by using multilocus genotypes. Proceedings of the National Academy of Sciences, USA 94:9197–9201.

Ritland, K. 2005. Multilocus estimation of pairwise relatedness with dominant markers. Molecular Ecology 14:3157–3165.

Robinson, J. G. 1988a. Demography and group structure in wedge-capped capuchin monkeys, *Cebus olivaceus*. Behavior 104:202–232.

Robinson, J. G. 1988b. Group size in wedge-capped capuchin monkeys *Cebus olivaceus* and the reproductive success of males and females. Behavioral Ecology and Sociobiology 23: 187–197.

Roeder, A. D., Archer, F. I., Poinar, H. N. and Morin, P. A. 2004. A novel method for collection and preservation of feces for genetic studies. Molecular Ecology Notes 4:761–764.

Rudran, R. 1979. The demography and social mobility of a red howler (*Alouatta seniculus*) population in Venezuela. In J. F. Eisenberg (ed), *Vertebrate Ecology of the Northern Neotropics* (pp. 107–126). Washington, DC: Smithsonian Institution Press.

Rudran, R. and Fernandez-Duque, E. 2003. Demographic changes over thirty years in a red howler population in Venezuela. International Journal of Primatology 24:925–947.

Rumiz, D. I. 1990. *Alouatta caraya*: Population density and demography in northern Argentina. American Journal of Primatology 21:279–294.

Savage, A., Giraldo, L. H., Soto, L. H. and Snowdon, C. T. 1996. Demography, group composition, and dispersal in wild cotton-top tamarin (*Saguinus oedipus*) groups. American Journal of Primatology 38:85–100.

Shimooka, Y., Campbell, C., Di Fiore, A., Felton, A. M., Izawa, K., Link, A., Nishimura, A., Ramos-Fernandez, G., and Wallace, R., (in press). Demography and group composition of *Ateles*. In: C. J. Campbell (ed.), Spider Monkeys: Behavior, Ecology and Evolution of the Genus *Ateles*. Cambridge University Press.

Silk, J. B. 2001. Ties that bond: The role of kinship in primate societies. In L. Stone (ed.), *New Directions in Anthropological Kinship* (pp. 71–92). Boulder, CO: Rowman and Littlefield.

Silk, J. B. 2002. Kin selection in primate groups. International Journal of Primatology 23:849–875.

Silk, J. B. 2006. Practicing Hamilton's rule: Kin selection in primate groups. In P. M. Kappeler and C. P. van Schaik (eds.), Cooperation in Primates and Humans: Mechanisms and Evolutions, (pp. 25–46). New York: Springer.

Smith, K., Alberts, S. C. and Altmann, J. 2003. Wild female baboons bias their social behavior toward paternal half-sisters. Proceedings of the Royal Society of London, B 270:503–510.

Stevenson, P. R. 2002. Frugivory and Seed Dispersal by Woolly Monkeys at Tinigua National Park, Colombia. Unpublished Ph.D. thesis,Stony Brook, NY.

Stevenson, P. R., Quiñones, M. J. and Ahumada, J. A. 1994. Ecological strategies of woolly monkeys (*Lagothrix lagotricha*) at Tinigua National Park, Colombia. American Journal of Primatology 32:123–140.

Strier, K. B. 1987. Demographic patterns in one group of muriquis. Primate Conservation 8:73–74.

Strier, K. B. 1990. New World primates, new frontiers: Insights from the woolly spider monkey, or muriqui (*Brachyteles arachnoides*). International Journal of Primatology 11:7–19.

Strier, K. B. 1991. Demography and conservation of an endangered primate, *Brachyteles arachnoides*. Conservation Biology 5:214–218.

Strier, K. B. 1994a. Brotherhoods among atelins: Kinship, affiliation, and competition. Behavior 130:151–167.

Strier, K. B. 1994b. The myth of the typical primate. Yearbook of Physical Anthropology 37: 233–271.

Symington, M. M. 1987. Sex ratio and maternal rank in wild spider monkeys: When daughters disperse. Behavioral Ecology and Sociobiology 20:421–425.

Symington, M. M. 1988. Demography, ranging patterns, and activity budgets of black spider monkeys (*Ateles paniscus chamek*) in the Manu National Park, Peru. American Journal of Primatology 15:45–67.

Taberlet, P., Griffin, S., Goossens, B., Questiau, S., Manceau, V., Escaravage, N., Waits, L. and Bouvet, J. 1996. Reliable genotype of samples with very low DNA quantities using PCR. Nucleic Acids Research 24:3189–3194.

Valderrama Aramayo, M. X. C. 2002. Reproductive Success and Genetic Population Structure in Wedge-Capped Capuchin Monkeys. Unpublished Ph.D. thesis,New York.

van de Casteele, T., Galbusera, P. and Matthysen, E. 2001. A comparison of microsatellite-based pairwise relatedness estimators. Molecular Ecology 10:1539–1549.

van Schaik, C. P. 1989. The ecology of social relationships amongst female primates. In V. Standen and R. Foley (eds.), *Comparative Socioecology: The Behavioral Ecology of Humans and Other Mammals* (pp. 195–218). Oxford: Blackwell Scientific Publications.

Vigilant, L., Hofreiter, M., Siedel, H. and Boesch, C. 2001. Paternity and relatedness in wild chimpanzee communities. Proceedings of the National Academy of Sciences, USA 98: 12890–12895.

Vitalis, R. 2002. Sex-specific genetic differentiation and coalescence times: Estimating sex-biased dispersal rates. Molecular Ecology 11:125–138.

Wallman, J., Hoelzer, G. A. and Melnick, D. J. 1996. The effects of social structure, geographical structure, and population size on the evolution of mitochondrial DNA: I. A simulation model. Computer Applications in the Biosciences 12:481–489.

Wang, J. 2002. An estimator for pairwise relatedness using genetic markers. Genetics 160: 1203–1215.

Wang, J. 2004. Estimating pairwise relatedness from dominant genetic markers. Molecular Ecology 13:3169–3178.

Waser, P. M. and Jones, W. T. 1983. Natal philopatry among solitary mammals. Quarterly Review of Biology 58:355–390.

Watts, D. and Mitani, J. 2001. Boundary patrols and intergroup encounters among wild chimpanzees. Behavior 138:299–327.

Weir, B. S., Anderson, A. D. and Helper, A. B. 2006. Genetic relatedness analysis: Modern data and new challenges. Nature Reviews: Genetics 7:771–780.

Weir, B. S. and Cockerham, C. C. 1984. Estimating F-statistics for the analysis of population structure. Evolution 38:1358–1370.

Widdig, A., Bercovitch, F. B., Streich, W. J., Sauermann, U., Nürnberg, P. and Krawczak, M. 2004. A longitudinal analysis of reproductive skew in male rhesus macaques. Proceedings of the Royal Society of London, Series B 271:819–826.

Widdig, A., Nuernberg, P., Krawczak, M., Streich, W. J. and Berkovitch, F. 2002. Affiliation and aggression among adult female rhesus macaques: A genetic analysis of paternal cohorts. Behavior 139:371–391.

Widdig, A., Nürnberg, P., Krawczak, M., Streich, W. J. and Bercovitch, F. B. 2001. Paternal relatedness and age proximity regulate social relationships among adult female rhesus macaques. Proceedings of the National Academy of Sciences, USA 98:13769–13773.

Wilson, E. O. 1975. Sociobiology: The New Synthesis. Cambridge, MA: The Belknap Press.

Wilson, J. F. and Erlandsson, R. 1998. Sexing of human and other primate DNA. Biological Chemistry 379:1287–1288.

Wimmer, B. and Kappeler, P. M. 2002. The effects of sexual selection and life history on the genetic structure of redfronted lemur, *Eulemur fulvus rufus*, groups. Animal Behavior 64:557–568.

Wrangham, R. W. 1980. An ecological model of female-bonded primate groups. Behavior 75: 262–300.

Wright, S. 1931. Evolution in Mendelian populations. Genetics 16:97–159.

Wright, S. 1943. Isolation by distance. Genetics 23:114–138.

Wright, S. 1965. The interpretation of population structure by F-statistics with special regard to systems of mating. Evolution 19:395–420.

Zhang, D.-X. and Hewitt, G. M. 1996. Nuclear integrations: Challenges for mitochondrial DNA markers. Trends in Ecology and Evolution 11:247–251.

Chapter 10
Predation Risk and Antipredator Strategies

Stephen F. Ferrari

10.1 Introduction

For most animal species, predation exerts a fundamental selective pressure on morphological traits and behavior patterns and, ultimately, the ecological characteristics of a taxon (Cheney and Wrangham 1987; Stanford 2002). The platyrrhines are no exception here, especially as the comparatively small size of most genera makes them relatively vulnerable to a wide range of prospective predators. This is reflected in specific patterns of antipredator or "predator sensitive" (Miller and Treves 2007) behavior, which vary systematically among genera in complexity and intensity, and are quite clearly the result of specific selective pressures.

One attribute common to all platyrrhines is a high degree of specialization for an arboreal way of life, a primitive trait within the primates which provides a measure of protection from some types of predator, but also imposes body size constraints. The forest canopy generally offers good cover and is relatively inaccessible to the more terrestrial predators, especially those of larger size. Large-bodied cats such as jaguar (*Panthera onca*) and puma (*Puma concolor*) may be able to climb trees, but are too heavy to reach the outermost branches, and will normally be unable to move between treetops. Obviously, this restricted mobility places substantial limitations on their ability to prey on platyrrhines in the arboreal milieu.

The mobile substrate of the forest canopy also facilitates detection of the approach of other animals, as well as hindering pursuit. As a three-dimensional environment, the canopy offers a much greater variety of escape routes and hiding places in comparison with open ground. However, these advantages may be offset, at least partially, by the risks of injury, or even death, from falls.

Despite the fundamental importance of predation pressure as a selective mechanism in platyrrhine evolution, the analysis of underlying patterns is limited by the scarcity of reports on predation events. For a majority of platyrrhine species, in fact, there are no published reports whatsoever, and even where data are available, they

S.F. Ferrari (✉)
Department of Biology, Federal University of Sergipe, Sergipe, Brazil
e-mail: ferrari@pq.cnpq.br, ferrari@pitheciineactiongroup.org

P.A. Garber et al. (eds.), *South American Primates,* Developments in Primatology:
Progress and Prospects, DOI 10.1007/978-0-387-78705-3_10,

invariably refer to a small handful of records for a given species, or even genus. This considerably limits the potential for a more reliable analysis of taxon- or habitat-specific patterns, for example. Despite these sampling problems, a careful review of the available data provides a number of insights into predation risk and patterns in the platyrrhines, and their influence on the evolution of antipredator strategies.

10.2 Predators

10.2.1 Predator Species

Platyrrhines are vulnerable to three main types of predator – raptors (order Falconiformes), mammalian carnivores (order Carnivora), and snakes (order Serpentes) – and practically all recorded predation events have involved species representing one of these groups. One exception is Boinski's (1987) recorded attack of toucans (*Ramphastos swainsoni*: order Piciformes) on infant squirrel monkeys. A second exception is the predation of titi monkeys (*Callicebus* spp.) by another platyrrhine, the capuchin, *Cebus* spp. (Lawrence 2003; Sampaio and Ferrari 2005). Capuchins hunt small mammals such as squirrels systematically at many sites (e.g., Galetti 1990; Rose et al. 2003), but these two records appear to be the only observations of successful attacks on monkeys.

A second primate – *Homo sapiens* – is responsible for the predation of large numbers of platyrrhines in some regions (*see* Peres, this volume), although this is a relatively recent phenomenon in the context of their evolutionary history, and is not directly relevant to most of the questions discussed here. Rose et al. (2003) also report on the attempted capture of capuchins by spectacled caiman, *Caiman crocodilus*, although this presumably represents a very exceptional situation.

A characteristic shared by almost all other known – or potential – predators of platyrrhine monkeys is their solitary foraging behavior. Among the terrestrial Neotropical carnivores, only coatis (*Nasua* spp.) form large social groups, but to date, there have been no reports of attacks on platyrrhines. In fact, their principal predator-prey relationship with the platyrrhines is as prey, given that capuchins are known to prey regularly on *Nasua* nestlings at a number of sites (Rose et al. 2003).

On the other hand, tayras (*Eira barbara*) will sometimes forage in pairs or small family groups and, as diurnal, arboreal frugivores, they may come into relatively frequent contact with platyrrhines in the wild. In fact, a large proportion of the published accounts of encounters between platyrrhines and carnivores involve this species (Moynihan 1970; Galef et al. 1976; Defler 1980; Buchanan-Smith 1990; Phillips 1995; Stafford and Ferreira 1995; Asensio and Gomes-Marin 2002; Camargo and Ferrari 2007), although their attacks are almost invariably unsuccessful.

By contrast, all platyrrhines are highly social, and cooperative behavior is a fundamental characteristic of the antipredator strategies of most species. In fact, platyrrhines will often cooperate to confront predators many times their own size with remarkable efficiency. However, some genera are more discrete, and tend to avoid predators, rather than confront them.

10.2.2 Human Impact

In addition to the direct influence of hunting pressure on primate populations, the impact of human colonization on Neotropical forest ecosystems also affects predator populations, either directly, though hunting, or indirectly, through deforestation and habitat fragmentation. *A priori*, there is a general tendency for the larger-bodied predators to become (even more) rare or extinct, whereas smaller-bodied species may be less affected, or even benefited, under some conditions, which may include the reduction of competition from larger-bodied forms.

Over the short term, habitat fragmentation generally results in a "refuge" effect, in which forest remnants become crowded with animals – which will likely be weak and disoriented – from the surrounding area, supporting relatively high temporary densities of predators. Peetz et al. (1992) recorded a situation of this type, in which a group of red howlers (*Alouatta seniculus*) displaced by the flooding of the Guri reservoir was apparently attacked repeatedly by jaguar.

Over the long term, the persistence of alpha predators, such as jaguar and harpy eagles in the fragmented landscape may be determined by a complex of factors, including the size and configuration of remaining habitat fragments, and the characteristics of the surrounding matrix. These species are absent from much of the Brazilian Atlantic Forest, for example, where fragments are typically of 100 hectares or less in size, and often widely spaced.

This reduction in the abundance of potential predators and thus, predation pressure, may contribute to the relatively high population densities often recorded for some species – in particular howlers (e.g., Mendes 1989; Ferrari et al. 2003; Ludwig et al. 2007) – in fragments. On the other hand, these populations may become relatively vulnerable, not only because of their density and isolation, but possibly also through increasing naïveté to the risks of predation, in particular from terrestrial carnivores. This is reinforced by the relative abundance of reports of predation by carnivores in the context of habitat fragmentation (*see* below). Anthropogenic habitat disturbance may thus alter predation patterns significantly in a number of different ways.

10.3 Platyrrhine Morphology

10.3.1 Body Size

A fundamental line of defense against predation for any animal is its body size and, in general, the larger the animal, the smaller the number of potential predator species it has. Platyrrhines vary in adult body weight from just over 0.1 kg in the pygmy marmoset, *Cebuella pygmaea*, to more than 10 kg in the largest atelids, the muriqui (*Brachyteles*) and the woolly monkey (*Lagothrix*). This range of body size has a fundamental influence on inter-taxon patterns of predation and antipredator behavior. For the purposes of the present discussion, the platyrrhines can be divided into four body size classes (Table 10.1).

Table 10.1 Body size classes for the platyrrhine genera (genera according to Rylands et al. 2000; van Roosmalen and van Roosmalen 2003)

Miniature (< 0.7 kg)	Small (0.7–2 kg)	Medium (2–5 kg)	Large (>5 kg)
Callibella	*Aotus*	*Cacajao*	*Alouatta*
Callimico	*Callicebus*	*Cebus*	*Ateles*
Callithrix	*Saimiri*	*Chiropotes*	*Brachyteles*
Cebuella		*Pithecia*	*Lagothrix*
Leontopithecus			*Oreonax*
Mico			
Saguinus			

The advantages of large size may be offset by other considerations for arboreal animals such as platyrrhines, given the physical limitations of this lifestyle. In the miniature callitrichids, the lack of size may be more than compensated for by an increase in agility and social cohesion (*see* below). By contrast, body size almost certainly imposes certain limitations on the potential for predation by raptors, which normally attack in flight. Even the largest Neotropical raptors, female harpy eagles (*Harpia harpyja*) are unable to fly carrying prey of more than 5–6 kg in weight, and most other species are restricted to quarry of smaller size. Obviously, the large platyrrhines (Table 10.1) enjoy a degree of protection – a smaller number of potential raptor species – on the basis of body size alone.

By contrast, immature callitrichids are potentially vulnerable to practically all raptor species, and many other birds, such as great kiskadees, *Pitangus sulphuratus*, a common species throughout the Neotropics, or toucans, *Ramphastos* spp. (*see* Boinski 1987). In general, however, larger bodied raptors tend to prey on larger-bodied platyrrhines (Table 10.2). Most of the records of predation on smaller-bodied platyrrhines involve smaller species of raptor, i.e., falcons and hawks, rather than eagles. It may be, in fact, that the relatively large size of the feet and talons (up to 10 cm in length in adult females) of eagles such as *Harpia* proscribes the efficient capture of animals as small as callitrichids. In addition, these monkeys typically occupy relatively dense vegetation, which may impose further limitations on the potential for capture by the larger-bodied raptors. One notable exception is Over-sluijs Vasquez and Heymann's (2001) observation of the predation of infant tamarins by *Morphnus guianensis*, the second largest Neotropical raptor (*see* also Ford and Boinski 2007).

Obviously, body size also imposes specific limits on the potential of terrestrial predators to attack platyrrhines of different sizes. Once again, however, whereas small-bodied predators may be physically unable to confront large platyrrhines, larger-bodied predators may be far less adept at the capture of the smallest monkeys, given not only their reduced size and increased agility, but also the complexity of their preferred forest habitats. In other words, whereas a marmoset might be easy prey for a jaguar on open ground, it may be almost impossible to capture within the confines of the arboreal matrix.

This is reflected in the data on predation by mammals (Tables 10.3 and 10.4), where, once again, there is a tendency for the larger platyrrhines to be attacked

Table 10.2 Records of predation on platyrrhines by raptors (Falconiformes) and other birds (Piciformes). Here and in Tables 10.3–10.5, the platyrrhine species are listed in ascending order of body size (see Table 10.1)

Platyrrhine species	Age (sex) class	Raptor	Habitat	Type of record	Reference
Callithrix jacchus	Subadult (male)	Collared forest falcon (*Micrastur semitorquatus*)	Forest fragment	Direct observation	Alonso and Langguth (1989)
Callithrix penicillata	Adult (male and female), juvenile	Ornate hawk eagle (*Spizaetus ornatus*)	Savanna	Nest remains	Greco et al. (2004)
Saguinus fuscicollis	Juvenile	Bicolored hawk (*Accipiter bicolor*)	Continuous forest	Direct observation	Terborgh (1983)
	Adult	*S. ornatus*	Continuous forest	Direct observation	Robinson (1994)
	Infant	Guyanan crested eagle (*Morphnus guianensis*)	Continuous forest	Direct observation	Oversluijs Vasquez and Heymann (2001)
Saguinus imperator	Adult	*S. ornatus*	Continuous forest	Direct observation	Terborgh (1983)
Saguinus midas	Adult	Harpy eagle (*Harpia harpyja*)	Continuous forest	Nest remains	Ford and Boinski (2007)
Saguinus mystax	Infant	*M. guianensis*	Continuous forest	Direct observation	Oversluijs Vasquez and Heymann (2001)
Saguinus nigricollis	Unspecified	Barred forest falcon (*Micrastur ruficollis*)	Continuous forest	Direct observation	Izawa (1978)
Saimiri oerstedi	Infant	Chesnut-mandibled toucan (*Ramphastos swainsoni*)	Continuous forest	Direct observation	Boinski (1987)
Saimiri sciureus	Unspecified*	*S. ornatus*	Continuous forest	Direct observation	Boinski (1987)
	Adult	*M. semitorquatus*	Continuous forest	Perch remains	Boinski (1987)
	Adult	*S. ornatus*	Continuous forest	Direct observation	Robinson (1994)
		H. harpyja	Continuous forest	Nest remains	Ford and Boinski (2007)
Pithecia pithecia	Unspecified	*H. harpyja*	Continuous forest	Nest remains	Rettig (1978)

Table 10.2 (continued)

Platyrrhine species	Age (sex) class	Raptor	Habitat	Type of record	Reference
	Young-adult	*H. harpyja*	Continuous forest	Nest remains	Ford and Boinski (2007)
Cebus capucinus	Adult (female)	*H. harpyja*	Island	Direct observation	Touchton, Hsu and Palleroni (2002)
Cebus spp.	Unspecified	*H. harpyja*	Continuous forest	Nest remains	Fowler and Cope (1964)
	Unspecified	*H. harpyja*	Continuous forest	Direct observation, nest remains	Rettig (1978)
	Unspecified	*H. harpyja*	Continuous forest	Nest remains	Izor (1985)
	Young-adult	*H. harpyja*	Continuous forest	Nest remains	Ford and Boinski (2007)
Chiropotes chiropotes	Unspecified	*H. harpyja*	Continuous forest	Nest remains	Rettig (1978)
Chiropotes utahickae	Adult (male)	*H. harpyja*	Continuous forest	Direct observation	Martins *et al.* (2005)
Alouatta belzebul	Adult (female)	*H. harpyja*	Continuous forest	Direct observation	Pers. Obs.
Alouatta palliata	Juvenile (male and female), adult (female)	*H. harpyja*	Island	Direct observation	Touchton, Hsu and Palleroni (2002)
Alouatta seniculus	Adult	*H. harpyja*	Continuous forest	Direct observation	Peres (1990)
	Adult	*H. harpyja*	Continuous forest	Direct observation	Sherman (1991)
	Unspecified	*H. harpyja*	Continuous forest	Nest remains	Rettig (1978)
	Unspecified	*H. harpyja*	Continuous forest	Nest remains	Izor (1985)
	Young-adult	*H. harpyja*	Continuous forest	Nest remains	Ford and Boinski (2007)
Ateles paniscus	Juvenile	*M. guianensis*	Continuous forest	Direct observation	Julliot (1994)
	Juvenile	*H. harpyja*	Continuous forest	Nest remains	Ford and Boinski (2007)
Lagothrix lagothricha	Unspecified	Black-and-chesnut eagle (*Oroaetus isidori*)	Unspecified	Direct observation	C. Lehman (1959) cited by Defler (2004)

**"a small monkey...probably a squirrel monkey" (Boinski 1987, p. 398).

by larger-bodied predators. By contrast, the few records of predation by snakes (Table 10.5) involve relatively small-bodied platyrrhines and mostly large-bodied boid snakes. While a sample this small should be treated with caution, it does indicate a distinct pattern of predation with regard to body size, in comparison with raptors and carnivores.

Finally, body size has an opposite effect in the case of human predation, given that hunters invariably prefer larger-bodied species, and will rarely, if ever, target species such as callitrichids. This reinforces the overall effects of modern human colonization of the forest, based on extensive deforestation and habitat fragmentation. Peres (1997), for example, reported that populations of woolly (*Lagothrix* spp.) and spider (*Ateles* spp.) monkeys faced decimation from hunting pressure long before the advance of significant levels of deforestation on agricultural frontiers in the western Amazon. These two genera – together with *Brachyteles* and *Oreonax* – are especially vulnerable to such pressure because of their relatively slow growth and maturation rates. Despite suffering similar pressures, howlers (*Alouatta* spp.) tend to be more tolerant, and are often relatively abundant in anthropogenic landscapes.

10.3.2 Pelage Coloration

Camouflage is a second basic antipredator strategy, which may also compensate partially for body size limitations, although it is important to remember that most snakes – especially boids – do not depend on visual cues to locate their prey. While some platyrrhines present relatively cryptic pelage (Rowe 1996), others are surprisingly colorful, especially considering their small size. Many of the marmosets of the genus *Mico*, and tamarins (*Leontopithecus* and *Saguinus*) present somewhat vivid coloration patterns which appear to contradict the supposition that predation pressure has molded pelage characteristics.

The most cryptic coloration for mammalian pelage appears to be the agouti or multibanded pattern (Hershkovitz 1968), in which the alternate light-dark banding of the hairs results in a fuzzy visual effect, which may be especially effective in complex environments such as that of the tropical forest. Among the platyrrhines, the best example of agouti coloration is the pygmy marmoset, *Cebuella pygmaea*, which presents this pattern uniformly over the whole of the body.

Despite the ornate facial pelage of some species, the Atlantic Forest marmosets (*Callithrix* spp.) are also relatively cryptic in coloration, with the general agouti of the dorsum merging into an annular pattern on the rump and tail. In the common (*Callithrix jacchus*) and pencil-ear (*Callithrix penicillata*) marmosets, the overall coloration is predominantly grayish, and the animals are relatively well camouflaged when clinging to vertical trunks, a typical posture in these species, in particular when feeding on gum. Coincidentally – or otherwise – these two species are found throughout an extensive area of the dry Caatinga and Cerrado woodland biomes of eastern and central Brazil, where they often occupy relatively open habitats, in contrast with the denser vegetation of the Atlantic Forest, to which the remaining species are endemic.

Table 10.3 Observed predation of platyrrhines by mammals (Mammalia)

Platyrrhine species	Age (sex) class	Predator	Habitat	Type of record	Reference
Callithrix geoffroyi	Adult	Margay (*Leopardus wiedii*)	Fragmented forest	Vestigial*	Passamani et al. (1997)
	Adult	Oncilla (*Leopardus tigrina*)	Fragmented forest	Vestigial*	Passamani et al. (1997)
Saguinus geoffroyi	Unspecified	Tayra (*Eira barbara*)	Unspecified	Prey carried in predator's mouth	Moynihan (1970)
Callicebus dubius†	Female	Capuchin (*Cebus* sp.)	Unspecified	Direct observation	Lawrence (2003)
Callicebus moloch	Infant	Tufted capuchin (*Cebus apella*)	Reservoir island	Direct observation	Sampaio and Ferrari (2005)
Alouatta palliata	Subadult male	Jaguar (*Panthera onca*)	Zoological garden	Direct observation	Cuarón (1997)
Alouatta seniculus	Adult (n = 5)	*P. onca*‡	Reservoir island	Vestigial	Peetz et al. (1992)
Ateles belzebuth	Adult male	*P. onca*	Continuous forest	Direct observation	Matsuda and Izawa (2008)

*Inferred from observation of predators following groups from which reintroduced animals disappeared and their remains were located by radio telemetry;
†Identified as *Callicebus brunneus* by the author, but probably *Callicebus dubius*, according to van Roosmalen, van Roosmalen and Mittermeier (2002);
‡Inferred from "Strong circumstantial evidence..." (Peetz et al. 1992, p. 226).

Table 10.4 Records of predation on platyrrhines by felids based on fecal analyses

Platyrrhine species	Predator	Habitat	Reference
Callithrix jacchus	Jaguarundi (Herpailurus yagouaroundi)	Scrub forest	Olmos (1993)
Aotus trivirgatus	Jaguar (Panthera onca)	Pantanal	Schaller (1983)
Callicebus personatus	Ocelot (Leopardus pardalis)	Forest fragment	Bianchi (2001)
Callicebus torquatus	Margay (Leopardus wiedii)	Continuous forest	Defler (2004)
Cebus apella	L. pardalis	Forest fragments	Bianchi (2001)
Cebus capucinus	Puma (Puma concolor)	Continuous forest	Chinchilla (1997)
Alouatta caraya	Panthera onca	Pantanal	Schaller (1983)
	Puma concolor	Fluvial island	Ludwig et al. (2007)
Alouatta guariba	L. pardalis	Forest fragments	Bianchi (2001)
	L. pardalis	Forest fragment	Miranda et al. (2005)
Alouatta palliata	Panthera onca	Continuous forest	Chinchilla (1997)
	Puma concolor	Continuous forest	Chinchilla (1997)
Alouatta pigra	Panthera onca	Large fragment	Novak et al. (2005)
	Puma concolor	Large fragment	Novak et al. (2005)
Ateles chamek	Panthera onca	Continuous forest	Emmons (1987)
Ateles geoffroyi	Panthera onca	Large fragment	Novak et al. (2005)
	Puma concolor	Large fragment	Novak et al. (2005)
	Puma concolor	Continuous forest	Chinchilla (1997)
Brachyteles arachnoides	Panthera onca	Large fragment	Olmos (1994)
Brachyteles hypoxanthos	L. pardalis	Forest fragment	Bianchi (2001)

By contrast, many of the Amazonian marmosets of the genus *Mico* are among the most extravagantly colored of all platyrrhines. For example, the pelage of both the white (*Mico leucippe*) and the gold-and-white (*Mico chrysoleuca*) marmoset is virtually unpigmented, while the face is bright pink, making these animals relatively conspicuous against the backdrop of the forest canopy. This pattern appears to contradict significantly the conclusion that predation pressure has selected for cryptic coloration.

However, the brown marmoset (*Mico melanurus*), which is found as far south as the Chaco scrublands of Paraguay, is predominantly dark brown in color – principally on the dorsal surface – with a deeply pigmented face and ears. This species may in fact be parapatric with *C. penicillata* in the Cerrado of the upper Araguaia river. As in *Callithrix*, then, the species with apparently the most cryptic coloration is distributed in the most open habitats, where they are presumably more visible to prospective predators, in particular raptors.

All other callitrichids not only inhabit rainforest ecosystems, but most species also exhibit a strong preference for secondary and edge habitats, where the vegetation is relatively dense. *Saguinus* presents variation in coloration patterns similar to that of *Mico*, perhaps exemplified most clearly by the saddle-back tamarin *Saguinus fuscicollis* (Hershkovitz 1977: Plate III), which includes subspecies ranging from all white (*Saguinus fuscicollis melanoleucus*) to a general agouti pattern (*Saguinus fuscicollis fuscicollis*). The fact that these two subspecies can be found on opposite

Table 10.5 Records of predation on platyrrhines by snakes (Serpentes)

Platyrrhine species	Age (sex) class	Predator	Habitat	Type of record	Reference
Callithrix aurita	Infant	Pit viper (Bothrops jararaca)	Fragment	Direct observation	Corrêa and Coutinho (1997)
Saguinus mystax	Adult (female)	Anaconda (Eunectes murinus)	Fluvial island	Direct observation	Heymann (1987)
Leontopithecus rosalia	Adult (female)	Boa constrictor (Boa constrictor)	Fragment	Radio-collared animal in predator's stomach	Kierulff et al. (2002)*
Callicebus discolor	Adult	B. constrictor	Flooded forest	Direct observation	D.F. Cisneros-Heredia et al. (2005)
Cebus capucinus	Juvenile (male)	B. constrictor	Dry forest	Direct observation	Chapman (1986)
Chiropotes utahickae	Adult (female)	B. constrictor	Fragment (island)	Direct observation	Ferrari et al. (2004)

*M.C. Kierulff (pers. comm.) reports at least two other predation events involving Boa constrictor.

banks of the Juruá river suggests that the difference in their pelage characteristics has been determined by factors other than predation pressure.

A number of other callitrichids are found within the western Amazonian range of *S. fuscicollis*. Curiously, some of these species – in particular the emperor tamarin, *Saguinus imperator* – exhibit some of the most extravagant pelage of any platyrrhine, whereas others, such as the pygmy marmoset, and Goeldi's monkey (*Callimico goeldii*), both of which present a single, relatively drab coloration pattern throughout their extensive geographic ranges, which is similar in size to that of *S. fuscicollis*.

The evolution of highly cryptic coloration in *Cebuella* can easily be accounted for by its reduced body size and implicit increase in predation risk. The use of relatively miniscule home ranges – often a single tree – may also be a factor. By contrast, *Callimico* is as large, or larger than the sympatric *Saguinus* species, although differences in habitat preferences – *Callimico* appears to be a bamboo forest specialist – may also be relevant in this case.

The small-bodied (Table 10.1) platyrrhines can be divided into two groups here: the relatively cryptic titis (*Callicebus*) and owl monkeys (*Aotus*), and the brightly colored squirrel monkeys (*Saimiri*). Most titi species are uniformly brown or grayish, although some species present a two-tone pattern, in which the dorsal and ventral surfaces present different shades of the same color, or even distinct colors. Species of the *Callicebus torquatus* group have distinct white or yellow markings on the neck and hands, which are not normally visible when the animal is in a typical quadrupedal posture. Similarly, in the two-tone species, the relatively less cryptic color is normally located on the ventrum, where it is less visible to predators. In many species of the *Callicebus moloch* and *Callicebus cupreus* species groups, for example, the grayish tones of the dorsum contrast with the reddish or yellowish coloration of the neck and ventrum. Presumably, at least some of these markings function in intraspecific communication, although their distribution on the body appears to be consistent with the influence of predation pressure.

Owl monkeys are uniformly grayish in color, except for the members of the "red-neck" group (cf. Hershkovitz 1983), in which the ventral surface of the neck, chest, and abdomen is russet. The general pattern is remarkably similar to that of the titis, in particular *C. moloch* (in the case of red-necked owl monkeys), and presumably has been subject to similar selective pressures, at least at some time in their history. The obvious difference is that present-day owl monkeys are essentially nocturnal, although they may be relatively visible in moonlight and during crepuscular periods.

The third small-bodied platyrrhine – *Saimiri* – presents an intriguing contrast to this discussion, with its almost extravagant coloration pattern, in which yellow and white details contrast brightly with black and greenish tones. This is not the only contrast between the two groups, however, and the key to this difference in coloration appears to be behavioral. In fact, the two groups represent opposite extremes within the Platyrrhini in many behavioral characteristics, such as modal group size – the smallest and largest, respectively – and ranging behavior. Whereas both *Aotus* and *Callicebus* typically forage discretely (or nocturnally) in dense vegetation, squirrel monkeys range relatively widely, foraging intensively for insects and major fruit sources (Terborgh 1983). In addition to being more conspicuous in general, squirrel monkeys face interference competition from other frugivores (including potential predators, such as tayras) in feeding patches, and also form systematic associations with capuchins (*Cebus* spp.), which are potential predators. Given these differences in their ecology, then, predator pressure has almost certainly favored the evolution of strategies other than crypticity, such as increased group size (*see* below).

At the next step up in body size, the smallest of the medium-bodied platyrrhines – *Pithecia* – is also the most cryptically colored, with the possible exception of the males of the white-faced saki, *Pithecia pithecia*, which present a distinctive white facial mask contrasting with a dark gray or black body. With the exception of *Pithecia albicans*, which is buffy-colored, the basic saki color pattern is a mottled gray, which is effectively a variation on the agouti pattern. Like titis and owl monkeys, sakis are known for their relatively cryptic behavior, and are notoriously difficult to observe in the wild.

Practically all other medium- to large-bodied platyrrhines are essentially monochromatic, and more often than not, relatively drab in color, with either dark brown, gray or black predominating in most species. With their bright red faces, and relatively indiscreet pelage, the bald uacaris (*Cacajao calvus* and *Cacajao rubicundus*) are a notable exception here. These species are also unusual in their degree of specialization for the occupation of the flooded whitewater forests of western Amazonia, ecosystems known for their reduced biological diversity, in particular of terrestrial animals (Queiroz 1995).

While bald uacaris may face fewer predator species than most other platyrrhines, it is unclear to what extent this – rather than other ecological, physiological and behavioral factors – may have contributed to their unusual coloration. Unfortunately, *Cacajao* is one of the least studied platyrrhine genera, and there are no known records of predation.

An intriguing question remains here. If the risk of predation is inversely related to body size, and cryptic coloration is an important antipredator strategy, why are most of the apparently least cryptic species among the smallest-bodied? The answer appears to lie in specific details of each taxon's behavior and ecology, including habitat use, as we shall see below.

10.4 Antipredator Behavior

10.4.1 Habitat Use

Whereas their arboreal lifestyle may impose certain limits on body size, a curious, but universal phenomenon within the platyrrhines is the preference of the larger-bodied species for the higher forest strata, where the substrate is invariably less stable and robust. This pattern is immortalized in Fleagle's (1988, p. 117) diagram of the vertical distribution of seven sympatric species in Surinam. In addition to the overall, community-wide pattern, the same tendency can be observed within body size classes (and taxonomic families), given that the slightly larger spider monkeys (*Ateles paniscus*) are seen just above the howlers (*Alouatta seniculus*), and the larger-bodied *Chiropotes chiropotes* and the smaller *Pithecia pithecia* are found in the uppermost and lowest strata, respectively.

This pattern is further emphasized by the systematic foraging associations typically formed between sympatric Amazonian species of capuchins (*Cebus* spp.) and squirrel monkeys (*Saimiri*), and also tamarins, *Saguinus* (Terborgh 1983; Heymann and Buchanan-Smith 2000; Porter 2001). In the case of the tamarins, the smaller-bodied species (*Saguinus fuscicollis*) invariably occupies lower levels in the forest than its sympatric congener. The smaller species also forages in a distinct fashion, involving the frequent manual investigation of substrates, often on vertical trunks, in contrast with the more typical "leaf-gleaning" technique of the larger tamarins. Squirrel monkeys also typically forage in leafy substrates, in lower strata than those in which capuchins tend to forage destructively for prey hidden in woody substrates.

Obviously, the vertical distribution of a species within the forest is related to a number of different aspects of its ecology – including niche separation – but the association with body size seems to be too systematic not to be related in some way to predation pressure. If, as argued above, agility is the main line of defense against carnivores, and body size limits the potential for predation by raptors, the larger, less agile monkeys are likely to be safer in the higher strata, which are less accessible to carnivores. By contrast, the smaller-bodied species will be comparatively well protected from attack by raptors in the lower strata, where they are normally able to detect and avoid the approach of carnivores.

The moot question here is to what extent the use of different forest strata is influenced by predation pressure, and a causality dilemma is implicit here, i.e., do large platyrrhines use the higher strata because they are large, or are they large because they use the higher strata? In most cases, the relationship between body size and vertical spacing is consistent with the ecological adaptations of the taxon. For specialized frugivores such as spider monkeys, for example, larger body size correlates with the larger feeding patches provided by canopy trees, and the relatively long distances traveled during foraging. Large size is advantageous for folivores such as howlers, in particular because of digestive constraints, and the uppermost layers of the forest canopy also provide the most productive source of foliage. By contrast, insectivory is negatively correlated with body size, and the smallest, most insectivorous platyrrhines (callitrichids) typically forage for relatively large prey, especially orthopterans, in the denser, lower forest strata.

Obviously, while larger species may be relatively less vulnerable to predation by raptors by virtue of their body size, they are not totally immune to such risks (Table 10.2). As foraging raptors rely on visual cues to locate their prey, predation pressure would likely select for cryptic coloration in the species occupying the upper canopy and, as we have seen, the larger-bodied platyrrhines tend to be more cryptic than many of the smaller, more agile species, which forage lower down in the forest.

10.4.2 Group Size

As for body size, there is considerable variation in social organization and group size among the different platyrrhine genera, ranging from small, nuclear family groups in *Aotus* and *Callicebus* (Table 10.6), to agglomerations of over one hundred individuals in *Saimiri*. As we have seen, these extremes of group size also correlate inversely with those in the relative discretion of pelage coloration and behavior, hinting at the influence of predation pressure on complementary traits.

For many genera, the primary determinant of group size is unlikely to be predation pressure. In *Aotus* and *Callicebus*, in particular, parental care seems to be the key to obligatory monogamy and, by default, small group size (Wright 1986; Fernandez-Duque 2007). In its turn, small group size also complements characteristics such as cryptic coloration and behavior, which once again hints at a causality dilemma, i.e., cryptic traits as a possible precursor of small group size, and even monogamy, rather than vice versa.

In the callitrichids, increased reproductive output has led to the evolution of a cooperative breeding system, which incorporates additional adult group members, with group size typically peaking at 10–15 members, depending on the genus. Group size is normally limited by the systematic dispersion of adults of both sexes, in addition to breeding restrictions in females. While ecological constraints may be important here, predation pressure could also have had a key role in determining limits on group size, in particular in the tiny *Cebuella*, and presumably also in the slightly larger, but poorly known *Callibella*.

Squirrel monkeys are unlike all other small-bodied platyrrhines in a number of different ways, most notably – in the present context – in their social behavior, as well as many aspects of their ecology. The formation of relatively large social groups appears to be an integral component of their general foraging strategy, and in this scenario, large groups may compensate for body size in a number of ways, including protection against predators (increased vigilance, mobbing) and interference competition. By seeking safety in numbers, squirrel monkeys are presumably relieved of the behavioral and morphological constraints of crypticity.

With the obvious exception of *Saimiri*, group size tends to be larger in the larger-bodied platyrrhines, ranging from what may be small extended family groups

Table 10.6 Typical social group formation in the different platyrrhine genera

Nuclear family (2–7 members)	Extended family (4–20 members)	Harem (4–20 members)	Large bands (>20 members)
Aotus	*Callibella**	*Alouatta*	*Ateles*
Callicebus	*Callimico*	*Cebus*	*Brachyteles*
	Callithrix		*Cacajao*
	Cebuella		*Chiropotes*
	Leontopithecus		*Lagothrix*
	Mico		*Oreonax**
	Saguinus		*Saimiri*
	Pithecia		

* Unknown, but inferred from genera of the same subfamily.

similar to those of callitrichids in *Pithecia* to troops of dozens of individuals in genera such as *Ateles*, *Brachyteles*, *Cacajao*, *Chiropotes*, and *Lagothrix* (and presumably also *Oreonax*). As in *Aotus* and *Callicebus*, the comparatively small groups in *Pithecia* complement the relatively cryptic coloration and behavior of this genus.

In the larger pitheciids and atelids, the typical fission-fusion mode of social organization points to foraging strategies as a primary determinant of group size. Troops tend to be more cohesive during periods of resource abundance, which suggests that these monkeys preferentially form larger groups. Larger groups are more visible, however, which may reinforce selection pressure on pelage coloration.

The two other genera in this body size range – *Alouatta* and *Cebus* – are intermediate in group size, a characteristic which is presumably related primarily to their essentially polygynous breeding systems, i.e., groups with multiple adult females, but one or few adult males. In most other aspects of their behavioral ecology, however, these genera occupy opposite extremes of the variation observed in the Platyrrhini, including diet, foraging strategies, activity rates, and ranging behavior, and in particular their response to predators (passive vs. active, *see* below). Once again, the functional link between predation pressure and group size appears to be tenuous in these genera.

10.4.3 Behavioral Strategies

Platyrrhines also vary considerably in their behavioral response to predators. In broad terms, reactions are either passive (avoidance, hiding or fleeing) or active (mobbing or monitoring), and vary systematically within a species according to the type of predator (Caine 1993; Kirchhof and Hammerschmidt 2006). Responses to a given type of predator may also vary among genera, reflecting other morphological and behavioral differences. As some responses appear to be learned (Ferrari and Lopes 1990; Barros et al. 2002), it seems likely that local differences may arise according to shifts in the variables that may determine the frequency of contact with predators, such as their diversity and abundance, and habitat structure.

There appears to be least variation in the response to raptors, which presumably reflects the lack of viable alternative strategies. The only effective way to avoid predation by a raptor is to detect the bird in time to take appropriate evasive action. Intragroup cooperation is a fundamental component of anti-raptor strategies, and specific alarm calls – short, sharp whistles – are a standard characteristic of platyrrhine behavior. In the smaller species, alarm calls may be elicited by virtually any object, including leaves and airplanes, passing overhead, reflecting a "better safe than sorry" strategy.

On hearing an alarm call, an individual will freeze or take cover instantaneously, often replying simultaneously with its own calls. This reaction is observed in infants as soon as they are able to move independently, and is, apparently, an instinctive behavior pattern (which may be reinforced by contact with adults during the dependency phase). This conclusion is supported by my own observations of a wild-born orphaned infant dusky titi (*Callicebus brunneus*), which would react in typical freeze-hide fashion to flying objects, or even imitations of alarm calls, even though it had had no further contact with conspecifics after approximately its first month of life, i.e., when it was still physically dependent on its parents.

This universal, and apparently instinctive response to aerial predators presumably reflects the intensity of the selective pressure exerted by raptors on platyrhine populations. As argued above, attacks by raptors are particularly effective because of the predator's lack of contact with the forest substrate, which minimizes the potential for detection. It seems likely that undetected attacks are almost invariably fatal. The ample influence of predation by raptors is reflected in the relatively large number of recorded events, and the variety of platyrrhine species involved (Table 10.2), even when sampling problems are taken into consideration, as discussed below.

Whereas raptors attack in a predictable fashion, terrestrial predators pose a more complex threat, not least because of the diversity of taxa involved. In addition to these predators, platyrrhines share the forest canopy with a variety of other vertebrates, with which they may interact in a number of different ways, in particular as predators themselves. Reacting to a potential prey animal with anti-predator behaviors would obviously be non-adaptive, reinforcing the importance of predator recognition. In some cases, there may be a fine line between a prey animal – a lizard, for example – and a potential predator, i.e., a snake.

Clearly, then, efficient defense against terrestrial predators requires considerable behavioral flexibility, in marked contrast with the instinctive response to aerial predators. The evolution of complex, learned responses involving intragroup cooperation may have been made more viable here for two main reasons – relatively reduced predation risk in comparison with raptors, and positive selective pressure arising from factors such as increased foraging efficiency.

In their detailed study of antipredator strategies in the buffy-headed marmoset (*Callithrix flaviceps*), Ferrari and Lopes (1990) recorded distinct behavioral strategies in response to carnivores and snakes. The former were mobbed aggressively with loud "tsak-tsak" vocalizations, and the group was observed putting to flight carnivores as large as tayras, which are more than an order of magnitude larger, in terms of body weight, than an adult marmoset. On the other hand, snakes were approached with extreme caution and continuous, low-volume intragroup communication calls. The primary objective in this case appeared to be the assessment and monitoring of the potential threat, and the safe withdrawal of the marmoset group. Intuitively, it would seem reasonable to assume that mobbing a snake with aggressive movements and vocalizations would be more likely to provoke or facilitate an attack rather than prevent one.

In each case, the behavior of the youngest group members was consistently different from that of the adults, and presumably reflected their lack of experience. This reinforces the idea of a learned, rather than an instinctive response, which is also supported by the reaction of the same orphaned dusky titi mentioned above – now a subadult – when presented with a live (non-venomous) snake. The titi was not intimidated in any way by the snake, and in fact, its response was to grab it in an apparent attempt to investigate a novel object.

Different platyrrhines are likely to react in different ways to terrestrial predators, of course, depending in particular on their relative vulnerability, but also other aspects of their behavior and ecology. In fact, the callitrichids' aggressive mobbing of carnivores is a "bluff" type of strategy, considering that a monkey this small would be no match for even the smallest of Neotropical carnivores. Capuchins, by contrast, are much larger and more robust, and appear to mob virtually any potential predator, including snakes (Chapman 1986; Perry et al. 2003) and human observers, in an aggressive fashion, which included branch-shaking and throwing objects. Similar behavior is exhibited by atelines, such as spider monkeys, although their behavior contrasts absolutely with that of the closely related howlers, which generally avoid rather than confront potential dangers.

Howler groups typically respond to potential predators by taking cover in dense vegetation, where they may remain motionless for periods of anything up to a number of hours (pers. obs.), although under some circumstances – e.g., Asensio and Gomes-Marin (2002) – they may mob predators. In the case of an actual attack by tayras, by contrast, an adult male *Alouatta belzebul* watched inert (Camargo and Ferrari 2007). As howlers typically spend a relatively large proportion of their daily activity period at rest, this strategy may be both dictated and made feasible by the howlers' lifestyle, which is sustained by a relatively low-quality, folivorous diet.

Titis are also relatively discrete in their reaction to predators, although in this case, group size, rather than metabolic constraints, may be the key factor. *Callicebus* groups normally contain only two adults, and a relatively large proportion of immature members. In addition, as one adult will, more often than not, be caring for a dependent infant, confrontation with predators would likely not only be ineffective, but risky. These same considerations would also apply to *Aotus*, and its nocturnal habits would probably further reinforce cryptic, rather than confrontational strategies.

An additional, fundamentally important aspect of the antipredator strategies of small-bodied platyrrhines, in particular (but also larger species such as capuchins – Di Bitetti et al. 2000), is their utilization of sleeping sites. Callitrichids enter into a state of torpor during the night, during which they are extremely vulnerable to predation (Franklin et al. 2007), and the choice and use of sleeping sites appears to have been molded by intense selective pressures (Ferrari and Lopes 1990; Heymann 1995; Day and Elwood 1999). Sites are typically well hidden (dense vegetation, tree holes) or relatively inaccessible (isolated emergents), and are often used in a "systematically" variable way, primarily by the use of different sites on consecutive nights, at least where appropriate sites are available.

10.5 Patterns of Predation

10.5.1 Sampling Problems

Despite their importance from both an ecological and an evolutionary viewpoint, predation events are rare and unpredictable, and have been recorded too infrequently for most species to permit more than a superficial analysis of pattern. In addition to this basic sampling problem, two factors hinder a more systematic analysis of tendencies within the Platyrrhini. One is the "polarization" of field studies toward the body size extremes, i.e., the comparatively much larger set of data available for the atelids and callitrichids, in comparison with the small- to medium-sized cebids and pitheciids. Theoretically, this bias could be compensated for by calculating a standardized rate of predation per unit of time spent monitoring per individual, but few studies provide precise information on contact time.

The other major sampling problem is the influence of anthropogenic habitat disturbance – including hunting pressure – on the populations of both platyrrhines and their predators. While any natural population is subject to some variation in population density, anthropogenic disturbance tends to cause major fluctuations that are normally rare in the wild, except in the context of natural disasters or epidemics. Habitat fragmentation may also alter predator behavior and efficiency, in a number of different, often contrasting ways, so there is no simple rule of thumb for assessing the influence of anthropogenic factors, which may affect almost all ecosystems, to a greater or lesser extent.

In addition, many of the records involving raptors and felids (Tables 10.2 and 10.4) are derived from studies of predator, rather than primate species. This biases the record in a number of ways, not least because a given species of predator will tend to attack primates of a specific body size range. A majority of the records for raptors are derived from studies of harpy eagles, for example, which tend to prey on medium- or large-bodied primates. In addition, where the evidence is indirect – i.e., remains found below nests or in scats – it is virtually impossible to determine whether the predator captured the primate, or scavenged its remains. This problem is presumably most relevant to carnivores, which are essentially terrestrial.

While an attempt was made to conduct the most thorough revision of the literature possible, it seems inevitable that some published reports have been overlooked. Ironically, while some authors have recognized the importance of predation events, and have published specific notes on their observations, others have concealed their records in more general texts, which demand laborious interpretation. In many of these cases, it is not exactly clear whether the author has observed a simple encounter, a thwarted attack or an actual predation event.

For example, Terborgh (1983) reports attacks on platyrrhines by a variety of raptor species at Manu in Peru, but only two actual predation events (Table 10.2). Many of the species mentioned – such as *Morphnus guianensis* – are certainly capable of capturing primates, but there is a substantial difference between this potential and *de facto* predation, especially when samples sizes are as small as they are here. A number of misunderstandings have even been perpetuated in the literature. Many authors cite Galef et al. (1976) as evidence of the predation of platyrrhines by tayras, for example, but this paper only mentions two reported attacks. While tayras do appear to attack primates relatively frequently in comparison with other carnivores, they seem to be successful only very rarely, and Galef Jr.'s paper only serves to reinforce this pattern.

Some older reports also present obscure or contradictory information. For example, a number of recent papers refer to Beebe's (1925) record of the predation of a capuchin by a margay, but Oliveira (1998) disputes the identification of the felid, given that the reported body weight was more than twice that of an adult margay, and more compatible with that of an ocelot. As it was not a direct observation of predation, and would thus add little to the overall analysis, this and other, similar records have been omitted here.

Obviously, this set of problems exacerbates that of the greatly reduced data set. While observations have increased substantially in the two decades since the major review of Cheney and Wrangham (1987), there are simply no records of predation whatsoever for the vast majority of platyrrhine species, including those of the Atelidae and Callitrichidae. Nevertheless, the data set now available (Tables 10.2, 10.3, 10.4, and 10.5) does permit the tentative evaluation of certain patterns, such as taxon-specific trends.

10.5.2 Rates and Trends

While predator-oriented data should be treated with caution for the analysis of general patterns, some studies of *Harpia* (Cope and Flower 1964; Rettig 1978; Izor 1985; Ford and Boinski 2007) and felids (Chinchilla 1997; Bianchi 2001; Bianchi and Mendes 2007) indicate that platyrrhines may constitute a relatively important category of prey, at least for some populations, and imply much higher rates of predation than suggested by records from primate-oriented studies. In addition, Boinski (1987) has estimated that 50% of infant *Saimiri oerstedi* are lost to raptors before reaching six months of age, based on observed attacks and disappearances. Presumably, predators will be responsible for the demise of most platyrrhines in the wild, including those incapacitated by disease or injury, or made vulnerable by inexperience, senility, or social isolation. In ecologically stable populations, death rates should generally be similar to birth rates – otherwise, the population would grow *ad infinitum* – but whereas births are observed in most long-term field studies, confirmed records of deaths (rather than disappearances) are disproportionately rare.

Even though the disappearance of some individuals, i.e., infants, is traditionally considered to represent mortality rather than migration, the cause is invariably unknown, or at best inferred, so these records add little to this analysis. Even when remains are found – which is rare, presumably because most attacks occur in the forest canopy – the chances of identifying the predator are minimal, although some useful evidence has been collected from beneath the nests of raptors (Table 10.2).

Even if the multitude of potential sampling problems is taken into consideration, one clear pattern does emerge – predation by snakes (Table 10.5) appears to be relatively rare in comparison with raptors and mammals. Despite the small number of records here, the data point to a number of trends. One is an apparent tendency for the predation of species (or individuals) of relatively small body size. In addition, most recorded events occurred in fragmented or edge habitats, and at least three took place on or near the ground, the main exception here being the bearded saki, which was attacked in the canopy. The predation of a tamarin by an anaconda (Heymann 1987) is emblematic here, given that this snake is aquatic, and presumably rarely, if ever, explores the forest canopy.

Reports of unsuccessful attacks – rather than encounters – by snakes also appear to be relatively rare in comparison with the two other main groups of predators, with two notable exceptions, recorded by Shahuano Tello et al. (2002) and Perry et al. (2003), in which groups of *Saguinus mystax* and *Cebus capucinus*, respectively, cooperated to repel attacks by boas. Defler (1979) also reports on *B. constrictor* attacking *Cebus albifrons*. Overall, then, all but one of recorded attacks have involved boids, which use primarily thermosensing to detect their prey (as does the pit viper). Once again, however, it is possible that attacks by snakes are relatively discrete, and thus less likely to be recorded during field work unless successful. Until further evidence is compiled, then, predation by snakes appears to be a relatively infrequent phenomenon, which typically involves the smaller-bodied platyrrhines, and generally occurs in the lower strata of the forest, often in marginal habitats.

A priori, there is a relatively extensive list of records of predation by other mammals (Tables 10.3 and 10.4), although the bulk of this evidence is derived from predator-oriented studies, including, ironically, that of Sampaio and Ferrari (2005). Most of the remaining records (Peetz et al. 1992; Cuarón 1997; Passamani et al. 1997) include either circumstantial evidence or exceptional circumstances (extreme environmental impact, radio collars), and should be considered as evidence of the capacity of carnivores to prey on platyrrhines rather than a reflection of natural predation rates. This leaves Matsuda and Izawa's (2008) record as the only direct observation of predation by a carnivore under natural conditions, although Moynihan (1970) provides an anecdotal report, and Camargo and Ferrari (2007) did conclude that their observed attacks by tayras on *Alouatta belzebul* would have been fatal if not for the unintentional intervention of the observers.

All the other studies (Table 10.4) are based on the analysis of fecal samples, rather than the observation of attacks, which makes them open to the possibility of scavenging, although this is refuted by some authors, such as Olmos (1994), who claims that the absence of arthropod larvae in the feces analyzed in his study contradicts this conclusion. In other studies, the scavenging hypothesis appears to be challenged by the frequency with which primate prey was ingested. In Bianchi's (2001) study of ocelot scats at the Caratinga Biological Station, the remains of primates – primarily *Alouatta guariba* – were observed in more than a quarter of all specimens, and contributed more than half of the estimated biomass ingested by *Leopardus pardalis* at this site. While it is certainly possible that some of these individuals were scavenged, it seems unlikely that all of them were. On the contrary, Bianchi (2001) suggests that the occurrence of *L. pardalis* at Caratinga is only viable because of the high densities of *A. guariba* in this forest, and their frequent use of the lower forest strata, and in particular the ground, when traveling through the highly fragmented habitat of this reserve, or drinking water.

Once again, despite the limitations of the data set, a number of patterns are apparent. One is a body size trend. In contrast with snakes, callitrichids are rare here, whereas atelids are abundant, in terms of both records and species. The only report of the predation of a callitrichid by a carnivore under natural conditions appears to be that of Moynihan (1970). Bianchi (2001) collected 20 *L. pardalis* scats with remains of primates, twelve of which contained *Alouatta* and three *Brachyteles*, and many other much smaller-bodied mammals, reptiles, and even invertebrates, but no evidence of the predation of *Callithrix*, despite the relative abundance of marmosets at both study sites. In addition, the relative lack of data for smaller-bodied felids such as *Leopardus wiedii* and *Leopardus tigrina* does not appear to be the result of an absence of studies, but rather the lack of primates in their diets.

The overall pattern is thus one of relatively large felids preying on relatively large-bodied platyrrhines. However, this trend also appears to be at least partly related to habitat variables, given that the majority of records refer to fragmented or discontinuous habitats, and disproportionately few are from the Amazon. While all felids are capable of climbing trees, the larger ones, in particular, are essentially terrestrial, and their ability to prey on primates may depend on a specific set of circumstances, perhaps best exemplified by Peetz et al. (1992) observation of the systematic predation of howlers by jaguar on a newly formed reservoir island.

One carnivore that does frequent the forest canopy regularly is the tayra (*Eira barbara*), and there are a relatively large number of records of encounters and attacks involving this species (Galef et al. 1976; Defler 1980; Buchanan-Smith 1990; Phillips 1995; Stafford and Ferreira 1995; Asensio and Gomes-Marin 2002; Camargo and Ferrari 2007). However, whereas tayras may have ample opportunity to attack primates, they do not appear to be very efficient predators, although Camargo and Ferrari (2007) did record circumstantial evidence of the short-term decimation of a population of *Alouatta belzebul*, in a situation similar, albeit on a different scale, to that recorded by Peetz et al. (1992). Bianchi (2001) may have also recorded a similar situation at Caratinga, where one-fifth of ocelot scats contained the remains of howlers. In all three cases, high primate population densities and habitat fragmentation were presumably essential prerequisites.

The relative lack of records involving one other arboreal mammal – the capuchin – is perhaps the most surprising pattern here. Of all mammals, capuchins would presumably have the best and most frequent opportunities to prey on other platyrrhines, but there is little evidence that they explore such chances on a regular basis. While it is true that relatively few field studies of capuchins are available, this is not the case for the callitrichids, ostensibly their most likely potential prey, at least in terms of body size. However, capuchins not only range over much wider areas than most callitrichids, but are also far from discrete in their movements through the forest, so, on the rare occasions that they cross paths, the callitrichids will normally have ample opportunity to avoid coming into close proximity.

Their typical foraging association with squirrel monkeys nevertheless brings capuchins into contact with large numbers of potential prey, although they appear to ignore this potential resource. This may in part be related to the relative difficulty of capturing squirrel monkeys, but may also reflect the long-term benefits of association for the capuchins, which would almost certainly be lost if they assumed the role of predators.

While some capuchins may engage in coordinated hunts of squirrels (Fragaszy, Visalberghi and Fedigan 2004), the overall paucity of reports suggests that they are relatively inefficient predators of other mammals. A decisive difference between squirrels and platyrrhines may be the behavioral complexity of the latter, and in particular, their social organization. This may explain why recorded predation by capuchins is limited to titi monkeys (Lawrence 2003; Sampaio and Ferrari 2005), which live in small family groups.

The third and final predator group – the raptors – presents a very different overall pattern, in particular with regard to body size, considering that the bulk of the records (including unsuccessful attacks: e.g., Defler 1979; Terborgh 1983; Eason 1989; Heymann 1990; Gilbert 2000; Miranda et al. 2006) involve atelids and callitrichids. Records for platyrrhines of intermediate size, such as *Cebus*, *Chiropotes*, and *Pithecia*, are only available from predator-oriented studies, which nevertheless indicate that these medium-sized primates – rather than atelids – are a staple prey for at least some, but not all (Galetti and Carvalho 2002) harpy eagles. The predominance of observed attacks on atelids and callitrichids may thus be primarily a by-product of the polarization of platyrrhine field studies, as discussed above.

An additional contrast here is the predominance of data from relatively well-preserved Amazonian sites. This is almost certainly due in part to the local extinction of large raptors such as *Harpia* and *Morphnus* throughout much of the other, more impacted Neotropical biomes, in particular the Brazilian Atlantic Forest. However, other, smaller raptor species do persist in most areas, and if anything, a bias toward the smaller-bodied platyrrhines – in particular the callitrichids, which are generally more abundant in disturbed habitats – might be expected, but this is not the case.

Whereas habitat fragmentation may make platyrrhines more vulnerable to terrestrial carnivores, then – presumably by bringing them into closer proximity – it may have a neutral, or even an opposite effect with regard to aerial predators. This may be especially true in the relatively dense vegetation of edge, secondary, and altered habitats, which may reduce the effectiveness of raptor foraging strategies. One other possibility is that the predation of monkeys is predominantly an activity of the larger raptors, such as *Harpia*, *Morphnus* and *Spizaetus*, which are, of course, rare or absent from most fragmented landscapes. This is certainly supported by the data (Table 10.2), even if the sampling problems discussed above are taken into account.

Overall, then, the evidence suggests that raptors are the primary predators of platyrrhines under natural conditions, and this is reflected in the more universal, instinctive strategies that have evolved in response to this type of predator. Predation by terrestrial species appears to be relatively rare, on the other hand, and more dependent on random or fortuitous circumstances. Most platyrrhines are probably also better able to deal with potential threats from terrestrial predators through a variety of behavioral strategies, often based on intragroup cooperation. However, data from a number of predator-oriented studies indicate that carnivores may be important predators of platyrrhines in fragmented or relatively open habitats, in particular where primate population densities are relatively high.

10.6 Summary

Predation pressure is a fundamental driving force in primate evolution, and has exerted a clear influence on many morphological and behavioral traits of the platyrrhines, including body size, pelage coloration, social organization, and habitat use. The strictly arboreal lifestyle of these primates is a primary determinant of their vulnerability, and a key factor in the evolution of antipredator strategies. Platyrrhines are susceptible to three main groups of predators: raptors (Falconiformes), mammals (Carnivora), and snakes (Serpentes). Each of these groups poses specific risks to different types of platyrrhines, and demands specific antipredator strategies. Whereas predation by raptors is prevented primarily by vigilance and avoidance, terrestrial predators are generally deterred by monitoring and mobbing. The principal problem for the interpretation of predation patterns in the platyrrhines is the scarcity of records, although the data now available do permit the cautious identification of certain trends. While predation by raptors appears to be relatively frequent, for example, records of successful attacks by snakes are rare. There have also been relatively few observations of predation by carnivores, although there is a growing body of indirect evidence (scat analyses) which suggests that large felids may be important predators of larger bodied platyrrhines, at least under certain specific conditions, such as anthropogenic habitat fragmentation, which may alter the primates' behavior and vulnerability. By contrast, predation by raptors is relatively uncommon under these conditions. Perhaps surprisingly, the sum of the evidence does not suggest that smaller-bodied platyrrhines are at greater risk of predation than larger-bodied species, except for an apparent tendency for snakes (predominantly boids) to prey on smaller-bodied platyrrhines. On the other hand, predation by carnivores appears to involve predominantly larger-bodied species. Overall, while a large number of new records have been collected over the past few years, the data set is still too small to permit a more definitive analysis of predation patterns.

Acknowledgments I am grateful to CNPq (Process no. 307506/2003-7) for their long-term support of my research, and to Alfredo Cuarón, Cecília Kierulff, Diego Cisneros-Heredia, Liza Veiga, Nayara Cardoso, Rita Bianchi, and Susan Ford for their help with data. I also thank Nancy Caine and Eckhard Heymann for their comments on the original manuscript.

References

Alonso, C., and Langguth, A. 1989. Ecologia e comportamento de *Callithrix jacchus* (Primates: Callitrichidae) numa ilha de floresta atlântica. Rev. Nord. Biol. 6:105–137.

Asensio N., and Gomes-Marin, F. 2002. Interspecific interaction and predator avoidance behavior in response to tayra (*Eira barbara*) by mantled howler monkeys (*Alouatta palliata*). Primates 43:339–342.

Barros, M., Boere, V., Mello Jr., E. L., and Tomaz, C. 2002. Reactions to potential predators in captive-born marmosets (*Callithrix penicillata*). Int. J. Primatol. 23:443–454.

Beebe, W. 1925. Studies of a tropical jungle: one quarter of a square mile of jungle at Kartabo, British Guiana. Zoologica 6:1–193.

Bianchi, R. C. 2001. Dieta da jaguatirica *Leopardus pardalis* (Linnaeus, 1758) em Mata Atlântica. Masters thesis, Universidade Federal do Espírito Santo, Vitória.

Bianchi, R. C., and Mendes, S. L. 2007. Ocelot (*Leopardus pardalis*) predation on primates in the Caratinga Biological Station, southeastern Brazil. Am. J. Primatol. 69:1–6.

Boinski, S. 1987. Birth synchrony in squirrel monkeys (*Saimiri oerstedi*): a strategy to reduce neonatal predation. Behav. Ecol. Sociobiol. 21:393–400.

Buchanan-Smith, H. 1990. Polyspecific association of two tamarin species, *Saguinus labiatus* and *Saguinus fuscicollis*, in Bolivia. Am. J. Primatol. 22:205–214.

Caine, N. G. 1993. Flexibility and co-operation as unifying themes in *Saguinus* social organization and behavior: the role of predation pressures. In A. B. Rylands (ed.), *Marmosets and Tamarins: Systematics, Behavior, and Ecology* (pp. 200–219). Oxford: Oxford University Press.

Camargo, C. C., and Ferrari S. F. 2007. Interactions between tayras (*Eira barbara*) and red-handed howlers (*Alouatta belzebul*) in eastern Amazonia. Primates 48:147–150.

Chapman, C. A. 1986. *Boa constrictor* predation and group response in white-faced cebus monkeys. Biotropica 18:171–172.

Cheney D. L., and Wrangham, R. W. 1987. Predation. In B. B. Smuts, D. L. Cheney, R. M. Seyfarth, R. W. Wrangham, and T. T. Struhsaker (eds.), *Primate Societies* (pp. 227–239). Chicago: Chicago University Press.

Chinchilla, F. A. 1997. La dieta del jaguar (*Panthera onca*), el puma (*Felis concolor*) y el manigordo (*Felis pardalis*) (Carnívora: Felidae) en el Parque Nacional Corcovado, Costa Rica. Rev. Biol. Trop. 45:1223–1229.

Cisneros-Heredia, D. F., León-Reyes, A., and Seger, S. 2005. *Boa constrictor* predation on a titi monkey, *Callicebus discolor*. Neotrop. Primates 13:11–12.

Corrêa, H. K. M., and Coutinho, P. E. G. 1997. Fatal attack of a pit viper, *Bothrops jararaca*, on an infant buffy-tufted ear marmoset (*Callithrix aurita*). Primates 38:215–217.

Cuarón, A. D. 1997. Conspecific aggression and predation: costs for a solitary mantled howler monkey. Folia Primatol. 21:100–105.

Day, R. T., and Elwood, R. W. 1999. Sleeping site selection by the golden-handed tamarin *Saguinus midas midas*: the role of predation risk, proximity to feeding sites, and territorial defence. Ethology 105:1035–1051.

Defler, T. R. 1979. On the ecology and behavior of *Cebus albifrons* in eastern Colombia: I Ecology. Primates 20:475–490.

Defler, T. R. 1980. Notes on interactions between the tayra (*Eira barbara*) and the white-fronted capuchin (*Cebus capucinus*). J. Mammal. 61:156.

Defler, T. R. 2004. Primates of Colombia. Bogotá: Conservation International.

Di Bitetti, M. S., Vidal, E. M. L., Baldovino, M. C., and Benesovsky, V. 2000. Sleeping site preferences in tufted capuchin monkeys (*Cebus apella nigritus*). Am. J. Primatol. 50:257–274.

Eason, P. 1989. Harpy eagle attempts predation on adult howler monkey. Condor 91:469–470.

Emmons, L. H. 1987. Comparative feeding ecology of felids in a neotropical rainforest. Behav. Ecol. Sociobiol. 20:271–283.

Fernandez-Duque, E. 2007. Aotinae: social monogamy in the only nocturnal haplorhines. In C. J. Campbell, A. Fuentes, K. C. MacKinnon, M. Panger, and S. K. Bearder (eds.), *Primates in Perspective* (pp. 139–154). New York: Oxford University Press.

Ferrari, S. F., Iwanaga, S., Ravetta, A. L., Freitas, F. C., Sousa, B. A. R., Souza, L. L., Costa, C. G., and Coutinho, P. E. G. 2003. Dynamics of primate communities along the Santarém-Cuiabá highway in southern central Brazilian Amazônia. In L. K. Marsh(ed.), *Primates in Fragments* (pp. 123–144). New York: Kluwer Academic.

Ferrari, S. F., and Lopes, M. A. 1990. Predator avoidance behavior in the buffy-headed marmoset, *Callithrix flaviceps*. Primates, 31:323–338.

Ferrari, S. F., Pereira, W. L. A., Santos, R. R., and Veiga, L. M. 2004. Fatal attack of a *Boa constrictor* on a bearded saki (*Chiropotes satanas utahicki*). Folia Primatol. 75:111–113.

Fleagle, J. G. 1988. Primate Adaptation and Evolution. San Diego: Academic Press.

Ford, S. M., and Boinski, S. 2007. Primate predation by harpy eagles in the Central Surinam Nature Reserve. Am. J. Phys. Anthropol., Suppl. 44:109.

Fowler, J. M., and Cope, J. B. 1964. Notes on the harpy eagle in British Guiana. Auk 81:257–273.

Fragaszy, D. M., Visalberghi, E., and Fedigan, L. M. 2004. The Complete Capuchin: the Biology of the Genus *Cebus*. Cambridge: Cambridge University Press.

Franklin, S. P., Hankerson, S. J., Baker, A. J., and Dietz, J. M. 2007. Golden lion tamarin sleeping-site use and pre-retirement behavior during intense predation. Am. J. Primatol. 69:325–335.

Galef Jr., B. G., Mittermeier, R. A., and Bailey, R. C. 1976. Predation by the tayra (*Eira barbara*). J. Mammal. 57:760–761.

Galetti, M. 1990. Predation on the squirrel, *Sciurus aestuans* by capuchin monkeys, *Cebus paella.* Mammalia 54:152–154.

Galetti, M., and Carvalho Jr., O. 2002. Sloths in the diet of a harpy eagle nestling in eastern Amazonia. Wilson Bul. 112:535–536.

Gilbert, K. A. 2000. Attempted predation on a white-faced saki in the Central Amazon. Neotrop. Primates 8:103–104.

Greco, M. V., Andrade, M. A., Carvalho, G. D. M., Carvalho-Filho, E. P. M., and Carvalho, C. E. 2004. *Callithrix penicillata* na dieta de *Spizaetus ornatus* (Aves: Accipitridae) em área de cerrado no Estado de Minas Gerais. In S. L. Mendes, and A. G. Chiarello (eds.), A Primatologia no Brasil – 8, (pp. 155–160). Santa Teresa: Sociedade Brasileira de Primatologia.

Hershkovitz, P. 1968. Metachromism or the principle of evolutionary change in mammalian tegumentary colors. Evolution 22:556–575.

Hershkovitz, P. 1977. Living New World Monkeys, Platyrrhini, with an Introduction to Primates, Volume 1. Chicago: Chicago University Press.

Hershkovitz, P. 1983. Two new species of night monkeys, genus *Aotus* (Cebidae, Platyrrhini): a preliminary report on *Aotus* taxonomy. Am. J. Primatol. 4:209–243.

Heymann, E. W. 1987. A field observation of predation on a moustached tamarin (*Saguinus mystax*) by an anaconda. Int. J. Primatol. 8:193–195.

Heymann, E. W. 1990. Reactions of wild tamarins, *Saguinus mystax* and *Saguinus fuscicollis* to avian predators. Int. J. Primatol. 11:327–337.

Heymann, E. W. 1995. Sleeping habits of tamarins, *Saguinus mystax* and *Saguinus fuscicollis* (Mammalia; Primates; Callitrichidae), in north-eastern Peru. J. Zool. (Lond.) 237: 211–226.

Heymann, E. W., and Buchanan-Smith, H. 2000. The behavioral ecology of mixed-species groups of callitrichine primates. Biol. Rev. 75:169–190.

Izawa, K. 1978. A field study of the ecology and behavior of the black-mantle tamarin (*Saguinus nigricollis*). Primates 19:241–274.

Izor, R. J. 1985. Sloths and other mammalian prey of the harpy eagle. In G. G. Montogomery (ed.), *The Evolution and Ecology of Armadillos, Sloths, and Vermilinguas* (pp. 343–346). Washington DC: Smithsonian Institution Press.

Julliot, C. 1994. Predation of a young spider monkey (*Ateles paniscus*) by a crested eagle (*Morphnus guianensis*). Folia Primatol. 63:75–77.

Kierulff, M. C., Raboy, B. E., Oliveira, P. P., Miller, K., Passos, F. C., and Prado, F. 2002. Behavioral ecology of lion tamarins. In D. G. Kleiman and A. B. Rylands (eds.), *Lion Tamarins: Biology and Conservation* (pp. 157–187). Washington DC: Smithsonian Institution Press.

Kirchhof, J., and Hammerschmidt, K. 2006. Functionally referential alarm calls in tamarins (*Saguinus fuscicollis* and *Saguinus mystax*) – evidence from playback experiments. Ethology 112:346–354.

Lawrence, J. M. 2003. Preliminary report on the natural history of brown titi monkeys (*Callicebus brunneus*) at the Los Amigos Research Station, Madre de Díos, Peru. Am. J. Phys. Anthropol. Suppl. 36:136.

Ludwig, G., Aguiar, L. M., Miranda, J. M. D., Teixeira, G. M., Svoboda, W. K., Malanski, L. S., Shiozawa, M. M., Hilst, C. L. S., Navarro, I. T., and Passos, F. C. 2007. Cougar predation of black-and-gold howlers on Mutum Island, southern Brazil. Int. J. Primatol. 28:39–46.

Martins, S. S., Lima, E. M., and Silva Júnior, J. S. 2005. Predation of a bearded saki (*Chiropotes utahicki*) by a harpy eagle (*Harpia harpyja*). Neotrop. Primates, 13:7–10.

Matsuda, I., and Izawa, K. 2008. Predation of wild spider monkeys at La Macarena, Colombia. Primates 49:65–68.

Mendes, S. L. 1989. Estudo ecológico de *Alouatta fusca* (Primates, Cebidae) na Estação Biológica de Caratinga, MG. Rev. Nord. Biol. 6:71–104.

Miller, L. E., and Treves, A. 2007. Predation on primates: past studies, current challenges, and directions for the future. In C. J. Campbell, A. Fuentes, K. C. MacKinnon, M. Panger and S. K. Bearder (eds.), *Primates in Perspective* (pp. 525–543). New York: Oxford University Press.

Miranda, J. M. D., Bernardi, I. P., Abreu, K. C., and Passos, F. C. 2005. Predation on *Alouatta guariba clamitans* Cabrera (Primates, Atelidae) by *Leopardus pardalis* (Linnaeus) (Carnivora, Felidae). Rev. Brasil. Zool. 22:793–795.

Miranda, J. M. D., Bernardi, I. P., Moro-Rios, R. F., and Passos, F. C. 2006. Anitpredator behavior of Brown howlers attacked by black hawk-eagle in southern Brazil. Int. J. Primatol. 27: 1097–1101.

Moynihan, M. 1970. Some behavior patterns of platyrrhine monkeys. II. *Saguinus geoffroyi* and some other tamarins. Smithson. Contr. Zool. 28:1–77.

Novak, A. J., Main, M. B., Sunquist, M. E., and Labisky, R. F. 2005. Foraging ecology of jaguar (*Panthera onca*) and puma (*Puma concolor*) in hunted and non-hunted sites within the Maya Biosphere Reserve, Guatemala. J. Zool. 267:167–178.

Oliveira, T. G. 1998. *Leopardus wiedii*. Mam. Spec. 579:1–6.

Olmos, F. 1993. Notes on the food habits of Brazilian caatinga carnivores. Mammalia 57:126–130.

Olmos, F. 1994. Jaguar predation on muriqui *Brachyteles arachnoides*. Neotrop. Primates 2:16.

Oversluijs Vasquez, O. M. R., and Heymann, E. W. 2001. Crested eagle (*Morphnus guianensis*) predation on infant tamarins (*Saguinus mystax* and *Saguinus fuscicollis*, Callitrichinae). Folia Primatol. 72:301–303.

Passamani, M., Mendes, S. L., Chiarello, A. G., Passamani, J. A., and Laps, R. R. 1997. Reintrodução do sagüi-da-cara-branca (*Callithrix geoffroyi*) em fragmentos de Mata Atlântica no Sudeste do Brasil. In S. F. Ferrari and H. Schneider (eds.), *A Primatologia no Brasil – 5* (pp. 119–128). Belém: Sociedade Brasileira de Primatologia.

Peetz, A., Norconk, M. A., and Kinzey, W. G. 1992. Predation by jaguar on howler monkeys (*Alouatta seniculus*) in Venezuela. Am. J. Primatol. 28:223–228.

Peres, C. A. 1990. A harpy eagle successfully captures an adult male red howler monkey. Wilson Bul. 102:560–561.

Peres, C. A. 1997. Primate community structure at twenty western Amazonian flooded and unflooded forests. J. Trop. Ecol. 13:381–405.

Perry, S., Manson, J. H., Dower, G., and Wikberg, E. 2003. White-faced capuchins cooperate to rescue a groupmate from a *Boa constrictor*. Folia Primatol. 74:109–111.

Phillips, K. 1995. Differing responses to a predator (*Eira barbara*) by *Alouatta* and *Cebus*. Neotrop. Primates, 3:45–46.

Porter, L. M. 2001. Dietary differences among sympatric Callitrichinae in northern Bolivia: *Callicmico goeldii*, *Saguinus fuscicollis* and *S. labiatus*. Int. J. Primatol. 22:961–992.

Queiroz, H. L. 1995. Preguiças e Guaribas: os Mamíferos folívoros arborícolas do Mamirauá. Brasília: CNPq.

Rettig, N. L. 1978. Breeding behavior of the harpy eagle (*Harpia harpyja*). Auk 95:629–643.

Robinson, S. K. 1994. Habitat selection and foraging ecology of raptors in Amazonian Peru. Biotropica 26:443–458.

van Roosmalen, M. G. M., van Roosmalen, T., and Mittermeier, R. A. 2002. A taxonomic review of the titi monkeys, genus *Callicebus* Thomas, 1903, with the description of two new species, *Callicebus bernhardi* and *Callicebus stephennashi*, from Brazilian Amazonia. Neotrop. Primates 10(Suppl):1–52.

van Roosmalen, M. G. M., and van Roosmalen, T. 2003. The description of a new marmoset genus, *Callibella* (Callitrichinae, Primates), including its molecular phylogenetic status. Neotrop. Primates 11:1–10.

Rose, L. M., Perry, S., Panger, M. A., Jack, K., Manson, J. H., Gros-Louis, J., MacKinnon, K. C., and Vogel, E. 2003. Interspecific interactions between *Cebus capucinus* and other species: data from three Costa Rican sites. Int. J. Primatol. 24:759–796.

Rowe, N. 1996. The Pictorial Guide to the Living Primates. East Hampton: Pogonias Press.

Rylands, A. B., Schneider, H., Langguth, A., Mittermeier, R. A., Groves, C. P., and Rordiguez-Luna, E. 2000. An assessment of the diversity of New World primates. Neotrop. Primates 8: 61–93.

Sampaio, D. T., and Ferrari, S. F. 2005. Predation of an infant titi monkey (*Callicebus moloch*) by a tufted capuchin (*Cebus apella*). Folia Primatol. 76:113–115.

Schaller, G. B. 1983. Mammals and their biomass on a Brazilian ranch. Arquiv. Zool. 31:1–36.

Shahuano Tello, S. N., Huck, M., and Heymann, E. W. 2002. *Boa constrictor* attack and successful group defence in moustached tamarins, *Saguinus mystax*. Folia Primatol. 73:146–148.

Sherman, P. T. 1991. Harpy eagle predation on a red howler monkey. Folia Primatol. 56:53–56.

Stafford, B. F., and Ferreira, M. F. 1995. Predation attempts on callitrichids in the Atlantic coastal rain forest of Brazil. Folia Primatol. 65:229–233.

Stanford, C. B. 2002. Avoiding predators: expectations and evidence in primate antipredator behavior. Int. J. Primatol. 23:741–757.

Terborgh, J. 1983. Five New World Monkeys: A Study in comparative Ecology. Princeton: Princeton University Press.

Touchton, J. M., Hsu, Y.-C., and Palleroni, A. 2002. Foraging ecology of reintroduced captive-bred subadult harpy eagles (*Harpia harpyja*) on Barro Colorado Island, Panama. Ornitol. Neotrop. 13:1–15.

Wright, P. C. 1986. Ecological correlates of monogamy in *Aotus* and *Callicebus*. In J. G. Else and P. C. Lee (eds.), *Primate Ecology and Conservation* (pp. 159–167). Cambridge: Cambridge University Press, Cambridge.

Chapter 11
Mechanical and Nutritional Properties of Food as Factors in Platyrrhine Dietary Adaptations

Marilyn A. Norconk, Barth W. Wright, Nancy L. Conklin-Brittain, and Christopher J. Vinyard

11.1 Introduction

Platyrrhines face a vast array of potential food resources in the Neotropics. Ecological challenges associated with finding, ingesting, masticating, and digesting foods are influenced by food availability and accessibility. Food availability is influenced by seasonal variation in forest productivity, fruiting synchrony, and crop size (e.g., Stevenson 2001; Chapman et al. 2003, but *see* Milton et al. 2005). Accessibility, on the other hand, is related to such factors as fruit and seed size, the ability to breach mechanically challenging tissues, to tolerate secondary chemical compounds, and to balance nutrient intake. Our goal in this chapter is to examine the diversity of platyrrhine responses to this second variable – gaining access to and processing foods.

All platyrrhine genera include fruit in their diets, but the annual percentage of fruit intake ranges widely from 8% in *Cebuella* to 86% in *Ateles* (Table 11.1). A wide variety of other resources including exudates, fungi, leaves, flowers, nectar and insect or vertebrate prey make up the balance, or at times the bulk, of annual diets. Some particularly interesting feeding behaviors seen in platyrrhines signal the evolution of specific adaptations. These include the ability to extract and digest plant resources such as gums by *Cebuella* and *Callithrix* (Nash 1986; Power and Oftedal 1996), fungi by *Callimico* (Porter 2001; Porter and Garber 2004; Hanson et al. 2006; Rehg 2006), and seeds by the pitheciins (van Roosmalen et al. 1988; Ayres 1989; Kinzey and Norconk 1990; Kinzey 1992; Peetz 2001; Norconk and Conklin-Brittain 2004). Although gums, seeds and fungi are ingested by other primate species [especially lemurs (Nash 1989; Hemingway 1998) and colobines (Waterman and Kool 1994; Kirkpatrick 1998)], they are used very intensively by these platyrrhines, composing either a majority of their diet during a single season, a subset of the annual diet, or are routinely and extensively used throughout the year.

M.A. Norconk (✉)

Department of Anthropology and School of Biomedical Sciences, Kent State University, Kent, OH 44242, USA

e-mail: mnorconk@kent.edu

P.A. Garber et al. (eds.), *South American Primates,* Developments in Primatology: Progress and Prospects, DOI 10.1007/978-0-387-78705-3_11,

© Springer Science+Business Media, LLC 2009

Table 11.1 Average annual percentages of plant parts and insects in the diets of 16 platyrrhine genera

Primate genus	Exudates (%)	Arils (%)	Fungi (%)	Fruit pulp (%)	Flowers (%)	Seeds (%)	Whole fruit (%)	Young leaves (%)	Mature leaves (%)	Insects (%)
Cebuella	60			8						30
Callithrix	45			16						39
Leontopithecus	9			53	7					25
Saguinus	10			35				3		45
Callimico	1		29	29						41
Saimiri				25	5			10		60
Cebus				47		8		8		33
Aotus				45	14			41		
Callicebus				59	4	27		6		4
Pithecia		2		16	2	61	11	5		3
Cacajao				18	6	67				
Chiropotes				42	1	51				4
Alouatta				34	9			38	16	
Lagothrix				64	4	1	8	6		9
Brachyteles				27	11	5		42	9	
Ateles				86	3			11		

Sources for dietary data: Ayres 1986; Strier 1991; Norconk 1996; Palacios et al. 1997; Peetz 2001; Porter 2001; NRC 2003; Di Fiore 2004; Wallace 2005. Species listed by body size from small to large. These data are composites for each genus and do not necessarily equal 100%.

Although considerable work has been done on platyrrhine feeding strategies, we plan to further examine these food accessibility issues taking a slightly different approach. We devised three variables that could be compared among 16 genera of platyrrhines. Each of the variables was compiled from both field and laboratory studies and each represents considerable built-in complexity. The three variables are (1) the morphology of the masticatory apparatus, (2) the mechanical protection of plant foods opened or ingested by platyrrhines, and (3) the nutrient composition of foods ingested and dietary intake in the form of metabolizable energy. We approach this chapter from diverse backgrounds and hope that by combining our efforts we will contribute to a more comprehensive understanding of the dietary strategies of the modern platyrrhine radiation.

11.2 General Characteristics of Platyrrhine Diets

Platyrrhines provide a nice test of Kay's body mass threshold model. All of them are frugivores to some extent, thus one would expect to find a shift between higher proportions of protein-rich insects to leaves at one kilogram of body mass (Kay 1984). Most of the smallest-bodied platyrrhines balance diets of exudates with fruit and insects and fit the model well with *Callimico* unique in its dietary combination of insects with fruit and fungi (Porter 2001). The percentage of insects in the diet is much reduced in larger-bodied platyrrhines ($< 10\%$) and the ateline diets are relatively high in their leaf portions. However, two platyrrhine genera do not appear to fit Kay's model. *Cebus* spp. have a higher intake and *Aotus* spp. have a lower intake of insects than expected, based on their body mass. The well-known extractive foraging strategy of *Cebus* enhances their reliability of access to protein-rich animal prey (e.g., Fragaszy 1986; Janson and Boinski 1992). The explanation for the low intake of insects in *Aotus* spp. is perhaps related to the inability to quantify insect eating in this nocturnal primate (Fernandez-Duque 2007). Seeds predominate in the diets of the saki-uacari group (pitheciines) and are a relatively rarely used resource among the rest of the platyrrhines. In the sakis and uacaris, various categories of fruit (e.g., fruit pulp, seeds, and whole fruit) make up more than 75% of the diet. Ultimately, a revision of Kay's model may be necessary as more nutritional data are collected to include better documentation of protein levels in wild foods and wild primate diets. Kay (1984) suggested that a dietary shift occurred along a continuum of body mass in primates – from protein derived from insects in primates weighing less than 1kg to protein derived from leaves in larger primates. Nutritional studies suggest that protein is both a ubiquitous resource in tropical plants, and an unpredictable one – particularly for fruit. Fruit pulp is often found to be relatively low in protein (e.g., 9.5%) (on the basis of dry matter, DM) compared to the average for leaves (22% DM) (Conklin-Brittain et al. 1998: Table III), but not always. *Capparis muco* fruit pulp eaten by white-faced sakis is 18.7% (DM) crude protein (Norconk and Conklin-Brittain 2004: Table I), but at this point we do not know if this is a relatively common or a rare occurrence. With the incorporation of other variables, such as longer termed studies and nutritional data, we will inevitably expand the limited dimensionality of Kay's original paradigm.

11.3 Materials and Methods: Our Approach in this Chapter

We begin our examination of how the diversity of platyrrhine feeding strategies relates to accessing and processing foods by reviewing variation in platyrrhine masticatory apparatus form, dietary mechanical properties, digestion and nutrient intake. Following these reviews, we attempt to integrate summary data from each review to explore potential interrelationships. Finally, we highlight where future research can further our understanding of how platyrrhines access and process selected resources.

11.3.1 Morphometric Sample of the Masticatory Apparatus

In order to review masticatory apparatus functional morphology, morphometric data on platyrrhine skulls were compiled from either unpublished measurements taken on museum specimens or from published species means. For the unpublished museum data, wild-shot individuals were sampled preferentially. These measurements were either taken with calipers or from video analysis following the methods outlined in Spencer and Spencer (1995).

11.3.2 Food Properties Sample

The review of feeding ecology and food mechanics are based primarily on data collected at Turtle Mountain in the Iwokrama reserve in central Guyana, South America from October 1999 to December 2000 (Wright 2004). Plant tissues were categorized as: fruit mesocarp, epicarp, seed coat (or endocarp), exocarp-mesocarp (adhering epicarp-mesocarp), endosperm, pod, seed (whole), fruit (whole), leaf (leaf lamina), spadix (spathes and fruiting spadices of aroid epiphytes), stem (of flowers & leaves), petal (of flowers), flower reproductive (non-petal or stem flower parts), gum (exudate) or aril following van Roosmalen (1984). Detailed feeding data were recorded when possible with particular attention paid to the sequence of oral/manual food processing. The position of the food along the dental arcade was also recorded when detailed observations could be made and the frequency of distinct processing techniques and sequences was calculated.

While in the field, samples of food tissues were collected and their fracture toughness was measured. The portable universal tester used in the field was designed specifically for testing the physical properties of foods processed by primates (Darvell et al. 1996; Lucas et al. 2001), and its use is well established in field studies of primate dietary ecology.

11.3.3 Nutritional Sample

The nutritional sample was drawn from several sources. (1) Fifty-seven plant species collected from Lago Guri, Venezuela, from 1991 to 1995 during long-term studies of *Pithecia pithecia* and *Chiropotes satanas* (Kinzey and Norconk 1993; Norconk 1996; Norconk and Conklin-Brittain 2004; unpublished data). (*See* Norconk and Conklin-Brittain 2004 for methods used in the nutritional analysis). Additional smaller data sets were also compiled for sympatric primates, *Alouatta seniculus*, and *Cebus olivaceus* in Lago Guri. (2) Fifty-five plant species, including both leaves and fruits, collected in Belize (Silver et al. 2000), during a thesis project studying the diet of black howler monkeys (*Alouatta pigra*). (3) Forty-five plant species also collected in Venezuela (Castellanos and Chanin 1996), for a study of spider monkey (*Ateles belzebuth*) fruits and arils. (4) Sixteen plant species, mainly fruits, and two insects from Isla Barro Colorado in Panama (Hladik et al. 1971). (5). Parts of several smaller datasets, especially for flowers, exudates, fungi, and insects (Gaulin and Craker 1979; Nash 1984; Garber 1988; Brown and Zunino 1990; Oftedal 1991; Smith 2000; Hanson et al. 2006).

Using this nutritional sample, we generated an estimated metabolizable energy density (kcal/100 g of diet dry weight) for each major food source by first computing total nonstructural carbohydrates *(TNC) = 100 – % insoluble fiber – % lipids – % protein – % ash.* Estimated metabolizable energy (ME) was calculated using the commonly used physiological fuel values of 4 kcal/g for carbohydrates (TNC) and protein, and 9 kcal/g for lipids (NRC 2003; Conklin-Brittain et al. 2006). We estimated metabolizable energy density for the diets of 16 platyrrhine genera, ranked the genera from highest to lowest and compared that to their body weights. Empirically determined ME, where energy lost through feces, urine, and respiration is subtracted from energy gained by intake, is more accurate, but this method has been used for very few wild primates (Altmann 1998; Conklin-Brittain et al. 2006; Miller et al. 2006).

11.4 Mechanical Assessment of Platyrrhine Masticatory Apparatus Form

The masticatory apparatus becomes involved in the feeding process as foods are brought into the mouth and mechanically reduced before being passed into the rest of the gastrointestinal tract for nutrient extraction. For the part of this process involving the masticatory apparatus, feeding can be somewhat arbitrarily broken down into ingestion, followed by mastication, and finally swallowing (Hiiemae 2000).

Ingestion involves using the teeth to forcibly extract and/or separate potential foods for subsequent chewing (Hiiemae and Crompton 1985). Ingestion may be the most variable of these three processes as it can pose minimal to significant mechanical challenges to an animal. Ingestive activities have been variably described as incision, harvesting, biting, husking, cropping, gouging, tearing, breaching, scraping and/or piercing in South American primates (Eaglen 1986; Rosenberger 1992; Anapol and Lee 1994; Vinyard et al. 2003; Wright 2005). After bringing a bite-sized piece of food into the mouth, chewing or mastication begins. Chewing typically involves consecutive, rhythmic patterns of jaw movement and loading during which foods are mechanically broken down between the upper and lower postcanine teeth (Hiiemae 1978). Swallowing is the most coordinated and likely stereotypical of these three processes. We do not consider swallowing in this chapter.

The mechanical demands placed on the masticatory apparatus during ingestion and mastication are broadly divisible into those related to force production and dissipation versus those associated with moving the jaw and tongue. While the forces generated at the bite point are paramount to successful food breakdown, the ability to efficiently produce these bite forces and successfully resist internal loads in the skull during feeding also impact the form of the platyrrhine masticatory apparatus. Most functional analyses of masticatory apparatus form among platyrrhines focus on the generation of these external bite forces and the dissipation of the resulting internal loads (Bouvier 1986a; Daegling 1992; Anapol and Lee 1994; Wright 2005). Research on jaw and tongue movements during feeding (Vinyard et al. 2003) lags behind functional studies of the force-related demands on the platyrrhine masticatory apparatus.

The structural and mechanical properties of an animal's diet as well as how it chooses to manipulate this diet must be fundamentally linked to the mechanical demands placed on the masticatory apparatus during ingestion and mastication. This overarching conclusion is bolstered by observations that the range of jaw movements and magnitude of jaw loads during chewing are influenced by these structural and mechanical properties as well as the relative position of a chewing cycle in a chewing sequence (Hiiemae and Kay 1973; Luschei and Goodwin 1974; Hiiemae 1978; Hylander 1979a; Hylander et al. 1987, 2000; Chew et al. 1988; Agrawal et al. 1998; Vinyard et al. 2006). Thus, we should expect any relationship between function and masticatory apparatus form to be predicated on dietary properties and feeding behavior. Our goal for the remainder of this section is to examine the functional consequences of variation in masticatory apparatus form among platyrrhines.

11.4.1 Jaw Forms Linked to Force Production

We can initially approximate a bite force anywhere along the tooth row using a static beam model in either a sagittal or frontal view (e.g., Hylander 1975; Smith 1978). In sagittal projection, bite force can be estimated as:

$$\text{Bite force}(Fb) = (muscle\ moment\ arm(IN) * muscle\ force(Fm))/bite\ point\ load\ arm(OUT)$$

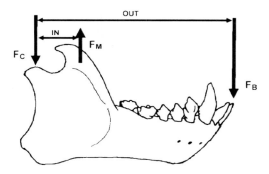

$F_B = IN \cdot F_M / OUT$

Mechanical Advantage for Jaw Muscles in generating F_B = IN / OUT

Fig. 11.1 A static beam model of vertical bite force at the incisors. The mechanical advantage of the jaw-closing muscles is estimated as the ratio of the perpendicular distance from the line of action of the jaw-closing muscle (arrow labeled F_m) to the temporomandibular joint (IN-lever) divided by the distance from the joint to the incisal bite point (OUT-lever). The sum of the vertical force created by the jaw-closing muscles is represented by F_m. (The location of the jaw-closing muscle line of action is arbitrarily placed for the purpose of illustration). Abbreviations: F_m = vertical force generated by the jaw-closing muscles, F_b = vertical bite force, F_c = vertical reaction force at the condyle, IN = the in-lever or moment arm for the jaw-closing muscles and OUT = the out-lever or load arm for biting at the incisors

(Fig. 11.1). This static bite force estimate is divisible into the force generated by the combined contraction of the jaw adductors (Fm) and a leverage component describing the mechanical advantage (IN/OUT) of the muscle (IN) and bite point (OUT) moment arms. Morphological changes affecting either of these components may influence an animal's ability to generate bite force during ingestion or mastication. Unfortunately, we know next to nothing about variation in bite forces among platyrrhines.

One reason we know so little about bite forces in platyrrhines is that we lack a comprehensive functional analysis of their jaw-closing muscles. The existing information on platyrrhine jaw muscles focuses on descriptive morphology (e.g., Ross 1995) or descriptions combined with muscle weights (Starck 1933; Schumacher 1961; Turnbull 1970; Cachel 1979). Taylor and Vinyard (2004) provide the only functional analysis of jaw-muscle architecture in platyrrhines. They compared masseter architecture in gouging and non-gouging callitrichines, but lacked the taxonomic breadth to be informative across platyrrhines. Based on the limited data available at the time, Bouvier and Tsang (1990) suggested that platyrrhines do not differ markedly from catarrhines in their relative jaw-muscle configurations.

In contrast to data on muscle architecture, we know more about how variation in skull form impacts the mechanical advantage of the platyrrhine masticatory apparatus. Most recently, Wright (2005) compared mechanical advantage for the jaw-closing muscles at the incisors, canines and M_2s across ten platyrrhine species. This analysis focused on *Cebus* spp. finding that they tended to exhibit the highest mechanical advantage for the jaw-closing muscles with the exception of the medial pterygoid. *Chiropotes satanas* tended to exhibit the next highest mechanical advantage followed by *Pithecia pithecia*, *Lagothrix lagotricha* and *Ateles paniscus*. *Alouatta seniculus*, *Callicebus* spp. and *Aotus trivirgatus* exhibited the least mechanical advantage among these species. Anapol and Lee (1994) estimated masseter and temporalis lever arms for eight platyrrhines. Their analysis focused more on variation among lever arm lengths noting that the masseter lever arm showed relatively less variation than the temporalis lever arm.

11.4.2 Jaw Forms Linked to Force Production: A Reanalysis

We can extend these previous analyses of jaw-muscle mechanical advantage by both adding species and incorporating an estimate of the relative force contributed by each of the three main jaw-closing muscles. We measured moment arms for the temporalis, masseter and medial pterygoid, similar to Wright (2005), for 22 platyrrhine species. We then scaled these moment arms by the percentage of the total jaw-adductor muscle weight each muscle represents based on the platyrrhines ($n = 4$) measured by Turnbull (1970). Individual moment arms as well as the average moment arm scale close to or slightly below isometry relative to incisor, canine and molar biting moment arms (Table 11.2). Similar scaling patterns are observed when regressed on body mass. The only scaling comparison to deviate from isometry is the negative allometry of the masseter and average muscle moment arm relative to biting at M^1.

Relative mechanical advantage among platyrrhine species trends toward a size-related decrease in biting leverage, particularly for biting along the postcanine dentition (but see Pirie 1976). Figure 11.2 shows that smaller platyrrhines tend to have greater mechanical advantage on average than larger species for biting at M_1 ($r = -0.57$; $P = 0.006$). A similar, but weaker, trend is observed in an analogous comparison for biting at the incisors ($r = -0.41$; $P = 0.06$). A size-related trend is not observed for biting at the canines ($r = -0.23$; $P = 0.29$). These size-correlated trends appear strongest in callitrichines; although, it is unclear if this pattern is directly related to masticatory or ingestive functions in this group. In summary, larger platyrrhines may start out with a size-correlated disadvantage for producing bite forces during mastication or ingestion.

Superimposed on this size-related trend are several differences in mechanical advantage that correlate with variation in diet. Among the non-callitrichines, *Cebus apella* possesses the highest leverage for biting at M_1 (Fig. 11.2) as well as the canines and incisors (data not shown). *Chiropotes satanas* and *Cacajao melanocephalus* have the next highest advantage for M_1 biting followed by *Pithecia pithecia* and *Cebus albifrons*. These results support previous observations that these "hard-object" feeders tend to have relatively greater mechanical advantage (Anapol and Lee 1994; Wright 2005), particularly during anterior tooth use. After the pithecines and *Cebus* spp., there are a group of primates with intermediate mechanical

Table 11.2 Scaling of moment arms for the jaw-closing muscles versus biting load arms and body mass among platyrrhines

Jaw-Muscle Moment Arm[1]	Incisor Load Arm		Canine Load Arm		M₁ Load Arm		Body Mass^{1/3}	
	LS/RMA Slope (95% CI)[2]	R[3]	LS/RMA Slope (95% CI)	R	LS/RMA Slope (95% CI)	R	LS/RMA Slope (95% CI)	R
Temporalis	0.98 (±0.117)	0.969	1.00 (±0.120)	0.968	0.94 (±0.117)	0.966	0.98 (±0.213)	0.906
	1.01 (±0.117)		1.03 (±0.120)		0.97 (±0.117)		1.08 (±0.213)	
Masseter	0.96 (±0.055)	0.992	0.97 (±0.057)	0.992	0.91 (±0.069)	0.987	0.97 (±0.153)	0.947
	0.97 (±0.055)		0.98 (±0.057)		0.92 (±0.069)		1.02 (±0.153)	
Medial Pterygoid	0.92 (±0.123)	0.962	0.94 (±0.126)	0.961	0.88 (±0.117)	0.957	0.90 (±0.225)	0.883
Scaled Jaw-Muscle Average[4]	0.96 (±0.123)	0.995	0.98 (±0.126)	0.995	0.92 (±0.125)	0.991	1.02 (±0.225)	0.883
	0.96 (±0.043)		0.98 (±0.046)		0.92 (±0.057)		0.90 (±0.225)	
	0.96 (±0.043)		0.98 (±0.046)		0.93 (±0.057)		1.02 (±0.225)	

[1] All regressions and correlations are based on 22 species' means (n = 22). (see Fig. 11.1 for measurement and species' descriptions).
[2] Top line is least-squares regression (LS) of a muscle moment arm on a moment arm for various bite points. Bottom line is reduced-major axis regression (RMA). The 95% confidence interval for the slope estimate is in parentheses. All regressions are significant α = 0.05.
[3] Pearson's product moment correlation. All correlations are significant α = 0.05.
[4] Calculated as average of the three muscle moment arm after taking into account relative percentage of total jaw-adductor mass across platyrrhines. [= (*Temporalis**0.52 + *masseter**0.33 + *medial pterygoid**0.17)/3].

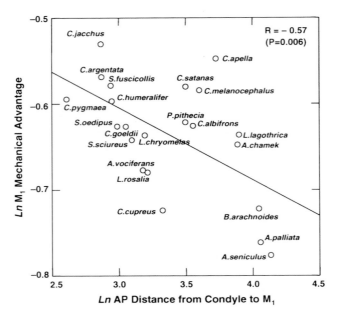

Fig. 11.2 Plot of mechanical advantage for biting at M_1 versus the distance from the condyle to M_1 among platyrrhines. M_1 mechanical advantage is estimated as: $((TempMA^*0.52)+(MassMA^*0.31)+(MedPtery^*0.17))/Condyle\text{-}M_1$ distance. TempMA is measured as the distance from the back of the condyle to the tip of the coronoid process. MassMA is the distance from the back of the glenoid fossa to the tip of the anterior attachment of the masseter at the root of the zygoma. MedPtery is the AP distance from the back of the glenoid to the midline of the pterygoid plates. Each of these muscle moment arms is multiplied by the relative percentage that muscle contributes to overall jaw-adductor weight in four platyrrhines measured by Turnbull (1970). This scaling provides an initial estimate of relative force contribution, based on weight, from each of the jaw-closing muscles. Condyle-M_1 is the AP distance from the back of the glenoid to M_1

leverage that are classically described as "frugivorous." More folivorous species, such as *Alouatta palliata, A. seniculus* and to a lesser extent *B. arachnoides*, show the lowest mechanical advantage for M_1 biting (Wright 2005). In contrast to the relatively higher leverage in *Cebus* spp. and pitheciines, leverage improvement is not related to folivory in platyrrhines. It is interesting to speculate based on these observations that while breaking down pliant, tough leaves may require significant mechanical work at the molars, manifested as repetitive crack propagation, it may not necessarily involve generating extremely high bite forces during chewing.

11.4.3 Jaw Forms Linked to Load Resistance

The platyrrhine masticatory apparatus likely experiences its largest internal loads during the power strokes of incision and mastication when foods are mechanically fractured and/or reduced. The morphological bottom line for resisting these loads is that bigger is better. In some cases, bigger in a certain direction (i.e., a specific shape) provides improved load resistance ability. In other situations, larger in magnitude regardless of direction, offers increased load resistance. Previous in vivo analyses of living primates indicate that the mandibular condyles, corpora and symphyses resist significant loads during mastication and ingestion (e.g., Hylander 1979a,b, 1984, 1985; Hylander et al. 1987, 1998).

Multiple studies have translated Hylander's in vivo strain data into expected morphological differences among platyrrhines that differ in diet and/or feeding behavior. We can summarize the morphological variation in load resistance ability across platyrrhines by combining shape measures (i.e., shape ratios) of the mandibular condyle, corpus and symphysis in a multivariate principal components analysis (PCA) (Table 11.3; Fig. 11.3). The first component of this PCA explains approximately 52% of the variation in these length and width shapes among platyrrhines. Furthermore, all variables have positive loadings suggesting that this component can be initially interpreted as a jaw robusticity factor. With the exception of anterior-posterior condyle length, each variable is significantly correlated with its first component score suggesting that most of these shape measures are contributing to this linear estimate of jaw robusticity.

C. satanas and *C. melanocephalus* have the largest scores along the first component suggesting these taxa have relatively robust mandibles linked to their ingesting mechanically challenging seeds (e.g., Bouvier 1986a; Anapol and Lee 1994; Kinzey 1992; but see Marriog et al. 2004) (Fig. 11.3). The two *Cebus* species have the next highest scores supporting earlier work that members of this genus have relatively robust jaws (Kinzey 1974; Bouvier 1986a; Cole 1992; Daegling 1992;

Table 11.3 Principal components analysis (PCA) for shapes related to load resistance in platyrrhine mandibles[1]

Shape Variable[2]	Component 1[3]	Component 2	Component 3	Component 4
Corpus Depth	0.439/**0.78**	0.246/0.29	0.495/0.39	−0.485/−0.34
Corpus Width	0.449/**0.80**	−0.022/−0.03	−0.564/−0.44	−0.442/−0.31
Symphysis Length	0.467/**0.83**	0.072/0.08	0.046/0.04	0.741/**0.52**
Symphysis Width	0.361/**0.64**	−0.538/−0.62	−0.366/−0.29	0.060/0.04
Condyle Length	0.102/0.18	−0.752/−0.87	0.515/0.41	−0.079/−0.06
Condyle Width	0.498/**0.88**	0.279/0.32	0.189/0.15	0.104/0.07
Eigenvalues[4]	3.14/52.3%	1.35/22.4%	0.62/10.3%	0.49/8.2%

[1] PCA was performed on the correlation matrix for shape variables. Corpus depth (SI) and breadth (ML) are measured at M_1. Symphysis length (primarily SI) and width (AP) are measured following Hylander (1985). Condyle length (AP) and width (ML) are measured from the articular surface of the joint.

[2] Shape variables were created by dividing each measure by the distance from the condyle to M_1.

[3] The first value represents the eigenvectors for each component. The second value is the correlation between the original variable and its component score. Bold correlations are significant at $\alpha = 0.05$.

[4] Eigenvalues are reported first followed by the percentage of total variation explained by that component.

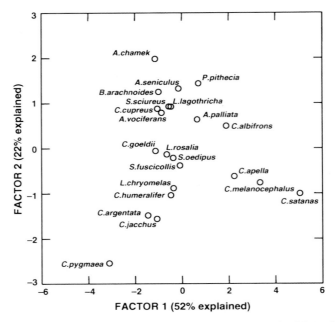

Fig. 11.3 Plot of factors one and two for principal components analysis (PCA) of mandibular shapes across 22 platyrrhine species. See Table 11.3 for descriptions of measurements

Anapol and Lee 1994; Wright 2005). The third pitheciine, *P. pithecia*, and the two folivorous *Alouatta* species are intermediate between these robust forms and the remaining platyrrhine species. The position of *P. pithecia* supports arguments that it is the less robust member of this seed-eating clade (Kinzey 1992; Anapol and Lee 1994). We can hypothesize based on the position of *Alouatta* along this component that folivory is correlated with modest jaw robusticity among platyrrhines. The left half of component one is occupied by the remaining primarily frugivorous, insectivorous and/or gummivorous platyrrhines. The tree-gouging marmosets (i.e., *Callithrix* and *Cebuella*) have relatively gracile jaws among these platyrrhines (Vinyard et al. 2003; Vinyard and Ryan 2006) suggesting that this behavior may not involve relatively large bite forces.

Variation along the second component is primarily contrasting differences in a-p condylar and symphyseal length shapes among platyrrhines. The tree-gouging marmosets along with *C. apella*, *C. melanocephalus* and *C. satanas* possess relatively elongated condyles and symphyses. Previous morphological analyses suggest that anteroposterior condyle length may be more important in facilitating jaw opening ability (i.e., wide gapes) or load resistance at wide gapes (Smith et al. 1983; Bouvier 1986a,b; Vinyard et al. 2003). If a-p condyle length is unrelated to loads or related to load resistance only in this specific mechanical context, then this might help explain its lack of strong contribution to component one and emphasis in component two. Vinyard et al. (2003) suggest that marmosets have several morphological features of their masticatory apparatus, including anteroposteriorly elongated condyles (a measure directly correlated with curvature and hence rotational ability) that facilitate wide jaw gapes during gouging. Similar work has not been done in pitheciines. We also speculate that the elongated symphyses of these taxa may relate to both improved load resistance ability (e.g., Bouvier 1986a) as well as the need to house the enlarged and procumbent anterior teeth possessed by several of these taxa.

11.4.4 Dental Morphology Linked to Feeding

The teeth play a pivotal role in the mechanics of food breakdown as they provide the points of contact between the masticatory apparatus and foods. Thus, the shape of these contacts (i.e., occlusal morphology) and their spatial distribution (related to tooth size) fundamentally affect how foods break down during feeding. It is not surprising then that primatologists have paid considerable attention to the teeth in functional studies linking platyrrhine masticatory apparatus form to feeding behaviors (Zingeser 1973; Kinzey 1974, 1992; Kay 1975; Rosenberger and Kinzey 1976; Hershkovitz 1977; Rosenberger 1978, 1992; Eaglen 1984; Teaford 1985; Greenfield 1992; Martin et al. 2003; Spencer 2003; Wright 2005). This is an extensive body of work and we provide only a synopsis here.

11.4.5 Postcanine Teeth – Chewing

We can initially, albeit imperfectly, divide the toothrow into the postcanine versus anterior teeth based on basic functional roles. The postcanine teeth, particularly the molars, are used in food reduction during chewing, while the anterior teeth are typically employed during ingestion of food bites. Relative molar areas for 16 platyrrhine genera show *Ateles* and several predominantly fruit-eating/insect-eating callitrichines have relatively small molar areas compared to more leaf-eating and seed-eating platyrrhines (Fig. 11.4) (Zingeser 1973; Pirie 1978; Kanazawa and Rosenberger 1989; Rosenberger 1992; Anapol and Lee 1994). The dedicated seed eaters, *Chiropotes* and *Cacajao*, are intermediate in relative molar area (Fig. 11.4), while *Pithecia* exhibits relatively larger molar areas than these two seed predators (Anapol and Lee 1994). The relatively large molar areas of *Cebus* support interpretation that these species ingest and masticate relatively hard and tough foods (Anapol and Lee 1994; Wright 2005). *Callicebus* has the largest relative molar areas among platyrrhines (Fig. 11.4).

Lucas (2004) develops an excellent series of arguments linking molar occlusal morphology to mechanical properties of foods. In short, fruit eaters are hypothesized to have relatively rounded cusps providing broad opposing surfaces for bursting the cell walls of small packets of fruit flesh. Primary leaf eaters are expected to have opposing blades on upper and lower occlusal surfaces that assist in propagating cracks through tough, flat leaves. Insect eaters are predicted to have sharp blades on

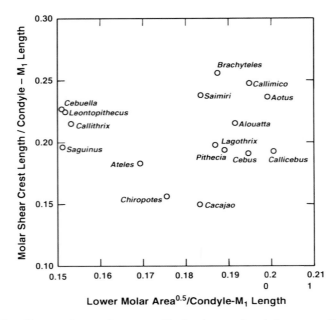

Fig. 11.4 Plot of lower molar area shape versus M_1 shearing crest length shape across 16 platyrrhine genera. Both measures were divided by an estimate of the chewing load arm to create mechanical shape variables. Lower molar area is the sum of individual molar area estimates (length x width) for $M_1 - M_3$ (M_1 and M_2 in callitrichids lacking an M_3). Molar area data are from Kanazawa and Rosenberger (1989), Rosenberger (1992) and Lucas et al. (1986). Molar shearing crest length is estimated as the sum of 6 shearing crest lengths on the M_1 based on data from Anthony and Kay (1993) and Meldrum and Kay (1997). Descriptions of the measured crests are provided in Kay (1977). To increase the number of platyrrhines represented in this plot, species data were averaged at the genus level. Thus, some generic estimates include different measures from separate species. As such, these comparisons only provide a first-approximation of platyrrhine interrelationships and should be used as preliminary evidence for future comparisons among species

their occlusal surfaces to aid in propagating cracks through cuticle. Seed predators should have rounded cusps that fit in opposing basins providing a mortar and pestle effect on these stress-resistant food items.

One simple way of summarizing occlusal morphology compares relative lengths of "shearing" crests along the molar occlusal surface; a technique developed by Kay and colleagues (Kay 1977; Kay and Covert 1984; Covert 1986; Anthony and Kay 1993; Meldrum and Kay 1997; Kirk and Simons 2001). Species consuming large percentages of leaves and insects should have relatively higher values and hence more blade-like crests, while more frugivorous and gramnivorous species should have lower values linked to more rounded cusps (Kay 1975; Lucas 2004). Using previously published data (Anthony and Kay 1993; Meldrum and Kay 1997), the seed predators *Cacajao* and *Chiropotes* along with the highly frugivorous *Ateles*, exhibit the least developed shearing crests relative to M_1 moment arm length (Kinzey 1992; Rosenberger 1992) (Fig. 11.4). Alternatively, the folivorous *Brachyteles* exhibits the most developed shearing crest lengths (Zingeser 1973).

Alouatta while having relatively longer crests than most genera (Rosenberger and Kinzey 1976), does not exhibit an extreme degree of cresting among platyrrhines. This underlines the generalist nature and broad dietary capabilities of howlers (Milton 1980; Kinzey 1997; Di Fiore and Campbell 2007). Similarly, the frugivorous *Lagothrix* and hard-object feeding *Pithecia* and *Cebus* only show moderately less cresting than many platyrrhines.

11.4.6 Anterior Teeth – Ingestion

Platyrrhines use their incisors and canines in as wide a range of ingestive behaviors as any primate clade. Thus, it is not surprising to see a broad range of platyrrhine anterior tooth morphologies (Rosenberger 1992) linked to this behavioral diversity. Rosenberger (1992) provides an excellent, detailed review of the functional morphology of platyrrhine anterior teeth. We rely heavily on this review. Rosenberger (1992) demonstrates that pitheciines have tall, mediolaterally (ml) narrowed and buccolingually (bl) broad lower incisors and robust canines that facilitates whittling down or reducing the exocarp and harvesting seeds from hard and tough fruit pericarps (Kinzey and Norconk 1990; Kinzey 1992; Rosenberger 1992; Anapol and Lee 1994) Tree-gouging marmosets, *Callithrix* and *Cebuella*, have modified their lower anterior teeth to form a sharp wedge facilitating their biting into tree barks to elicit exudate flow (Coimbra-Filho and Mittermeier 1977; Rosenberger 1978, 1983, 1992; Sussman and Kinzey 1984; Nash 1986; Garber 1992; Natori and Shigehara 1992). *Cebus apella* tends to have robust anterior teeth linked to aggressive ingestion of a broad range of potential foods including hard and tough objects (Eaglen 1984; Rosenberger 1992; Anapol and Lee 1994; Wright 2005). The more frugivorous atelines (*Ateles* spp. and *Lagothrix* spp.) tend to have broad spatulate incisors, thought to be related to peeling and ingesting fruits (Eaglen 1984; Rosenberger 1992; Anthony and Kay 1993). Alternatively, atelines that ingest a higher proportion of leaves, *Alouatta* spp. and *Brachyteles* spp., have comparatively reduced incisors that researchers hypothesize is related to a reduced mechanical loading of the anterior teeth in leaf ingestion (Zingeser 1973; Eaglen 1984; Rosenberger 1992; Anthony and Kay 1993). Kinzey (1974) remarked that the very wide incisors of *Aotus* were heavily worn with a flat wear pattern, but dietary information and feeding ecology data are still rather poor for night monkeys (Fernandez-Duque 2007).

11.5 Mechanical Properties of Fruit Ingested by Platyrrhines

11.5.1 Assessing Food Toughness

In a study of the mechanical properties of foods processed by six platyrrhine primates (*Alouatta seniculus*, *Ateles paniscus*, *Cebus apella*, *Cebus olivaceus*, *Chiropotes satanas* (cf. *sagulatus*), *Pithecia pithecia*) in Guyana, South America,

Wright (2004) measured dietary toughness and related this to the way in which these primate species processed selected plant foods and their constituent tissues. The plant diets of these species were divided into the fourteen aforementioned tissue types (Section 11.3.2) and compared using the percentage of each tissue category in the diet (used as an estimate of processing frequency), the average toughness of processed tissues, and the maximum toughness of processed tissues (used as an estimate of peak performance). Fruit and other plant parts were often a composite of tissues that were opened with the anterior teeth and chewed with the cheek teeth. Ranks were calculated separately for plant tissues that were opened with the anterior dentition versus those tissues that were masticated with the postcanine dentition. The nonparametric Friedman and Kruskal-Wallis (Zar 1999) tests were used to compare the diets of these species. These statistics provide a ranked score of dietary demand for each primate species. The ability to incorporate multiple sympatric species in the analyses placed the dietary profile of a single species in the context of the entire community.

The most demanding masticated tissues (i.e., maximum toughness values) were processed by *Alouatta seniculus* and *Cebus apella* (Table 11.4). Although *A. seniculus* and *C. apella* were comparably ranked, their ranks were equal for different reasons. *Cebus apella* masticated (*Astrocaryum vulgare*; palm fruit, $10,909\,\mathrm{Jm^{-2}}$) (Table 11.4) and breached (*Dimorphandra conjugata*; pod, $8,585\,\mathrm{Jm^{-2}}$) (Table 11.5) the tissues with the greatest maximum toughness, despite the fact that the majority of its diet had relatively low toughness values. Out of 436 sampled trees in the *C. apella* habitat, 26 (6.2%) were *D. conjugata* (Wright 2005). This tree species fruited only once during the 14-month study period. This suggests that fallback resources are playing a strong role in shaping the masticatory adaptations of *C. apella*. *A. seniculus* frequently ate tough leaves, but was capable of breaching a single exceedingly tough seed coat (*Catostemma fragrans*). Out of 665 sampled trees within the *A. seniculus* habitat, one (0.23%) was *C. fragrans*. Additionally, this tree fruited

Table 11.4 Percentage of feeding bouts, average toughness ($\mathrm{Jm^{-2}}$), maximum toughness ($\mathrm{Jm^{-2}}$) and species with the maximum toughness values for food items that were masticated by individuals in six primate species in Guyana

Species	N[1]	% Feeding Bout	Average R	Maximum R	Species for Maximum Value
Ateles paniscus	20	51	470	1765	Bignoniaceae (unknown)
Alouatta seniculus	20	77	731	2639	Mimosoideae (Fabaceae): *Mora excelsa*
Cebus apella	22	63	669	10909	Areaceae: *Astrocaryum vulgare*
Cebus olivaceus	32	58	390	2729	Annonaceae: (unknown)
Chiropotes satanas	8	57	389	1031	Lecythidaceae: *Eschweilera sagotiana*
Pithecia pithecia	5	47	309	825	Connaraceae: *Connarus lambertii*

[1] N = number of plant tissue specimens tested.

Table 11.5 Percentage of feeding bouts, average toughness (Jm^{-2}), maximum toughness (Jm^{-2}) and species with the maximum toughness values for food items opened or breached specimens by individuals in six primate species in Guyana

Species	N[1]	% Feeding Bout	Average R	Maximum R	Species for Maximum Value
Ateles paniscus	19	49	839	2139	Polygalaceae: *Moutabea guianensis*
Alouatta seniculus	10	33	1381	7902	Bombacaceae: *Catostemma fragrans*
Cebus apella	13	37	1111	8584	Mimosoideae (Fabaceae): *Dimorphandra conjugata*
Cebus olivaceus	24	42	1042	3449	Tiliaceae: *Apeiba enchinata*
Chiropotes satanas	6	43	1385	2773	Caesalpinioideae (Fabaceae): *Eperua grandiflora*
Pithecia pithecia	5	53	1336	4329	Mimosoideae (Fabaceae): *Inga bourgoni*

[1] N = number of plant tissue specimens tested.

only once during the 14 months. Thus, both frequent use (i.e., leaves) and fallback resources (hard fruit) may play a role in shaping this species masticatory system. The sakis also placed high dietary demands on their anterior dentition. In the case of *Pithecia*, this involved breaching tough seed tissues, whereas fruit pericarps were the toughest tissues breached by *C. satanas* (cf. *sagulatus*). Those species that processed tough tissues with either the anterior dentition or cheek teeth also exhibited marked seasonal shifts in diet. These shifts include a higher percentage of leaves in the diet of *A. seniculus*, an increase in the percentage of embedded insect foraging or palm fruit exploitation in *C. apella*, and increased consumption of legume seeds in *P. pithecia*.

The importance of seasonal changes in dietary emphasis, from brittle to tough plant tissues in *C. apella* and *P. pithecia,* and tough to brittle tissues in *A. seniculus* suggests that masticatory features often identified as 'specializations' may actually facilitate broadening the dietary niche. These features permit the annual exploitation of a broad array of plant tissues that vary widely in toughness, and also may account for variation in the size of geographic ranges. For example, *C. apella*, *A. seniculus*, and *P. pithecia* have larger geographic ranges than *C. olivaceus*, *A. paniscus*, and *C. satanas* (cf. *sagulatus*). Although many factors play a role in the ability of a species to colonize and exploit new habitats, the ability to exploit a wide array of demanding plant and animal tissues appear to be a critical factor.

11.5.2 Food Size and Shape

Food size has a tremendous impact on masticatory function, yet it is infrequently reported in studies of primate dietary ecology. Gape changes the orientation and location of forces at the temporomandibular joint (TMJ) and the angle of contact of the tooth cusps relative to the food substrate. Thickness is arguably the primary factor influencing ingestion technique. Plant food thickness ranges from sheet-like leaves at one extreme to tree trunks at the other. Primates that exhibit relatively wide gapes are those that (1) fix their upper anterior dentition and then use the lower anterior dentition as a plane to strip away layers of bark and wood in the case of the marmosets, (2) open or scrape relatively thick/large fruits in the case of the pitheciines and atelines, or (3) open the mouth widely to place relatively large fruits or seeds on the postcanine dentition, particularly the premolars, to permit the application of relatively high muscle forces, as in the case of *Cebus* spp., particularly *C. apella*. While it is clear that marmosets exploit the thickest items when feeding (i.e., tree trunks for their exudates) and howler monkeys exploit the thinnest foods (i.e., leaves), it is less clear how the diets of "frugivores" vary according to fruit size.

We compiled average fruit thickness and breadth data to compare the dimensions of ingested fruits among *A. paniscus*, *C. apella*, *C. olivaceus*, *C. satanas* (cf. *sagulatus*), and *P. pithecia*. BW compiled data on spider monkeys and two *Cebus* species. Fruit sizes were taken from van Roosmalen (1985) for fruit species that these primates were observed to exploit in Guyana, South America. Data for the pitheciines were collected by MN at Lago Guri, Venezuela. It appears from observations of saki feeding behavior that the greatest fruit dimension is avoided or bypassed during processing. For example, fruit pods are held so that incisive forces are directed perpendicular to the food's long axis (i.e., similar to how humans eat an ear of corn). This feeding behavior eliminates any influence of pod length on jaw gape, but either pod width or breadth may influence the maximum gape used in this behavior. Thus, the results for these two dimensions are shown separately. For fruit breadth, the primate species are arrayed in ascending order according to fruit size (Fig. 11.5). *C. olivaceus* exploited the narrowest fruits followed by *A. paniscus*, *P. pithecia*, *C. apella* and *C. satanas* (cf. *sagulatus*). This, accords well with the findings for condylar length. Findings for fruit thickness (Fig. 11.6) are comparable, with only *C. apella* and *P. pithecia* trading positions.

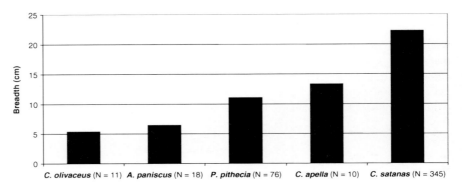

Fig. 11.5 Comparison of the breadth of fruits consumed by six Guiana Shield primates

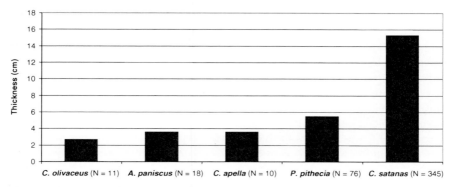

Fig. 11.6 Comparison of the thickness of fruits consumed by six Guiana Shield primates

11.6 Characteristics of the Platyrrhine Gut and Digesta Retention

Most platyrrhines have a generalized gut with some modification in the hindgut (Chivers and Hladik 1980; Lambert 1998). Enlargement of the colon and/or cecum increases the probability that fermentation will improve digestion of harder-to-digest foods, such as dietary fiber and complex sugars. There have been a number of in vivo estimates of digesta retention using indigestible markers, beginning with Milton's (1984a) study of 14 primate species. The measurement of time to first appearance (TFA) of a marker, also referred to as transit time, has been the most commonly used method of estimating digesta retention in primates to date. However, Lambert (1998), and especially Van Soest et al. (1983), noted that mean retention time (MRT) would more accurately estimate retention capability. A major problem of TFA is that the precise time of day the marker is fed to an animal affects the TFA much more than the MRT. Nevertheless, we are obliged to use TFAs here because they are currently available for more primate species than are MRTs (in addition to Milton 1984a 1998; see Power 1996; Power and Oftedal 1996 for callitrichines: Edwards and Ullrey 1999 for *Alouatta* spp.; and Norconk et al. 2002 for *Pithecia pithecia*).

Platyrrhines apparently move digesta through their gut relatively quickly, often within three to eight hours (Lambert 1998). A short TFA suggests that most platyrrhines do not digest much of the cell wall fraction, which can be considerable even in fruit (see Section 11.6.1). Instead, their strategy (at least during active periods) is to move the indigestible ballast of digesta (e.g., seeds, dietary fiber, and chitin) through the gut relatively quickly to make room for more easily digestible foods (Foley and Cork 1992; Power and Oftedal 1996). Only a few platyrrhines, such as *Alouatta* spp. and *Pithecia* spp., appear to retain digesta for longer than their daily period of activity (i.e., longer than 9 to 10 hours. Compare these data with the much longer TFAs in catarrhines documented in Lambert 1998). Interestingly, some platyrrhines adjust the intake of difficult-to-digest food items to occur before a long sleeping or resting period (Chapman and Chapman 1991; Heymann and Smith 1999). By ingesting fibrous foods late in the day, they may both improve digestion and nutrient extraction (at least some platyrrhines are known to defecate only during waking hours (Milton 1998; Norconk et al. 2002)), and shift the energy expended in food search during active periods to digestion during resting periods.

In platyrrhines, the smallest-bodied members of three of the clades have relatively longer TFAs than their larger-bodied relatives. Power and Oftedal (1996) showed experimentally that *Cebuella* and *Callithrix* slowed transit time on a diet of gum arabic compared with TFA on a baseline (non-gum) diet whereas no difference in TFA was found with three larger-bodied tamarins (*Saguinus fuscicollis*, *S. oedipus*, and *Leontopithecus rosalia*). The tamarins, unlike the marmosets, also exhibited reduced digestibility on the gum diet. *Cebuella*, who is most dependent on gum in the wild, had the longest TFA (6.3 hours: Power and Oftedal 1996) of the four callitrichines tested. Among pitheciines, *Pithecia monachus* (Milton 1984a) and *P. pithecia* (Norconk et al. 2002) have TFAs in excess of 15 hours, compared with the estimated five hours for the two larger genera, *Chiropotes* and *Cacajao* (Milton 1984a). Finally, body mass is similar among the atelines, but *Brachyteles*, the largest-bodied of the group, has a transit time of about eight hours (Milton 1984a), compared with the 25- to 32-hour average transit time for particulate markers in the smaller-bodied *Alouatta* spp. (Edwards and Ullrey 1999). The ability to use more ubiquitous resources such as leaves and gums can also influence home range size and energy invested in travel. *Cebuella*, *Pithecia*, and *Alouatta* all have relatively small home ranges and shorter daily paths than sympatric close relatives, *Saguinus*, *Chiropotes*, and *Ateles*, respectively (Milton 1988; Strier 1992; Soini 1993; Di Fiore and Campbell 2007; Norconk 2007).

Gut adaptations and/or intake of fibrous foods (especially the soluble fibers in gums) may result in increased retention of digesta and improve the competitive abilities of these platyrrhines by giving them access to more ubiquitous resources (i.e., gums in the case of the callitrichines and leaves for *Alouatta* and *Pithecia*). The mechanisms might seem counter-intuitive in that high fiber diets are generally considered useful to increase passage rate and decrease constipation in humans. Nevertheless, with the proper gut adaptations, soluble fibers are easily retained (Foley and Cork 1992); whereas insoluble fibers are generally assumed to increase passage rate (e.g., wheat bran). However, finely ground insoluble fibers do not cause laxation in humans (Wrick et al. 1983). Thus, chewing food very thoroughly may increase digesta retention in hind-gut fermenters like the Platyrrhines.

11.7 Nutritional Characteristics of Platyrrhine Diets

11.7.1 Difficulties Comparing Nutritional Characteristics of Plant Parts Ingested by Platyrrhines

The data set in Table 11.6 is a summary of the nutrient composition of at least 128 plant species, with essentially no overlap and very few unidentified species.

Table 11.6 Estimated nutritional values of food items from: Hladik et al. 1971; Gaulin and Craker 1979; Garber 1984, 1993; Nash 1986; Brown and Zunino 1990; Oftedal 1991; Castellanos and Chanin 1996; Power 1996; Silver et al. 2000; Norconk and Conklin-Brittain 2004 and unpublished data; and Hanson et al. 2006. Total non-structural carbohydrates (TNC) = $100 - \%NDF - \%lipids - \%protein - \%ash$. Estimated metabolizable energy (ME) was calculated using the general physiological fuel values of 4 kcal/g for TNC and protein, and 9 kcal/g for lipids (NRC 2003; Conklin-Brittain, Knott and Wrangham 2006)

Food (N)	% NDF	% Lipid	% CP	% Ash	% TNC**	kcals/100g
Exudates (3)	0*	0	18.7	2.9	78.4***	388.4
Arils & Palms (9)	29.7	34.8	7.5	2.5	25.5	445.2
Fungi (4)	74.9*	1.3	9.6	3.3	10.9***	93.7
Fruit Pulp (61)	25.7	4.3	7.6	4.5	58.3	299.0
Whole Fruit (33)	41.8	5.9	8.4	5.6	38.3	240.0
Flowers (18)	44.1	2.3	16.8	6.7	30.1	208.4
Seeds (35)	35.5	15.4	8.9	2.5	37.6	324.6
Young Leaves (34)	51.4	1.7	20.1	6.5	20.3	176.6
Mature Leaves(16)	58.3	1.5	14.4	8.0	17.7	141.7
Insects [†] (4)	32.8	16.4	45.3	2.8	2.7	339.5

[†] The category 'Insects' is a combination of adults and immatures.

* NDF= neutral-detergent fiber or total insoluble fibers. Exudates contain substantial quantities of soluble fiber, which is usually not measured by NDF. In the case of fungi, however, soluble fibers are contaminating the NDF (see Hanson et al. 2006), elevating the value considerably.

** TNC = total non-structural carbohydrates include starch, mono- and disaccharide sugars, and soluble fibers. The soluble fibers are fermented, giving 3 kcal/g fermented material, as opposed to the 4 kcal/g of digested sugar or starch. However, we are assigning 4 kcal/g of the total TNC because we have no data indicating how much is fiber and how much is starch and sugars (Conklin-Brittain, Knott and Wrangham, 2006).

*** These numbers are either mostly soluble fiber in exudates, or in the fungi, artificially low because most of the soluble fiber stayed in the NDF.

Sites from Venezuela, Belize and Panama are the most strongly represented. Not all of the datasets used, however, were as complete as the Lago Guri, Venezuela data. For example, none of the howler monkey datasets reported lipid content, not even for the fruits. Studies of leaf-eating monkeys generally assume that the lipid intake is very low, so it has never been measured, and hence there has never been a study of howler monkeys designed to test this assumption. As a consequence of this, using the small number of leaves in the Lago Guri data, and comparing that to a very large dataset from Kibale Forest, Uganda (Conklin-Brittain unpub. data), we have assigned approximate lipid values, one for leaves, one for flowers, and one for fruit. We chose values that may be lower than reality because fat content so heavily influences energy content (i.e., 9 kcal/g for fat versus 4 kcal/g carbohydrates and protein), and we did not want to artificially elevate the ME values. We wanted to keep this data set purely Neotropical, and therefore we used the African data only to reassure ourselves that we had chosen reasonable values.

In addition, different laboratory methods have been available historically (i.e., during the time span of these reports), making it difficult to combine all of these data, especially with respect to fiber analysis. Thus, the most historical data set (Hladik et al. 1971) has had a conversion factor applied to "update" the fiber values

(National Research Council (NRC) 2003: Table 3.2, pp. 65–66). On the one hand, this is worthwhile because this study adds species not repeated in other reports. On the other hand, conversion factors generally blur the detailed complexity that may exist.

A different conversion factor was applied to fiber values reported by Castellanos and Chanin (1996). They published the largest and most complete nutrient analysis of spider monkey diets to date, but the fiber values are lower than we expected. We strongly suspect that they reported acid-detergent values rather than neutral-detergent values, but there is insufficient detail in the methods to determine if this is indeed the case. Consequently we applied the conversion factor recommended to convert ADF values to NDF values, in NRC (2003: Table 3.2, pp. 65–66) because acid-detergent values cannot be used in the calculations of energy content (Conklin-Brittain et al. 2006).

11.7.2 Nutritional Characteristics of Plant Parts Ingested by Platyrrhines

The report that exudates have zero fiber is not accurate (Table 11.6); this is a reflection of the method used to quantify fiber. Neutral detergent only extracts insoluble fiber, and exudates like gums are mostly or completely soluble fibers (complex non-starch polysaccharides or NSP). The procedure for assessing Total Dietary Fiber (TDF) measures insoluble and soluble fractions separately, but it is very expensive and rarely used except on human foods. In the system we are reporting here, the percentage of NDF, protein, lipid and ash are subtracted from 100% to calculate percentage of total nonstructural carbohydrates (TNC) and the soluble fibers (NSP) are therefore included in the TNC by default. This means that the physiological fuel value of 4 kcal/g of carbohydrate is probably too high, but we do not know how much of the TNC is starch and simple sugars versus NSP (i.e., soluble fiber) for these wild plant exudates. The NSP are digested through fermentation (Nash 1986; Lambert 1998) in the large intestine (or cecum) and thus the physiological fuel value is at most 3 kcal/g of NSP (Conklin-Brittain et al. 2006). Thus the ME value is probably a slight overestimate for the callitrichine genera.

Ingestion of arils by platyrrhines is underestimated in our summary of feeding behavior (Table 11.1) because primate ecologists tend to lump arils with fruit or seeds. However, arils are widely known for their lipid-rich qualities (e.g., *Virola* spp. and palm fruit) (Aguiar et al. 1980; Howe and Vande Kerckhove 1981; Moermond and Denslow 1985; Forget 1991). Fruit-producing arils are often characterized as bird-dispersed fruits (e.g., Janson 1983), but they also figure prominently in *Ateles* (Russo 2005; Russo et al. 2005), *Pithecia pithecia* (Norconk and Conklin-Brittain 2004) and *Callicebus torquatus* diets (Palacios et al. 1997).

The fungi values used in Table 11.6 are exceptionally high in NDF. According to Hanson et al. (2006) the analyses used were not the traditional NDF procedure because there were severe filtering difficulties. A method equivalent to the TDF method was used and as a result all fibers, soluble and insoluble, are contained within, and therefore elevate, the value listed here as "NDF". The NDF usually does not include soluble fibers. Fungi were equivalent to fruit pulp in terms of crude protein. This suggests that fungi could be a seasonal protein substitute for ripe fruit for *Callimico* (Porter 2007), but ME values of fungus were by far the lowest in our sample. Hanson et al. (2006) conclude that digestion trials are needed to better understand the nutritional value of fungi.

Since NDF acts as a feeding deterrent (Wrangham et al. 1998), adding ballast but not nutritional value for nonspecialists, it is not surprising that fruit pulp is the most widely used platyrrhine food type (Table 11.1). Fruit pulp is among the least fibrous of foods and has the second highest (and most accessible) total digestible carbohydrate component (Table 11.6). Whole fruit, where both pulp and seeds are chewed up and digested, can have considerably more fiber than the fruit pulp in this data set (Wrangham et al. 1993).

Flowers are moderately high in fiber too, and their protein content is as high as leaves. Nectar can be an important dry season resource (Garber 1988, 1993; Terborgh 1983), but the nectar in flowers would have to be sampled separately if it is important to differentiate it from the rest of the flower. The TNC fraction, where you would expect to find the nectar sugars, is lower here than that for fruit pulp and even whole fruit. The ME of nectar from resources like *Symphonia globulifera* would perhaps resemble exudates more closely than the flowers in this sample.

Seeds are more energy dense than are fruit, flowers or leaves. It is not surprising that most of the larger (> 1 kg body weight) primates eat seeds. Seeds ranked relatively high in ME due to their high lipid levels, but are challenging resources since they are often protected mechanically (see above; Kiltie 1982; Lucas et al. 2000) or chemically (e.g., Waterman and Kool 1994; Guimarães et al. 2003). They are also relatively low in protein.

Young leaves in this summary are somewhat, but not dramatically, different from mature leaves. This sample represents mature leaves that are actually eaten, so it is not surprising that they would be nutritionally similar to young leaves. Overall both leaf types are not very energetically dense.

Our small insect sample includes adults and immature stages combined. Soft-bodied insects, such as caterpillars, are likely to have a much higher ME value (Milton 1984b) and are important seasonal resources for some platyrrhines (Veiga and Ferrari 2006). Insects, as with exudates and fungi, need more nutrient analyses performed on a greater diversity of species.

11.7.3 Estimating Metabolizable Energy (ME) Intake in Platyrrhines

We have intensely scrutinized the data from the literature and summarized it according to food type in Table 11.6 and consumer species in Table 11.7. To create Table 11.7, the percentage of time spent feeding on a given food type (Table 11.1) was multiplied by the amount of each nutrient in each food type (Table 11.6), giving

Table 11.7 Nutrient and energy densities of the diets consumed by 16 platyrrhine genera and rank of platyrrhine genera based on the estimated energy density of their diets. These weighted averages were calculated by multiplying the percentage of each food type in the diet (Table 11.1) times the nutrient values for each food type (Table 11.6). Body mass estimates are averaged from Smith and Jungers (1997: Table 1)

Primate genus	Body mass (kg)	NDF[1] g/100 g of diet	Lipid g/100 g of diet	CP[2] g/100 g of diet	Ash g/100 g of diet	TNC[3] g/100 g of diet	Estimated ME[4] kcal/100 g of diet	Rank
Cebuella	0.12	11.9	5.3	25.4	2.9	54.5	367	1
Callithrix	0.37	16.9	7.1	27.3	3.1	45.6	355	2
Saguinus	0.48	25.3	8.9	25.5	3.3	36.9	330	3
Callimico	0.48	42.6	8.4	23.8	3.4	21.8	258	13
Leontopithecus	0.59	24.9	6.6	18.2	3.8	46.5	318	5
Saimiri	0.81	33.4	11.2	31.9	3.8	19.6	307	8
Aotus	0.93	38.8	3.0	14.0	5.7	38.6	237	14
Callicebus	1.05	30.9	7.6	10.6	4.1	46.9	298	10
Pithecia	2.23	35.4	12.1	10.4	3.5	38.7	305	9
Chiropotes	2.86	30.6	10.4	9.7	3.4	45.9	316	6
Cebus	2.96	29.8	8.8	20.8	3.8	36.7	310	7
Cacajao	3.05	31.1	11.2	8.3	2.9	46.5	320	4
Alouatta	6.32	41.6	2.5	14.0	5.9	35.9	223	16
Lagothrix	7.68	27.9	5.1	11.6	4.3	51.1	296	11
Ateles	8.56	29.1	4.0	9.2	4.8	52.9	284	12
Brachyteles	8.84	40.4	3.0	14.1	5.6	36.9	231	15

[1] NDF = neutral detergent fiber
[2] CP = crude protein
[3] TNC = total nonstructural carbohydrates
[4] ME = metabolizable energy.

the estimated grams of each nutrient in the diet for each primate genus (Table 11.7). Using these values, the metabolizable energy as kcal per 100 g of diet was calculated using the physiological fuel values 4 kcal/g carbohydrate and protein and 9 kcal/g lipid as discussed above.

Classic energetics studies state that the smaller the mammal the higher the energy requirements per kg of body mass (Blaxter 1989, pp. 123–133). In addition, and because of gut size restrictions in small animals, they need a higher caloric density in their food compared to larger mammals (Robbins 1993). Our values listed in Tables 11.1, 11.6, and 11.7 are the result of averaging values from various study sites and plant foods from all around South and Central America. Nevertheless, our data sets comply with these over-arching principles regarding body mass and caloric density of food (regressing kcal/100 gm of diet against body mass, Table 11.7; $r^2 = 0.27$, $p = 0.041$). Consequently we feel confident that our summaries in Tables 11.6 and 11.7 are reasonable.

The ability to eat higher fiber diets is supposed to increase with increasing body size; however, we did not find that to be true using this data set ($r^2 = 0.11$, $p = 0.20$). On the other hand, removing *Callimico* because they eat so much fungus and the fiber analysis of fungus was problematic, and removing *Aotus* because their intake of insects (a somewhat low fiber food) was perhaps underestimated, the rest of the genera follow the rule that fiber concentration increases with body size ($r^2 = 0.31$; $p = 0.04$).

Using our nutritional sample as representative of Neotropical plants, we found that *Cebuella* and *Callithrix*, the smallest-bodied platyrrhines, had the highest ME intakes due to the very high proportion of exudates in their diets (Table 11.1). The two closely related species, *Saguinus* and *Leontopithecus*, ranked lower than the marmosets, apparently due to higher intake of fruit pulp. A better year-round estimate of nectar intake (for *Saguinus*) and estimating insect intake separately for soft-bodied and hard-bodied insects may increase the ME estimate for tamarins. Nectar intake of 22.1 and 30.6% for a dry season month for *Saguinus fuscicollis* and *S. mystax*, respectively (Garber 1988:101), suggests that ME is underestimated for these tamarins. Closely related *Callimico* shows a distinctly different pattern than the marmosets and tamarins, ranking near the bottom in ME due exclusively to their high ingestion of fungi. Unlike insects that provide a mix of total digestible carbohydrates depending on NDF and lipid values, fungi analyzed by Hanson et al. (2006) appear to be uniformly low in lipids and protein, and high in NDF. High molar shearing crests may facilitate reduction of fungus particle size and improve the potential for digestion of *Callimico*'s key fallback food in the dry season (Hanson et al. 2006; Porter 2007).

Pithecia, ranked below the two other pitheciines that ingest a higher proportion of seeds. Presently, dietary studies suggest that *Pithecia* has a more diverse diet than *Chiropotes* or *Cacajao*, which means a lower proportion of seeds. The estimate of lipid intake for *Pithecia* is the highest for the platyrrhines based on combined seed and aril intake (Table 11.1), but the sakis also have a high fiber intake which reduces their ME estimate. The sister group to the pitheciines, *Callicebus*, ranked just below *Pithecia*. Interestingly, their diets appear to be quite similar if the intake of seeds (high in *Pithecia*; low in *Callicebus*) is exchanged with the intake of fruit pulp (low in *Pithecia*; high in *Callicebus*). Both ingest more young leaves than *Cacajao* and *Chiropotes*, but due to the relatively few species that are well represented in dietary studies it is difficult to capture the diversity in *Callicebus* spp.

The atelines all rank in the lower third in terms of ME estimates. The general prescription for a diet high in ME is one that is high in lipids and low in NDF. Fruit pulp, the major component of *Ateles* diets, is relatively low in both NDF and lipids. *Ateles* and *Lagothrix* both have relatively high intakes of fruit pulp, but ingestion of insects by *Lagothrix* appears to be the factor pushing them just above *Ateles* in ME rank. *Brachyteles* and *Alouatta* rank last among the platyrrhines due to their high intake of young leaves and perhaps a fallback strategy of ingesting some mature leaves (although the differences between the two in our composite database were not great). There are three possible problems with the ateline data. First, the ME estimate for *Ateles* would be more accurate with an improved estimate of the contribution of arils to their diets. Second, a recent study of *Brachyteles* (Talebi et al. 2005) suggests that fruit intake is higher than was reported in earlier studies (Milton 1984c; Strier 1991). The ME we obtained for *Brachyteles* is based on a diet that is about half leaves and half fruit, (i.e., close to that of *Lagothrix*). Third, the relatively low apparent protein value for *Ateles* is a reflection of the balance of intake that is weighted heavily toward ripe fruit. While reports of *Ateles* spp. diets are somewhat variable (leaves ranging 7–17%: Di Fiore and Campbell 2007; Table 10.3) our choice of a representative intake of 11% (Table 11.1) is not low for many studies. Thus, our finding could indicate a protein value that is lower than expected, but accurate for the animal, or it could be related to a broader life history strategy, which for *Ateles* spp. is among the slowest of the non-hominoid primates (Strier 2006; Di Fiore and Campbell 2007).

Saimiri and *Cebus* are closely ranked in the middle of the ME values just below the callithrichines and larger-bodied saki-uacaris (Table 11.7). The insect component of their diets was the highest of any non-callitrhichine and the dietary figures we used for *Saimiri* suggest that they have the highest intake of insects among platyrrhines. We suspect that their rank in terms of ME may be higher with a better estimate of the insect component of their diet. The dietary composition we used for *Aotus* had a relatively high intake of young leaves, four times that of *Saimiri*, but probably a lower estimate of insects than is realistic (Fernandez-Duque 2007). As a result, the position of *Aotus* in the ranking near the bottom may be too low.

11.8 Integrating Morphology, Dietary Properties and Nutritional Data

To explore potential relationships among masticatory morphology, dietary properties and nutrient intake in platyrrhines, we created a single summary variable from each dataset that attempts to capture the overall diversity across the clade. In some cases, we calculated this descriptive index for all available species. However to facilitate comparing the three datasets, we averaged species data at the generic level. All 16 extant platyrrhine genera are represented in the morphological and nutritional datasets, while only six genera have available data on dietary mechanical properties.

We combined several measures of masticatory apparatus form to build a biomechanical robusticity index for platyrrhines. Similar to Anapol and Lee (1994), we averaged z-scores for 10 relative measures of the masticatory apparatus related to bite force production, load resistance and dental function to generate a robusticity score for a species (Fig. 11.7). The two pitheciines, *C. satanas* and

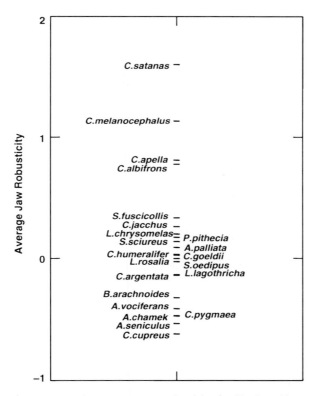

Fig. 11.7 Plot of average masticatory apparatus robusticity for 22 platyrrhine species. To estimate average jaw robusticity, we first calculated z-scores for 10 variables related to masticatory apparatus bite force production, load resistance and dental function: (1) M_1 biting efficiency, (2) symphysis length shape, (3) symphysis width shape, (4) corpus depth shape, (5) corpus width shape, (6) condyle width shape, (7) lower incisor area shape, (8) maxillary canine volume shape, (9) lower molar area shape, and (10) shearing crest length shape. We used the absolute values of z-scores for the final two measures (#9, #10) given that extremes represent mechanical solutions to different challenges posed by primate diets (see text for discussion). Shapes were created by dividing by the moment arm for biting at the M_1 or incisors. Z-score values for these measures were averaged to obtain an estimate of average jaw robusiticty in a species. In some species, data were unavailable for all 10 measures. We included species that had 6 or more of these dimensions in their average. Data for lower incisors was taken from Rosenberger (1992) and canine volumes from Thoren et al. (2006). Sources for molar dimensions (#9 and #10) are provided in Fig. 11.4

C. melanocephalus, and the two cebids, *C. apella* and *C. albifrons*, exhibit the largest average scores for this masticatory apparatus index (Fig. 11.7). While this is not unexpected, given previous research, it is surprising that the more folivorous *Alouatta* spp. and *B. arachnoides* are not differentiated from other fruit- and insect-eating platyrrhines. These results suggest that the larger size of these animals may provide them sufficient performance abilities in chewing leaves and/or their diet may not be as tough as previously thought (Teaford et al. 2006).

We represented the dietary mechanical properties dataset using the mean of the average toughness estimates taken from masticated (Table 11.4) and breached (Table 11.5) items, respectively. While there are inter-specific differences in toughness values between masticated and breached items, their average is adopted as a broad measure of overall loading experienced in the masticatory apparatus during feeding. The dataset from Turtle Mountain was supplemented with dietary toughness data from a two-month study of common marmosets (*Callithrix jacchus*) at Estação Ecológica do Tapacurá Pernambuco, Brasil (Vinyard et al. n.d.).

The metabolizable energy estimate (ME) from Table 11.7 represents variation in dietary nutrition among platyrrhine genera. Because this dataset is negatively correlated with body mass among platyrrhines, we examined both the absolute ME and the residual ME values from regression on body mass.

11.8.1 Comparisons of Metabolizable Energy (ME), Overall Masticatory Apparatus Shape and Average Dietary Toughness

Metabolizable energy (ME) shows little association with the jaw robusticity index (Fig. 11.8). When comparing residual ME, relative to body mass, (data not shown) a similar pattern is evident with the main difference being that the five callitrichid genera tend to have reduced relative values with respect to the remaining platyrrhines. The three genera with the most robust jaws (*Chiropotes*, *Cacajao* and *Cebus*) tend to have intermediate absolute ME estimates. Alternatively, they have higher relative ME estimates. This increase in relative ME may suggest that evolutionary changes in these three genera might have involved increasing jaw robusticity to provide the mechanical capacity for accessing structurally challenging, but energy-rich foods in their respective environments.

Dietary toughness and jaw robusticity show little association among the six platyrrhine genera represented here (Fig. 11.9). The small sample for dietary toughness precludes any definitive statements regarding the potential relationship between jaw shapes and dietary properties. While *Alouatta* and *Callithrix* have the highest average toughness values, they arrive at the top for different reasons. *Alouatta* has the highest average toughness during mastication. *Callithrix* breaches foods, specifically tree barks, with high toughness values. Marmosets may circumvent the high toughness of barks and wood fibers by planing off layers of tissue rather than cutting through fibers. The genera with relatively robust jaws tend to exhibit

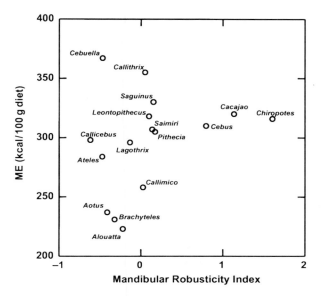

Fig. 11.8 Plot of metabolizable energy (kcal/100 g diet) versus the mandibular robusticity index for 16 platyrrhine genera

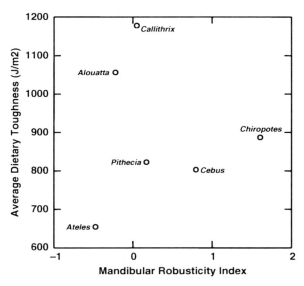

Fig. 11.9 Plot of average dietary toughness (J/m²) versus the mandibular robusticity index for 6 platyrrhine genera

only intermediate toughness values. Collectively, this suggests that jaw robusticity may not share a particularly strong association with average dietary toughness. Jaw robusticity may be more closely associated with infrequent use of fallback or temporally limited resources. In this context, it is worth mentioning that if maximum values are compared, then *Cebus* diets are among the toughest foods sampled among these six genera (Wright 2004).

Average dietary toughness and metabolizable energy also do not show a consistent pattern across these six genera (Fig. 11.10). With the exception of *Alouatta*, there is a potential direct association between these two variables for the remaining genera. However, this pattern is contingent on the exudativorous *Callithrix* and as such really requires additional species to verify the direction of the trend. Hill and Lucas (1996) found that fiber (NDF) correlated well with toughness for Japanese macaques' (*Macaca fuscata*) leaf foods. One would expect that as fiber content (or toughness) increases, metabolizable energy would decrease. Hence it is odd to see the trend in Fig. 11.10, where as toughness increases, so does the kcals/100 g of diet (except for the *Alouatta*). Unfortunately NDF cannot be used to double check this trend (e.g., by regressing NDF against our factorially calculated ME), because the NDF value was used in part of the calculation of the ME, so they are not statistically independent values. While it remains possible that South American taxa may show similar intra-specific patterns as seen in *M. fuscata*, we need more data using an independent toughness measure to evaluate this relationship.

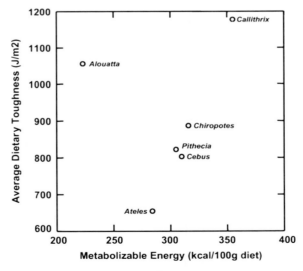

Fig. 11.10 Plot of average dietary toughness (J/m^2) versus metabolizable energy (kcal/100 g diet) for 6 platyrrhine genera

11.8.2 Synthesizing Nutrition, Jaw form and Food Properties

The extant platyrrhines represent an infraorder of at least four primate radiations, some of them dating from the early to middle Miocene (Setoguchi and Rosenberger 1987; Kay et al. 1998; Fleagle and Tejedor 2002). Compared with the more recent expansion of the cercopithecoid monkeys, platyrrhines exhibit considerable variation, in body mass and a variety of specializations, for extracting resources. As such, it is difficult to capture the diversity of the group with a few summary variables. Nevertheless, top-down analyses can be an effective means to illuminate patterns and point out deficiencies in the data. Based on our summary measures of nutritional intake and masticatory apparatus, only the larger bodied pitheciines (*Cacajao* and *Chiropotes*) had both relatively high ME intake and robust jaws. The callitrichines showed nearly the opposite pattern of having a high ME intake (by virtue of the proportion of gums in their diet) and relatively low jaw robusticity.

We are careful to point out that a poor fit between these two variables as we constructed them does not signify that nutritional and mechanical data are unrelated in platyrrhines. First, these variables could be correlated in certain species or clades. Second, other specific nutritional and mechanical variables, rather than our summary measures, may be related across platyrrhines. Finally, examining patterns at the generic level may have obscured important relationships among species. Discovering whether these potential relationships exist will require additional data and future analyses. Therefore, our results do not preclude the co-evolution of masticatory form and dietary nutrition in platyrrhines. Based on these data, however, any co-evolution linking dietary nutrition and jaw robusticity appears to follow species- or clade-specific patterns that likely differ throughout the infraorder.

We also failed to demonstrate clear relationships between dietary mechanical properties and either jaw robusticity or metabolizable energy. Our attempt to compare food properties across platyrrhines is somewhat premature because of the lack of available data. Therefore, we reserve any conclusions regarding these potential relationships. Data are needed both for additional species and other mechanical properties. Describing the mechanical variation in diets must be a priority for future studies of platyrrhine feeding adaptations. Data on food mechanics should also be coupled with detailed analyses of fruit availability and abundance in order to assess the role of fallback or temporally limited resources in shaping diet, ranging, and masticatory anatomy among platyrrhines. Fallback resources are often aseasonal resources and widely available (e.g., bark, leaves, tree exudates), but are used for only a subset of the year. Fallback foods are not considered to be preferred resources (i.e., not taken in the relative abundance in which they occur) whereas seasonally exploited resources, even if difficult to access, are taken disproportionately to their abundance. This difference is critical for differentiating the feeding strategies of primates.

Having stated these caveats, we provide a final assessment of our analysis. Platyrrhines illustrate an incredible diversity of morphology and diet in Primates. They are similar in that all are essentially arboreal and ingest fruit when it is available. Plant diversity in the Neotropics has enabled the evolution of at least four radiations – they occur in highest densities in Amazon Basin forests near the equator where plant diversities are also highest (Peres and Janson 1999). Despite the high level of diversity, we identified two patterns that help to define the platyrrhines. Within-clade (subfamily) variation tends to involve either ME or mandibular robusticity while the other variable(s) remain relatively stable (Fig. 11.8).

Cebus and *Saimiri* are very different in social behavior and in the components of their diets, but multiple species are sympatric and some (*C. apella* and *S. sciureus*) form mixed-species groups (Terborgh 1983; Podolsky 1990). From our analysis, differences between these genera lie in mandibular and dental robusticity while metabolic intake is very similar. Among pitheciines, *Pithecia* spp. have a broader geographic and habitat distribution than *Chiropotes* and *Cacajao*. By adding the fourth member of this clade (*Callicebus*) the pitheciines span the entire continuum of diversity in platyrrhine mandibular robusticity (Fig. 11.8). *Chiropotes* and *Cacajao* are largely allopatric, but both are sympatric in some parts of their range with *Pithecia,* and with both *Pithecia* and *Callicebus* in the Amazon Basin. *Pithecia* and *Callicebus* are the smaller-bodied generalists that add leaves and insects to seeds to form their primary diet. As in the cebines, the range of ME variation in the pitheciines is very low.

While allometric differences help explain variation in the cebine and pitheciine radiations, a different pattern emerges in the atelines. Variation in mandibular robusticity among atelines is low, but with a considerable range of variation in dietary ME (Fig. 11.8). Yet, the four genera present a dispersed cluster with relatively low ME and low robusticity. A similar pattern is demonstrated within callitrichines clustering in the high ME, low mandibular robusticity quadrant. In this clade, the smallest species are the most specialized – the opposite pattern of the pitheciines.

11.9 Future Directions in Studying Platyrrhine Feeding Adaptations

It is commonplace for review chapters to call for improvements in data collection and analysis. We are no different and we think that these statements serve to remind us that more precise data will further our efforts in studying platyrrhine feeding adaptations. With respect to morphometric analyses of jaw mechanics, we argue that (1) additional in vivo studies are needed to validate morphometric proxies of mechanical abilities and (2) morphometric measures can be improved to provide more precise estimates of mechanical ability (e.g., Deagling 2007).

Analyses of dietary mechanical properties in the field are still in their relative infancy. We examined data on dietary toughness in this chapter, but the technology is currently available to measure other properties such as stiffness, hardness and friction in primate diets (Lucas et al. 2001). In particular, it would be helpful to document variation among platyrrhines for two fragmentation indices, $(E^*R^{0.5})$ and $(R/E)^{0.5}$ (Agrawal et al. 1997; Lucas et al. 2002; Lucas 2004). These indices describe stress-limited and displacement-limited patterns of food breakdown, respectively, and would help to document expected morphological features that might be associated with a particular species diet. We anticipate that documenting variation in food properties will provide some of the most important advances in our understanding of platyrrhine jaw and tooth form over the next decade.

The relationship between primate feeding ecology and an interest in nutrition is gaining ground with more researchers recognizing the relevance of not only quantifying intake, but also collecting relevant food parts for nutritional analysis. Of particular importance are seasonal variation in food availability, the use of fallback resources, and the intake of nutrients and feeding deterrents. In the past, primate research has practiced the conventional wisdom that for herbivores and omnivores it is more important to balance nutrients and avoid secondary plant compounds than to determine the energy content of foods. However, increasing evidence shows that reproduction in primates, and hence fitness, is dependent on overall energy intake (Knott 2001, 2005; Emery-Thompson et al. 2007). The feeding selectivity that we see in herbivorous and omnivorous primates is still meant to achieve optimal energy intake, although initially the consumer needs to balance nutrients and avoid digestion inhibitors. The practical impact of this is that complete analyses have to be performed on all foods in order to continue making progress with this theory. To continue using factorial estimates of metabolizable energy as the nutrition variable, a complete laboratory analysis consists of protein, lipid, total ash, and neutral-detergent fiber (NDF). Total dietary fiber (TDF) might be preferred over NDF for the gummivorous species, because it quantifies the soluble fiber as well as the insoluble fiber.

Finally, we urge further integration of phylogenetic, morphological, nutritional, dietary mechanical, and ecological research methods in studying platyrrhines. Even though we saw little evidence for associations among these different data sets in platyrrhines, we still consider the integration of these otherwise disparate research agendas as a great source of potential advancement in our understanding of platyrrhine feeding adaptations. When possible, we recommend collecting all four kinds of data in single species or population. As part of this integration, traditionally lab-based techniques need to be taken to the field and joined with behavioral ecology research on free-ranging platyrrhines (Wright 2005; Williams et al. 2008). In addition, non-invasive captive animal nutritional work comparing in vivo digestibilities and dietary properties, as well as the effects on digestibility of different particle sizes resulting from different tooth morphologies may help identify where these different areas of research can work together. Through increasing our precision and integration, we hope to further our understanding of both the range and patterns of adaptations for accessing foods among platyrrhine primates.

Acknowledgments We wish to thank Paul Garber, Alejandro Estrada, César Bicca-Marques, Eckhard Heymann and Karen Strier for inviting us to combine our expertises and think broadly about platyrrhine feeding adaptations, and Alfie Rosenberger for critical comments on the manuscript. We thank curators and staff members at the Field Museum of Natural History, National Museum of Natural History, American Museum of Natural History, Natural History Museum of London, Natural History Museum, Basel, and Museum National d'Histoire Naturelle for access to skeletal collections. Marilyn A. Norconk wishes to thank the co-authors, particularly Chris Vinyard, for continuing to work on the chapter while she was in the field. She is also grateful to Luis Balbás and members of Estudios Básicos, EDELCA-Guri for logistical support in Venezuela, and students especially Terry Gleason, Brian Grafton, Jason Brush, and Suzanne Walker. Barth Wright would like to thank the support staff, rangers and student assistants of the Iwokrama Reserve, Guyana for logistical support in the field. Funding for this research was supported by NSF (SBR 98-07516, SBR-9701425, BCS-094666, BCS-0412153, BCS-9972603), The Wenner Gren Foundation for Anthropological Research #6138, The Leakey Foundation, The National Geographic Society, and Sigma Xi.

References

Agrawal, K. R., Lucas, P. W., Prinz, J. F., and Bruce, I. C. 1997. Mechanical properties of foods responsible for resisting food breakdown in the human mouth. Arch. Oral Biol. 42:1–9.

Agrawal, K. R., Lucas, P. W., Bruce, I. C., and Prinz, J. F. 1998. Food properties that influence neuromuscular activity during human mastication. J. Dent. Res. 77:1931–1938.

Aguiar, J. P. L., Marinho, H. A., Rebelo, Y. S., and Shrimpton, R. 1980. Aspectos nutritivos de alguns frutos da Amazônia. Acta Amazonica 10:755–758.

Altmann, S. A. 1998. Foraging for Survival: Yearling Baboons in Africa. Chicago: University of Chicago Press.

Anapol, F., and Lee, S. 1994. Morphological adaptations to diet in platyrrhine primates. Am. J. Phys. Anthropol. 94:239–261.

Anthony, M. R. L., and Kay, R. F. 1993. Tooth form and diet in Ateline and Alouattine primates: reflections on the comparative method. Am. J. Sci. 293-A:356–382.

Ayres, J. M. 1986. Uakaris and Amazonian Flooded Forest. PhD Dissertation. University of Cambridge.

Ayres, J. M. 1989. Comparative feeding ecology of the Uakari and Bearded Saki, *Cacajao* and *Chiropotes*. J. Hum. Evol. 18:697–716.

Blaxter, K. L. 1989. Energy Metabolism in Animals and Man. New York: Cambridge University Press.

Bouvier, M. 1986a. A biomechanical analysis of mandibular scaling in Old World monkeys. Am. J. Phys. Anthropol. 69:473–482.

Bouvier, M. 1986b. Biomechanical scaling of mandibular dimension in New World Monkeys. Int. J. Primatol. 7:551–567.

Bouvier, M., and Tsang, S. M. 1990. Comparison of muscle weight and force ratios in New and Old World monkeys. Am. J. Phys. Anthropol. 82:509–515.

Brown, A. D., and Zunino, G. E. 1990. Dietary variability in *Cebus apella* in extreme habitats: evidence for adaptability. Folia Primatol. 54:187–195.

Cachel, S. M. 1979. A functional analysis of the primate masticatory system and the origin of the anthropoid post-orbital septum. Am. J. Phys. Anthropol. 50:1–18.

Castellanos, H. G., and Chanin, P. 1996. Seasonal differences in food choice and patch preference of long-haired spider monkeys (*Ateles belzebuth*). In M. A. Norconk, A. L. Rosenberger and P. A. Garber (eds.), *Adaptive Radiations of Neotropical Primates* (pp. 451–466). New York: Plenum.

Chapman, C. A., and Chapman, L. J. 1991. The foraging itinerary of spider monkeys: when to eat leaves? Folia Primatol. 56:162–166.

Chapman, C. A., Chapman, L. J., Rode, K. D., Hauck, E. M., and McDowell, L. R. 2003. Variation in the nutritional value of primate foods: among trees, time periods, and areas. Int. J. Primatol. 24:317–333.

Chew, C. L., Lucas, P. W., Tay, D. K. L., Keng, S. B., and Ow, R. K. K. 1988. The effect of food texture on the replication of jaw movements in mastication. J. Dent. 16:210–214.

Chivers, D. J., and Hladik, C. M. 1980. Morphology of the gastrointestinal tract in primates: comparisons with other mammals in relation to diet. J. Morphol. 166:337–386.

Coimbra-Filho, A. F., and Mittermeier, R. A. 1977. Tree-gouging, exudate-eating and the "short tusked" condition in *Callithrix* and *Cebuella*. In D. G. Kleiman (ed.), *The Biology and Conservation of the Callitrichidae* (pp. 105–115). Washington DC: Smithsonian Institution Press.

Cole, T. M. 1992. Postnatal heterochrony of the masticatory apparatus in *Cebus apella* and *Cebus albifrons*. J. Hum. Evol. 23:253–282.

Conklin-Brittain, N. L., Wrangham, R. W., and Hunt, K. D. 1998. Dietary response of chimpanzees and cercopithecines to seasonal variation in fruit abundance. II. Macronutrients. Int. J. Primatol 19:971–998.

Conklin-Brittain, N. L., Knott, C. D., and Wrangham, R. W. 2006. Energy intake by wild chimpanzees and orangutans: methodological considerations and a preliminary comparison. In G. Hohmann, M. M. Robbins and C. Boesch (eds.), *Feeding ecology in apes and other primates: ecological, physical and behavioral aspects* (pp. 445–471). New York: Cambridge University Press.

Covert, H. H. 1986. Biology of early Cenozoic primates. In D. R. Swindler and J. Erwin (eds.), *Comparative Primate Biology, Systematics, Evolution, and Anatomy* (pp. 335–359). New York: Alan R. Liss.

Daegling, D. J. 1992. Mandibular morphology and diet in the genus *Cebus*. Int. J. Primatol. 13: 545–570.

Darvell, B. W., Lee, P. K. D., Yuen, T. D. B., and Lucas, P. W. 1996. A portable fracture toughness tester for biological materials. Meas. Sci. Tech. 7:954–962.

Deagling, D. J. 2007. Morphometric estimation of torsional stiffness and strength in primate mandibles. Am. J. Phys. Anthropol. 132:261–266.

Di Fiore, A. 2004. Diet and feeding ecology of woolly monkeys in a western Amazonian rain forest. Int. J. Primatol. 25:767–801.

Di Fiore, A., and Campbell, C. J. 2007. The Atelines: Variation in ecology, behavior, and social organization. In C. J. Campbell, A. Fuentes, K. C. MacKinnon, M. Panger and S. K. Bearder (eds.), *Primates in Perspective* (pp. 155–185). Oxford UK: Oxford University Press.

Eaglen, R. H. 1984. Incisor size and diet revisited: The view from a platyrrhine perspective. Am. J. Phys. Anthropol. 64:263–275.

Eaglen, R. H. 1986. Morphometrics of the anterior dentition in strepsirhine primates. Am. J. Phys. Anthropol. 71:185–201.

Edwards, M. S., and Ullrey, D. E. 1999. Effect of dietary fiber concentration on apparent digestibility and digesta passage in non-human primates. II. Hindgut and forgut fermenting folivores. Zoo Biol. 18:537–549.

Emery-Thompson, E., Kahlenberg, S. M., Gilby, I. C., and Wrangham, R. W. 2007. Core area quality is associated with variance in reproductive success among female chimpanzees at Kibale National Park. Anim. Behav. 73:501-512.

Fernandez-Duque, E. 2007. Aotinae: social monogamy in the only nocturnal haplorrhines. In C. J. Campbell, A. Fuentes, K. C. MacKinnon, M. Panger and S. K. Bearder (eds.), *Primates in Perspective* (pp. 139–154). Oxford UK: Oxford University Press.

Fleagle, J. G., and Tejedor, M. F. 2002. Early platyrrhines of southern South America. In W. C. Hartwig (ed.), *The Primate Fossil Record* (pp. 161–173). Cambridge UK: Cambridge University Press.

Foley, W. J., and Cork, S. J. 1992. Use of fibrous diets by small herbivores: How far can the rules be "bent"? Trends Ecol. Evol. 7:159–162.

Forget, P.-M. 1991. Comparative recruitment pattern of two non-pioneer tree species in French Guiana. Oecologia 85:434–439.

Fragaszy, D. M. 1986. Time budgets and foraging behaviors in wedge-capped capuchins (*Cebus olivaceus*): age and sex differences. In D. Taub and F. King (eds.), *Current Perspectives in Primate Social Dynamics* (pp. 159–174). New York: Van Nostrand Press.

Garber, P. A. 1988. Foraging decisions during nectar feeding by tamarin monkeys (*Saguinus mystax* and *Saguinus fuscicollis*, Callitrichidae, Primates) in Amazonian Peru. Biotropica 20: 100–106.

Garber, P. A. 1992. Vertical clinging, small body size, and the evolution of feeding adaptations in the Callitrichinae. Am. J. Phys. Anthropol. 88:469–482.

Garber, P. A. 1993. Feeding ecology and behavior of the genus *Saguinus*. In A. B. Rylands (ed.), *Marmosets and Tamarins: Systematics, Behavior, and Ecology* (pp. 273–295). Oxford UK: Oxford University Press.

Gaulin, S. J. C., and Craker, L. E. 1979. Protein in vegetation and reproductive tissues of several Neotropical species. J. Agri. Food Chem. 27:791–795.

Greenfield, L. O. 1992. Relative canine size, behavior and diet in male ceboids. J. Hum. Evol. 23:469–80.

Guimarães Jr., P. R., José, J., Galetti, M., and Trigo, J. R. 2003. Quinolizidine alkaloids in *Ormosia arborea* seeds inhibit predation but not hoarding by agoutis (*Dasyprocta leporina*). J. Chem. Ecol. 29:1065–1072.

Hanson, A. M., Hall, M. B., Porter, L., and Lintzenich, B. 2006. Composition and nutritional characteristics of fungi consumed by *Callimico goeldii* in Pando, Bolivia. Int. J. Primatol. 27:323–346.

Hemingway, C. 1998. Selectivity and variability in the diet of Milne-Edwards' sifakas (*Propithecus diadema edwardsi*): Implications for folivory and seed-eating. Int. J. Primatol. 19: 355–377.

Hershkovitz, P. 1977. Living New World Monkeys (Platyrrhini). Chicago: University of Chicago Press.

Heymann, E. W., and Smith, A. C. 1999. When to feed on gums: temporal patterns of gummivory in wild tamarins, *Saguinus mystax* and *Saguinus fuscicollis* (Callitrichinae). Zoo. Biol:18: 459–471.

Hiiemae, K. M. 1978. Mammalian mastication: A review of the activity of the jaw muscles and the movements they produce in chewing. In P. M. Butler and K. A. Joysey (eds.), Development, Function and Evolution of Teeth (pp. 361–398). London: Academic Press.

Hiiemae, K. M. 2000. Feeding in mammals. In K. Schwenk (ed.), *Feeding* (pp. 411–448). New York: Academic Press.

Hiiemae, K. M., and Crompton, A. W. 1985. Mastication, food transport, and swallowing. In M. Hildebrand, D. B. Bramble, K. Liem and D. Wake (eds.), *Functional Vertebrate Morphology* (pp. 262–290). Cambridge: Harvard University Press.

Hiiemae K. M., and Kay R. F. 1973. Evolutionary trends in the dynamics of primate mastication. In M. R. Zingeser (ed.), *Craniofacial Biology of Primates*, Vol. 3: Symp. Fourth Int. Cong. Primatology, (pp. 28–64). Basel: S. Karger.

Hill, D. A., and Lucas P. W. 1996. Toughness and fiber content of major leaf foods of wild Japanese macaques (*Macaca fuscata yakui*) in Yaushima. Am. J. Primatol. 38:221–231.

Hladik, C. M., Hladik, A., Bousset, J., Valdebouze, P., Viroben, G., and Delort-Laval, J. 1971. Le regime alimentaire des primates de L'île de Barro-Colorado (Panama). La Terre et la Vie. 1:25–117.

Howe, H. F., and Vande Kerckhove, G. A. 1981. Removal of wild nutmeg (*Virola surinamensis*) crops by birds. Ecol. 62:1093–1106.

Hylander, W. L. 1975. The human mandible: Lever or link? Am. J. Phys. Anthropol. 43:227–242.

Hylander, W. L. 1979a. The functional significance of primate mandibular form. J. Morphol. 160:223–240.

Hylander, W. L. 1979b. Mandibular function in *Galago crassicaudatus* and *Macaca fascicularis*: An in vivo approach to stress analysis of the mandible. J. Morphol. 159:253–296.

Hylander, W. L. 1984. Stress and strain in the mandibular symphysis of primates: A test of competing hypotheses. Am. J. Phys. Anthropol. 61:1–46.

Hylander, W. L. 1985. Mandibular function and biomechanical stress and scaling. Am. Zool. 25: 315–330.

Hylander, W. L., Johnson, K. R., and Crompton, A. W. 1987. Loading patterns and jaw movements during mastication in *Macaca fascicularis*: A bone-strain, electromyographic, and cineradiographic analysis. Am. J. Phys. Anthropol. 72:287–314.

Hylander, W. L., Ravosa, M. J., Ross, C. F., and Johnson, K. R. 1998. Mandibular corpus strain in primates: Further evidence for a functional link between symphyseal fusion and jaw-adductor muscle force. Am. J. Phys. Anthropol. 107:257–271.

Hylander, W. L., Ravosa, M. J., Ross, C. F., Wall, C. E., and Johnson, K..R 2000. Symphyseal fusion and jaw-adductor muscle force: an EMG study. Am. J. Phys. Anthropol. 112:469–492.

Janson, C. 1983. Adaptation of fruit morphology to dispersal agents in a Neotropical forest. Science 219:187–189.

Janson, C. H., and Boinski, S. 1992. Morphological and behavioral adaptations for foraging in generalist primates: the case of the cebines. Am. J. Phys. Anthropol. 88:483–498.

Kanazawa, E., and Rosenberger, A. L. 1989. Interspecific allometry of the mandible, dental arch, and molar area in anthropoid primates: functional morphology of masticatory components. Primates 30:543–60.

Kay, R. F. 1975. The functional significance of primate molar teeth. Am. J. Phys. Anthropol. 43:195–215.

Kay, R. F. 1977. The evolution of molar occlusion in the Cercopithecidae and early catarrhines. Am. J. Phys. Anthropol. 46:327–352.

Kay, R. F. 1984. On the use of anatomical features to infer foraging behavior in extinct primates. In P. S. Rodman and J. G. H. Cant (eds.), *Adaptations for Foraging in Nonhuman Primates: Contributions to an Organismal Biology of Prosimians, Monkeys and Apes* (pp. 21–53). New York: Columbia University Press.

Kay, R. F., and Covert, H. H. 1984. Anatomy and behavior of extinct primates. In D. J. Chivers, B. A. Wood and A. Bilsborough (eds.), Food Acquisition and Processing in Primates (pp. 467–508). New York: Plenum Press.

Kay, R. F., Johnson, D., and Meldrum, D. J. 1998. A new pitheciin primate from the middle Miocene of Argentina. Am J. Primatol. 45:317–336.

Kiltie, R. A. 1982. Bite force as a basis for niche differentiation between rain forest peccaries (*Tayassu tajacu* and *T. pecari*). Biotropica 14:188–195.

Kinzey, W. G. 1974. Ceboid models for the evolution of hominoid dentition. J. Hum. Evol. 3:193–203.

Kinzey, W. G. 1992. Dietary and dental adaptations in the Pitheciinae. Am. J. Phys. Anthropol. 88:499–514.

Kinzey, W. G. 1997. New World Primates: Ecology, Evolution and Behavior. New York: Aldine de Gruyter, Inc.

Kinzey, W. G., and Norconk, M. A. 1990. Hardness as a basis of fruit choice in two sympatric primates. Am. J. Phys. Anthropol. 81:5–15.

Kinzey, W. G., and Norconk, M. A. 1993. Physical and chemical properties of fruit and seeds eaten by *Pithecia* and *Chiropotes* in Surinam and Venezuela. Int. J. Primatol. 14:207–227.

Kirk, E. C., and Simons, E. L. 2001. Diets of fossil primates from the Fayum Depression of Egypt: a quantitative analysis of molar shearing. J. Hum. Evol. 40:203–229.

Kirkpatrick, C. 1998. Ecology and behavior in snub-nosed and douc langurs. In N. G. Jablonski (ed.), Ecology and Behavior in Snub-nosed and Douc Langurs (pp. 155–190). Singapore: World Scientific Publishing, Singapore.

Knott, C. 2001. Female reproductive ecology of the apes: implications for human evolution. In P. T. Ellison (ed.), Reproductive Ecology and Human Evolution (pp. 429–463). New York: Aldine de Gruyter.

Knott, D. D. 2005. Energetic responses to food availability in the great apes: implications for hominin evolution. In D. K. Brockman and C. P. van Schaik (eds.), Seasonality in Primates: Studies of Living and Extinct Human and Non-human Primates (pp. 351–378). New York: Cambridge University Press.

Lambert, J. E. 1998. Primate digestion: Interactions among anatomy, physiology, and feeding ecology. Evol. Anthropol. 7:8–20.

Lucas. P. W. 2004. Dental Functional Morphology: How Teeth Work. Cambridge: Cambridge University Press.

Lucas, P. W., Corlett, R. T., and Luke, D. A. 1986. Postcanine tooth size and diet in anthropoid primates. Z. Morphol. Anthropol. 76:253–276.

Lucas, P.W., Turner, I.M., Dominy, N.J., and Yamashita, N. 2000. Mechanical defenses to herbivory. Ann. Bot. 86:913–920.

Lucas, P. W., Beta, T., Darvell, B. W., Dominy, N. J., Essackjee, H. C., Lee, P. K. D., Osorio, D., Ramsden, L., Yamashita, N., and Yuen, T.D.B. 2001. Field kit to characterize physical, chemical, and spatial aspects of potential primate foods. Folia Primatol. 72:11–25.

Lucas, P. W., Prinz, J. F., Agrawal, K. R., and Bruce, I. C. 2002. Food physics and oral physiology. Food Qual. Pref. 13:203–213.

Luschei, E. S., and Goodwin, G. M. 1974. Patterns of mandibular movement and jaw muscle activity during mastication in the monkey. J. Neurophysiol. 37:954–966.

Marroig, G., De Vivo, M., and Cheverud, J.M. 2004. Cranial evolution in sakis (*Pithecia*, Platyrrhini) II: evolutionary processes and morphological integration. J. Evol. Biol. 17: 144–155.

Martin, L. B., Olejniczak, A. J., and Maas, M. C. 2003. Enamel thickness and microstructure in pitheciin primates, with comments on dietary adaptations of the middle Miocene hominoid *Kenyapithecus*. J. Hum. Evol. 45:351–367.

Meldrum, D. J., and Kay, R. F. 1997. *Nuciruptor rubricae*, a new pitheciin seed predator from the Miocene of Colombia. Am. J. Phys. Anthropol. 102:407–427.

Miller, K. E., Bales, K. L., Ramos, J. H., and Dietz, J. M. 2006. Energy intake, energy expenditure, and reproductive costs of female wild golden lion tamarins (*Leontopithecus rosalia*). Am. J. Primatol. 68:1037–1053.

Milton, K. 1980. The foraging strategy of howler monkeys. New York: Columbia University Press.

Milton, K. 1984a. The role of food-processing factors in primate food choice. In P. S. Rodman and J. G. H. Cant (eds), *Adaptations for Foraging in Nonhuman Primates: Contributions to an Organismal Biology of Prosimians, Monkeys, and Apes* (pp. 249–279). New York: Columbia University Press.

Milton, K. 1984b. Protein and carbohydrate resources of the Maku Indians of north-western Amazonia. Am. Anthropol. 86:7–27.

Milton, K. 1984c. Habitat, diet, and activity patterns of free-ranging woolly spider monkeys (*Brachyteles arachnoides* E. Geoffroy 1806). Int. J. Primatol. 5:491–514.

Milton, K. 1988. Foraging behavior and the evolution of primate cognition. In A. Whiten and R. Byrne (eds.), *Machiavellian Intelligence: Social Expertise and the Evolution of Intellect in Monkeys, Apes and Humans* (pp. 285–305). Oxford UK: Oxford University Press.

Milton, K. 1998. Physiological ecology of howlers (*Alouatta*): Energetic and digestive considerations and comparison with the Colobinae. Int. J. Primatol. 19:513–548.

Milton, K., Giacalone, J., Wright, S. J., and Stockmayer, G. 2005. Do frugivore population fluctuations reflect fruit production? Evidence from Panama. In J. L. Dew and J. P. Boubli (eds.), Tropical Fruits and Frugivores: The Search for Strong Interactors (pp. 5–35). Dordrecht, The Netherlands: Springer.

Moermond, T. C., and Denslow, J. S. 1985. Neotropical avian frugivores: patterns of behavior, morphology, and nutrition, with consequences for fruit selection. Ornithol. Mono. 36:865–897.

Nash, L.T. 1984. Observations on the ecology and behavior of *Galago senegalensis* at the ADC Mutara Ranch, 29 July – 15 August, 1984. Rep. Inst. Prim. Res., Natl. Mus. Kenya.

Nash, L.T. 1986. Dietary, behavioral, and morphological aspects of gummivory in primates. Yrbk. Phys. Anthropol. 29:113–137.

Nash, L.T. 1989. Galagos and gummivory. J. Hum. Evol. 4:199–206.

National Research Council 2003. Feeding ecology, digestive strategies, and implications for feeding programs in captivity. In Committee on Animal Nutrition, ad hoc committee on Nonhuman primate nutrition, Board on Agriculture and Natural Resources (eds.), *Nutrient Requirements of Nonhuman Primates* (pp. 5–40). Washington D.C.: The National Academies Press.

Natori, M., and Shigehara, N. 1992. Interspecific differences in lower dentition among eastern-Brazilian marmosets. J. Mammal. 73:668–671.

Norconk, M. A. 1996. Seasonal variation in the diets of white-faced and bearded sakis (*Pithecia pithecia* and *Chiropotes satanas*) in Guri Lake, Venezuela. In M. A. Norconk, A. L. Rosenberger and P. A. Garber (eds.), Adaptive Radiations of Neotropical Primates (pp. 403–423). New York: Plenum.

Norconk, M. A. 2007. Sakis, uakaris, and titi monkeys: behavioral diversity in a radiation of seed predators. In C. J. Campbell, A. Fuentes, K. C. MacKinnon, M. Panger, and S. K. Bearder (eds.), *Primates in Perspective* (pp. 123–138). Oxford UK: Oxford University Press

Norconk, M. A., and Conklin-Brittain, N. L. 2004. Variation on frugivory: the diet of Venezuelan white-faced sakis. Int. J. Primatol. 25:1–26.

Norconk, M. A., Oftedal, O. T., Power, M. L., Jakubasz, M., and Savage, A. 2002. Digesta passage and fiber digestibility in captive white-faced sakis (*Pithecia pithecia*). Am. J. Primatol. 58: 23–34.

Oftedal, O.T. 1991. The nutritional consequences of foraging in primates: the relationship of nutrient intakes to nutrient requirements. Phil. Trans. Roy. Soc. Lond., Ser. B. 334:161–170.

Palacios, E., Rodríguez, A., and Defler, T. R. 1997. Diet of a group of *Callicebus torquatus lugens* (Humboldt 1812) during the annual resource bottleneck in Amazonian Colombia. Int. J. Primatol. 18:503–522.

Peetz, A. 2001. Ecology and social organization of the bearded saki *Chiropotes satanas chiropotes* (Primates: Pitheciinae) in Venezuela. Ecol. Mono. 1:1–170.

Peres, C. A., and Janson, C. H. 1999. Species coexistence, distribution and environmental determinants of Neotropical primate richness: a community-level zoogeographical analysis. In J. G. Fleagle, C. Janson and K. E. Reed (eds.), Primate Communities (pp. 55–74). Cambridge UK: Cambridge University Press.

Pirie, P. L. 1976. Allometry in the Masticatory Apparatus of Primates. Ph.D. dissertation, Ohio State. University, Columbus, OH.

Pirie, P. L. 1978. Allometric scaling in the postcanine dentition with reference to primate diets. Primates 19:583–591.

Podolsky, R. D. 1990. Effects of mixed-species association on resource use by *Saimiri sciureus* and *Cebus apella*. Am. J. Primatol. 21:147–158.

Porter, L. 2007. The Behavioral Ecology of Callimicos and Tamarins in Northwestern Bolivia. Pearson Prentice Hall, Upper Saddle River, NJ.

Porter, L. M. 2001. Dietary differences among sympatric Callitrichinae in northern Bolivia: *Callimico goeldii, Saguinus fuscicollis* and *S. labiatus*. Int. J. Primatol. 22:961–992.

Porter, L. M., and Garber, P. A. 2004. Goeldi's monkeys: a primate paradox? Evol. Anthropol. 13:104–115.

Power, M. L., and Oftedal, O. T. 1996. Differences among captive callitrichids in the digestive responses to dietary gum. Am. J. Primatol. 40:131–144.

Power, M. L. 1996. The other side of callitrichine gummivory. In M. A. Norconk, A. L. Rosenberger, and P. A. Garber (eds.), *Adaptive Radiations of Neotropical Primates* (pp. 97–110). New York: Plenum.

Rehg, J. A. 2006. Seasonal variation in polyspecific associations among *Callimico goeldii, Saguinus labiatus*, and *S. fuscicollis* in Acre, Brazil. Int. J. Primatol. 27:1399–1428.

Robbins, C. T. 1993. Wildlife Feeding and Nutrition. 2nd eition. Academic Press, San Diego.

Rosenberger, A. L. 1978. Loss of incisor enamel in marmosets. J. Mammal. 59:207–208.

Rosenberger, A. L. 1983. Aspects of the systematics and evolution of the marmosets. In M. T. de Mello (ed.), A Primatologia no Brasil (pp. 159–180). Brasilia, Brazil: Sociedade Brasileira de Primatologia.

Rosenberger, A. L. 1992. Evolution of feeding niches in New World monkeys. Am. J. Phys. Anthropol. 88:525–562.

Rosenberger, A. L., and Kinzey, W. G. 1976. Functional patterns of molar occlusion in platyrrhine primates. Am. J. Phys. Anthropol. 45:281–298.

Ross, C. F. 1995. Allometric and functional influences on primate orbit orientation and the origins of Anthropoidea. J. Hum. Evol. 29:201–227.

Russo, S. E., Campbell, C. J., Dew, J. L., Stevenson, P. R., and Suarez, S. A. 2005. A multi-forest comparison of dietary preferences and seed dispersal by *Ateles* spp. Int. J. Primatol. 26: 1017–1038.

Russo, S. E. 2005. Linking seed fate to natural dispersal patterns: factors affecting predation and scatter-hoarding of *Virola calophylla* seeds in Peru. J. Trop. Ecol. 21:243–253.

Schumacher, G.H. 1961. Funktionelle Morphologie der Kaumuskulatur. Jena, Germany: Gustav Fischer.

Setoguchi, T., and Rosenberger, A. 1987. A fossil owl monkey from La Venta, Colombia. Nature 326:692–694.

Silver, S. C., Ostro, L. E. T., Yeager, C. P., and Dierenfeld, E. S. 2000. Phytochemical and mineral components of foods consumed by black howler monkeys (*Alouatta pigra*) at two sites in Belize. Zoo Biol. 19:95–109.

Smith, A. C. 2000. Composition and proposed nutritional importance of exudates eaten by saddleback (*Saguinus fuscicollis*) and mustached (*Saguinus mystax*) tamarins. Int. J. Primatol. 21:69–83.

Smith, R. J. 1978. Mandibular biomechanics and temporomandibular joint function in primates. Am. J. Phys. Anthropol. 49:341–50.

Smith, R.J., and Jungers, W.L. 1997. Body mass in comparative primatology. J. Hum. Evol. 32:523–559.

Smith, R. J., Petersen, C. E., and Gipe, D. P. 1983. Size and shape of the mandibular condyle in primates. J. Morphol. 177:59–68.

Soini, P. 1993. The ecology of the pygmy marmoset, *Cebuella pygmaea:* some comparisons with two sympatric tamarins. In A. B. Rylands (ed.), *Marmosets and Tamarins: Systematics, Behavior, and Ecology* (pp. 257–261). Oxford UK: Oxford University Press.

Spencer, M. A., and Spencer, G. S. 1995. Technical note: Video-based three-dimensional morphometrics. Am. J. Phys. Anthropol. 96:443–453.

Spencer, M. A. 2003. Tooth-root form and function in platyrrhine seed-eaters. Am. J. Phys. Anthropol. 122:325–335.

Starck, D. 1933. Die Kaumuskulatur der Platyrrhinen. Gegen. Morphol. Jahr. 72:212–285.

Stevenson, P. R. 2001. The relationship between fruit production and primate abundance in Neotropical communities. Biol. J. Linn. Soc. 72:161–178.

Strier, K. B. 1991. Diet in one group of woolly spider monkeys, or muriquis (*Brachyteles arachnoides*). Am. J. Primatol. 23:113–126.

Strier, K. B. 1992. Ateline adaptations: behavioral strategies and ecological constraints. Am. J. Phys. Anthpol. 88:515–524.

Strier, K. B. 2006. Primate Behavioral Ecology. Pearson Allyn and Bacon, Boston, MA.

Sussman, R. W., and Kinzey, W. G. 1984. The ecological role of the Callitrichidae: a review. Am. J. Phys. Anthropol. 64:419–449.

Talebi, M., Bastos, A., and Lee, P. C. 2005. Diet of southern muriquis in continuous Brazilian Atlantic forest. Int. J. Primatol. 26:1175–1187.

Taylor, A. B., and Vinyard, C. J. 2004. Comparative analysis of masseter fiber architecture in tree-gouging (*Callithrix jacchus*) and nongouging (*Saguinus oedipus*) callitrichids. J. Morphol. 261:276–285.

Teaford, M. F. 1985. Molar microwear and diet in the genus *Cebus*. Am. J. Phys. Anthropol. 66:363–370.

Teaford, M. F., Lucas, P. W., Ungar, P. S., and Glander, K. E. 2006. Mechanical defenses in leaves eaten by Costa Rican howling monkeys (*Alouatta palliata*). Am. J. Phys. Anthropol. 129: 99–104.

Terborgh, J. 1983. Five New World Primates: a study in comparative ecology. Princeton University Press, Princeton, NJ.

Thoren, S., Lindenfors, P., and Kappeler, P.M. 2006. Phylogenetic analyses of dimorphism in primates: Evidence for stronger selection on canine size than on body size. Am. J. Phys. Anthropol. 130:50–59.

Turnbull, W. D. 1970. Mammalian masticatory apparatus. Fieldiana, Geology 18:148–356.

van Roosmalen, M. G. M. 1984. Subcategorizing foods in primates. In D. J. Chivers, B. A. Wood and A. Bilsborough (eds.), Food Acquisition and Processing in Primates (pp. 167–175). New York: Plenum Press.

van Roosmalen, M. G. M. 1985. Habitat preferences, diet, feeding strategy and social organization of the black spider monkey (*Ateles paniscus paniscus* Linnaeus 1758) in Surinam. Acta Amazon. 15(3/4, suppl):1–238.

van Roosmalen, M. G. M., Mittermeier, R. A., and Fleagle, J. G. 1988. Diet of the northern bearded saki (*Chiropotes satanas chiropotes*): A neotropical seed predator. Am. J. Primatol. 14:11–35.

Van Soest, P. J., Uden, P., and Wrick, K.F. 1983. Critique and evaluation of markers for use in nutrition of humans and farm and laboratory animals. Nutr. Rep. Int. 27:17–28.

Veiga, L. M., and Ferrari, S. F. 2006. Predation of arthropods by southern bearded sakis (*Chiropotes satanas*) in eastern Brazilian Amazonia. Am. J. Primatol. 68:209–215.

Vinyard, C. J., and Ryan, T. M. 2006. Cross-sectional bone distribution in the mandibles of gouging and non-gouging platyrrhines. Int. J. Primatol. 27:1461–1490.

Vinyard, C. J., Wall, C. E., Williams, S. H., and Hylander, W. L. 2003. Comparative functional analysis of skull morphology of tree-gouging primates. Am. J. Phys. Anthropol. 120:153–170.

Vinyard, C. J., Wall, C. E., Williams, S. H., Johnson, K. R., and Hylander, W. L. 2006. Masseter electromyography during chewing in ring-tailed lemurs (*Lemur catta*). Am. J. Phys. Anthropol. 130:85–95.

Vinyard, C. J., Wall, C. E., Williams, S. H., Mork, A. L., Garner, B. A., Melo, L. C. O., Valença Montenegro, M. M., Valle, Y. B. M., Monteiro da Cruz, M. A., Lucas, P. W., Schmitt, D., Taylor, A. B., and Hylander, W. L. in press. The evolutionary morphology of tree gouging in marmosets. In S. M. Ford, L. M. Porter and L.C. Davis (eds.), *The Smallest Anthropoids*: The Marmoset/Callimico Radiation. New York: Springer.

Wallace, R. B. 2005. Seasonal variations in diet and foraging behavior of *Ateles chamek* in a southern Amazonian tropical forest. Int. J. Primatol. 26:1053–1075.

Waterman, P., and Kool, K. M. 1994. Colobine food selection and plant chemistry. In A. G. Davies and J. F. Oates (eds.), *Colobine monkeys: Their Ecology, Behavior and Evolution* (pp. 251–284). Cambridge UK: Cambridge University Press.

Williams, S. H., Vinyard, C. J., Glander, K. E., Deffenbaugh, M., Teaford, M., and Thompson, C. L. 2008. A preliminary report on a telemetry system for assessing jaw-muscle function in free-ranging primates. Int. J. Primatol.

Wrangham, R. W., Conklin, N. L., Etot, G., Obua, J., Hunt, K. D., Hauser, M. D., and Clark, A. P. 1993. The value of figs to chimpanzees. Int. J. Primatol. 14:243–256.

Wrick, K. L., Robertson, J. B., Van Soest, P. J., Lewis, B. A., Rivers, J. M., Roe, D. A., and Hackler, L.R. 1983. The influence of dietary fiber source on human intestinal transit and stool output. J. Nutr. 113:1464–1479.

Wright, B. W. 2004. Ecological Distinctions in Diet, Food Toughness, and Masticatory Anatomy in a Community of Six Neotropical Primates in Guyana, South America. Ph.D. Dissertation. Urbana-Champaign: University of Illinois.

Wright, B. W. 2005. Craniodental biomechanics and dietary toughness in the genus *Cebus*. J. Hum. Evol. 48:473–492.

Wright, K. W. 2005. Interspecific and Ontogenetic Variation in Locomotor Behavior, Habitat Use and Postcranial Morphology in *Cebus apella* and *Cebus olivaceus*. Ph.D. Dissertation. Northwestern University.

Zar, J. H. 1999. Biostatistical Analysis. New York: Prentice Hall.

Zingeser, M. R. 1973. Dentition of *Brachyteles arachnoides* with reference to Alouttine and Atelinine affinities. Folia Primatol. 20:351–390.

Chapter 12
Neutral and Niche Perspectives and the Role of Primates as Seed Dispersers: A Case Study from Rio Paratari, Brazil

Kevina Vulinec and Joanna E. Lambert

Niche and neutral perspectives have quite different implications for how one should manage natural resources and craft conservation strategies. A unified theory of communities that judiciously blends both perspectives is needed if ecologists are to understand the processes governing biodiversity at a fundamental level and then apply this understanding to the urgent problem of maintaining diversity in our rapidly changing world.
(Holt, 2006; 533)

12.1 Introduction

Community ecologists are currently evaluating the degree to which neutrality is applicable to understanding and interpreting species assemblages, richness, abundance, and functional traits. A basic tenet of so-called neutral models is that species are roughly equivalent in their dispersal and recruitment abilities and that species traits have limited influence on the presence or abundance of that species in a community (Hubbell 2005; Holt 2006). Since species are relatively equal, with equivalent dispersal and recruitment abilities, ecological space is filled at random from a wide pool of potential recruits (Hubbell 2005). In the case of animal-dispersed plant species, differences in species assemblages are the result of dispersal limitation – i.e., the failure of seeds to arrive at appropriate sites (Schupp et al. 2002) – rather than inherent, evolved trait differences in competitive and dispersal ability.

Others view community structure and the end product of seed dispersal from the perspective of evolved niche space and emphasize directional selection of fruit/seed traits resulting from consistent primary dispersal into safe sites. This equilibrium, or so-called "niche", perspective deflates the importance of stochasticity in systems, and instead emphasizes the role of species competition and coexistence and their impact on co-evolution and dependence among species (Morin 1999). With regard

K. Vulinec (✉)
Department of Agriculture and Natural Resources, Delaware State University, Dover, DE 19901-2277, USA
e-mail: kvulinec@desu.edu

P.A. Garber et al. (eds.), *South American Primates,* Developments in Primatology: Progress and Prospects, DOI 10.1007/978-0-387-78705-3_12,
© Springer Science+Business Media, LLC 2009

to seed dispersal, it has been argued that plant species distributions are produced largely as a function of primary dispersal, which selects for "disperser syndromes" identified on the basis of both plant attributes (e.g., fruit/seed size, shape, fruit pulp color, etc.) and the dispersers themselves (e.g., "bird syndrome", "mammal syndrome"; Janson 1983; Gautier-Hion et al. 1985). Janson (1983), for example, concludes that fruit can be characterized as either a "Bird fruit" or "Mammal fruit" according to co-varying traits selected for by large guilds of birds, volant mammals, or primates; "monkey-dispersed fruits" were identified as being greater than 14 mm in width, brown, green, white or yellow with a protective pericarp. Similarly, Gautier-Hion et al. (1985) suggest in Gabon, fruits dispersed by primates tend to be red or multi-colored, fleshy, or dehiscent with arillate seeds weighing 5–50 g. Julliot (1996b), too, has defined a "primate syndrome" whose fruits are characterized by being middle-to large-sized and pulpy, yellow or orange in color, and having an indehiscent, hard pericarp and a few well-protected large seeds.

Taking this equilibrium, co-evolutionary argument one step further – literally – some authors have argued that two-phased dispersal systems ("diplochory") occur in forests (Vander Wall and Longland 2004; Andresen 2005). In Phase One of this system, seeds are swallowed by fruit-eating mammals such as primates ($1°$ dispersal), and then secondarily dispersed in Phase Two ($2°$ dispersal) by either seed-hoarding rodents, which harvest them from herbivore dung, or by dung beetles, which then transport and bury the herbivore's dung. According to Vander Wall and Longland (2004), the selective advantage to the first phase of diplochory is that seeds are removed from parent plants, so that seedlings have a better chance of surviving, while the selective advantage of the second phase is that seeds are moved to hidden sites, where they have a greater chance of avoiding the detection of seed predators.

Despite differences in neutral and niche models, they are not necessarily dichotomous; they can be theoretically integrated and viewed as different ends of a continuum of impact on communities (Gravel et al. 2006; Holt 2006; Scheffer and van Nes 2006). For example, Gravel et al. (2006) have recently modeled an open system in which each species has its own unique set of resource requirements, but the likelihood of that species' recruitment into an assemblage is dependent on its abundance in the metacommunity. These authors demonstrate that without immigration from an external source such as nearby intact forest, interspecific competition structures niche space; however, with immigration (i.e., an influx of recruits) – neutrality emerges. The likelihood of neutrality increases as a function of total species richness, because as species richness increases, species abundance tends to decrease; with decreasing abundance, stochastic events are more likely to impact the probability of dispersal and recruitment (Gravel et al. 2006).

In this chapter, we consider the degree to which neutralizing events may influence plant community structure. We tackle this by considering variance in the quantity and quality (*sensu* Schupp 1993) of Phase One ($1°$ dispersal) and Phase Two ($2°$ dispersal) seed dispersal by primates and dung beetles based on these animals' population and community structures. We reason that each step in the seed dispersal process has the potential of a positive return for the plant, but that each step is also characterized by high variance (for example primates can serve as both seed predators and seed dispersers of the same plant) that can swamp directional selection pressure and effectively neutralize competitive interactions.

We present a case study near the Rio Paratari located in the State of Amazonas, Brazil, an area that is home to at least seven primate species (Vulinec et al. 2006). We present data (collected over 3 months during two years) from systematic transect surveys on abundance, biomass, ranging, and habitat use by six of the seven primates in the area. Data are also presented on dung beetle spatial and temporal abundance and seed burying behavior, as are data on the demography and physiognomy of the local plant community. We focus on primate-dung beetle interactions because this association is ubiquitous in the Neotropics (Vulinec et al. 2006); primates represent a significant portion of the seed disperser biomass in the region (Peres 1997), and the interaction of primates with secondarily dispersing dung beetles has a considerable impact on seed dissemination and recruitment (Feer 1999; Vulinec 2000; Andresen 2001). We demonstrate that at Rio Paratari there is: (i) high within and between species variance in primate abundance across habitats and seasons; (ii) high variance in dung beetle abundance across seasons and among traps within habitats and, (iii) no correlation between primate habitat use (as measured by distribution and biomass) and forest structure. We structure our paper around the following posits and questions:

The quantity component of Phase One dispersal refers to the total number of seeds dispersed from a plant (Schupp 1993). Since previous studies demonstrate a positive relationship between primate biomass and the number of seeds dispersed in tropical forests (Lambert and Garber 1998), we ask: *what is the density and biomass of fruit-eating primates at Rio Paratari?*

(1) The quality component of Phase One dispersal in part relates to the arrival of seeds to particular, "safe" sites in forests (Schupp 1993). The patterns in which seeds are primarily dispersed are strongly influenced by the patterns in which primates use their habitat, and habitat use can be measured by quantifying primate distribution via transect surveys (Julliot 1996a; Beckman and Muller-Landau 2007). We thus ask: *do particular primate species restrict their patterns of range use to particular parts of the forest?*

(2) As with Phase One, the quantity component of Phase Two seed dispersal relates to the total number of seeds secondarily dispersed, which in turn relates to the abundance of secondary dispersers. Dung beetles are thought to be particularly important secondary dispersers in the Neotropics (Vulinec et al. 2006). Thus, we ask: *Does abundance and species composition of dung beetles as secondary dispersers vary significantly in space and time?*

(3) Similar to the quality component of Phase One dispersal, the quality component of Phase Two in part relates to the arrival of seeds to particular, "safe" sites in forests (Schupp 1993). Thus, we ask: *are there dung beetle burial behaviors that may exert selection on fruit traits? Or, do these behaviors interrupt or dilute the effects of directional selection?*

(4) Finally, as current demography and distribution of plant species reflect, in part, the history of primary and secondary dispersal in a system (Garber 1986), we

ask: *is there evidence that presence and biomass of primates correlates with tree community structure or demography in Rio Paratari?*

12.2 Methods

12.2.1 Site

Data were collected during December 2001, July 2002, and August 2002 around the riverine region of the Rio Paratari, a whitewater tributary of the Amazon, between Rio Madeira and Rio Purus. The region is largely seasonally inundated várzea habitat; those areas not flooded are described as lowland moist tropical rainforest (Ayres 1993). Some parts of this region are farmed for manioc via slash-and-burn. In general, after forests are cut and burned, crops are grown for approximately three years and then left fallow. Local people extract numerous resources from the forest including Brazil nuts, various fruits, and occasionally timber. In addition, hunting is common in this area and has had a significant effect on the presence and density of primate populations (*see* below) (Vulinec, pers. obs.).

12.2.2 Primate Surveys (Quantity and Quality Component of Phase One Dispersal)

Survey data were collected in intact primary forest, in an area of > 50-year secondary natural re-growth from the manioc plantings, and along riverbanks and flooded forest channels. Surveys were conducted during three seasons: low water rainy season (December), high water dry season (July), and medium-depth water dry season (August). We conducted primate surveys using standard methods described in Vulinec et al. (2006). One or two people walked these transects at approximately 1 km/h. Canoe transects (with two people plus boat operator) were conducted along the main branch of the river during December and the main branch of the river plus flooded forest during July and August. We collected data on: primate species, number of individuals (seen or heard), time of day, location on transect, and observer distance. Infants were not included in counts. We walked or canoed line transects for 16 days in primary forest, 18 days in secondary forest, and 14 days along rivers in canoes. We used an index of sighting rates (number of primate groups encountered per 10 km walked) to estimate relative abundance (Skoropa 1988; Peres 1997). We calculated density using a 50% cut-off rule to select the sighting distance (Chapman et al. 1988). This method compensates for observer differences, as sightings at the farthest distances are excluded from the census if they occur infrequently (Peres 1997). We used observer- to-animal distance because perpendicular distance is known to underestimate transect width for forest primates (Chapman et al. 1988). Additionally, we estimated distances by sight after practice with measured distances, and we calculated density as the number of individuals sighted within the truncated sighting distance divided by the area sampled, i.e., the length of the transect multiplied by the truncated distance (Chapman et al. 1988).

To estimate biomass, we used published average weights for each species (Emmons 1997; Peres 1997) times our calculated densities and reported this as kg/km^2. To examine ranging and habitat use patterns over space and time, we classified locations as a point every 100–150 m and numbered these as 1–15 in secondary forest, 16–29 in primary forest, and 30–32 along the riverbank and in flooded forest (#33). The primary forest transect and secondary forest transect were separated by 10 km. Although we surveyed by canoe over 5 km of riverside and 2 km of flooded forest, we did not have a way to distinguish each point as we did on terra firme. Thus, we pooled all sightings for riverbank into three points and all those in flooded forest into one for display on the graph of number of individuals (Fig. 12.1). The separation of points among habitats is therefore not evenly spaced, whereas the points within primary and secondary are relatively evenly spaced. We used total number of sightings, total number of individuals, and biomass at each point over the survey period to examine occurrence in all locations. There are limitations to the transect data because of the lack of precise points in flooded and riverside habitats, nevertheless, total distance and width of the transect allowed us to calculate primate density and biomass. We pooled all riverside and flooded forest transects into one habitat for ANOVA of biomass across seasons and habitats. We surveyed roughly equivalent areas in all three habitats, primary forest, secondary forest, and flooded/riverside forest (0.16 km^2, 0.18 km^2, and 0.14 km^2, respectively).

Hunting has been and remains heavy; the larger primates such as the howling monkey (*Alouatta seniculus*) and the common woolly monkey (*Lagothrix lagothricha*) are reported by locals to have been common in the past but are now rare (Vulinec et al. 2006). Given the recent extinction or decline of these primate species, their density and biomass cannot be assessed. The most commonly encountered monkey was the bare-eared squirrel monkey (*Saimiri ustus*). Other relatively common primates at the site include saki monkeys (*Pithecia irrorata*), titi monkeys (*Callicebus caligatus*), brown capuchins (*Cebus apella*), and saddleback tamarins (*Saguinus fuscicollis*).

We report density and biomass in the primary forest and secondary forests (in this analysis only, inundated forest was included with secondary forest). In the case of location versus month and species, neither number of sightings nor number of individuals was a normally distributed variable, and therefore we used Kruskal-Wallis nonparametric ANOVA to compare the number of sightings, biomass, and the number of individuals of all species pooled and each species separately by date (December, July, August) and location (Points 1–33).

12.2.3 Dung Beetle Abundance and Seed Burial (Quantity and Quality Components of Phase Two Dispersal)

We surveyed dung beetles with 20 baited pitfall traps set along the same route as primate transects (Vulinec 2000). Ten traps were in secondary forest and 10 in primary forest. These traps were at least 100 m apart (Larsen and Forsyth 2005) and baited with approximately 25 cc of human dung. Traps separated by 100 m are considered

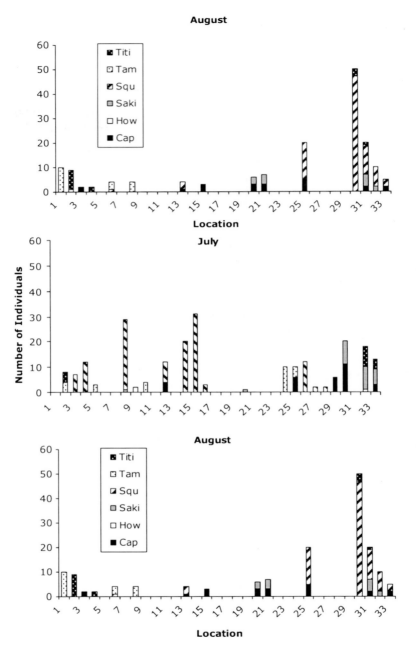

Fig. 12.1 Number of primate individuals seen (total over all survey periods) of six species of primates. Species codes: Titi = *Callicebus caligatus* Tam = *Saguinus fuscicollis*, Squ = *Saimiri ustus*, Saki = *Pithecia irrorata*, How = *Alouatta seniculus*, Cap = *Cebus apella*. Numbers on x-axis represent locations along transects: 1–15 in secondary forest, 16–29 in primary forest, and 30–32 along the riverbank and in flooded forest (#33)

independent in that beetles captured in one trap would not affect the number of individuals captured in the next trap (Larsen and Forsyth 2005). We set traps for 15 24-h periods during December, July, and August for a total of 300 trap-days. Dung beetles are classified into guilds by their dung burial behavior (Halffter and Edmonds 1982; Doube 1990; Vulinec 2002; Vulinec et al. 2006). In this study, we used a modified classification from Vulinec (2002) of three key guilds important to seed burial. Beetle species that bury no seeds are classified as "N." These are generally small beetles or those that do not bury dung, such as most *Eurysternus* species. Those species that only bury small seeds are classified as "S" and are rollers, forming balls and rolling them away from the site of deposition. The two most common species in this category are *Canthon aequinoctialis* and *Canthon fulgidus*. The tunnelers are classified as "LS" signifying that they bury both small and large seeds directly under the dung pat without relocation. Examples of these beetles are *Dichotomius* and *Oxysternon* spp.

We were particularly interested in the stochasticity of dung beetle community structure through time. We used a Chi-square test to examine the proportions of each guild over all 15 collections. We considered a collection to be 10 traps set out for 24 h. On most dates two collections were made in different areas separated by approximately 5 km. Because biomass is an important character in the study of dung beetle seed burial (Radtke et al. 2007; Gardner et al. 2008), we also plotted biomass of key guilds over time. We additionally plotted average abundance per trap of four important beetle species at each trap over all collecting days to demonstrate variability within space and time.

We conducted an experiment to examine the variation associated with secondary dispersal of seeds by dung beetles. Because different beetle communities are active during the day and night, seeds deposited in dung at night or during the day are manipulated differently. To examine this variability, we imbedded 10 3 × 3 mm beads in 50 cc dung pats (human dung). Beads were used as seed surrogates to eliminate the effects of seed predators and fungi. Previous work has demonstrated no difference between the handling of beads and seeds by dung beetles (Vulinec 2000). We threaded the beads with 60 cm of sewing thread for later recovery. We set out 10 dung pats during the daytime and 10 pats overnight. Pats were 20 m from each other, and pats and beads were set out at dusk and picked up at dawn. Another set of 10 pats was set out during daylight and picked up before dusk. In addition, we estimated the percentage of dung pats buried (# dung pats remaining after 24 h/# set out) over the entire experiment. This experiment was conducted over one 24-h session in 2001 and four 24-h sessions in 2004.

12.2.4 Vegetation and Forest Structure Survey

We opportunistically collected fruit that had fallen on the ground during the December 2001 surveys. These fruits were examined for signs of primate feeding and were identified to genus (and species when possible) using Ribeiro et al. (1999). This collection was intended as a preliminary survey of local fruit and is not an exhaustive or systematic collection.

To examine the potential relationship between primate occurrence and vegetation forest structure, we surveyed vegetation in eight 10×50 m quadrats, four quadrats in areas of high primate activity and four in areas of low primate activity. Primate activity level was based on results from our line transect survey (Vulinec et al. 2006). We classified areas as "low primate activity" if there was a low number of individuals counted in this quadrat during the transects (points 6, 14, 18, 22) and areas of "high primate activity" those where number of individual primates was high (points 4, 9, 15, 26). Squirrel monkeys, capuchins, and tamarins were seen often in secondary forest, while squirrel monkeys were rarely recorded in the primary forest. Saki and titi monkeys were not seen often in either primary and secondary transects. Four quadrats were in primary forest and four were in 50-year secondary regrowth (two each in areas of high monkey activity and two each in low). We counted all trees within the quadrat, and measured diameter at breast height (1.3 m high) of all trees larger than 10 cm. Smaller trees were classified as saplings (over 1 m) or seedlings (under 1 m). Trees greater than 10 cm dbh were classified as adults for the purposes of this study, although we could not verify this classification with fruiting data. We did not examine tree species, but only vegetation structure and phenology. Although this approach limits our data, we did not have access to the expertise to identify plants at this time. Relative soil density was measured with a penetrometer. Four readings at each 10×10 m subquadrat were taken and averaged. We then averaged the total for the quadrat from these readings. We measured canopy cover with a densitometer through four readings in the cardinal directions taken at the center of each 10×10 m subquadrat, and averaged over all five 10×10 m subquadrats. We then compared abundance of adult trees, saplings, seedlings, dbh of adults and the ratio of seedlings/adults, saplings/adults, between areas of high activity and areas of low activity. We also compared soil density and canopy cover in the same areas. Soil density was measured using a penetrometer at 10 random locations in primary forest and 10 in secondary forest. At each location, we took an average of three readings within one m^2. Canopy cover was estimated using a hemispherical gridded densitometer and scored as number of shaded grid sections out of total grid. We took four readings at each cardinal direction in each 10×10 subplot and averaged the percent shaded. We used these averages to calculate canopy cover for the entire plot in the analysis.

12.3 Results

12.3.1 Primate Distribution and Abundance

Primate abundance varied seasonally and spatially, although spatial variation was not statistically significant (Fig. 12.1). This is likely to result from the large variance in primate ranging patterns. Group sightings per 10 km walked did not vary significantly between the two most abundant primates, *Cebus apella* and *Saimiri ustus*, (Vulinec et al. 2006; Table 12.1). However, of these two species, squirrel monkeys showed much higher densities per km^2 than other species, particularly in secondary forest (Vulinec et al. 2006). Squirrel monkeys occurred at twice the density of the next most abundant primate, brown capuchin monkeys (48.4 squirrel monkey individuals/km^2 vs. 23.7 capuchin individuals/km^2). Standard deviation of number of individuals seen per transect in all species was nearly equal to the mean (Table 12.1; Vulinec et al. 2006) suggesting that number of individuals are randomly distributed, or extremely variable (Zar 1999; Horwich et al. 2001). The number of sightings and number of individuals did not vary by season or by species (Kruskal-Wallis tests) with the exception of *C. caligatus*, which was counted significantly more often in August than December or July (Kruskal-Wallis $= 10.18, df = 2$, $P = 0.006$). However, squirrel monkey troop size varied with season, being significantly larger during periods of low water – (6.20 mean number of individuals/troop in December vs. 10.28 mean number of individuals/troop during July and August; $F_{1.75} = 8.21, P = 0.005$), but not with habitat (riverside, 8.28 mean individuals/troop versus forest, 6.28 mean individuals/troop; $F_{1.75} = 2.96, P = 0.09$). Total biomass/km^2 did not vary between primary and secondary forest ($t = -0.75$; $df = 8$; $P = 0.48$). Biomass of individual species did not vary with either season or location using Kruskal-Wallis one-way analysis of variance on ranks. The biomass of titi monkeys approached significance with season ($H = 44.69, df = 32$, $P = 0.07$), while all other species P-values fell between 0.13 and 0.75 for both season and location. This result is not due to evenness of distributions, but rather the large variance in biomass among locations and over seasons. While frugivorous squirrel monkeys were the most abundant primate, and other seed-dispersing primates were common in the area, principally or frequently seed-destroying primates, such as saki monkeys and capuchins, were also abundant (Table 12.1). However, average troop size of the larger species was relatively low (Table 12.1), compared to troop sizes in less hunted areas (Bennett et al. 2001; Vulinec, unpub.).

Table 12.1 Primate sighting rates (SR: sightings/10 km walked), population density (# individ./km^2) in primary forest (DP), population density in secondary forest (DS), biomass (kg/km^2) in primary forest (BP), biomass in secondary forest (BS), mean number individuals/transect (MI), standard deviation of individuals/transect (SD), and average troop size (TS) of seven primate species that occur in the Rio Paratari area. *Lagothrix lagotricha* was seen once in the area

Species	SR	DP	DS	BP	BS	MI	SD	TS
Saguinus fuscicollis	0.8	3.3	3.2	1.3	1.3	5.9	5.2	5.9
Saimiri ustus	4.2	6.9	41.5	15.3	91.4	12.1	11.8	6.7
Callicebus caligatus	0.9	0	6.4	0	6.7	4.0	2.7	2.7
Pithecia irrorata	1.3	2.3	4.2	3.0	9.2	3.6	2.9	2.6
Cebus apella	2.2	17.1	6.6	47.6	31.2	6.1	5.1	4.7
Alouatta seniculus	0.2	0	1.3	0	20.4	2.4	1.7	2.4
Lagothrix lagotricha	*	*	*	*	*	*	*	*
Total	--	29.6	63.2	67.2	160.25	7.2	8.1	5.1

12.3.2 Dung Beetles

Dung beetle guilds varied extensively in abundance over time (Fig. 12.2). The average number of beetles/trap in the three guilds varied significantly with collection ($X^2 = 382.2$, $df = 28$, $P < 0.001$). Non-seed burying dung beetles dominated the biomass in most pitfall collections. This dominance is reflected in the fact that 39% of simulated dung pats set out were left unburied. The biomass of four abundant beetle species representing two seed-dispersing guilds varied considerably from trap to trap over the collecting period and among traps set 100 m of each other (Fig. 12.3). These four species exhibited the largest biomass/trap of species that disperse seeds. Each of the four species we highlight here manipulates and buries seeds in a different manner. Both *Canthon* species are rollers and bury only small seeds in a scattered and shallow pattern. *Canthon fulgidus* is diurnal; *C. aequinoctialis* is nocturnal (Vulinec 2002). *Oxysternon conspicillatum* is a large, diurnal tunneler, which buries dung and both small and large seeds directly under the site of deposition. *Dichotomius boreus* is a large nocturnal tunneler that also buries both large and small seeds (Vulinec 1999, 2002). It is clear that seeds imbedded in dung at this site are most likely to be left on the ground at the site of deposition due to the abundance of guild N, which buries no seeds (Fig. 12.2). In addition, 17% of buried seeds were buried at a depth greater than 10 cm, potentially too deep for the germination of most seeds (Shepherd and Chapman 1998).

We found that of those seeds moved and buried by dung beetles, the beetle species arriving first has the greatest influence on whether a seed is buried under the dung, moved a short distance away, buried deeply, or buried shallowly. Seeds are expected to have a better chance of being buried if they are deposited in August when there is a higher proportion of S and LS beetle guilds (Fig. 12.2). They also have a better chance of being buried away from dung deposition sites if deposited in primary forest, which decreases the chance of detection by rodent seed predators

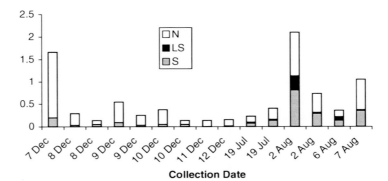

Fig. 12.2 Average dung beetle guilds biomass per trap by collection date. Duplicate dates represent two 10-trap collections at sites separated by 3 km. Guild codes: N = beetle species that bury no seeds; LS = Beetle species that bury both large and small seeds; S = Beetle species that bury only small seeds. Location codes: Pr = primary forest; Se = secondary forest

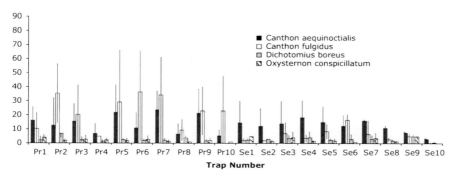

Fig. 12.3 The abundance of four key dung beetle species over 15 total days of collections during December, July, and August. Bars are average abundance in each of 20 pitfall traps. Error bars are ± 1 SD. Trap codes: Pr = Primary forest, Se = Secondary forest

(Fig. 12.3). This pattern is likely due to the higher abundance of *Canthon fulgidus*, a diurnal roller. Hence, secondarily dispersed seeds are subjected to varied movement and burial, depending on whether they are dispersed diurnally, nocturnally, in primary forest, secondary forest, or by the stochasticity of which beetle species arrives first to dung. In our study, seeds deposited during the day are rarely buried, but are instead moved up to 0.5 m from the site of deposition (Fig. 12.4). Seed burial was significantly deeper if it occurred during the night rather than the day, but distance moved was significantly greater during the day (Kruskal-Wallis analysis on ranks, Dunn's method of pairwise comparisons: $H = 101.08, df = 3, P < 0.001$). Seeds deposited during the day are moved horizontally a longer distance but at a more shallow depth or not buried at all. Overall, seeds were moved an average distance of 7.5 cm ($SD = 2.9$) and buried an average depth of 5.4 cm ($SD = 0.6$). Sixty percent of 7 mm beads set out in the experiment were buried; however, only 34% were buried at a depth of between 2 and 10 cm. Seeds deposited during the night were buried deeper but were not moved horizontally (Fig. 12.4). Seeds moved even a short distance from the original site of deposition are less likely to be found by rodents (Estrada and Coates-Estrada 1991). Seeds buried deeply may also not be discovered by rodents, but have less chance of germinating (Shepherd and Chapman 1999).

12.3.3 Plant Demography and Vegetation Structure

Our preliminary collection of fruits included 21 species of plants fruiting during December. Of these, nine had teeth marks from primates (Table 12.2). We noted whether these species have been recorded in the literature as primate-, bird-, and/or bat-dispersed.

Two-way ANOVAs, using primate occurrence and habitat as factors, showed no significant effect of primate presence on the ratio of saplings and seedlings to adult trees ($F_{1,12} = 1.34$, $P = 0.27$). However, habitat (primary vs. secondary forest)

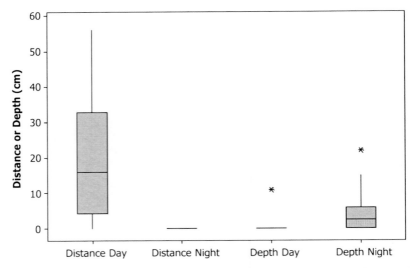

Fig. 12.4 Box plot summarizing bead burial by dung beetles. Distance, depth (both in cm) and day versus night are represented by boxes. The central line is the median; the shaded box encompasses the 25–75 percentile. Error bars represent the 10th to 90th percentile; outliers are represented by points

was significantly different, with secondary forest having a higher ratio ($F_{1,12} = 26.85$, $P < 0.001$; Fig. 12.5). There was no significant interaction between ratios in areas with high primate use versus random areas with little or no primate presence ($F_{1,12} = 0.87$, $P = 0.37$). There were significantly more numbers of seedlings and saplings in secondary forest than in primary forest ($F_{1,4} = 29.68$, $P = 0.006$), as would be expected. However, there was no effect of monkey occurrence in these areas ($F_{1,4} = 1.26$, $P = 0.33$), and no interaction between these factors ($F_{1,4} = 0.10$, $P = 0.77$). Average diameter at breast height of trees > 10 cm did not show a significant relationship with habitat, monkey presence, or an interaction between the two ($F_{1,4} = 0.007$, $P = 0.94$; $F_{1,4} = 1.23$, $P = 0.33$; $F_{1,4} = 0.17$, $P = 0.70$).

Canopy cover did not differ between plots where primates were abundant versus plots where primates were uncommon ($F_{1,4} = 0.81$, $P = 0.42$). There was a nearly significantly higher degree of canopy cover in primary forest compared to secondary forest ($F_{1,4} = 5.13$, $P = 0.09$), but no interaction effect ($F_{1,4} = 0.73$, $P = 0.44$). Additionally, soil density did not vary between primary and secondary forest ($t = 2.12$, $df = 16$, $P = 0.76$).

12.4 Discussion

Based on our preliminary sample of primates censused, the *quantity* component of Phase One dispersal appears to be highly affected by the abundance and biomass of frugivorous vertebrates in the locality (Table 12.1; Fig. 12.1). At Rio Paratari, primates represent the greatest component of the non-volant vertebrate frugivore

Table 12.2 List of plants with fruit fall below parent tree. Columns show those with primate teeth marks observed in this study, those known to be dispersed by primates, birds, bats, and other mammals (e.g., rodents, ungulates, tapirs). 1 = Howe & Miriti 2004; 2 = Ribeiro et al. 1999; 3 = Feer 1999; 4 = Link and Di Fiore 2006; 5 = Wenny 2000b; 6 = Julliot 1997; 7 = Suarez 2006

Plant	Primate Teeth Marks	Primate Dispersed	Bird Dispersed	Bat Dispersed	Other Mammals	Ref.
Ficus sp.	x	x	x	x	x	2
Diospyros sp.	x	x				2
Duguetia surinamensis	x	x				3
Ocotea sp.	x	x				5
Pouteria sp.	x	x				6
Couratari stellata	x					
Duguetia flagellaris	x					
Gnetum sp.	x					
Virola mollissima	x					7
Cecropia purpurascens		x	x	x		1
Compsoneura sp.		x	x			2
Neea sp.		x				4
Duguetia stelechantha						
Hevea guianensis						
Protium nitidifolium						
Rinorea amapensis						
Rinorea macrocarpa						
Sloanea sp.						
Virola theiodora						
Virola venosa						
Xylopia benthamii						

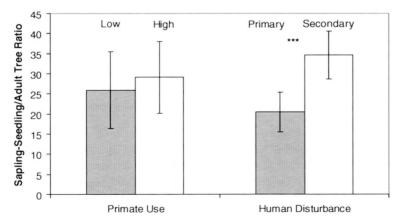

Fig. 12.5 Comparison of areas of low and high primate activity and primary and secondary forest in sapling + seedling/adult tree density

coterie, with the largest biomass per area (K. Vulinec, unpub.). Other mammalian seed dispersers recorded at this location [paca (*Agouti paca*), red-rumped agouti (*Dasyprocta leporina*), tapir (*Tapirus terrestris*), white-lipped peccary (*Tayassu pecari*), collared peccary (*Tayassu tajacu*), coatimundi (*Nasua nasua*), tayra (*Eira barbara*), and gray brocket deer (*Mazama gouazoubira*)] are heavily hunted, and have low population densities (K. Vulinec and D. Mellow, unpub. data from line transects and camera traps). Indeed, village elders reported a decrease in all large mammals in the area, particularly larger primates over the past 20 years (K. Vulinec, pers. obs.). As such, plants may be currently experiencing local dispersal limitation and/or smaller primates may have a greater relative affect on local plant community structure than they did historically (Stoner et al. 2007). Nonetheless, given their high abundance relative to other primary dispersers, primates overall are likely having a proportionately large impact on total seed removal, seed predation, and dissemination. Although, it is important to note that without census data on other frugivores, this remains a preliminary observation.

How primate distribution and habitat use impacts a measure of the *quality* component of Phase One – arrival at suitable microsites – is not clear. For example, based on recent surveys, all primate species inhabiting the riverine region of the Rio Paratari tended to use secondary forest more than primary rainforest (Vulinec et al. 2006), suggesting that primate-dispersed seeds are more likely to arrive in secondary growth than in mature forest. However, a disjuncture exists in the relationship between primate habitat use and density of seedlings and saplings. While we did find a higher ratio of seedling/saplings to adult trees in secondary growth versus primary forest, we found no association between areas in which primates are active and those where they were not, in either habitat type (Fig. 12.5). Sampling error cannot be ruled out. However, we speculate that this may be because in this region of Amazonia, vertebrate-dispersed fruit species exhibit a peak in fruiting phenology during the same season when primates are most widely distributed (Ayres 1993; this study; Fig. 12.1). At Rio Paratrai, primates also forage along riverbanks and in flooded forest during periods of inundation, where seeds are dispersed into water and secondarily dispersed or consumed by fish (Kubitzki and Ziburski 1994). During seasons of high fruit availability, plants are subjected to the greatest diversity of primate and non-primate seed dispersers (Ayres 1993; Kubitzki and Ziburski 1994; Vulinec et al. 2006). Thus, although primates are expected to contribute heavily to the quantity component of Phase One seed dispersal, as in other analyses (e.g., Herrera 1985; Howe and Westley 1988; Fischer and Chapman 1993; Lambert and Garber 1998; Chapman et al. 2002) the effects of directional selection by any single taxon on plant fruit and seed traits is likely to be minimal.

The Phase Two dispersers in this system include at least two large caviomorph rodent species (red-rumped agouti and paca; K. Vulinec and D. Mellow, unpub.), an uncounted number of small rodent and ant species, and 34 dung beetle species trapped during this study. These dung beetle species were classified into three guilds based on how they manipulate seeds deposited in dung: none, only small seeds, or both large and small seeds (Vulinec 2002; Vulinec et al. 2006). The dung beetle community differed among trap sites, habitat, and season (Fig. 12.2). There was substantial variation in dung beetle distributions daily and spatially, even between traps separated by 100 m (Larsen and Forsyth 2005; Fig. 12.3). The four most important species in terms of effective seed burial varied in their distributions among locations, and this variation has a large impact on whether a seed is buried sufficiently deeply, too deeply, shallowly, or left on the surface (Fig. 12.3). During August, beetles that buried seeds were more abundant than at other times of the year; however this abundance did not correspond with periods of high fruit availability (Ayres 1993; K. Vulinec, unpub).

Whether a seed is dispersed diurnally or nocturnally (e.g., some howler monkey defecations occur in the early morning before dawn; Julliot 1996a) had a major influence on whether that seed was buried, moved, or both buried and moved: seeds deposited in dung during the day were rarely buried, but were moved as much as 0.5 m from site of deposition, and seeds deposited during the night were almost never moved, but were buried (Fig. 12.4). Burial was rarely more than 10 cm, an optimal depth for germination of many seeds (Shepherd and Chapman 1998). However, most dung was left unburied due to the prevalence of the non-seed burying beetle guild. Seeds deposited in primary forest had a better chance of being buried shallowly and at a short distance away from the site of deposition due to the abundance of *Canthon fulgidus*. If dung beetles are a major component of a system of diplochory, they are likely to have both positive and negative effects on seed dispersal because a particular guild of beetles would treat similarly sized seeds the same way, regardless of plant species or the seed germination requirements (deposition depth and suitable site) thus reducing any selective force on seeds for size (Estrada and Coates-Estrada 1991; Shepherd and Chapman 1998). Most of the seed burial by dung beetles occurs during the night, rather than during the peak hours of primate movement.

Plant species respond to the integrated selective pressure of their seed dispersers and seed predators; i.e., the sum of selective pressures exerted by the entire coterie of fauna that exploit a species' seeds (Chapman and Chapman 2002; Russo 2003). The suggestion that primate frugivore behavior creates directional selection on fruit traits to the point of giving rise to a "primate syndrome" (Snodderly 1979; Julliot 1996a,b, 1997) ignores the variability in fruit handling and ranging behaviors (within and between species) and the fate of seeds after primary dispersal. Indeed, as demonstrated by a number of authors in both the Neotropics (e.g., Lambert and Garber 1998; Norconk et al. 1998) and Paleotropics (e.g., Asia: Corlett and Lucas 1990; Africa: Kaplin and Moermond 1998; Lambert and Garber 1998; Lambert 1999), even a single primate species in a given habitat can exhibit an array of seed-handling behaviors when exploiting the same seed species – (seed dropped, swallowed, defecated) as a function of season, fruit availability, social context, and fruit pulp ripeness. The variance in seed handling behaviors becomes even greater when evaluating the impact of several primate species, and greater still in an evaluation of the entire coterie of animals that exploit either the pulp or seeds of a plant species (Lambert 2002). It is thus more likely that selection on fruits, in most cases, has promoted traits that serve to attract a large number of dispersers. Combining two dispersers (in a proposed diplochorous system) is even less likely, given the coordinated pathway a seed must take, to be dispersed to an appropriate site: to first be detected by primary disperser rather than a seed predator, then chosen by the disperser, swallowed whole, transported away from parent plant, defecated, found incidentally in dung by dung beetles, buried directly or rolled away and buried, buried between 2–10 cm (depending on seed size), buried in an appropriate location, and not preyed on at any time during the dispersal cycle.

In general, relatively few studies have demonstrated strong directional selection on quantitative traits. In an analysis of 63 studies on the strength of selection on quantitative traits, Kingsolver et al. (2001) concluded that strong directional selection was uncommon. Jordano (1995) found that directional selection gradients on fruit traits indicated weak or erratic selection effects on maternal phenotypes, but that selection on individual seed phenotypes such as size characters was much more significant. Fruit traits (e.g., color, pulp volume) will be affected mainly by primary dispersers. Seeds traits, on the other hand, may be subjected to selection from other factors. Primate endozoochory is thought to exert selection on plants for smaller seeds, while several studies have found that larger seeds germinate at a higher rate than small seeds (Shepherd and Chapman 1998; van Ulft 2004; Pizo et al. 2006; but see Wenny 2000a). Dung beetles are likely to exert pressure in both directions, as smaller seeds will have a greater likelihood of being buried, but less potential for germination at greater depths. Thus, competing pressures may work at opposite ends of the character spectrum, and rather than exerting directional selection, lead to diluted and unpredictable selection on plant traits.

Results presented here offer a starting point to suggest that – as discussed and demonstrated by others (e.g., Holt 2006; Gravel et al. 2006; Scheffer and van Nes 2006) – community assemblages and character evolution are impacted both by stochastic events (Neutral Model) and species interactions and coexistence (Niche Model). In metacommunities with high species richness, the impact of stochastic events on community structure, species interactions, and trait evolution will be greater as a function of decreasing species abundance and smaller population sizes. It has been well documented that smaller populations also are more vulnerable to population decline and extinction; this is particularly true with large species and/or with species with slow life histories (e.g., large mammals such as primates, large-seeded tree species of primary forest; Cowlishaw and Dunbar 2000). Given extreme anthropogenic disturbance and its impact on decreasing primate – and other primary seed dispersers' – population sizes in the Rio Paratari area and throughout the Neotropics, stochastic events will be increasingly influential on species assemblages, biomass, and interactions. This suggests that forest managers will hence increasingly have to implement conservation tactics on a local, case by case basis.

12.5 Summary

Neutral and niche models are currently being tested to further understanding of species assemblages and functional traits. Some argue that plant traits coevolved with specific primary dispersers and that in some plant species diplochorous dispersal (both specific primary and secondary dispersers) is optimal for maximizing seed survival. Here, we evaluate how primary (primates) and secondary (dung beetles) dispersers impact seed dispersal in a case study of central Amazonia. Specifically, we asked: what is the density and biomass of primates at this location? Do particular primate species restrict their ranging to certain habitats? Does abundance and species composition of dung beetles vary in space and time? Are there dung beetle behaviors that may exert selection or dilute the effects of selection on fruit traits? Finally, is there evidence that presence and biomass of primates correlates with tree community structure or demography in Rio Paratari? Our data represent a first step in addressing these questions, and indicate that during the season of highest fruit availability plants are subjected to the greatest diversity of primate and non-primate seed dispersers. We also found the most effective seed-burying dung beetle species varied in their distributions among seasons resulting in a large impact on depth of seed burial. During August (the season of lowest fruit availability), beetles that buried seeds were more abundant than at other times of the year. Dung deposited at night had a better chance of being buried (with seeds imbedded) than during the day. We found a higher ratio of seedling/saplings to adult trees in secondary growth versus primary forest, as would be expected, but we found no association between areas of primate presence and absence. From this case study, we conclude that the suggestion that primate frugivore behavior creates directional selection on fruit traits to the point of giving rise to a "primate syndrome" or a system of obligate diplochory, ignores the variability in ranging behaviors and the fate of seeds after primary dispersal. It is more likely that selection on fruits, in most cases, has promoted traits that serve to attract a large number of dispersers. The anthropogenic impact on primates and other primary seed dispersers at this site implies that stochasticity in ecological and evolutionary interactions will be increasingly important, thus, conservation tactics may necessarily need to be implemented on a local, case by case basis.

Acknowledgments   We would like to thank the following organizations for their support: Delaware State University, The Center for Field Research (Earthwatch), and the Women's International Science Collaboration (AAAS and NSF). In Brazil, we are grateful to O Conselho Nacional de Desenvolvimento Científico e Tecnológico (CNPq), Instituto Nacional de Pesquisas da Amazônia (INPA), and the people of the villages surrounding Rio Paratari. We are especially thankful to Claudio R. V. da Fonseca of INPA. We are particularly grateful to Geoff Bland who worked tirelessly and with delightful insight for several field seasons, Maureen Leahy for her superb help with behavioral data collection and shuffling, and Sam Kahn for excellent field forays and jokes during work in the time of malaria. We are grateful to Dave Mellow, Betty Ferster, Roger Masse, Júlio César Bicca-Marques, Paul Garber, Alejandro Estrada, and two anonymous reviewers for valuable comments on the manuscript. We also thank over 40 Earthwatch volunteers for their help. Additionally, this project could not have been done without the continual aid and friendship of our Brazilian field assistants, Marcelo Beniti, Marzinho Souza, and their families.

References

Andresen, E. 2001. Effects of dung presence, dung amount and secondary dispersal by dung beetles on the fate of *Micropholis guyanensis* (Sapotaceae) seeds in Central Amazonia. J. Trop. Ecol. 17:61–78

Andresen, E. 2005. Interacción entre primates, semillas, y escarabajos coprófagos en bosque húmidos tropicales: un caso de diplochoria. Univesidad y Ciencia Numero Especial 002, 73–84.

Ayres, J. M. 1993. As matas da Várzea do Mamirauá. MCT-CNPq/Sociedade Civil Mamirauá. Brasilia.

Beckman N. G., and Muller-Landau H. C. 2007. Differential effects of hunting on pre-dispersal seed predation and primary and secondary seed removal of two Neotropical tree species. Biotropica 39:328–339.

Bennett, C. L., Leonard, S., and Carter, S. 2001. Abundance, diversity, and patterns of distribution of primates on the Tapiche River in Amazonian Peru. Am. J. Primatol. 54:119–126.

Chapman, C. A., and Chapman, L. J. 2002. Plant-animal coevolution: Is it thwarted by spatial and temporal variation in animal foraging? In D. Levey, W. R. Silva, and M. Galetti (eds.), Seed Dispersal and Frugivory: Ecology, Evolution, and Conservation (pp. 275–290). CAB International Press, Wallingford, Oxfordshire, UK.

Chapman, C. A., Chapman, L. J., Cords, M., Gauthua, M., Gautier-Hion, A., Lambert, J. E., Rode, K. D., Tutin, C. E. G., and White, L. J. T. 2002. Variation in the diets of Cercopithecus Species: Differences within forests, among forests, and across species. In M. Glenn and M. Cords (eds.), The Guenons: Diversity and Adaptation in African Monkeys, (pp. 319–344). Plenum Press, New York.

Chapman, C. A., Fedigan, L. M., and Fedigan, L. 1988. A comparison of transect methods of estimating population densities of Costa Rican primates. Brenesia 30:67–80.

Cowlishaw G., and Dunbar, R. 2000. Habitat Disturbance. In: G. Cowlishaw and R. Dunbar (eds.), Primate Conservation Biology (pp. 191–241). Chicago: The University of Chicago.

Corlett, R. T., and Lucas, P. W. 1990. Alternative seed-handling strategies in primates: seed-spitting by long-tailed macaques (Macaca fascicularis). Oecologia 82:166–171.

Doube, B. M. 1990. A functional classification for analysis of the structure of dung beetle assemblages. Ecol. Entomol. 15:371–383.

Emmons, L. H. 1997. Neotropical rainforest mammals: a field guide. 2nd Edition. The University of Chicago Press, Chicago.

Estrada, A., and Coates-Estrada, R. 1991. Howler monkeys (Alouatta palliata), dung beetles (Scarabaeidae) and seed dispersal: ecological interactions in the tropical rain forest of Los Tuxtlas, Mexico. J. Trop. Ecol. 7:459–474.

Feer, F. 1999. Effects of dung beetles (Scarabaeidae) on seeds dispersed by howler monkeys (Alouatta seniculus) in the French Guianan rain forest. J. Trop. Ecol. 15:129–142.

Fischer, K., and Chapman, C. A. 1993. Frugivores and fruit syndromes: Differences in patterns at the genus and species levels. Oikos 66:472–482.

Garber, P. A. 1986. The ecology of seed dispersal in two species of callitrichid primates (Saguinus mystax and Saguinus fuscicollis). Am. J. Primatol. 10:155–170.

Gardner, T. A., Hernández, M. I. M., Barlow, J., and Peres, C. 2008. Understanding the biodiversity consequences of habitat change: the value of secondary and plantation forests for neotropical dung beetles. Journal of Applied Ecology doi: 10.1111/j.1365-2664.2008.01454.x

Gautier-Hion, A., Duplantier, J.-M., Quris, R., Feer, F., Sourd, C., Decoux, J.-P., Dubost, G., Emmons, L., Erard, C., Hecketsweiler, P., Moungazi, A., Roussilhon, C., and Thiollay, J.-M. 1985. Fruit characters as a basis of fruit choice and seed dispersal in a tropical forest vertebrate community. Oecologia 65:324–337.

Gravel, D., Canham, C. D., Beaudet, M., and Messier, C. 2006. Reconciling niche and neutrality: the continuum hypothesis. Ecol. Let. 9:399–409.

Halffter, G., and Edmonds, W. D. 1982. The Nesting Behavior of Dung Beetles (Scarabaeinae): An Ecological and Evolutive Approach. Pub. Instit. Ecol., Mexico City, Mexico.

Herrera, C. M. 1985. Determinants of plant-animal coevolution: The case of mutualistic dispersal of seeds by vertebrates. Oikos 44:132–141.

Holt, R. D. 2006. Emergent neutrality. TREE 21:531–533.

Horwich, R. H., Brockett, R. C., James, R. A., and Jones, R. C. 2001. Population structure and group productivity of the Belizean Black Howling Monkey (Alouatta pigra): implication for female socioecology. Primate Rep. 61:47–65.

Howe, H. F., and Miriti, M. N. 2004. When seed dispersal matters. Bioscience 651–660.

Howe, H. F., and Westley, L. C. 1988. Ecological Relationship of Plants and Animals. Oxford University Press, Oxford.

Hubbell, S. P. 2005. Neutral theory in community ecology and the hypothesis of functional equivalence. Funct. Ecology 19:166–172.

Janson, C. H. 1983. Adaptation of fruit morphology to dispersal agents in a neotropical forest. Science 219:187–189

Jordano, P. 1995. Frugivore-mediated selection on fruit and seed size: birds and St. Lucie's cherry, *Prunus mahaleb*. Ecology 76:2627–2639.

Julliot, C. 1996a. Fruit choice by red howler monkeys (*Alouatta seniculus*) in a tropical rain forest. Am. J. Primatol. 40:261–282.

Julliot, C. 1996b. Seed dispersal by red howling monkeys (*Alouatta seniculus*) in the tropical rain forest of French Guiana. Int. J. Primatol. 17:239–258.

Julliot, C. 1997. Impact of seed dispersal by red howler monkeys, *Alouatta Seniculus*, on the seedling population in the understory of tropical rain forest. Ecology 85:431–440.

Kaplin, B. A., and Moermond, T. 1998. Variation in seed handling by two species of forest monkeys in Rwanda. Am. J. of Primatol. 45:83–101.

Kingsolver, J. G., Hoekstra, H. E., Hoekstra, J. M., Berrigan, D., Vignieri, S. N., Hill, C. E., Hoang, A., Gibert, P., and Beerli, P. 2001. The strength of phenotypic selection in natural populations. Amer. Nat. 157:245–261.

Kubitzki, K., and Ziburski, A. 1994. Seed dispersal in flood plain forests of Amazonia. Biotropica 26:30–43.

Lambert, J. E. 2002. Exploring the link between animal frugivory and plant strategies: the case of primate fruit-processing and post-dispersal seed fate. In D. Levey, W. R. Silva and M. Galetti (eds.), Seed Dispersal and Frugivory: Ecology, Evolution, and Conservation, (pp. 365–379). CABI Publishing, Wallingford, Oxfordshire, UK.

Lambert, J. E. 1999. Seed handling in chimpanzees (*Pan troglodytes*) and redtail monkeys (*Cercopithecus ascanius*): implications for understanding hominoid and cercopithecine fruit processing strategies and seed dispersal. Am. J. Phys. Anth. 109: 365–386.

Lambert, J. E., and Garber, P. A. 1998. Evolutionary and ecological implications of primate seed dispersal. Am. J. Primatol. 45:9–28.

Larsen, T. H., and Forsyth, A. 2005. Trap spacing and transect design for dung beetle biodiversity studies. Biotropica 37:322–325.

Link, A., and Di Fiore, A. 2006. Seed dispersal by spider monkeys and its importance in the maintenance of neotropical rain-forest diversity. J. Trop. Ecol. 22:235–246.

Morin, P. 1999. Productivity, intraguild predation, and population dynamics in experimental food webs. Ecology 80:752–760.

Norconk, M. A., Grafton, B. W., and Conklin-Brittain, N. L. 1998. Seed dispersal by neotropical seed predators. Am. J. Primatol. 45:103–126.

Peres, C. A. 1997. Primate community structure at twenty Amazonian flooded and unflooded forests. J. Trop. Ecol. 13:381–405.

Pizo, M. A., Von Allmen, C., and Morellato, L. P. C. 2006. Seed size variation in the palm *Euterpe edulis* and the effects of seed predators on germination and seedling survival. Acta Oecologica 29:311–315.

Radtke, M., da Fonseca, C. V., and Williamson, B. G. 2007. The old and young Amazon: dung beetle biomass, abundance, and species diversity. Biotropica 39:725–730.

Ribeiro, J. E. L. da S., Hopkins, M. J. G.,Vicentini, A., Sothers, C., Costa, A., M. A. da S., De Brito, J. M., de Souza, M. A. D., Martins, L. H. P., Lohmann, L. G., Assunção, P. A. C. L., Pereira, E. da C., da Silva, C. F., Mesquita, M. R., and Procópio, L. C. 1999. Flora da Reserva Ducke: Guia de identificação das plantas vasculares de uma floresta de terra-firma na Amazônia Central. Instituto Nacional de Pesquisas de Amazônia, Manaus.

Russo, S. E. 2003. Responses of dispersal agents to tree and fruit traits in *Virola calophylla* (Myristicaceae): implications for selection. Oecologia 136:80–87.

Scheffer, M., and van Nes, E. H. 2006. Self-organized similarity, the evolutionary emergence of groups of similar species. PNAS 103:6230–6235.

Schupp, E. W. 1993. Quantity, quality and the effectiveness of seed dispersal by animals. Plant Ecology 107–108:15–29.

Schupp, E. W., Milleron, T., and Russo, S. E. 2002. Dissemination limitation and the origin and maintenance of species-rich tropical forests. In D. J. Levey, W. R. Silva and M. Galetti (eds.), *Seed Dispersal and Frugivory: Ecology, Evolution and Conservation* (pp. 19–33). CABI Publishing, Wallingford, Oxfordshire.

Shepherd, V., and Chapman, C. A. 1998. Dung beetles as secondary seed dispersers: Impact on seed predation and germination. J. Trop. Ecol. 14:199–216.

Skoropa, J. P. 1988. The Effect of Selective Timber Harvesting on Rain-Forest Primates in Kibale Forest, Uganda. Ph.D. Thesis, University of California, Davis.

Snodderly, D. M. 1979. Visual discrimination encountered in food foraging by a neoptropical primate: implications for the evolution of color vision. In E. H. Burtt Jr. (ed.), *The Behavioral Significance of Color* (pp. 237–279). New York, Garland Press.

Stoner, K. E., Riba-Hernández, P., Vulinec, K., and Lambert, J. E. 2007. The role of mammals in creating and modifying seedshadows in tropical forests and some possible consequences of their elimination. Biotropica 39:316–327.

Suarez, S. A. 2006. Diet and travel costs for spider monkeys in a nonseasonal, hyperdiverse environment. Int. J. Primatol. 27:411–436.

Vander Wall, S. B., and Longland, W. S. 2004. Diplochory and the evolution of seed dispersal. Trends Ecol. Evol. 19:155–161.

van Ulft, L. H. 2004. The effect of seed mass and gap size on seed fate of tropical rain forest tree species in Guyana. Plant Biol. (Stuttg.) 6:214–221.

Vulinec, K. 1999. Dung beetles, monkeys, and seed dispersal in the Brazilian Amazon. Ph. D. Dissertation. University of Florida.

Vulinec, K. 2000. Dung beetles (Coleoptera: Scarabaeidae), monkeys, and conservation in Amazonia. Fl. Entomol. 83:229–241.

Vulinec, K. 2002. Dung beetle communities and seed dispersal in primary forest and disturbed land in Amazonia. Biotropica 34:297–309.

Vulinec, K., Lambert, J., and Mellow, D. J. 2006. Primate and dung beetle communities in secondary growth rainforests: Implications for conservation of seed dispersal systems. Int. J. Primatol. 27:855–879.

Wenny, D. G. 2000a. Dispersal of a high quality fruit by specialized frugivores: high quality dispersal? Biotropica 32:327–337.

Wenny, D. G. 2000b. Seed dispersal, seed predation, and seedling recruitment of a neotropical montane tree. Ecol. Monog. 70:331–351.

Zar, J. H. 1999. Biostatistical Analysis. Prentice Hall, New Jersey, USA.

Chapter 13
The Use of Vocal Communication in Keeping the Spatial Cohesion of Groups: Intentionality and Specific Functions

Rogério Grassetto Teixeira da Cunha and Richard W. Byrne

13.1 Introduction

In group-living species, especially ones occupying low visibility habitats or in which group members tend to spread over large distances, there must be some signaling mechanism(s) to ensure that the group ultimately remains as a spatially cohesive unit. The very nature of the problem precludes the use of tactile and visual communication, while the slower rate of diffusion of chemicals in the air or water makes impractical the use of chemical communication. Thus, the acoustic channel is expected to be the mode of choice.

Primates have been particularly well studied in that respect – besides being an intensively studied order, they are par excellence the mammalian taxon that exemplifies diurnal group living in poor visibility (forest) habitats. Vocalizations with some contact role have been found in almost all species studied (*see*, for example, Mitani and Nishida 1993; Rendall et al. 2000; Uster and Zuberbühler 2001; Oda 2002; Mendes and Ades 2004; Range and Fischer 2004). However, poor conceptual and nomenclatural standardization has been the norm for vocalization studies, and calls with a variety of names have been proposed to perform a wide range of functions, all traceable to a "contact" or "cohesion" role (*see* Table 13.1 for a non-exhaustive review of Neotropical primates' studies). These include: maintain contact at close (visual) range (Epple 1968; Pook 1977); maintain contact at intermediate ranges and/or in situations likely to lead to separation (traveling/foraging) (Byrne 1981; Harcourt et al. 1993); maintain contact at a distance/regain contact (Daschbach et al. 1981; Palombit 1992; Halloy and Kleiman 1994); coordinate or initiate and direct group travel (Boinski 1991; Boinski 1993); monitor the position of others (Caine and Stevens 1990); attract others in particular situations (Dittus 1988; Mitani and Nishida 1993).

Richard W. Byrne (✉)
School of Psychology, University of St. Andrews, St. Andrews, UK
email: rwb@st-andrews.ac.uk

P.A. Garber et al. (eds.), *South American Primates,* Developments in Primatology: Progress and Prospects, DOI 10.1007/978-0-387-78705-3_13,
© Springer Science+Business Media, LLC 2009

Table 13.1 Review of studies with Neotropical primates in which a given call could be described as a contact, cohesion, isolation or lost call by its explicit labeling, the description of the contexts in which it is emitted and/or the proposed function[1]

Species	Call name	Context and/or function	Type of study[2]	Reference
Callimico goeldii	Rhythmical calls	Isolation	C	a
	Monosyllabic calls	Disturbances, but "may serve as contact calls"		
C. goeldii	Long-distance location calls	"when interindividual distances were estimated to exceed 20 m"	W/C	b
	Long-distance contact calls	"in the case where a vocalize attempted to maintain contact with other individuals over a long distance"; "emitted by a group member who was lost to all other animals"		
	Short-distance location calls	"when the vocalizer attempted to maintain contact with other individuals . . . within an interindividual distance of approximately 10 m"		
Callithrix geoffroyi; *C. jacchus*;	Faint phee	"Undisturbed and in close visual contact"	C	a
Mico argentatus	Heterotypical faint high-pitched notes	"close visual and bodily contact"		
	Longer phee calls	"When losing visual contact with each other"		
	Phee cries	Isolated animals		
	Twitters	Loose visual contact and isolation		
C. jacchus	Short whirr; whirr; short broken whirr, broken whirr	From close visual contact to visual separation ; rarely when isolated	C	c
	Shrilling calls	Isolation		
Cebuella pygmaea	Closed mouth trill	Moving through the environment and sometimes when able to see each other	C	d
	Quiet trill	Cohesion or contact over short distances		
	J-call	Dispersion/visually isolated; Contact at larger distances		
M. argentatus	Contact calls	Intra-group cohesion; coordinate group activities; loose contact	W	e
Leontopithecus rosalia	Pe calls	Close visual contact	C	a
	Whee calls	When animals lose visual contact		
	Heterotypical sequences culminated in pü pü pü when in high excitement	Isolated animals		
	Twitters	When losing visual contact		

Table 13.1 (continued)

Species	Call name	Context and/or function	Type of study[2]	Reference
L. rosalia	One-phrase long call (wah-wah)	"Promote group travel"	W	f
	Two –phrase long call (wah–wah and descending whine)	Isolation from group members		
L. rosalia	Two –phrase long call	"to locate and to recognize group members, which are near them but out of visual contact"	W	g
Saguinus fuscicollis	Short calls	General activity; before congregating in groups; "spatially locating an individual . . . stray from the group", and also aids group cohesion	C	h
	Soft long call	Same contexts and function, but higher arousal		
S. fuscicollis	Chee, chip, chee-chip, multiple	From close visual contact to visual separation ; rarely when isolated	C	c
	Shrilling calls	Isolation		
S. geoffroyi; S. oedipus	Te calls	Undisturbed and in close visual contact	C	a
	Variations on te	Loose visual contact and disturbance		
	Monosyllabic calls given in isolation, trills	Isolation		
S. labiatus	Slide call	Social monitoring of group members; Group cohesion and coordination	C	i
S. mystax	Long call	Isolated animals, responded by its troop.	C	j
S. oedipus	Combination long calls	Socially isolated animals	C	k
	Large initially modulated whistle	"Given greater than .6m from other animals in low arousal situations"		
Cebus albifrons	Caw	"Probably serves as a contact call"	C	l
C. capucinus	Raucous squawk	Distant contact call	C	l
C. capucinus	Arrawh	"Exclusively produced by individuals separated by great distances from other group members"	W	m
	Adult trill	Start, lead, and change direction of group travel		
C. capucinus	Lost call	Isolation/separation from group members	W	n

Table 13.1 (continued)

Species	Call name	Context and/or function	Type of study[2]	Reference
C. olivaceus	Arrawh	"Single individual moving about and (…) apparently trying to locate its troop"	W	o
C. olivaceus	Huh	"Movement (…) was blocked or (…) the individual acted as if it did not know where to go"		
C. olivaceus	Arrawh	Isolated animals; animals lagging during group progression or moving in a different direction	W	p
	Huh	Intermediate distances and several contexts, often within visual contact; may work to maintain a characteristic spacing		
C. olivaceus	Chirps	Foraging/moving	W	q
Saimiri oerstedii	Peeps	Increasing distance;	W	r
	Twitters	Initiate/guide troop movement		
S. sciureus	Squeak	"contact call when left alone"	C	l
S. sciureus	Isolation peep	Lost of visual contact/large separation from the group; "may help a separated animal find its way back to the group"	C	s
	Peep	"…Associated with contact and attention"		
	Chirp	"Contact call for short to medium distances"		
Callicebus moloch	Whistles and trills	"May be uttered by individuals … "lost" or isolated"	C/W	t
C. moloch	Chirrup	Captive animals when isolated		
	Chirrup	"Contexts in which locating and recognizing group members foster group cohesion"	W	u
	Moan	Promotes approach of the mate to the caller		
Chiropotes albinasus; C. sagulatus	Whistle	"When group dispersion is relatively large (the animals are spread over an area of 1/4ha)"	W	v
Alouatta palliata	Deep metallic cluck	"Initiates progression, controls its direction and rate and coordinates the animals of the clan"	W	w
A. palliata	Notes similar to clucking	"Maintain auditory contact between the individuals of the clan"	W	x
A. palliata	Whimper	Troop progression	W	y
	Wrah-ha calls	Mothers of dependent infants when separated from the troop		

Table 13.1 (continued)

Species	Call name	Context and/or function	Type of study[2]	Reference
A. palliata	Broad band contact call	Group movements and sexual activity	W	z
Ateles geoffroyi	Whinny	Several contexts. In some circumstances elicit vocal response and/or approach from distant individuals. It is a localization signal which promotes cohesion through	W	aa
A. geoffroyi	Whinny	"serves a contact function by providing information to listeners about the identity and location of the caller"	W	ab
Brachyteles hypoxanthus	Staccatos and neighs with large proportion of short elements	Short-range vocal exchanges; "Co-ordination of nearby individuals"	W	ac
	Neighs with more long elements	Long-range vocal exchanges		

[a]Epple (1968), for C. jacchus see also Schrader and Todt (1993), Norcross and Newman (1997; 1999), Norcross et al. (1999); [b]Masataka (1982); [c]Pook (1977); [d]Pola and Snowdon (1975), Snowdon and Cleveland (1980), Snowdon and Hodun (1981); [e]Veracini (2002); [f]Halloy and Kleiman (1994); [g]Sabatini & Ruiz-Miranda (2008); [h]Moody & Menzel (1976); [i]Caine and Stevens (1990); [j]Snowdon and Hodun (1985); [k]Cleveland and Snowdon (1982), see also Jordan, Weiss, Hauser, and McMurray (2004); [l]Andrew (1963); [m]Boinski (1993); [n]Digweed, Fedigan and Rendall (2007); [o]Oppenheimer and Oppenheimer (1973); [p]Robinson (1982); [q]Robinson (1984); [r]Boinski (1991); [s]Winter, Ploog and Latta (1966), see also Masataka and Symmes (1986), Symmes, Newman, Talmage-Riggs, and Lieblich (1979), Lieblich and Symmes (1980), Smith, Newman, Hoffman, and Fetterly (1982); [t]Moynihan (1966); [u]Robinson (1979a), see also Robinson (1981); [v]Ayres (1981); [w]Carpenter (1934); [x]Collias and Southwick (1952); [y]Baldwin and Baldwin (1976); [z]Jones (1998); [aa]Teixidor and Byrne (1999); [ab]Ramos-Fernández (2005); [ac]Mendes and Ades (2004).

[1]Here we follow the taxonomy of Rylands and Mittermeier (this volume), updating species names whenever in disagreement with such taxonomy and possible. Species are presented in alphabetical order within each family. Separate studies of the same species were treated in different lines.

[2]C – captivity/experimental studies; W – naturalistic studies; C/W – when the species was studied by the same author(s) both in captivity and in the wild.

13.2 To Inform or Not to Inform? That Is the Question

Some researchers working with contact calls have reported the impression of a call-and-answer system, or at least vocal exchanges (Byrne 1981; Snowdon and Hodun 1985; Dittus 1988; Caine and Stevens 1990; Sugiura 1993; Teixidor and Byrne 1999). That is, they have the feeling, sometimes not backed up with strong data, that the core of the group answers to the isolated or distant animal as if informing it on the location of the group, or that separated animals or sub-groups exchange calls back-and-forth regularly, as if mutually informing each other of their location. This interpretation has been seriously questioned by some authors, however. If animals are to answer others on purpose, they argue, then by necessity they would have to know that the other animal is lost, isolated or wants to be in contact. In turn, such intentional answering would imply that these animals have a theory of mind (Premack and Woodruff 1978), an ability monkeys have yet to be proven to possess (Tomasello and Call 1997). It is argued that when there seems to be a system of vocal exchange in place, the supposed responses are actually a consequence of the state of mind of the "responder" itself, not an answer to the first caller (Cheney et al. 1996; Cheney and Seyfarth 1999; Rendall et al. 2000; Seyfarth and Cheney 2003). That is, the calls are not given with the intent of maintaining contact or informing the whereabouts of the group to the separated animal(s), something we label the **personal-status hypothesis**.

This idea is received with some resistance by field workers, given the widespread impression of animals answering one another. And even among the original critics of the intentionality idea, the position is sometimes softened. Rendall and co-authors (2000) recognize that other mechanisms could produce vocal exchanges, and that there may be degrees of perception of the other's perspective, falling short of full theory of mind, although they do not develop these ideas in detail.

When analyzing this issue we draw on Dennett's intentionality system. Dennett (1983, 1987) proposes that the degree of intentionality in animal communication can be understood in terms of a series of levels. In the lowest, zero-order intentional system, the reaction is mechanistic and does not involve mental states, as when a thermostat responds to low temperature. First-order intentionality involves purposive, goal-directed behavior by the agent, as when a monkey *wants* to deter a competitor. In second-order intentionality, the agent takes account of the mental states of others in addition to its own: for example when a monkey plans to deter a competitor, dominant to itself, by deliberately misleading it into believing that the monkey itself has seen a predator, when it has not. Thus, in the view of the personal-status hypothesis, monkeys performing contact calls need only be zero-order intentional systems, emitting calls in mechanical response to environmental cues, or first-order-intentional systems, calling because they are themselves afraid of their own isolation. These authors have specifically criticized a second-order intentionality stance, in which monkeys uttering contact calls would be aware of the knowledge status of other fellow monkeys and would be calling in order to alter such status.

But these are not the only options available, and Byrne (2000) proposes a way out of this dilemma. He argues that contact calling may involve a genuine call-and-answer system, based on a rather more elaborate representation of the environment, but one that is entirely first-order intentional, without comprehension of others' mental states. In this case "...both signalers and hearers *want* to reunite... and *know* that they should call in order to do so swiftly. It is not necessary that they be able to understand what others know...contact calling may be a goal-directed tactic, learned or even perhaps hard-wired, employed flexibly and selectively, but without insight into its mechanism" (Byrne 2000, p. 507). We baptize this idea the **reunite hypothesis**.

An important aspect of contact calls that would favor the cohesive function in the manner proposed by the reunite hypothesis, or even be necessary for some of the predictions, is the individual distinctiveness of the signal. Individual distinctiveness in the acoustic structure of contact calls has already been found in several studies (Snowdon and Cleveland 1980; Macedonia 1986; Gautier and Gautier-Hion 1988; Rendall et al. 1996; Teixidor and Byrne 1999; Oda 2002). Furthermore, some of these studies have shown that individuals not only perceive such differences, but also respond to them in adaptive ways (e.g., Dittus 1988; Rendall et al. 1996 2000).

The two hypotheses above lead to different, testable predictions (*see* Table 13.2). The personal-status hypothesis predicts that upon hearing isolated animals calling, only those animals that are themselves also separated would call back. The reunite hypothesis, on the other hand, predicts that hearers can call back irrespective of their own state of separation from the group. Also, if "both signalers and hearers want to reunite", then one could go further and predict that the more closely two animals are associated (by kinship or friendship), the more likely it is for an answer to occur. Another possible prediction is that proximity with higher ranking animals is more desirable for a series of reasons (Ramos-Fernández 2005), and thus they should be answered more promptly by lower ranking one than the other way round. In the last two cases, note that the call would need to have cues for individual identity, something that seems to be the case with several contact calls (*see* above).

Table 13.2 Predictions of alternative hypotheses regarding the motivational basis underlying the answering behavior to contact calls

Personal-status	Reunite
Only if also separated/isolated	Calling does not depend on the state of separation
Dependent exclusively on the state of separation/isolation	Dependent on the degree of association by kinship and/or friendship with caller; Lower ranking more likely to answer to higher rank than the other way round.

13.2.1 The Moo Call of Black Howler Monkeys *Alouatta caraya*

In order to illustrate the different approaches to the question and the applicability of our framework, we attempt to compare these two hypotheses in a particular case, the *moo* call of the black howler monkey, *Alouatta caraya*. Note that our intention here is not a specific test of the hypotheses, but a broad discussion of some issues involved with them in a real-world context.

Until recently there were very few studies in which low amplitude vocalizations of howler monkeys had been described, and evidence for a contact call was scant. Jones (1998) briefly analyzed a so-called "broad band contact call" of female mantled howler monkeys (*Alouatta palliata*), given during group movements but also during sexual activity. She suggested that the call was associated with food in both contexts, but may also incite male-male competition, a somewhat broad proposal. In the only extensive repertoire analysis of a howler species published to date (Baldwin and Baldwin 1976), no unambiguous and specific adult contact call was described. Thus, individuals of *A. palliata* emitted *whimpers* on a variety of situations, including troop progressions; infants produced *caws* when they were lost from their mothers, but also in other situations; and mothers of dependent infants produced *wrah-ha* calls when separated from the troop. Finally, in a very brief report on *A. caraya* calls, the authors mentioned a vocalization (*cry*) emitted in stressful situations, such as when the caller was away from the group (Calegaro-Marques and Bicca-Marques 1995). Recently, based on a 19-month field work study on the vocal behavior of a group of habituated wild black howler monkeys living on a forest patch in the Brazilian Pantanal, we determined that a low amplitude vocalization, the *moo* call, served a contact function (da Cunha and Byrne, in preparation)

Moo calls are relatively long vocalizations (around 1 s), tonal, with a low fundamental, presenting just a few clear overtones, and in general with a typical convex frequency modulation but spanning a small frequency range (*see* Fig. 13.1). A broad noisy segment sometimes precedes this tonal vocalization, especially from infants. Fundamental frequency varies among individuals between 100 and 200 Hz. Frequency modulation can take other forms or even be absent, but the one depicted is the most common.

Several data led us to infer a contact function for this call. First of all, travelling, the behavioral state which has the highest risk of separation between individuals, showed the highest calling rates among the analyzed behavioral states. On the other hand, rates were significantly lower during feeding, a sedentary activity for howlers. We were also able to show that group spread was significantly greater than normal during periods of *moo* calling. Furthermore, spread decreased significantly after calling began, although baseline levels of group spread were usually not reached by the time calling ceased. (da Cunha and Byrne in preparation)

Since our field work was designed around other questions, we did not collect precise data that could allow us to test between the personal-status and the reunite hypotheses (below, we offer a detailed study design for those wanting specifically to pursue this and other questions related to contact calls). However, some ad libitum observations can offer hints. First of all, the field observer (RGTC) noted a mild

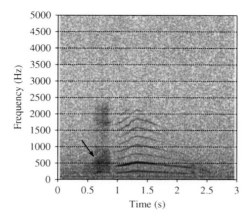

Fig. 13.1 Spectrogram of a *moo* call. Gaussian window. Time window: 0.04 s. Bandwidth = 32.5 Hz. Note the noisy component of the call

contagion effect exerted by the call, especially in more protracted sessions. But, more critically, *moos* were at times apparently answered by other *moos* from distant individuals or by a volley of *mutters* (another soft call), in which case several individuals would call nearly simultaneously. In neither case did it seem necessary for all the calling animals to be separated. In fact, having several individuals simultaneously out of reach of various others was quite an uncommon situation for these howler monkeys. At most, one animal or a small sub-group was isolated from the remainder of the group. Strangely, *mutters* are rather quiet calls, given mostly during feeding, that almost certainly do not carry far in the forest: they do not seem to offer any locational cue for the isolated animal. This seems to indicate that, though the "answerers" may desire to reunite, they do not always perform the most adequate behavior to that end. This interpretation would support the notion that the responders do not intend to inform the original callers. However, these data give little support for the notion that monkeys only answer when they are also separated from other individuals.

Occasionally, when a *moo* session was very prolonged, with calls from various individuals being emitted for quite some time, the alpha male would emit an *oodle*, a soft call type, often marking pauses in the middle of loud call sessions. This call, although not loud, could function as an acoustic beacon for the group if they are not too spread out. Additionally, on at least one occasion, animals resting in close proximity emitted a noticeably long sequence of calls, in the absence of only the alpha male. The male later reunited with the group, apparently without emitting calls himself. This confirms that separation is not a necessary condition for calling, but wanting to reunite with group members may be a sufficient one.

A more puzzling set of observations occurred when animals were in close proximity (e.g., with a group spread of 15m or less) yet still called. There were also situations in which there was a high rate of calling involving adults, and, although the group was more spread out, every animal was within visual reach of a couple of others. Another intriguing fact is that large group diameters were not necessarily followed by emission of *moos*, and on many occasions separated individuals regained contact without calling. These observations all seem to indicate that the status of the caller was not a necessary condition for *moo* calling, since animals that called were not necessarily separated from others: this runs against the personal-status hypothesis. Neither was the separation status of the caller sufficient to trigger calls or "answers", since separated animals were observed not to call at times.

All these apparently puzzling observations can be more easily accommodated under the reunite hypothesis. On this account, animals are expected to call more often (but not always) when they themselves are separated: the situation most likely to lead to a desire to reunite. The reunite hypothesis can potentially also explain calling between animals in closer proximity, within visual reach and calling from the core of the group when one or a few individuals are absent, since it is not the separation status but the desire to reunite that triggers calling.

13.3 Contact Calls, Yes, and so What?

Another issue, unrelated to the problem of intentionality, is that of the specific function played by a given contact call. As we have already noted, authors have labeled a range of vocalizations as contact or cohesion calls, though they apparently perform quite different *specific* functions. These run from maintaining contact at close or visual range up to regaining contact among widely separated individuals, from gaining access or proximity to an individual up to initiating group travel.

A first problem with respect to such variation, and one common to most vocalization studies, is that authors do not specify at which functional level they are working. In other words, authors may use the same word, "function" or "purpose", to imply quite different meanings. They may be referring to (1) the ultimate social consequence, (2) the immediate behavioral means of operation, or (3) the informational content of a call from the point of view of the receiver (we arrange these variants from more general to more specific levels). Note that there is another explanatory level when dealing with calls, which is the proximate mechanism through which a call operates. This level is located between the immediate behavioral means of operation and the informational content of a call. Since it has rarely been proposed as a "function" in the literature we will consider it alongside level (2). Consider, for example, a call that helps isolated animals to regain contact. This description refers to its immediate behavioral means of operation. While its social consequence may be to keep the spatial cohesion of the group, the proximate mechanism could be the exchange of calls between animals who wish to reunite. Finally, the informational content of the call could be the position of the group to the isolated individual and/or the emotional state of the caller.

An additional problem pertaining more specifically to the immediate behavioral means of operation of contact calls is that, in some cases, a given label may apply to a continuum. For example, "maintaining/regaining contact between group members" can be applied to calls given between animals that are separated in different degrees. The division into different categories (such as maintaining contact between close or distant group members) may be arbitrary.

Keeping these problems in mind, we will attempt here to propose a classification scheme for different functional explanations that could be grouped under the label "contact". The scheme was devised to help the study of contact calls by organizing the diversity of proposals into a common framework. This can both help to compare the studies that were already conducted, as well as organize future data collection aimed at providing data that could be used in a comparative way. We will stop at the level of the proximate mechanism, since the problem of the informational content of the call is a more specific (and thorny) one, with a whole body of literature related to the discussion of the referential vs. emotional content of vocalizations, or even if calls do have any informational content at all (Owren and Rendall 2001).

Starting at the first functional level (social consequence), we will only consider calls which function to retain the spatial cohesion of the group, thus following the majority of studies on contact calls, or at least their unstated assumptions. This restriction leaves out a few studies in which vocalizations were labeled as contact calls but were proposed to have a different function, namely gaining or maintaining access/proximity/physical contact to an individual. We take these to refer to the second functional level (immediate behavioral means of operation), with a completely different social consequence (first functional level), probably related to purely social issues. As examples, we have the contact chatter of *Leontopithecus rosalia* emitted in physical contact contexts (McLanahan and Green 1977), and the contact calls of *Macaca fuscata* and *Theropitecus gelada*, produced before grooming contacts (Masataka 1989), and "before or during positive social interactions" (Aich et al. 1987), respectively.

For a call that works in keeping the spatial cohesion of groups, there are three possible immediate behavioral means of operation. For each, there seems to be only one possible proximate mechanism (the next lower level of explanation).

13.3.1 Keeping Contact

In this category we include all calls that apparently function to keep regular contact, although they may be given by individuals in different states of separation. It encompasses calls given by animals at close (or visual contact) range, and calls given by animals when visibility between them is poor or not possible at all. The pigmy marmoset illustrates these possibilities, showing three different contact calls depending on the separation distance (Snowdon and Hodun 1981). Northern muriquis use different calls to keep contact with close and quite distant group members (Mendes and Ades 2004), something apparently useful for this fission-fusion species.

There are a number of situations in which it could be useful for animals to produce such calls even when at close ranges or in visual contact. One of these is in nocturnal species, for which visual contact is made even more difficult; indeed, categorization based on distance becomes somewhat arbitrary here. Another situation is when animals are concentrating on a given activity, such as when feeding or moving fast, and can maintain contact through calling without having to scan constantly for others, something that was observed to happen with the "monitoring" call of saddle-back tamarins (Caine and Stevens 1990).

In all these cases the most likely mechanism to accomplish the behavioral means of operation of "keeping in contact" is reciprocal (or antiphonal) calling, which then guides the actions of the individuals.

13.3.2 Regaining Lost Contact

Here we include those calls given in cases in which an animal/sub-group gets accidentally separated from others, usually described as being "lost" or isolated (a description that is somewhat contentious and hard to prove). We also incorporate in this category those calls used by fission-fusion species to help in guiding reunion of individuals. In these cases, the separation is not accidental but forms part of the normal behavioral repertoire of the species. As well as reciprocal calling, there is another theoretically possible proximate mechanism, in the light of the reunite hypothesis. If an animal wants to reunite with a separated member, before the individual itself wants to, then calling just by the first individual might then attract the attention of the separated one, who can then reunite without calling itself. The whinny of the spider monkey illustrates well this possibility, with calls not always being responded antiphonally, but sometimes just with an approach by another individual (Ramos-Fernández 2005).

13.3.3 Coordinating Group Travel

Vocalizations included in this class may be of two kinds. (A) Calls given just before the initiation of a group travelling session, with the aim of inducing other individuals to follow the leader. (B) Calls uttered in the middle of travelling bouts with the apparent intent of changing route direction (and thus changing the lead as well). In this latter case, we take it that animals do not call because they want to reunite or maintain continuous contact with other(s). Rather, we are talking about a situation in which only one individual calls in order to change the behavior of all others, and who do not need to call back in order for the vocalization to fulfill its role. As examples, we have the studies by Boinski, who found calls with such functions in white-faced capuchin monkeys, *Cebus capucinus* (Boinski 1993), and squirrel monkeys, *Saimiri oerstedi* (Boinski 1991). In this category, the mechanism is either a temporal association between a vocal signal and a change of behavior (start of travelling bout or change of direction) or constant calling by the leader to help group members to orient themselves.

13.4 The *Moo* Call: Applying the Scheme

In order to illustrate the application of the categorization presented above, we will analyze the possible immediate behavioral means of operation of the *moo* call of black howler monkeys. The first question to be addressed in this case is the relation between calling and the variables "travelling" and "group diameter". Group diameter tends to be large during travel, so if calling is associated with travel, it is bound to relate to large group diameters as well. (The reverse relationship does not hold, as the call may be specifically used in static contexts of large group spread.) However, the proximal cue for calling might be either the travelling context itself, or a high dispersion. In the latter case, the specific function could be keeping contact between dispersed members or regaining lost contact. When travelling itself is the primary cause of calling, the immediate function could be to keep contact with other animals or be more specifically related to co-ordination of group travel.

Anecdotal observations point to higher-than-average dispersion as the primary cause for calling. First, it was observed that animals did not call exclusively when travelling or about to start travelling. On a variety of occasions, animals (singles, small sub-groups or even the "core" of the group when one or more individuals were missing) gave *moo* calls when there was loss of visual contact between members of the group. The individual or party that joined the others would normally be the calling one, and the calls usually ceased after reunion. In a couple of cases, the separated animal(s) stopped calling after hearing a *moo* from another individual, and started travelling in the direction of the sound. Once, the alpha male was quite separated from the group ($+/-$ 150 m, a very large value for these howlers), kept emitting *moos* for at least 40 minutes at regular intervals. His erratic movements and apparent indecision gave the impression he was trying to re-locate the group. We have already mentioned the situation in which the core of the group gave *moo* calls for a long time in the absence of only the alpha male. All these observations point toward a function of regaining contact (category 2), and do not fit well into a regular exchange of calls as it happens in those cases where the function is keeping contact (category 1). Given the kinds of circumstances in which it was produced, an exclusive role in travel coordination (category 3) also does not seem to apply.

Sometimes, however, there was a high rate of calling involving adults, yet every animal was within visual reach of a couple of others, although not of the whole group. Examples include cases when there were two animals heading in very different directions, or part of the group started moving while others kept resting, or travelling sessions that led to a large group spread. The calls usually terminated when the group was more cohesive or when a clear travel session ensued. Those observations seem to indicate a function of keeping contact (category 1) or coordinating group travel (category 3). Such an interpretation is reinforced by the observation that travelling under very windy conditions, which impairs hearing the sounds made by other travelling animals, also seemed to trigger calling. However, if co-ordination is the role, it is definitely of a different nature from that described in squirrel and capuchin monkeys (Boinski 1991, 1993), as initiators and leaders were not the main or exclusive callers.

Thus, it seems that *moos* can perform two, or even all three of the proposed immediate behavioral roles. There are few species in which the *same* call performs the three different behavioral roles categorized above. In the majority of cases there is a call which is used for keeping contact, and when the animals are "lost" there is another call type. Guinea baboons are an exception (Byrne 1981), although splitting in sub-groups is part of their routine, while this is not true of howlers. The possibility that *moos* might have a broader range of functions than contact calls of most other species makes them an interesting target for further research, as does the fact that it is also produced by immature individuals in stressful situations, a completely different context and apparently unrelated to its contact role.

13.5 Proposed Study Designs: a Road Map to Contact Calls

We now propose some designs for future work, with three related aims in mind: testing whether a given vocalization is indeed a contact call (in the sense of functioning to retain the spatial cohesion of a group); elucidating the behavioral role and the particular mechanism of a contact call; and testing the two alternative hypotheses that could explain the motivation behind a proposed contact call.

13.5.1 Contact Call

In order to show a given vocalization is indeed a contact one, it seems obvious to begin by looking at spatial cohesion. Thus, one must have clear and detailed measures of group spread. For that end, instantaneous samples should be taken of the group diameter (understood as the maximum distance between any two group members) at regular intervals, in order to provide a baseline picture of spread. However, this is a crude measure and somewhat unrealistic, given that a group of primates is not spread out in an even way in all directions. It would therefore be interesting as well to register the distance on an axis 90 degrees to the first one, as done by Palombit (1992). The multiplication of the two provides an index of group spread which might reflect more accurately the larger variety of situations in the real world than mere maximum distances.

The next step is to record these measures focally at those times when an animal produces the vocalization suspected to be a contact call, and also at regular intervals after the call or the start of the session (if calls are produced in volleys, be it by a single individual or by many) and after the cessation of calling. These intervals should be smaller than the ones used for the baseline data collection, but the experience of the researcher with the species being studied will indicate the best intervals.

The idea behind all these measures is to compare the baseline values with the focal ones in order to test two predictions. First, if the call possesses some contact-related function, then its emission should be associated with group diameters larger than usual. (Larger group spreads are situations in which maintaining contact is presumably more necessary or in which it is more likely for animals to get separated from the group or to be already isolated.) One would also predict a relation between the cessation of a call (or a bout of calling) and a decrease in values of group diameter, perhaps returning to average (or lower) values. Alternatively we could expect to observe reductions of group diameter more often than increases after calling.

The measures above allow a quantification of group spread, but not of the separation between individuals or whether they were within visual reach or not. Ideally one should also register the distances of every individual in relation to all others at regular time intervals. Such data would provide useful baseline figures about the average separation between each pair of animals. Clearly, however, this is not feasible with a large group of animals. In this situation the researcher might be able to sample a sub-set of the group on each day, or the distance of focal animals to all visible animals, and alternate the focal animal between previously defined sampling periods. These distances could also be recorded when a focal individual calls, noting especially which animals are likely to be in the visual field of the caller. Data collection, analysis and predictions are similar to those described above. The baseline data could also be analyzed in a similar way as done by Palombit (1992), using multiple regression, with the number of calls in a given time period as the dependent variable and distances of interest as the independent ones.

However, even if a given call is not produced in situations of large group spread and does not lead to a decrease in group diameter, it might still function as a contact call. For example, it could be used during fast travel or travel through a low visibility habitat, when the group diameter is within normal values but the animals must maintain contact vocally to avoid losing track of others, or to avoid having to scan the environment constantly. The call could also have a behavioral role of coordinating group travel. In order to take account of these and other possibilities, we imagined some refinements in data collection.

To investigate visibility, since conditions vary hugely from one study site to the other, the simplest measure would be to record regularly—alongside the distance of each animal to the others—whether pairs are likely to be within sight, with yes/no sampling. If the call were indeed a contact one, we would predict that the more animals that are out of sight of each other, the more likely it is for the animals to call.

Where calling functions to coordinate travel, one would predict that calling should be more associated with travelling than other behavioral states—with the proviso that initiation of travel may fail to produce this association. To test this prediction, the most straightforward method would be to collect data continuously on focal animals for fixed periods of time, recording behavioral states, such as travelling, feeding, resting, etc. Afterwards, the researcher would have a sample of calling rates along different behavioral states for the group individuals, and could perform an appropriate statistical test of the prediction above. Additionally, it may be interesting to focus on the mechanics of "travelling" itself, which can be divided into leading, following, or travelling alone. If the call does have a role in co-ordination, then one would expect it to be associated with leading but not to following or travelling alone. Special attention could be paid to attempts to change group direction, a leading role performed by animals that are not at the front of the progression. A possible criterion is to register if a following animal starts moving in a direction that deviates more than, say, 30 degrees from the current one. If the call is associated to such attempts then it also has a role in travel co-ordination. If the researcher has a large sample of events, including for example other calls, displays, etc., then the prediction is that if the two events (calling and attempts) are indeed associated, transition probabilities between them should be significantly higher than expected by chance (see Robinson 1979b; Snowdon and Cleveland 1984 for examples of the use of the technique). If a call is associated with travelling, but is produced by many or most animals, then it will probably have a role of "keeping contact". Such an idea would be reinforced if they were emitted more often during fast travel or in locomotion through poorer visibility habitats.

The data collection suggested up to this point can also help to elucidate if a call functions to regain lost contact. In this case, the prediction is that the call should be given by those individuals that are separated or visually isolated from all or the majority of others. Here the researcher needs also appropriate criteria to define separation. We suggest two alternatives. One could use fixed criteria based on experience (for example if an animal is more than y meters from the majority of the group, it is considered to be separated). Alternatively, separated animals could be those whose distance in relation to the majority of others lies at the upper end of the distribution of baseline distances.

To address the specific possibility of initiation of group travel, one should note all attempts to get a group moving. A preliminary definition of an attempt could be: a situation whereby an animal starts travelling after a resting or feeding session, moving away from the nearest individual more than x meters (x, for example, might be a proportion of the diameter of troop dispersion). To differentiate between "true" initiation attempts on one hand and solitary movement and failed attempts on the other, it is necessary to note if the group started moving within y minutes of an attempt, and the azimuth taken with respect to the initial alignment of the initiator with the rest of the group (Boinski 1991; Boinski 1993). If the majority of the animals in the group start moving in the same direction of an attempt, it is deemed successful, otherwise is a failure (or a solitary movement). However, particularities of different species may make a universal criterion worthless, and the researcher should use one that is appropriate for his/her species. If the vocalization under study does have a role in initiation of group travel, then one would expect a temporal association between calling and initiation of travel. Another prediction is that successful initiation attempts should be more associated to the call under study than failed attempts/solitary movements. In both cases, the test of the prediction can be made using transition probabilities, as proposed above with respect to attempts to change travel direction.

Some possible refinements include noting, in successful initiation attempts, the time lag between the attempt and the last animal to start moving. A distribution of those latencies should provide a more objective evaluation of those cases in which there seems to be "resistance" to follow. Conflicting initiation attempts, leading to real or potential group splitting, may be especially revealing. In both cases, if the leader(s) make use of the call, one has further data that reinforces the notion it plays a role in initiation of group travel.

Showing that there is a system of call exchange would help in determining a call works in keeping contact between group members. Here one can also make use of transition probabilities since, if there is a call-and-answer system, transition probabilities between calls from different individuals should be higher than expected by chance.

13.5.2 Personal-Status × Reunite

Since the personal-status and the reunite hypotheses differ in the predicted effect of separation between individuals in their answering behavior, this can provide a test between the hypotheses. In the case of the personal-status hypothesis, one would expect answers only from animals that are themselves separated, while this is not necessary in the light of the alternative hypothesis. Note that, for this test, the researcher needs to have specific criteria for two parameters: answer and separation. To determine if a given call is an answer to a previous emission, transition probabilities could help once again. An "answer" should be a call that is produced after another with a transition probability significantly higher than expected by chance. Some authors, however, have used either a time criterion (Cheney et al. 1996; Jordan et al. 2004; Ramos-Fernández 2005) or considered answers as calls produced during an emission of another animal (Digweed et al. 2007). Regarding separation, there could be many ways to define it or to test if the degree of separation interferes with the probability of an answer. For example, one could compare the distances between an answerer and the nearest individual to it, to the baseline sample of "nearest individual" distances (compiled from the regularly collected data on the distances between each individual and all others). The personal-status hypothesis predicts that there would be a significant difference between the two samples, while this is not true for the reunite hypothesis.

However, reality may not be as simple as the test above suggests. Proponents of the personal-status hypothesis could still argue that, apart from distance itself, many other factors could also influence the "feeling" of separation and the likelihood of an answer. These factors could include degree of visibility, position of the animal within the cloud of group spread, kind of activity being carried out, among others. Thus further data should be collected to tease apart these possibilities, as we suggest below.

The personal-status hypothesis predicts that animals in the periphery of the group, or lagging behind in travel sessions, should exhibit a higher probability of answering to a contact call than centre ones, something not expected under the reunite hypothesis. Thus, the position of the animal within the "cloud" of group spread should also be noted. For stationary behaviors (resting/feeding), a centre/periphery criterion should suffice. For travelling, one should record the position of the animal in the line of progression, divided in thirds, for example, as done with baboons (Cheney et al. 1996), or even the position of every individual if species' characteristics allow. If visibility and species allow, a sketch of the group cloud, with distances indicated, would provide much useful data. We suggest collecting such kind of information at regular intervals and also focally, during call emission.

The reunite hypothesis predicts that answers should be more likely to be given to calls emitted by close kin, friends or higher ranking animals (*see* Table 13.2 above). The data collected (both regularly and focally during call emission) regarding the distance of every individual in relation to others is enough to test these predictions. Regarding visibility, using the data collected as suggested above, the prediction from the personal-status hypothesis is that the more animals who are out of an animal's sight the more likely it will answer the calls of others, and that animals with many others in sight should not answer at all. The alternative hypothesis would predict that the likelihood of answers is not influenced by the presence/absence of other individuals within sight, and that answers could be given in situations of full visibility of the majority of the group.

We acknowledge that the collection of all the data suggested above would be taxing, and in fact impossible to do all at once. Alternation between different kinds of data collection on different days may be needed.

13.6 Playback Experiments

Besides collecting behavioral data, a researcher can also conduct playback experiments. Although a powerful technique, its use in the field of primate vocal communication studies in the wild has been mostly restricted to loud or alarm calls (Cheney et al. 1996; Rendall et al. 2000; Ramos-Fernández 2005 are the exceptions with contact calls that confirm the rule). In this final section we suggest some designs useful for the issues discussed above, but the possibilities here are limited only by the creativity of the researcher. We assume the reader is familiar with all the care that must be taken to avoid pseudo-replication, and similar precautions, to ensure the playback mimics as natural a situation as possible.

To test between the personal-status and the reunite hypotheses an experiment could be used. The starting point is a sample of contact calls from all individuals in the group (note that we must use calls from the study group itself in order to avoid possible confounding effects due to group members recognizing calls from strangers, and behaving as if an unfamiliar individual is nearby). Then, when a given individual is isolated from the group, the researcher can employ two tactics (following Cheney et al. 1996; Rendall et al. 2000). Contact calls from this separated individual could be played to the core of the group. If the personal-status hypothesis is correct, then one should expect no answer at all, since none of the individuals in the core of the group are in any way separated. On the other hand, the reunite hypothesis predicts that answers will occur, and that these will be more likely the higher the degree of association of an individual to the separated one whose calls were played back. The second tactic is to play calls from core group members to the separated animal. The prediction under the personal-status hypothesis is that the isolated animal should answer irrespective of the animal whose call was played back, since calling is triggered only by its own state of separation. Alternatively, the reunite hypothesis predicts that answers will be more likely the higher the degree of association between the caller and the isolated animal. The designs of Ramos-Fernández's (2005) experiments are elegant ones to show the use of calls in regaining contact or keeping contact at large distances.

Regarding the possible behavioral roles that can be played by a contact call, playback experiments might also be useful. For example, if the call in question works in keeping contact between group members, the researcher can play back calls at a range of situations varying in their need for keeping contact. For example, calls (always from individuals that are out of sight of the majority or considering only the answers of those animals out of the sight of the "caller") at various group diameters. If the call works in keeping contact, then one expects that the higher the group diameter at the playback time, the more likely it is for animals to answer or the higher its intensity.

Another option is to test if a call works in initiation of group travel. For this, one must play calls from different individuals to the group during a resting session (again ensuring that the real animal whose call is going to be played is out of the sight of the majority). If the call does promote travel initiation then one would expect that the reaction (either in looking at the speaker direction or actually starting to move in its direction) should be proportional to the frequency of the "caller" as a travel leader. Thus calls from infants, juveniles or lower ranking animals would generally be ignored, while calls from high ranking adults should generate stronger reactions. Unfortunately, apart from initiation, it may not be possible to test if a call works in other aspects of travel co-ordination, since this would involve playing back at an ongoing travel session, a situation which poses very difficult logistic constraints.

13.7 Summary

The existence of a call-and-answer system in primate contact calls has been questioned on the grounds that it would need a theory of mind in order to operate. On this view, answers would occur depending only on the state of separation of the answerers themselves (personal-status hypothesis);

We suggest that the behavior might be explained by a first-order intentionality mechanism, with no theory of mind implied: both animals involved want to reunite and know that they should call for that end (reunite hypothesis);

Anecdotal observations on a contact call of black howler monkeys *Alouatta caraya*, the *moo*, indicates that the behavior of the animals conforms to the predictions of the reunite hypothesis;

Part of the great variety of functional proposals for contact calls in the literature stems from a lack of precision of authors with respect to the explanatory level they are working, which they usually label as the call function;

Three levels are commonly used in vocalization studies, and we added a fourth one for completeness (number 3 below). These are (from the more general to the more specific): (1) the social consequence of the call, (2) its immediate behavioral means of operation, (3) its proximate mechanism, or (4) the informational content of a call from the point of view of the receiver;

Applying this hierarchical categorization to contact calls, most authors restrict the usage of the term to calls having a social consequence of keeping the spatial cohesion of the group;

We identified three immediate behavioral means of operation to achieve that consequence: keeping contact; regaining lost contact, and coordinating group travel;

The proximate mechanism of "keeping in contact" is reciprocal calling, which guides the necessary actions of the individuals; to regain lost contact animals can also use reciprocal calling or a member at the core of the group might want to reunite with a separated member, and then call, guiding the isolated individual, with no counter-calling being necessary; to coordinate group travel, there must be an association between calling and a change of behavior, be it the initiation of a travel bout or a change of direction, or constant calling by the leader to help group members to orient themselves;

The *moo* call seems to operate through the behavioral means of keeping in contact, and regaining lost contact, and maybe in coordinating group travel as well;

Our proposed observational and experimental design can help to address both questions raised here, the alternative hypotheses about the possibility of a call-and-answer system in contact vocalizations, and the various explanatory levels for these calls.

Acknowledgments We would like to thank EMBRAPA for kindly allowing using the study area for the conduction of the research on black howler monkeys' vocalizations and for providing many logistic facilities. In particular, Emiko Kawakami de Resende and José Anibal Comastri Filho for speeding up the work permission, decreasing the fees, and providing access to logistical help, Gentil Cavalcanti and Marcos Tadeu, responsible for logistics at the farm, Sandra Santos for general help, and the cowboys and drivers for helping to sort out a myriad of small problems. Dr. Eckhard Heymann and an anonymous reviewer provided many insightful comments in an earlier version of the manuscript, which was greatly improved with their contributions. RGTC's Ph.D. study was funded by a CAPES studentship (n° 1373/99 4). Field work was partly funded by a Russell Trust Award from St. Leonard's College, University of St. Andrews. The help of Marcelo Oliveira Maciel (fieldwork assistant) was invaluable at all phases of data collection.

References

Aich, H., Zimmermann, E. and Rahmann, H. 1987. Social position reflected by contact call emission in Gelada baboons (*Theropithecus gelada*). Zeitschrift fur Säugertierkunde 52: 58–60.

Andrew, R. J. 1963. The origins and evolution of the calls and facial expressions of the primates. Behaviour 20: 1–109.

Ayres, J. M. 1981. Observações sobre a ecologia e o comportamento dos cuxiús (Chiropotes albinasus e Chiropotes satanas, Cebidae, Primates). Instituto Nacional de Pesquisas da Amazônia e Fundação Universidade do Amazonas, Manaus, Master's thesis.

Baldwin, J. D. and Baldwin, J. I. 1976. Vocalizations of howler monkeys (*Alouatta palliata*) in southwestern Panama. Folia primatologica 26(2): 81–108.

Boinski, S. 1991. The coordination of spatial position: a field study of the vocal behaviour of adult female squirrel monkeys. Animal Behaviour 41(1): 89–102.

Boinski, S. 1993. Vocal coordination of troop movement among white-faced capuchin monkeys, *Cebus capucinus*. American Journal of Primatology 30(2): 85–100.

Byrne, R. W. 1981. Distance vocalisations of Guinea baboons (*Papio papio*) in Senegal: an analysis of function. Behaviour 78: 283–312.

Byrne, R. W. 2000. How monkeys find their way: Leadership, coordination, and cognitive maps of African baboons. In S. Boinski and P. A. Garber (eds.), *On the move: How and why animals travel in groups* (pp. 491–518). Chicago: University of Chicago Press.

Caine, N. G. and Stevens, C. 1990. Evidence for a "monitoring call" in red-bellied tamarins. American Journal of Primatology 22(4): 251–262.

Calegaro-Marques, C. and Bicca-Marques, J. C. 1995. Vocalizações de *Alouatta caraya* (Primates, Cebidae). In S. F. Ferrari and H. Schneider (eds.), *A Primatologia no Brasil – 5*, (pp. 129–140). Belém: SBPr/UFPA.

Carpenter, C. R. 1934. A field study of the behaviour and social relations of howling monkeys (*Alouatta palliata*). Comparative Psychology Monographs 10(2): 1–168.

Cheney, D. L. and Seyfarth, R. M. 1999. Mechanisms underlying the vocalizations of nonhuman primates. In M. D. Hauser and M. Konishi (eds.), *The design of animal communication* (pp. 629–643). Cambridge: Bradford: MIT Press.

Cheney, D. L., Seyfarth, R. M. and Palombit, R. A. 1996. The function and mechanisms underlying baboon "contact" barks. Animal Behaviour 52(3): 507–518.

Cleveland, J. and Snowdon, C. T. 1982. The complex vocal repertoire of the adult cotton-top tamarin (*Saguinus oedipus oedipus*). Zeitschrift fur Tierpsychologie 58: 231–270.

Collias, N. and Southwick, C. 1952. A field study of population density and social organization in howling monkeys. Proceedings of the American Philosophical Society 96(2): 143–156.

Daschbach, N. J., Schein, M. W. and Haines, D. E. 1981. Vocalizations of the slow loris, *Nycticebus coucang* (Primates, Lorisidae). International Journal of Primatology 2(1): 71–80.

Dennet, D. C. 1983. Intentional systems in cognitive ethology: the "Panglossian paradigm" defended. Behavioral and Brain Sciences 6: 343–390.

Dennet, D. C. 1987. The Intentional Stance. Cambridge, Mass.: Bradford books, MIT Press.

Digweed, S. M., Fedigan, L. M. and Rendall, D. 2007. Who cares who calls? selective responses to the lost calls of socially dominant group members in the white-faced capuchin (*Cebus capucinus*). American Journal of Primatology 69(1): 1–7.

Dittus, W. 1988. An analysis of toque macaque cohesion calls from an ecological perspective. In D. Todt, P. Goedeking and D. Symmes (eds.), *Primate vocal communication* (pp. 31–50). Berlin: Springer-Verlag.

Epple, G. 1968. Comparative studies on vocalization in marmoset monkeys (Hapalidae). Folia primatologica 8: 1–40.

Gautier, J.-P. and Gautier-Hion, A. 1988. Vocal quavering: a basis for recognition in forest guenons. In D. Todt, P. Goedeking and D. Symmes (eds.), *Primate vocal communication* (pp. 15–30). Berlin: Springer-Verlag.

Halloy, M. and Kleiman, D. G. 1994. Acoustic structure of long calls in free-ranging groups of golden lion tamarins, *Leontopithecus rosalia*. American Journal of Primatology 32: 303–310.

Harcourt, A. H., Stewart, K. J. and Hauser, M. D. 1993. Functions of wild gorilla "close" calls. I. Repertoire, context, and interspecific comparison. Behaviour 124(1–2): 89–122.

Jones, C. B. 1998. A broad-band contact call by female mantled howler monkeys: implications for heterogeneous conditions. Neotropical Primates 6(2): 38–40.

Jordan, K., Weiss, D., Hauser, M. D. and McMurray, B. 2004. Antiphonal responses to loud contact calls produced by *Saguinus oedipus*. International Journal of Primatology 25(2): 465–475.

Lieblich, A. K., Symmes, D., Newman, J. D. and Shapiro, M. 1980. Development of the isolation peep in laboratory-bred squirrel monkeys. Animal Behaviour 29(1): 1–9.

Macedonia, J. M. 1986. Individuality in a contact call of the ringtailed lemur (*Lemur catta*). American Journal of Primatology 11(2): 163–179.

Masataka, N. 1982. A field study on the vocalizations of Goeldi's monkeys (*Callimico goeldii*). Primates 23(2): 206–219.

Masataka, N. 1989. Motivational referents of contact calls in Japanese monkeys. Ethology 80(1–4): 265–273.

Masataka, N. and Symmes, D. 1986. Effect of separation distance on isolation call structure in squirrel monkeys (*Saimiri sciureus*). American Journal of Primatology 10(3): 271–278.

McLanahan, E. B. and Green, K. M. 1977. The vocal repertoire and an analysis of the contexts of vocalizations in *Leontopithecus rosalia*. In D. G. Kleiman (ed.), *The biology and conservation of the Callitrichidae* (pp. 251–269). Washington, DC: Smithsonian Institution Press.

Mendes, F. D. C. and Ades, C. 2004. Vocal sequential exchanges and intragroup spacing in the Northern muriqui *Brachyteles arachnoides hypoxanthus*. Anais da Academia Brasileira de Ciencias 76(2): 399–404.

Mitani, J. C. and Nishida, T. 1993. Contexts and social correlates of long-distance calling by male chimpanzees. Animal Behaviour 45(4): 735–746.

Moody, M. I. and Menzel, E. W., Jr. 1976. Vocalizations and their behavioral contexts in the tamarin *Saguinus fuscicollis*. Folia primatologica 25(2–3): 73–94.

Moynihan, M. 1966. Communication in the titi monkey, *Callicebus*. Journal of Zoology 150: 77–127.

Norcross, J. L. and Newman, J. D. 1997. Social context affects phee call production by nonreproductive common marmosets (*Callithrix jacchus*). American Journal of Primatology 43(2): 135–143.

Norcross, J. L. and Newman, J. D. 1999. Effects of separation and novelty on distress vocalizations and cortisol in the common marmoset (*Callithrix jacchus*). American Journal of Primatology 47(3): 209–222.

Norcross, J. L., Newman, J. D. and Cofrancesco, L. M. 1999. Context and sex differences exist in the acoustic structure of phee calls by newly paired common marmosets (*Callithrix jacchus*). American Journal of Primatology 49(2): 165–181.

Oda, R. 2002. Individual distinctiveness of the contact calls of ring-tailed lemurs. Folia primatologica 73: 132–136.

Oppenheimer, J. R. and Oppenheimer, E. C. 1973. Preliminary observations of *Cebus nigrivittatus* (Primates: Cebidae) on the Venezuelan llanos. Folia primatologica 19(6): 409–436.

Owren, M. J. and Rendall, D. 2001. Sound on the rebound: bringing form and function back to the forefront in understanding nonhuman primate vocal signaling. Evolutionary Anthropology 10(2): 58–71.

Palombit, R. A. 1992. A preliminary study of vocal communication in wild long-tailed macaques (*Macaca fascicularis*): II. Potential of calls to regulate intragroup spacing. International Journal of Primatology 13(2): 183–207.

Pola, Y. V. and Snowdon, C. T. 1975. The vocalizations of pygmy marmoset *Cebuella pygmaea*. Animal Behaviour 23: 826–842.

Pook, A. G. 1977. A comparative study of the use of contact calls in *Saguinus fuscicollis* and *Callithrix jacchus*. In The biology and conservation of the Callitrichidae, ed. D. G. Kleiman, pp. 271-280. Washington, DC: Smithsonian Institution Press.

Premack, D. and Woodruff, G. 1978. Chimpanzee problem-solving: a test for comprehension. Science 202(4367): 532–535.

Ramos-Fernández, G. 2005. Vocal communication in a fission-fusion society: do spider monkeys stay in touch with close associates? International Journal of Primatology 26(5): 1077–1092.

Range, F. and Fischer, J. 2004. Vocal Repertoire of Sooty Mangabeys (Cercocebus torquatus atys) in the Taï National Park. Ethology 110: 301–324.

Rendall, D., Cheney, D. L. and Seyfarth, R. M. 2000. Proximate factors mediating "contact" calls in adult female baboons (*Papio cynocephalus ursinus*) and their infants. Journal of Comparative Psychology 114(1): 36–46.

Rendall, D., Rodman, P. S. and Emond, R. E. 1996. Vocal recognition of individuals and kin in free-ranging rhesus monkeys. Animal Behaviour 51(5): 1007–1015.

Robinson, J. G. 1979a. An analysis of the organization of vocal communication in the titi monkey *Callicebus moloch*. Zeitschrift fur Tierpsychologie 49: 381–405.

Robinson, J. G. 1979b. Vocal regulation of use of space by groups of titi monkeys *Callicebus moloch*. Behavioral Ecology and Sociobiology 5: 1–15.

Robinson, J. G. 1981. Vocal regulation of inter- and intragroup spacing during boundary encounters in the titi monkey, *Callicebus moloch*. Primates 22(2): 161–173.

Robinson, J. G. 1982. Vocal systems regulating within-group spacing. In C. T. Snowdon, C. H. Brown and M. R. Petersen (eds.), *Primate Communication* (pp. 94–116). New York: Cambridge University Press.

Robinson, J. G. 1984. Syntactic structures in the vocalizations of wedge-capped capuchin monkeys, *Cebus olivaceus*. Behaviour 90: 46–79.

Sabatini, V. and Ruiz-Miranda, C. R. 2008. Acoustical aspects of the propagation of long calls of wild *Leontopithecus rosalia*. International Journal of Primatology 29: 207–223.

Schrader, L. and Todt, D. 1993. Contact call parameters covary with social context in common marmosets, *Callithrix j. jacchus*. Animal Behaviour 46(5): 1026–1028.

Seyfarth, R. M. and Cheney, D. L. 2003. Signallers and receivers in animal communication. Annual Review of Psychology 54: 145–173.

Smith, H. J., Newman, J. D., Hoffman, H. J. and Fetterly, K. 1982. Statistical discrimination among vocalizations of individual squirrel monkeys (*Saimiri sciureus*). Folia primatologica 37: 267–279.

Snowdon, C. T. and Cleveland, J. 1980. Individual recognition of contact calls by pygmy marmosets. Animal Behaviour 28: 717–727.

Snowdon, C. T. and Cleveland, J. 1984. Conversations" among pygmy marmosets. American Journal of Primatology 7(1): 15–20.

Snowdon, C. T. and Hodun, A. 1981. Acoustic adaptations in pygmy marmoset contact calls: Locational cues vary with distances between conspecifics. Behavioral Ecology and Sociobiology 9: 295–300.

Snowdon, C. T. and Hodun, A. 1985. Troop-specific responses to long calls of isolated tamarins (*Saguinus mystax*). American Journal of Primatology 8(3): 205–213.

Snowdon, C. T. and Pola, Y. V. 1978. Responses of pigmy marmosets to synthesized variations of their own vocalizations. In D. J. Chivers and J. Herbert (eds.), *Recent advances in primatology* (pp. 811–813). London: Academic Press.

Sugiura, H. 1993. Temporal and acoustic correlates in vocal exchange of coo calls in Japanese macaques. Behaviour 124(3–4): 207–225.

Symmes, D., Newman, J. D., Talmage-Riggs, G. and Lieblich, A. K. 1979. Individuality and stability of isolation peeps in squirrel monkeys. Animal Behaviour 27(4): 1142–1152.

Teixidor, P. and Byrne, R. W. 1999. The "whinny" of spider monkeys: individual recognition before situational meaning. Behaviour 136(3): 279–308.

Tomasello, M. and Call, J. 1997. Primate Cognition. Oxford: Oxford University Press.

Uster, D. and Zuberbühler, K. 2001. The functional significance of Diana monkey "clear" calls. Behaviour 138(6): 741–756.

Veracini, C. 2002. Selected contact calls of a wild group of silvery marmosets (*Mico argentatus*, L. 1766). Folia primatologica 73: 333–334.

Winter, P., Ploog, D. and Latta, J. 1966. Vocal repertoire of the squirrel monkey (*Saimiri sciureus*), its analysis and significance. Experimental Brain Research 1: 359–384.

Chapter 14
Primate Cognition: Integrating Social and Ecological Information in Decision-Making

Paul A. Garber, Júlio César Bicca-Marques, and Maria Aparecida de O. Azevedo-Lopes

14.1 Introduction

Two major challenges that social animals face in exploiting their environment are the ability to locate ephemeral, widely scattered, and productive feeding sites and to obtain access to food resources also sought by other group members. This may include decisions concerning where to search, whom to follow, whom to avoid, an assessment of individual differences in competitive ability or dominance, and the costs and benefits of remaining in a food patch jointly occupied by others. Group foragers, therefore, are expected to solve problems of food acquisition by developing foraging strategies that integrate both social and ecological information (Giraldeau and Caraco 2000; Bicca-Marques and Garber 2005; Ottoni et al. 2005; Barrett and Henzi 2006; Bugnyar and Heinrich 2006).

Virtually all species of higher primates are gregarious foragers and live in stable social groups. In some primate species, group members exploit small isolated food patches or forage in a wide or dispersed front with nearest neighbors separated by distances of 15 m to several hundred meters (Chapman and Chapman 2000). Under these conditions, individual group members act principally as **searchers**[1], and encounter feeding sites as a result of their own search efforts (Barnard and Sibly 1981; Ranta et al. 1996; Giraldeau and Caraco 2000; Mottley and Giraldeau 2000). A searcher strategy may be associated with the concept of scramble competition (van Schaik 1989) and the advantages foragers gain through first access to a feeding site (DiBitetti and Janson 2001). Among primates that travel in a more cohesive unit or jointly exploit large and productive food patches, opportunities for information sharing may be a major benefit of social foraging, as certain group members direct their attention to the behavior of conspecifics in order

P.A. Garber (✉)
Department of Anthropology, University of Illinois at Urbana-Champaign, IL, USA
e-mail: p-garber@illinois.edu

[1] We adopted the term "searcher" as in Bicca-Marques (2003) instead of "finder" as in Bicca-Marques and Garber (2005) and Dominy et al. (2003) to avoid confusion with the term "finder's advantage" and because this term better describes a foraging strategy in which the searcher may or may not find food rewards as an outcome of its behavior.

P.A. Garber et al. (eds.), *South American Primates,* Developments in Primatology:
Progress and Prospects, DOI 10.1007/978-0-387-78705-3_14,
© Springer Science+Business Media, LLC 2009

to identify the location of potential feeding sites. These individuals are described as **joiners**, and co-feed or usurp food patches by taking advantage of the search effort (time and energy) of searchers. Joiners also may benefit by relying on the behavior of searchers to evaluate the level of predation risk at a feeding site. Whereas searchers rely largely on ecological information to locate food patches, joiners rely principally on social information (i.e., the sight or sound of conspecifics feeding or searching). In this regard, Giraldeau and Caraco (2000:152) define "all forms of exploitation of others' food discoveries or captures as kleptoparasitism." Joiners, therefore, are kleptoparasitic individuals who employ behavioral tactics such as agonism, stealth, appeasement, or cooperation to obtain food rewards (Giraldeau and Caraco 2000). Finally, under conditions in which foragers can simultaneously monitor the feeding behavior of others and search for food, individuals may act as **opportunists**, and more evenly distribute their time and energy budgets to both searching for food and kleptoparasitizing food from others depending on current social and ecological conditions (Vickery et al. 1991).

The costs and benefits of a searcher, joiner, or opportunist foraging strategy are dependent on a variety of factors including the productivity and distribution of food patches, the size of a finder's advantage, individual differences in social dominance and competitive ability, the number of group members adopting similar behavioral tactics, the cognitive capacity of individuals to store, recall, and integrate diverse sets of environmental information, and the reliability of past social experiences and ecological information as a basis for current foraging decisions (Barnard and Sibly 1981; Vickery et al. 1991; Ranta et al. 1996; Bicca-Marques and Garber 2005; Dubuc and Chapais 2007). Although there exist numerous studies of feeding ecology, dominance, and foraging behavior in nonhuman primates, little is known concerning the specific information primates use in decision-making, and the degree to which individual group members adopt searcher, joiner, or opportunist foraging strategies in response to changes in the size and availability of feeding sites and in response to the behavioral tactics of others (Garber 2000; DiBitetti and Janson 2001; Bicca-Marques and Garber 2005). It is important to stress that under a range of conditions, individuals living in cohesive social units are likely to benefit most by simultaneously searching for food and monitoring the behavior of others.

Figure 14.1 is a schematic representation of the kinds of information primates might encode and integrate in developing a set of "decision rules" to solve foraging problems. The concept of rule-based decision-making assumes that through processes of trial and error and various forms of social learning, individuals develop a set of expectations (hypotheses) and behavioral tactics (rules) and use these to solve current socioecological problems (Bugnyar and Heinrich 2006). This requires that an individual has the behavioral flexibility to generalize cause and effect relationships from one context to another by recognizing critical elements common to past and present situations, recall which behavioral tactic was most successful, and apply that tactic to the current problem (Watanabe and Huber 2006). If a given behavioral tactic fails to solve the problem, the individual is expected to incorporate that information and try a second or third behavioral tactic or decision rule that offered some probability of success in the past (Garber 2000; Hunt et al. 2006).

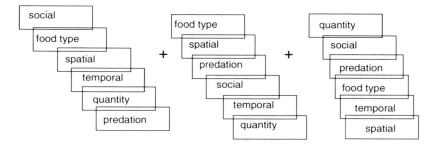

Using such amodel, a primate forager is expected to integrate several types of social and ecological information in making a foraging decision. This information is organized hierarchically with the most salient or important information for that decision listed first. In the first example, social information (presence, absence, or identity of co-feeders in a food patch) is more salient than food type (fruit vs. flower), food type is more salient than spatial information (distance traveled to this food patch), spatial information is more salient than temporal information (when patch was last visited), temporal information is more salient than quantity information (amount of food in the patch) and so on. Based on previous experience the forager uses this information to develop a particular decision rule. For example, enter a fruit feeding patch also occupied by an estrous female regardless of the size of that patch and when last visited. In the second example, the hierarchy of cues are different. In this case, food type is the most salient information because the forager needs a high energy resource (fruit) and distance to that patch and predation risk in that patch are more important in decision-making than the presence, absence, or identity of co-feeders. This information might be used to develop a decision-rule and applied in future feeding bouts. In the third example, expectations concerning the amount of food in the patch is the most salient information used by the forager.

Fig. 14.1 Representation of the hierarchy of information used by primates in developing decision rules

Moreover, we expect that the information used in developing behavioral tactics or rules is hierarchical, that is, under a given set of conditions certain information is more salient or more reliable than other information (Garber and Dolins 1996; Garber and Paciulli 1997; Bicca-Marques and Garber 2003, 2005). For example, based on previous experience, an individual might employ a foraging rule such as "if a distant tree had a large number of ripe fruits yesterday, revisit that tree and it will have a high probability of bearing ripe fruit today" (win-return rule in which expectations of food quantity or patch predictability may overshadow information on distance traveled to reach the feeding site; *see* Garber 1988). Some set of rules are likely to be relatively simple in that they integrate only limited amounts of social and ecological information, so-called "rules of thumb" (Bugnyar and Heinrich 2006). Such a behavioral rule for a subordinate individual might be to avoid any feeding site that is currently occupied by a more dominant female. Other rules are likely to be more complex, integrate several disparate types of information, require greater computational abilities, and be context specific. For example, if after a given period of time the quantity of food in a patch has not fallen below some critical level, then enter the patch because the dominant female is becoming satiated and will tolerate up to two additional foragers. Based on a common set of experiences in exploiting the same feeding sites or food types, all or most group members may develop a similar set of rules that are effective under a broad range of conditions. However, given differences in age, cognitive development, sex, visual or olfactory acuity, dominance status, and personal experience, individuals also may prioritize and integrate different information to effectively solve particular socioecological problems (Tomasello and Call 1997; Dominy et al. 2001; Reader and Laland 2002; Addessi and Visalberghi 2006).

In order to examine issues of primate cognition and decision-making, we conducted a series of controlled field experiments examining the ability of wild adult and immature black-chinned emperor (*Saguinus imperator imperator*) and Weddell's saddleback (*Saguinus fuscicollis weddelli*) tamarins to flexibly adopt alternative foraging tactics under conditions of changing food availability. Emperor (body weight = 435 ± 44 g, N = 40) and saddleback (355 ± 28 g, N = 17) tamarins live in relatively small, cohesive social units (emperor: 3–10 individuals, Bicca-Marques, Garber and Azevedo-Lopes, unpublished data; saddleback: 4–11 individuals, Digby et al. 2007), with most or all group members jointly exploiting a common set of feeding sites (Terborgh 1983; Bicca-Marques 2000; Garber and Bicca-Marques 2002). In addition, both species are characterized by a pattern of cooperative infant care-giving in which individuals closely monitor the actions of other group members, and coordinate behaviors associated with infant carrying, infant provisioning, and predator vigilance (Caine 1993; Garber 1997). In this regard, all members of a tamarin group are likely to have relatively equal access to the same social and ecological information (Bicca-Marques and Garber 2005). This has been referred to as group or public information (Valone 1989; Valone and Giraldeau 1993). Access to public information offers group members the opportunity to acquire new information or update current information by attending to the behavior of conspecific models or demonstrators (Coussi-Korbel and Fragaszy 1995; Ottoni et al. 2005).

Based on current socioecological and social foraging theory (Giraldeau and Caraco 2000; Bicca-Marques and Garber 2005) we made the following predictions:

Prediction 1 – At feeding sites in which the finder's advantage is low (large, nonmonopolizable, slowly depleting food patches) both dominant and subordinate tamarin group members are equally likely to adopt searcher, joiner, and/or opportunist foraging strategies and experience relatively equal feeding success.

Prediction 2 – At feeding sites in which the finder's advantage is relatively high (small, easily depletable, low quality, monopolizable food patches) dominant animals are expected to adopt a searcher strategy and subordinate animals are expected to experience significantly lower feeding success by joining at depleted patches.

Prediction 3 – At feeding sites of intermediate productivity, we expect that individuals in species characterized by lower levels of intragroup feeding-related agonism will adopt a broader range of searcher-opportunist-joiner foraging patterns and experience relatively equal feeding success, whereas in species characterized by higher levels of intragroup aggression or one in which subordinates are not tolerated as joiners regardless of the quality of the food patch, lower ranking individuals can more effectively increase their feeding success by always adopting a searcher strategy.

14.2 Methods

An experimental field study of foraging tactics and decision-making in wild emperor tamarins and wild saddleback tamarins was conducted at the Parque Zoobotânico (9°56'30"S, 67°52'08"W), Rio Branco, State of Acre, northwestern Brazil. The area is a 100 hectare protected reserve administered by the Federal University of Acre and characterized by a forest chronosequence dominated by secondary vegetation. There is a dry season from May through September (rainfall of the driest month: <20 mm) and a rainy season from October through April (rainfall of the wettest month: >300 mm; Deus et al. 1993). In addition to tamarins, this forest fragment also is inhabited by red titi monkeys (*Callicebus cupreus*), red-necked night monkeys (*Aotus nigriceps*), Bolivian squirrel monkeys (*Saimiri boliviensis boliviensis*), and pygmy marmosets (*Cebuella pygmaea*). Potential predators of primates at the study site include snakes and small birds of prey.

From March through August 2001 we collected data on one habituated group of 8–10 emperor tamarins and one habituated group of 11 saddleback tamarins. All group members were captured, tranquilized according to Santos et al. (1999), measured, weighed, and marked with colored collars for individual recognition.

The research design involved the construction of a feeding station located within the home range of the study groups. The feeding station contained eight visually identical feeding platforms (hereafter FP) located in a circular arrangement. Adjacent FPs were approximately 5 m apart and placed at a height of 1.5 m above the ground. The distribution of FPs was designed such that each individual could monitor the behavior of his/her group mates at all FPs. Feeding platforms were either baited with accessible bananas (reward platforms, RP) or inaccessible bananas inside wire mesh cages (non-reward platforms, NRP). All bananas (accessible or not) were tied to a screw installed on the platform to prevent the tamarins from removing the food reward. Feeding platforms were rebaited after all members of a study group left the area and no other group had arrived. A blind placed 5 m from the nearest FP was used to observe the behavior of the monkeys at the feeding station.

After a period of prebaiting in which all group members were observed to feed on the platforms, we initiated a series of six field experiments in which the amount and distribution of food available to the monkeys was systematically manipulated. Each experiment lasted 16 days (total 96 days of data collection). Depending on the protocol of the experiment, either two or all eight FPs contained a food reward. The amount of food on a platform was consistent during an experiment but varied across experiments from one unpeeled banana slice (about 13 g), one unpeeled whole banana (about 60 g), or three unpeeled whole bananas (about 190 g). Based on data provided by Goldizen et al. (1988), a wild saddleback tamarin can consume approximately 60 g of banana over a period of several hours while caught in a live-trap. Thus, patch quality in our experiments varied from poor (two RP containing 13 g of banana each = 26 g of banana) to very rich (eight RP containing three bananas each = 1,520 g of bananas; Table 14.1). In the poorest patch one or two individuals could consume all the available food in a short period of time, whereas the richest patch contained more than enough food to satiate all group members. In addition, the experimental conditions varied in terms of reward certainty. That is, food rewards could be more easily monopolized by dominant individuals when distributed on only two FPs than when food was present on all eight FPs. This difference in food quantity and monopolization enabled us to assess the effects of sex, age, and dominance on individual foraging strategies and access to resources.

Table 14.1 Food availability and distribution in each experimental condition

Experiment	# of reward platforms	Amount of food per reward platform	Total amount of food (grams)	Productivity
1	8	One banana	480	High
2	2	One banana	120	Medium
3	8	Three bananas	1,520	Highest
4	2	Three bananas	360	Medium
5	8	One banana slice	104	Low
6	2	One banana slice	26	Lowest

Given that non-reward platforms contained an inaccessible banana in a wire mesh cage, odor cues could not be used by the forager to locate food rewards. To eliminate the use of visual cues in identifying RPs versus NRPs, both accessible and inaccessible bananas were completely covered with an identical large leaf. During experiments in which there were only two RPs, the same FPs were baited with accessible bananas throughout the experiment. Thus, place was constant (predictable), and foragers could rely on spatial information to efficiently relocate baited feeding sites (*see* Bicca-Marques and Garber 2004).

Behavioral observations and data collection using the behavior sampling rule with continuous recording ("all occurrences"; Martin and Bateson 1993), began when tamarin vocalizations were heard or animals were visually detected within a distance of 10 meters from the feeding station. A visit to a platform was recorded when an individual was observed (1) sitting or standing on the platform and searching it for food, or (2) hanging on a substrate adjacent to the platform (no more than 1 m distant) and searching it for food (*see* Garber and Dolins 1996; Garber and Paciulli 1997). All platform visits were recorded.

Individual searching investment was defined as the number of times a given individual was the first group member to arrive at a previously non-inspected feeding platform (hereafter "inspection"). Assuming that all group members have an equal opportunity to look for food, individuals searching at a significantly greater frequency than expected, based on group size, are considered searchers (i.e., in a group of 10 individuals each group member is expected by chance to be the first to inspect a feeding platform 10% of the time). Individuals searching for food at a significantly lower frequency than expected are considered joiners, and those searching platforms at the frequency expected based on group size are considered opportunists. Binomial tests (Z) were used to compare observed and expected searching frequencies ("inspections") (Ayres et al. 2005). A level of confidence of 0.05 is used throughout this paper. The Bonferroni adjustment was not applied to provide an experiment-wise level of confidence because it is excessively conservative (*see* Gotelli and Ellison 2004) and because each individual's searching strategy was defined in comparison to the behavior of all other group members.

Repeated visits by an individual to the same platform during the same session (feeding station visit) were recorded and analyzed as an "individual single visit" (hereafter referred to as ISV; Bicca-Marques 2000; Bicca-Marques and Garber 2005). That is, if animal A visited platform 1 ten times in a given session, it was recorded as one visit to that platform. Foraging success or the ability of individuals to obtain a food reward was assessed by the expected or estimated size of the finder's advantage (based on the amount of food on a platform) as well as the percentage of ISVs to reward platforms that resulted in feeding.

The dominance status of group members was determined based on the frequency and distribution of dyadic agonistic interactions performed and received at feeding sites (*see* Janson 1985). A dominance index was calculated based on the proportion of agonistic interactions in which the individual initiated aggression (Lehner 1996). An individual was considered to have high social rank if he/she was the initiator of aggression more frequently than a recipient of aggression in dyadic comparisons. In contrast, low social rank was determined by a higher frequency of aggression received than performed.

14.3 Results

14.3.1 Dominance

Over the course of 96 days and six experimental conditions, emperor tamarins and saddleback tamarins were characterized by high levels of social tolerance and extremely low levels of aggression at baited feeding sites. During 2,784 platform visits we observed only 120 instances of within-group agonistic interactions among emperor tamarins. In the case of saddleback tamarins, 79 instances of agonistic interactions were observed during 3,235 platform visits. Overall, saddleback tamarins averaged 2.4 and emperor tamarins averaged 4.3 agonistic encounters per 100 platform visits (Table 14.2; this is consistent with data on low rates of agonistic behavior in natural, non-experimental field studies of wild tamarins, Heymann 1996; Garber 1997; and on greater male-female tolerance observed among saddleback tamarins in captive studies; Box et al. 1995). Nevertheless, differences in social

Table 14.2 Rates of agonistic behavior within groups of emperor and saddleback tamarins under changing conditions of food availability and patch productivity. Experiments are organized in decreasing order of food availability

	Saguinus imperator			*Saguinus fuscicollis*		
Experiment	# of FP visits	# of agonistic interactions	Rate of agonism (events/FP visit)	# of FP visits	# of agonistic interactions	Rate of agonism (events/FP visit)
3	529	15	0.028	534	12	0.022
1	439	25	0.056	545	5	0.009
4	478	40	0.084	481	24	0.049
2	449	32	0.071	626	32	0.051
5	604	6	0.009	734	6	0.008
6	285	2	0.007	315	0	0
Total	2,784	120	0.043	3,235	79	0.024

Table 14.3 Distribution of searcher (S), joiner (J), and opportunist (O) strategies in *Saguinus imperator* and *S. fuscicollis* relative to patch productivity, and individual social rank, age, and sex

Patch productivity (g)				1,520	480	360	120	104	26
Individual	Rank	Age	Sex	Exp 3	Exp 1	Exp 4	Exp 2	Exp 5	Exp 6
Saguinus imperator									
VRM	1	Ad	M	O	O	O	O	O	S
CAC	2	Ad	M	J	J	O	O	O	J
CAZ	3.5	Ad	M	S	O	O	S	J	J
CRO	3.5	Ad	F	J	S	–	O	–	–
COX	5	Ad	F	O	S	S	S	S	O
CLA	6	SA	M	J	J	O	J	J	J
CVE	7	SA	F	O	O	O	O	J	J
CBR	8	J	F	S	J	O	O	J	J
INF	*	I	?	O	–	–	–	S	S
Searchers =				2	2	1	2	2	2
Joiners =				3	3	0	1	4	5
Opportunists =				4	3	6	5	2	1

Total number of searchers = 11 (22.9%)
Total number of joiners = 16 (33.3%)
Total number of opportunists = 21 (43.8%)

Saguinus fuscicollis									
ROV	1	Ad	M	S	S	S	O	J	O
ROA	2	Ad	F	O	O	O	S	–	–
AZA	3	Ad	M	O	O	O	O	S	S
VBV	4	Ad	M	J	O	J	O	O	O
AMC	5	Ad	M	O	O	O	O	O	O
VRA	6	Ad	M	S	S	S	O	O	O
VEB	7	J	M	S	O	O	O	S	O
AZB	8	J	M	O	J	O	O	O	O
MAM	9	J	F	J	J	J	J	O	O
CIV	**	J	M	–	–	–	–	S	O
INF	*	I		J	–	J	–	J	O
Searchers =				3	2	3	1	3	1
Joiners =				3	2	3	1	2	0
Opportunists =				4	5	4	7	5	9

Total number of searchers = 13 (22.4%)
Total number of joiners = 11 (19.0%)
Total number of opportunists = 34 (58.6%)

Ad = adult, SA = subadult, J = juvenile, I = independent infant, M = male, F = female
* Infants were not assigned a social rank.
** The social rank of CIV was not evaluated because the individual was present in the group during two experiments only.

dominance were evident among group members, and this may have resulted in individual differences in foraging tactics and access to resources. Given that our emperor and saddleback tamarin study groups were similar in size and composition (*see* Table 14.3), characterized by a polyandrous mating pattern, and generally breed during the same period of the year (Terborgh 1983), differences in rates of agonism appear to represent subtle species-specific differences in social tolerance.

In both saddleback and emperor tamarins rates of agonistic interactions were highest during Experiments 2 and 4, when only two platforms each contained either one or three whole bananas, and lowest during Experiments 5 and 6, when either all or two platforms each contained only a banana slice (Table 14.1). The average number of tolerated or co-feeding group members also was affected by patch productivity. When patch productivity was medium to high (Experiments 3 and 4) emperor and saddleback tamarins averaged approximately 2.0–2.5 conspecifics feeding together on the same platform. However, when the amount of food on a platform was lowest (Experiments 5 and 6), the number of co-feeders averaged only 1.2–1.3 for both tamarin species.

Emperor tamarins were characterized by a relatively linear dominance hierarchy (Landau's Index of Linearity, h = 0.83) that was strongly affected by sex and age. All adult males were found to rank above all adult females, and the subadult male ranked above the subadult and juvenile females. Among saddleback tamarins, rank reversals were more common, and the dominance hierarchy was not linear (Landau's Index of Linearity, h = 0.41). The highest ranking saddleback tamarin group member was an adult male. However, the second highest ranking individual was the group's only adult female (Table 14.3).

14.3.2 Searcher, Joiner, and Opportunist Foraging Strategies

Individual tamarins exhibited considerable flexibility in using ecological and social information in foraging decisions. The most common strategy employed by both emperor and saddleback tamarins was an opportunistic foraging pattern (44% and 59%, respectively). In the case of emperor tamarins, the lowest-ranking adult male (CAZ) and the highest-ranking adult female (CRO) were found to adopt all three foraging strategies depending on the conditions of the experiment, whereas the second highest-ranking male (CAC) and the subadults were equally likely to be joiners or opportunists but not searchers. Other group members exhibited a more consistent foraging pattern. Adult female COX adopted a searcher strategy in most experiments, whereas the highest-ranking adult male (VRM) often adopted an opportunist strategy. Overall, emperor tamarin searchers tended to be adult females (five of 11 cases, p = 0.0392) and joiners tended to be subadult or juvenile group members (10 of 16 cases, p = 0.0374). The number of searchers in the group (two) remained relatively constant throughout the study, whereas the number of joiners increased and the number of opportunists decreased in the two poorest patches (Experiments 5 and 6, Table 14.3).

In saddleback tamarins, the dominant male (ROV) was the only group member to adopt searcher, joiner, and opportunist strategies. This highest ranking male, the second ranking adult male (AZA), the lowest ranking male (VRA), and a juvenile male (VEB) acted as searchers in at least two of the six experimental conditions. Saddleback joiners tended to be juveniles or infants (eight of 11 cases, p = 0.0364). Joiner strategies were most common in experiments in which patch productivity was higher (Experiments 3, 1, and 4) and absent in Experiment 6 when 13 g of banana were available on each of only two reward platforms (Table 14.3). Overall, in both tamarin species high-ranking individuals were not more likely to switch foraging strategies than were low-ranking group members.

14.3.3 Evidence of a Finder's Advantage

Although it was not possible to quantify the precise amount of food ingested by each tamarin visiting a reward platform, we were able to determine the number of individuals that fed successfully (obtained food) during each platform visit and used this as an indirect measure of a potential finder's advantage. In both emperor and saddleback tamarins, as the productivity of feeding sites decreased the percentage of successful individual foraging bouts also decreased (Table 14.4). Under conditions of highest food availability (Experiments 3 and 1) each tamarin group member,

Table 14.4 Individual percentage of successful feeding bouts in each experimental condition

Patch productivity (g)	1,520	480	360	120	104	26	
Experiment #	3	1	4	2	5	6	Mean ± s.d.
Saguinus imperator							
VRM (Ad M)	94	78	88	57	28	40	64 ± 27
CAC (Ad M)	97	96	67	74	31	0	61 ± 38
CAZ (Ad M)	96	95	87	33	14	0	54 ± 44
CRO (Ad F)	96	78	NA	77	NA	NA	84 ± 11*
COX (Ad F)	98	86	83	66	40	63	73 ± 21
CLA (SA M)	96	80	63	23	17	0	47 ± 39
CVE (SA F)	100	88	35	20	2	0	41 ± 43
CBR (J F)	93	77	62	33	9	0	46 ± 38
INF (I)	93	NA	72	NA	31	38	59 ± 29
Mean ± s.d.	96 ± 2	85 ± 8	70 ± 17	48 ± 23	22 ± 13	18 ± 25	
Saguinus fuscicollis							
ROV (Ad M)	95	80	76	35	4	0	48 ± 41
ROA (Ad F)	89	81	85	76	NA	NA	83 ± 6*
AZA (Ad M)	90	74	81	56	18	29	58 ± 29
VBV (Ad M)	100	87	80	53	13	0	56 ± 41
AMC (Ad M)	89	94	82	61	19	0	58 ± 39
VRA (Ad M)	86	85	78	68	21	9	58 ± 34
VEB (J M)	98	84	86	76	31	0	63 ± 38
AZB (J M)	94	84	70	37	4	13	50 ± 38
MAM (J F)	92	82	65	33	7	0	47 ± 39
CIV (J M)	NA	NA	NA	NA	28	8	18 ± 14**
INF (I)	36	NA	64	NA	5	0	26 ± 30
Mean ± s.d.	87 ± 18	83 ± 5	77 ± 8	55 ± 17	15 ± 10	6 ± 9	

Ad = adult, SA = subadult, J = juvenile, I = independent infant, M = male, F = female
* This individual was no longer in the group during both poorest patch experiments.
** This individual was present in the group during both poorest patch experiments only.

Table 14.5 Feeding success (average ± standard deviation % successful feeding bouts) of each foraging strategy under changing conditions of food availability

Patch productivity (g)	1520	480	360	120	104	26
Experiment #	3	1	4	2	5	6
Saguinus imperator						
Searchers	94.5	82.0	83.0	49.5	40.0	40.0
	± 2.1	± 5.6		± 23.3		
Joiners	96.3	84.3	–	23.0	11.0	–
	±0.5	±10.2			±7.9	
Opportunists	97.3	87.0	67.0	52.2	22.6	15.7
	±3.0	±8.5	±19.5	±25.0	±11.9	±31.5
Opportunists of high rank	94.0	86.5	80.6	69.3	29.5	0
		±12.0	±11.8	±10.7	±2.1	±0
Opportunists of low rank	99.0	88.0	53.3	26.5	9.0	21.0
	±1.4		±15.8	±9.1		±36.3
Saguinus fuscicollis						
Searchers	92.0	82.5	77.0	76.0	25.6	29.0
	±8.4	±3.5	±1.4		±6.8	
Joiners	96.0	83.0	72.5	33.0	4.0	–
	±5.6	±1.4	±10.6			
Opportunists	91.4	84.0	80.0	55.1	12.8	3.7
	±2.8	±7.3	±6.3	±15.1	±7.3	±5.3
Opportunists of high rank	91.3	80.6	83.0	48.0	13.0	0
	±3.2	±6.5	±2.8	±11.3		±0
Opportunists of low rank	91.5	89.0	79.3	60.5	12.7	5.0
	±3.5	±7.0	±8.3	±16.8	±8.5	±5.7

Data on infants were not included in these calculations. Cases in which a forager was carrying an infant were also omitted from these calculations.

except the infant saddleback, fed during most reward platform visits (Table 14.4), and all strategies resulted in relatively equal foraging success (>80%; Table 14.5). Therefore, the foraging success of both dominant and subordinate group members was high and relatively equal regardless of whether individuals principally used ecological and/or social information to locate feeding sites, lending support to Prediction 1.

In contrast, in the poorest food patches (Experiments 5 and 6), many group members experienced limited feeding success (Table 14.4), with searchers being more successful than joiners and opportunists (Table 14.5; emperor: $Z = 2.3647$, $n_s = 4$, $n_{jo} = 12$, $p = 0.0180$; saddleback: $Z = 2.8347$, $n_s = 4$, $n_{jo} = 16$, $p = 0.0046$). During Experiment 6 the potential of a finder's advantage and the ability to monopolize feeding sites were high, and only three of eight emperor tamarins (the highest ranking adult male, the lone adult female, and one infant) and four of ten saddleback tamarins (second and fifth ranking adult males and two low ranking juveniles) obtained access to a food reward (Table 14.4). Data from *S. fuscicollis* do not support Prediction 2. The dominant adult male (ROV) adopted either a joiner or an opportunist strategy and experienced minimal feeding success in these experiments, while subordinate individuals adopted either an opportunist or a searcher strategy and were able to obtain greater feeding success.

In the case of *S. imperator*, on the other hand, Prediction 2 is supported. The highest ranking adult male (VRM), the group's breeding and lone adult female (COX; adult female CRO was no longer in the group at the end of this study), and an independently moving infant were the only group members to act as searchers or opportunists and to have access to food rewards in Experiment 6. These same individuals plus the second highest ranking male (CAC) also were the only individuals to adopt one of these two strategies in Experiment 5 when each of all eight platforms contained 13 g of bananas.

Finally, the results of Experiment 4 support Prediction 3 (relating rank, less tolerant vs. more tolerant, and feeding success at sites of intermediate productivity) for both species. Emperor tamarins exhibited a higher level of feeding-related agonism than did saddleback tamarins, and *S. imperator* opportunists of lower rank experienced lower feeding success than searchers and opportunists of higher rank.

Data from Experiment 2 in which the amount of food available was only a third of that found in Experiment 4, however, do not consistently support Prediction 3. Among emperor tamarins, joiners and low ranking opportunists were found to have lower average feeding success than did searchers and high ranking opportunists, as expected. In addition, the lowest ranking adult male (CAZ) adopted a searcher strategy as predicted, however, he successfully fed during only 33% of his visits to reward platforms, and this was not consistent with Prediction 3. Among saddleback tamarins, individual feeding success showed greater variation than expected, ranging from 33–76% (Table 14.4). The only joiner (MAM) experienced the lowest feeding success in the group (33%) and the only searcher (ROA) and one opportunist (VEB) the highest (76%). However, both the highest ranking (ROV, 35%) and lowest ranking opportunists (AZB, 37%) had low feeding success suggesting that when individual saddleback tamarins arrived first at a platform containing a single banana they obtained a strong finder's advantage unrelated to rank. Linear regression analyses showed that, for each tamarin species, rank alone was not a reliable predictor of individual foraging success in any of the six experiments.

14.4 Discussion

A primary advantage of group foraging is that individuals may integrate ecological information with information obtained by monitoring the activities of conspecifics in order to more efficiently locate and exploit feeding sites (Giraldeau and Caraco 2000; Bicca-Marques and Garber 2005). In addition, social foragers may experience directly or observe the outcomes of dominance interactions and use this information to adjust their behavior in deciding when to feed and which food patches to visit. Benefits gained from observing other group members are influenced by several factors including age and attentiveness to conspecifics, spatial proximity to conspecifics, frequency with which conspecifics are encountered, predictability of actions or responses of different group members under a similar set of conditions, and the ability to understand cause and effect relationships between a conspecific's behavior, the foraging or social task, and successful and unsuccessful outcomes (Cambefort 1981; Galef 1991; Visalberghi 1994). Bugnyar and Heinrich (2006) refer to the manner in which animals adjust their actions in response to current social and ecological conditions as "situation-dependent use of behavior" (p. 373). They argue that such behavior may be explained by the use of sets of general and specific decision rules. For example, based on past experience a subordinate individual might employ a general foraging rule such as avoid feeding patches currently occupied by dominant animals. This rule would reduce the possibility of an aggressive encounter but depending on the size of the food patch, also result in limited access to food rewards. Individuals using such a behavioral rule are likely to be most successful when adopting a searcher strategy, if they have personal knowledge concerning the location and productivity of nearby feeding sites. Alternatively, this same individual might employ a decision rule of cautiously approaching a feeding site currently occupied by dominant conspecifics, and if not threatened or attacked, co-feed with them. Such a strategy involves some risk to the forager. Finally, a forager who has developed an affiliative bond with a particular individual or network of individuals might act as a joiner and enter a patch and co-feed with any of those individuals but avoid patches occupied by other individuals. An increased ability to identify and associate regularities in the behavioral patterns of others with specific conditions of the ecological and social environment is expected to result in more insightful and efficient decision-making (Watanabe and Huber 2006).

In this chapter we presented the results of a series of field experiments designed to examine the ability of wild Brazilian emperor (*S. imperator*) and saddleback (*S. fuscicollis*) tamarins to integrate social information (individual dominance status, and the behavior of conspecifics interacting, exploring, or feeding on a platform) and ecological information (the location, number, and productivity of reward platforms were constant or predictable during each experimental condition) in deciding where and when to feed. The configuration of our feeding station was analogous to the challenges primates face when exploiting a tree crown or food patch containing multiple feeding platforms each of which varied predictably in the size or density of the food reward. In our field experiments patch productivity ranged from low to medium to high.

Overall, we found that individual tamarins were sensitive to both social and ecological information, and flexibly altered their foraging strategy under conditions of changing food availability. The same individual tamarin might adopt a searcher, joiner, or opportunist strategy based on expectations concerning the size and monopolizability of the food reward, and the presence and social rank of particular conspecifics at a feeding site. Moreover, despite the fact that some age or sex classes were found to vary less than others in their use of particular search strategies (i.e. subadults were more likely to be joiners or opportunists, but not searchers, and emperor tamarin adult females tended to be searchers), data from a previous study (Bicca-Marques 2000; Bicca-Marques and Garber 2005) did not support age- or sex-based differences in cognitive ability in these New World primates.

In both tamarin species, most individuals adopted an opportunist foraging pattern to obtain food rewards. That is, during a given experimental condition the same

individual sometimes searched for food using ecological information and sometimes acted kleptoparasitically and located reward platforms by relying on the search
efforts of others. Although searchers must rely on ecological information to locate a
feeding site, to be successful they also may monitor the behavior and movements of
conspecifics in order to insure that they arrive first at a reward platform. At present,
the specific set of decision rules used by individual tamarins in selecting where
and when to feed remains unclear. There is evidence, however, that individuals integrated spatial, quantity, temporal, and social information in their foraging decisions.
In both saddleback and emperor tamarins individual feeding success varied significantly when the productivity of the patch was low or intermediate depending on
the foraging tactic employed. Therefore, foragers had opportunities to "test" the
success of alternative behavioral strategies or "hypotheses". For example in Experiments 2 and 4, when two platforms contained food rewards, only one emperor
tamarin adopted a joiner strategy, whereas in Experiments 1 and 3 when all eight
platforms contained food rewards, six individuals acted as joiners. It appears likely
that the ability of the same forager to go back and forth from a joiner, to a searcher
or opportunist strategy across experiments (Table 14.3) resulted from an ability to
combine social information on the tolerance of particular group members at feeding
sites, temporal information on recent feeding success, and ecological information
concerning expectations of the amount of food and number of reward platforms to
develop a successful set of foraging rules.

14.4.1 Test of Foraging Hypotheses

Given evidence for species differences in social dominance, and the fact that access
to food rewards was affected by the search strategies of others, we tested a series
of hypotheses relating social dominance, foraging tactics, and individual feeding
success. These predictions were (1) under conditions in which the finder's advantage was low, dominant and subordinate group members were equally likely to
adopt searcher, joiner, and/or opportunist foraging strategies and experience relatively equal feeding success, (2) under conditions in which the finder's advantage
was high, dominant animals would act as searchers and subordinates would have
decreased feeding success by joining at depleted patches, and (3) at feeding sites
of intermediate productivity, searchers, joiners, and opportunists would have relatively equal foraging success in more socially tolerant species (*Saguinus fuscicollis*),
whereas in a species with a higher level of food-related agonism (*Saguinus imperator*), subordinates would have higher feeding success if they acted as searchers.
Our results fully supported Prediction 1 and offered some support for Prediction
3 for both species, whereas Prediction 2 was supported for *S. imperator* but not
for *S. fuscicollis*. That is, under conditions in which the relative finder's advantage
was greatest (13 g of banana on each of eight platforms in Experiment 5 and 13 g of
banana on each of two platforms in Experiment 6 the highest ranking adult male saddleback tamarin was characterized by extremely) low feeding success (Table 14.4).

This interspecific difference may be related to the more tolerant, and less aggressive social interactions that characterize *S. fuscicollis*. Overall, individual foraging strategies in *Saguinus* spp. appear to be highly flexible, with group members integrating and tracking changes in both social and ecological information.

Analogous quantitative data on decision-making in wild New World primates are limited. DiBitetti and Janson (2001) report the results of an experimental field study in which wild capuchin monkeys (*Cebus nigritus*; formerly *Cebus apella nigritus*) foraging at large, nonmonopolizable food patches experienced similar feeding successes irrespective of individual dominance rank or foraging strategy, whereas searchers fed more successfully at small and/or monopolizable patches (*see* also Janson 1996). Garber and Dolins (1996), Garber and Paciulli (1997), and Garber (2000) conducted experimental field studies of decision-making in wild mustached tamarins (*Saguinus mystax* in Peru) and wild, white-faced capuchins (*Cebus capucinus* in Costa Rica) and found evidence that individuals integrated spatial, temporal, and quantity information in selecting feeding sites. In a separate study, Garber and Brown (2006) found that white-faced capuchins in Costa Rica attended to an array of two and three landmark cues (2-meter high pink and yellow poles separated by distances of several meters) to predict the location of baited feeding sites. Experimental field studies of *Callicebus cupreus*, *Aotus nigriceps*, *Saguinus fuscicollis*, and *Saguinus imperator* by Bicca-Marques and colleagues (Bicca-Marques 2000; Bicca-Marques and Garber 2003, 2004, 2005) support several of the results presented in the current study concerning the importance of social and ecological information in primate decision-making. Finally, Stone's (2007) experimental field study of risk-sensitive foraging in *Saimiri sciureus* indicates that wild Brazilian squirrel monkeys integrate information on predation risk, food abundance in the general environment, and food abundance on individual feeding platforms in their foraging decisions.

However, given species differences in relative brain size and patterns of brain growth and development, visual and olfactory acuity, manipulative ability, social tolerance and cooperative behavior, group cohesion, diet, and foraging techniques, individuals in different primate taxa are likely to have access to different sets of information and experiences from which to generate solutions to foraging problems (Tomasello and Call 1997; Whiten and Byrne 1997; Dominy et al. 2001; Reader and Laland 2002; Leigh 2004; Addessi and Visalberghi 2006). In this regard, Milton (2000) compared the foraging behavior of two ateline species. Spider monkeys (*Ateles geoffroyi*), which are ripe fruit specialists, live in fission-fusion societies and exploit large day and home ranges were compared with mantled howler monkeys (*Alouatta palliata*), which consume a larger proportion of leaves in their diet, live in more cohesive groups, and exploit small day and home ranges. Milton (2000) found that howlers traveled across 50% of their entire home range during an average 5 day period and states "the travel pattern shown by howlers should permit a troop to keep a fairly close eye on phenological activity within its total home range without the need for strong dependence on long-term memory" (p. 391). Moreover, given high levels of group cohesion during traveling, foraging, and feeding, new or inexperienced group members benefit significantly by taking advantage of group or public information concerning diet, the use of efficient travel routes, and the location of feeding sites. In contrast, spider monkeys appear to rely to a greater degree on ego-generated information, with individuals in different fission-fusion parties possessing both shared knowledge and private knowledge concerning the location of efficient travel routes and locally productive feeding sites (Milton 2000). In *Ateles* spp., males and females may form sex-segregated parties (DiFiore and Campbell 2007) and preferentially occupy and exploit different areas of the group's home range (Symington 1988; Chapman 1990). Females intensively exploit a smaller area and males range over a much larger area (Symington 1988; DiFiore and Campbell 2007). Moreover, it is reported that older female spider monkeys lead group movement, and appear to possess more detailed knowledge of the local availability and distribution of resources than either younger adult females or adult males (van Roosmalen 1985; but *see* Valero and Byrne 2007). Finally, spider monkey foraging parties frequently vocalize or produce loud calls. Although it remains unclear if the function of these calls is primarily to attract conspecifics (Chapman and Lefebvre 1990) or warn them away (van Roosmalen 1985), the calls provide other group members with ecological information concerning the size and location of the food patch and social information concerning the size and composition of the foraging party (Chapman and Lefebvre 1990; Milton 2000). Overall, it appears that the amount and type of information used by individual male and female spider monkeys in deciding which particular party or subgroup to join, who to follow, who to avoid, when to respond to food calls, which travel route to take to efficiently reach a nearby or distant feeding site, and whether public or private information is more reliable in a given context is likely to be very different than the information available to and used by male and female howler monkeys (Milton 2000).

Brachyteles represents another genus of atelines that includes a high proportion of leaves, along with ripe fruits in its diet (Strier et al. 1993). Data on a single muriqui (*Brachyteles hypoxanthus*) group over a 15-year period indicate an increase in group size from 23–27 to 57–63 (Dias and Strier 2003). When smaller, the group traveled and fed as a highly cohesive unit. As group size approached 60 animals, subgrouping became more common. Subgroups averaged 39 independently loco-moting individuals, and ranged in size from approximately 5–63 individuals (Dias and Strier 2003). On certain occasions subgroups were separated without visual and vocal contact for several days (Strier et al. 1993). Dias and Strier (2003) state, "the shift from the group's prior cohesion to its tendency to split up into subgroups may reflect an upper limit on the number of individuals that can coordinate their movements while spreading out during foraging" (p. 218). In this regard, the amount and type of public and private information available to individual muriquis in this group is likely to have changed considerably over time. Such a situation is probably not uncommon among primate species characterized by large variance in group size and offers an opportunity to examine factors that directly influence the manner in which individuals of different rank, age, sex, experience, and reproductive condition acquire and integrate social and ecological information.

In conclusion, along with natural field studies and controlled laboratory stud-ies, experimental field studies of wild primates offer an important methodological tool for examining individual and species differences in the ability to encode and integrate ecological information (e.g., spatial, temporal, quantity, olfactory, visual) and social information (e.g., dominance status, partner competency and reliability, friendships, food calls) in decision-making (the main limitation of experimental field studies seems to be the lack of control over the frequency, temporal distribution, and social contexts of visits to the artificial feeding sites by individual group members). We found that among emperor and saddleback tamarins, individuals of different age, sex, and dominance status flexibly responded to changing social and ecological conditions in deciding which platforms to visit and when, which group members to follow, and which to avoid. Although we were unable to identify the specific hierarchy of decision rules used by individual tamarin foragers, based on previous studies of tamarin cognitive ecology and the fact that individuals travel as a cohesive group, it is likely that all or most individuals were able to similarly estimate patch quality as well as the likelihood of obtaining access to resources in a patch using principally public information.(Garber 1988, 1989, 2000; Giraldeau and Caraco 2000; Bicca-Marques and Garber 2003, 2004, 2005). Public information represents an important type of socially transmitted learning that may facilitate the emergence of group-level behavioral "traditions" (Box and Russon 2004). For primates living in larger or less cohesive groups, we may expect more variance in foraging rules and decision-making as individuals come to rely more heavily on private or self-generated information (trial and error learning).

14.5 Summary

Group foragers face a series of ecological and social challenges associated with locating productive feeding sites and obtaining access to food resources also sought by others. Foraging efficiency, therefore, is influenced by decisions such as whom to follow, whom to avoid, and an assessment of the costs and benefits of locating a nearby or distant feeding patch singly, or remaining in a food patch jointly occupied by others. In this regard, nonhuman primates are likely to integrate spatial, temporal, quantity, and social information in developing behavioral rules or tactics to solve foraging problems. To examine questions concerning how the number and productivity of feeding sites influence patch choice in social foragers, we conducted a controlled experimental field study of decision-making in wild adult and immature emperor (*Saguinus imperator*) and saddleback (*Saguinus fuscicollis*) tamarins in Brazil. Our research design involved the construction of eight visually identical feeding platforms located in the home ranges of our study groups. The amount of food on a feeding platform and the number of baited platforms was varied systematically during six experimental conditions. We found that individual tamarins adopted a searcher (use ecological information), joiner (use social information), or opportunist (use either ecological and/or social information) foraging strategy depending on the conditions of the experiment. At productive sites, differences in individual feeding success were minimal and not influenced by the strategy adopted.

However, as the availability of food decreased at experimental patches, the foraging strategy adopted affected feeding success. When the potential for a finder's advantage was high (small and/or monopolizable feeding sites), searchers experienced greater feeding success than joiners and opportunists. Our results indicate that both tamarin species flexibly and effectively altered their use of ecological and social information in response to changes in food availability and distribution.

Acknowledgments We thank Alejandro Estrada, Ricardo Mondragon Ceballos and Charlie Janson for their constructive comments on an earlier version of this manuscript. PAG wishes to thank Sara and Jenni for teaching him about the ontogeny of primate foraging behavior as they have moved from searching for Pez dispensers to searching for cars. PAG also wishes to thank Chrissie McKenney for her love and patience. JCBM also thanks Cláudia, Gabriel, and Ana Beatriz for their permanent support. Supported by the Research Board and the Vice-Chancellor's Office of the University of Illinois at Urbana-Champaign.

References

Addessi, E., and Visalberghi, E. 2006. Rationality in capuchin monkey's feeding behavior? In S. Hurley and M. Nudds (eds.), *Rational Animals?* (pp. 313–329). Oxford: Oxford University Press.

Ayres, M., Ayres Jr., M., Ayres, D. L., and Santos, A. S. 2005. BioEstat 4.0. Belém: Sociedade Civil Mamirauá.

Barnard, C. J., and Sibly, R. M. 1981. Producers and scroungers: A general model and its application to captive flocks of house sparrows. Anim. Behav. 29:543–550.

Barrett, L., and Henzi, S. P. 2006. Monkeys, markets and minds: Biological markets and primate sociality. In P. M. Kappeler and C. P. van Schaik (eds.), *Cooperation in Primates and Humans: Mechanisms and Evolution* (pp. 209–232). Berlin: Springer-Verlag.

Bicca-Marques, J. C. 2000. Cognitive Aspects of Within-Patch Foraging Decisions in Wild Diurnal and Nocturnal New World Monkeys. PhD thesis, University of Illinois, Urbana. Available from: University Microfilms, Ann Arbor, MI: 9955588.

Bicca-Marques, J. C. 2003. Sexual selection and foraging behavior in male and female tamarins and marmosets. In C. B. Jones (ed.), *Sexual Selection and Reproductive Competition in Primates: New Perspectives and Directions* (pp. 455–475). Norman: American Society of Primatologists.

Bicca-Marques, J. C., and Garber, P. A. 2003. Experimental field study of the relative costs and benefits to wild tamarins (*Saguinus imperator* and *S. fuscicollis*) of exploiting contestable food patches as single- and mixed-species troops. Am. J. Primatol. 60:139–153.

Bicca-Marques, J. C., and Garber, P. A. 2004. The use of spatial, visual, and olfactory information during foraging in wild nocturnal and diurnal anthropoids: A field experiment comparing *Aotus*, *Callicebus*, and *Saguinus*. Am. J. Primatol. 62:171–187.

Bicca-Marques, J. C., and Garber, P. A. 2005. Use of social and ecological information in tamarin foraging decisions. Int. J. Primatol. 26:1321–1344.

Box, H. O., Röhrhuber, B., and Smith, P. 1995. Female tamarins (*Saguinus* – Callitrichidae) feed more successfully than males in unfamiliar foraging tasks. Behav. Proc. 34:3–12.

Box, H. O., and Russon, A. E. 2004. Socially mediated learning among monkeys and apes: some comparative perspectives. In L. J. Rogers and G. Kaplan (eds.), Comparative Vertebrate Cognition, pp. 97–140. New York: Kluwer Academic/Plenum Publishers.

Bugnyar, T., and Heinrich, B. 2006. Pilfering ravens, *Corvus corax*, adjust their behavior to social context and identity of competitiors. Anim. Cogn. 9:369–376.

Caine, N. G. 1993. Flexibility and co-operation as unifying themes in *Saguinus* social organization and behavior: The role of predation pressures. In A. B. Rylands (ed.), *Marmosets and Tamarins: Systematics, Behaviour, and Ecology* (pp. 200–219). Oxford: Oxford University Press.

Cambefort, J. P. 1981. A comparative study of culturally transmitted patterns of feeding habits in the chacma baboon, *Papio ursinus* and the vervet monkey, *Cercopithecus aethiops*. Folia Primatol. 36:243–263.

Chapman, C. A. 1990. Association patterns of spider monkeys: The influence of ecology and sex on social organization. Behav. Ecol. Sociobiol. 26:409–414.

Chapman, C. A., and Chapman, L. J. 2000. Determinants of group size in primates: The importance of travel costs. In S. Boinski and P. A. Garber (eds.), *On the Move: How and Why Animals Travel in Groups* (pp. 24–42). Chicago: The University of Chicago Press.

Chapman, C. A., and Lefebvre, L. 1990. Manipulating foraging group size: spider monkey food calls at fruiting trees. Anim. Behav. 39:891–896.

Coussi-Korbel, S., and Fragaszy, D. M. 1995. On the relation between social dynamics and social learning. Anim. Behav. 50:1441-1453.

Deus, C. E., Weigand Junior, R., Kageyama, P. Y., Viana, V. W., Ferraz, P. A., Nogueira-Borges, H. B., Almeida, M. C., Silveira, M., Vicente, C. R., and Andrade, P. H. 1993. Comportamento de 28 espécies arbóreas tropicais sob diferentes regimes de luz em Rio Branco, AC. Rio Branco: UFAC/PZ.

Dias, L. G., and Strier, K. B. 2003. Effects of group size on ranging patterns in *Brachyteles arachnoides hypoxanthus*. Int. J. Primatol. 24:209–221.

DiBitetti, M. S., and Janson, C. H. 2001. Social foraging and the finder's share in capuchin monkeys, *Cebus apella*. Anim. Behav. 62:47–56.

DiFiore, A., and Campbell, C. J. 2007. The atelines: variation in ecology, behavior, and social organization. In C. J. Campbell, A. Fuentes, K. C. MacKinnon, M. Panger and S. K. Bearder (eds.), *Primates in Perspective* pp. 155–185. New York: Oxford University Press.

Digby, L. J., Ferrari, S. F., and Saltzman, W. 2007. Callitrichines: The role of competition in cooperatively breeding species. In C. J. Campbell, A. Fuentes, K. C. MacKinnon, M. Panger and S. K. Bearder (eds.), *Primates in Perspective* (pp. 85–105). New York: Oxford University Press.

Dominy, N. J., Garber, P. A., Bicca-Marques, J. C., and Azevedo-Lopes, M. A. O. 2003. Do female tamarins use visual cues to detect fruit rewards more successfully than do males? Anim. Behav. 66:829–837.

Dominy, N. J., Lucas, P. W., Osorio, D., and Yamashita, N. 2001. The sensory ecology of primate food perception. Evol. Anthropol. 10:171–186.

Dubuc, C., and Chapais, B. 2007. Feeding competition in *Macaca fascicularis*: an assessment of the early arrival tactic. Int. J. Primatol. 28:357–367.

Galef Jr, B. G. 1991. Information centers of Norway rats: sites for information exchange and information parasitism. Anim. Behav. 41:295–301.

Garber, P. A. 1988. Foraging decisions during nectar feeding by tamarin monkeys (*Saguinus mystax* and *Saguinus fuscicollis*, Callitrichidae, Primates) in Amazonian Peru. Biotropica 20:100–106.

Garber, P. A. 1989. Role of spatial memory in primate foraging patterns: *Saguinus mystax* and *Saguinus fuscicollis*. Am. J. Primatol. 19:203–216.

Garber, P. A. 1997. One for all and breeding for one: cooperation and competition as a tamarin reproductive strategy. Evol. Anthropol. 5:187–199.

Garber, P. A. 2000. Evidence for the use of spatial, temporal, and social information by primate foragers. In S. Boinski and P. A. Garber (eds.), *On the Move: How and Why Animals Travel in Groups* (pp. 261–289). Chicago: The University of Chicago Press.

Garber, P. A., and Bicca-Marques, J. C. 2002. Evidence of predator sensitive foraging in small- and large-scale space in free-ranging tamarins (*Saguinus fuscicollis*, *Saguinus imperator*, and *Saguinus mystax*). In L. E. Miller (ed.), *Eat or be Eaten: Predator Sensitive Foraging in Primates* (pp. 138–153). Cambridge: Cambridge University Press.

Garber, P. A., and Brown, E. 2006. Use of landmark cues to locate feeding sitesin wild capuchin monkeys (*Cebus capucinus*): an experimental field study. In A. Estrada, P. A. Garber,

M. Pavelka and L. Luecke (eds.), *New Perspectives in the Study of Mesoamerican Primates* pp. 311–332. New York: Springer.

Garber, P. A., and Dolins, F. L. 1996. Testing learning paradigms in the field: evidence for use of spatial and perceptual information and rule-based foraging in wild moustached tamarins. In M. Norconk, A. L. Rosenberger and P. A. Garber (eds.), *Adaptive Radiation of Neotropical Primates* (pp. 201–216). New York, Plenum Press.

Garber, P. A., and Paciulli, L. M. 1997. Experimental field study of spatial memory and learning in wild capuchin monkeys (*Cebus capucinus*). Folia Primatol. 68:236–253.

Giraldeau, L. A., and Caraco, T. 2000. Social Foraging Theory. Princeton: Princeton University Press.

Goldizen, A. W., Terborgh, J., Cornejo, F., Porras, D. T., and Evans, R. 1988. Seasonal food shortage, weight loss, and the timing of births in saddle-back tamarins (*Saguinus fuscicollis*). J. Anim. Ecol. 57:893–901.

Gotelli, N. J., and Ellison, A. M. 2004. A Primer of Ecological Statistics. Sunderland: Sinauer.

Heymann, E. W. 1996. Social behavior of wild moustached tamarins, *Saguinus mystax*, at Estacion Biologica Quebrada Blanco, Peruvian Amazon. Am. J. Primatol. 38:101–113.

Hunt, G. R., Rutledge, R. B., and Gray, R. D. 2006. The right tool for the job: what strategies do wild New Caledonian crows use? Anim. Cogn. 9:307–316.

Janson, C. 1985. Aggressive competition and individual food consumption in wild brown capuchin monkeys (*Cebus apella*). Behav. Ecol. Sociobiol. 18:125–138.

Janson, C. H. 1996. Toward an experimental socioecology of primates: examples from Argentine brown capuchin monkeys (*Cebus apella nigritus*). In M. Norconk, A. L. Rosenberger, and P. A. Garber (eds.), *Adaptive Radiation of Neotropical Primates* (pp. 309–325). New York: Plenum Press.

Lehner, P. N. 1996. Handbook of Ethological Methods. 2nd edn. Cambridge: Cambridge University Press.

Leigh, S. R. 2004. Brain growth, life history, and cognition in primate and human evolution. Am. J. Primatol. 62:139–164.

Martin, P., and Bateson, P. 1993. Measuring Behaviour: An Introductory Guide. 2nd edn. Cambridge: Cambridge University Press.

Milton, K. 2000. Quo vadis? Tactics of food search and group movement in primates and other animals. In S. Boinski and P. A. Garber (eds.), *On the Move: How and Why Animals Travel in Groups* (pp. 375–417). Chicago: The University of Chicago Press.

Mottley, K., and Giraldeau, L. A. 2000. Experimental evidence that group foragers can converge on predicted producer-scrounger equilibria. Anim. Behav. 60:341–350.

Ottoni, E. B., Resende, B. D., and Izar, P. 2005. Watching the best nutcrackers: what capuchin monkeys (*Cebus apella*) know about others' tool-using skills. Anim. Cogn. 8:215–219.

Ranta, E., Peuhkuri, N., Laurila, A., Rita, H., and Metcalfe, N. B. 1996. Producers, scroungers and foraging group structure. Anim. Behav. 51:171–175.

Reader, S. M., and Laland, K. N. 2002. Social intelligence, innovation, and enhanced brain size in primates. Proc. Natl. Acad. Sci. USA 99:4436–4441.

Santos, F. G. A., Salas, E. R., Bicca-Marques, J. C., Calegaro-Marques, C., and Farias, E. M. P. 1999. Cloridrato de tiletamina associado com cloridrato de zolazepam na tranquilização e anestesia de calitriquídeos (Mammalia, Primates). Arq. Brasil. Méd. Vet. Zootec. 51:539–545.

Stone, A. I. 2007. Age and seasonal effects on predation-sensitive foraging in squirrel monkeys (*Saimiri sciureus*): a field experiment. Am. J. Primatol. 59:127–141.

Strier, K. B., Mendes, F. D. C., Rimoli, J., and Rimoli, A. O. 1993. Demography and social structure of one group of muriquis (*Brachyteles arachnoids*). Int. J. Primatol. 14:513–526.

Symington, M. M. 1988. Demography, ranging patterns, and activity budgets of black spider monkeys (*Ateles paniscus chamek*) in the Manu National Park, Peru. Am. J. Primatol. 15:45–67.

Terborgh, J. 1983. Five New World Primates: A Study in Comparative Ecology. Princeton: Princeton University Press.

Tomasello, M., and Call, J. 1997. Primate Cognition. New York: Oxford University Press.

Valero, A., and Byrne, R. W. 2007. Spider monkey ranging patterns in Mexican subtropical forest: do travel routes reflect planning? Anim. Cogn. 10:305–315.

Valone, T. J. 1989. Group foraging, public information, and patch estimation. Oikos 56:357–363.

Valone, T. J., and Giraldeau, L. A. 1993. Patch estimation by group foragers: what information is used? Anim. Behav. 45:721–728.

van Roosmalen, M. G. M. 1985. Habitat preferences, diet, feeding strategy and social organization of the black spider monkey (*Ateles paniscus paniscus* Linnaeus 1758) in Surinam. Acta Amazon. 15:1–238.

van Schaik, C. P. 1989. The ecology of social relationships amongst female primates. In V. Standen and R. A. Foley (eds.), *Comparative Socioecology* (pp. 195–218). Cambridge: Blackwell.

Vickery, W. L., Giraldeau, L. A., Templeton, J. J., Kramer, D. L., and Chapman, C. A. 1991. Producers, scroungers, and group foraging. Am. Nat. 137:847–863.

Visalberghi, E. 1994. Learning processes and feeding behavior in monkeys. In B. G. Galef Jr., M. Mainardi, and P. Valsecchi (eds.), *Behavioral Aspects of Feeding* (pp. 257–270). Chur: Harwood Academic.

Watanabe, S., and Huber, L. 2006. Animal logics: decisions in the absence of human language. Anim. Cogn. 9:235–245.

Whiten, A., and Byrne, R. W. 1997. Machiavellian Intelligence II: Extension and Evaluations. Cambridge: Cambridge University Press.

Part IV
Conservation and Management of South American Primates

Chapter 15
Impacts of Subsistence Game Hunting on Amazonian Primates

Benoit de Thoisy, Cécile Richard-Hansen, and Carlos A. Peres

15.1 Introduction

Human and nonhuman primates have coexisted as predator and prey in all major tropical forest regions for millennia. Coexistence has often been mystical, with widespread taboos and avoidance (Cormier 2006), but primates have usually contributed prominently to human diets. Drastic declines of several large-bodied primate species were but one consequence of human colonization of several major islands and continents, and hunting probably contributed to prehistorical extinctions of megafauna, such as the giant lemurs of Madagascar (e.g. *Palaeopropithecus ingens* and *Pachylemur insignis*) following the earliest human arrivals, 2,000 years ago (Burney 1999; Burney et al. 2004; Perez et al. 2005). In the New World, hunting of primates and other vertebrates has been reported since at least the earliest Mayan period (Vaughan 1993), and most likely drove several large-bodied taxa to extinction, such as the mega-*Brachyteles* of the Brazilian Atlantic forest (Cartelle and Hartwig 1996).

In contemporary times, monkeys and apes are still a key part of the traditional diets of most tropical forest dwellers. In many communities, socio-cultural and religious taboos may limit predation of some species, but consumption of primate meat remains frequent and widespread in most tribal and non-tribal territories throughout the humid tropics. In the Paleotropics, lemurs are hunted in Madagascar (Garcia and Goodman 2003) and game harvest severely threatens several species (Bollen and Donati 2006). Gibbons, pig-tailed macaques and white-fronted leaf monkeys are regularly consumed in Indonesian Borneo (Wadley et al. 1997), where current levels of hunting, combined with pressures on forest habitats, have resulted in dramatic declines of the endemic Hose's leaf monkey (Nijman 2004). In China, hunting pressure is one of the main drivers of local extinctions of western black-crested gibbons (Jiang et al. 2006). Isolated populations of primate species in India, including the threatened Nilgiri langur and the lion-tailed macaque are also hunted by local

B. de Thoisy (✉)
Kwata NGO, Association Kwata, "Study and conservation of Guianan Wildlife", French Guiana
e-mail: thoisy@nplus.gf

P.A. Garber et al. (eds.), *South American Primates,* Developments in Primatology: Progress and Prospects, DOI 10.1007/978-0-387-78705-3_15,
© Springer Science+Business Media, LLC 2009

communities for subsistence purposes (Madhusudan and Karanth 2002; Kumara and Singh 2004).

African primates face severe threats from widespread subsistence and commercial hunting in addition to habitat loss (Oates 1996; Cowlishaw 1999), exposing the so-called "bushmeat crisis" (Milner-Gulland et al. 2003). These threats are particularly acute in apes for which dramatic population losses due to commercial hunting are now evident (Walsh et al. 2003). Monkeys are also hunted by local rural communities for subsistence or commerce, resulting in unsustainable harvests in Côte d'Ivoire (Refish and Koné 2005), Zaïre and Congo (Wilkie et al. 1998), Cameroon (Muchaal and Ngandjui 1999), Tanzania (Fusari and Carpaneto 2000), and Mozambique (Fusari and Carpaneto 2006).

In Brazilian Amazonia, large primates are among the most threatened mammal species, and this is primarily driven by widespread hunting pressure (Peres 1990; Costa et al. 2005). However, ethnographic studies have rarely focused specifically on the relationships between humans and their platyrrhine primate prey, with a few notable exceptions (Shepard 2002; Cormier 2003). In many game harvest studies of Indian and *caboclo* settlements, primates are generally cited as one of the top-ranking orders of mammals in terms of numeric offtake (Peres 2000a, Jerozolimski and Peres 2003). Among 40 Amerindian groups for which reliable data on harvests were available, 30 hunted solely cebids, and only 10 hunted both cebids and callitrichids (Cormier 2006). In most cases, hunting is for meat and restricted to household subsistence purposes, but smaller monkeys can also be killed for ornamental purposes or captured as pets (Mittermeier 1991; Shepard 2002).

In this chapter, we first review the impacts of hunting on neotropical primate communities. Secondly, we describe the hunting patterns of Amazonian indigenous groups, and attempt to identify factors that regulate primate offtakes. To illustrate interactions between primate populations and hunting pressure exerted by traditional communities, we consider results from two cases studies based on a standardized series of line-transect surveys. The first includes an extensive network of over 70 forest sites censused throughout the lowland Amazon basin. The second used a similar methodology to census 41 French Guianan sites spread across the entire country. We present comparisons of primate richness and abundance between remote and hunted sites, as well as hunting practices in survey sites, and their observed impacts on primate populations. We also consider the future of Amazonian primates by exploring the relationship between traditional game harvest practices and population viability. Finally, we consider the need for further fieldwork to inform the imperative challenge to protect primate populations and their habitats.

15.2 Effects of Hunting on Primates

Some 230 of the 625 living primate species are threatened. Nearly one-third of the 202 Neotropical primate species are vulnerable, endangered or critically endangered, and hunting has been identified as a major cause of decline in the Northern

Muriqui *Brachyteles hypoxanthus* and the Brown spider monkey *Ateles hybridus brunneus*, two of the three South American species among the world's 25 most endangered primate species (Mittermeier et al. 2005). It is well established that species vulnerability to a well-identified extinction risk is inexorably related to species life history traits. Among the most common threats, primate species exhibiting low ecological flexibility are often vulnerable to forest habitat disturbance, whereas large-bodied species are more susceptible to hunting (Peres 1999; Isaac and Cowlishaw 2004).

Population declines and local extinctions in relation to direct human exploitation are widely reported in South and Central America, including Guyana (Sussman et al. 1995; Lehman 2000), Venezuela (Urbani 2006), Peru and Bolivia (Freese et al. 1982), French Guiana (de Thoisy et al. 2005), and Brazilian Amazonia (Peres 1990, 1997a, 1999, 2000a; Lopes and Ferrari 2000; Haugaasen and Peres 2005; Peres and Palacios 2007). Large cebids are usually the first target species, and consequently the most dramatically affected. For instance, a single family of rubber tappers in a remote forest site of western Brazilian Amazonia killed more than 200 woolly monkeys (*Lagothrix lagotricha*), 100 spider monkeys (*Ateles paniscus*), and 80 howlers (*Alouatta seniculus*) over a period of 18 months (Peres 1991). As dramatic as these figures are, they likely underestimate actual hunting-induced mortality. For instance, harvest estimates from market surveys do not include primates that are consumed in villages. In the Democratic Republic of Congo, 57.1% of primates are consumed in the villages and do not make it to the market, and in Liberia, primates were more valuable in rural than in urban areas (Lahm 1993; Colell et al. 1995). Also, interview results are often biased, since hunting and/or sale is officially prohibited in many areas where it occurs (Johnson 1996; Richard-Hansen and Hansen 2004; de Thoisy et al. 2005). Finally, animals lethally wounded by hunters in the forest often cannot be retrieved and are thus not included in village-based harvest estimates based on the number of carcasses intercepted. For example, this is particularly typical of all Amazonian atelines which often remain secured to the upper canopy by their prehensile tails, hence becoming inaccessible to hunters long after *rigor mortis* sets in (Peres 1991). Ohl et al. (2007) estimated that incorporating collateral mortality increased the impact of hunting by Matsigenka Indians on ateline primates in the Peruvian Amazon by 14–18%, depending on the species and hunting practices.

Hunted primate populations initially face a numerical reduction in reproductive individuals. Among large-bodied species, frugivores are frequently the most rapidly affected by harvests (Peres and Palacios 2007), perhaps because populations are more susceptible to fluctuations in resource supplies (Peres 1991; Ferrari et al. 1999). However, other biological or behavioral effects can also contribute to population declines. For instance, disturbance by hunters wielding fire-weapons may result in more cryptic behavior and less active foraging, which may have negative consequences to patterns of group dispersal (Johns and Skorupa 1987). Additional pressures on habitat and smaller group sizes may result in a break-down of social structure (Young and Isbell 1994), poor body condition (e.g., Olupot 2000) and direct demographic effects, including elevated infant mortality (Johns 1991) and decreased juvenile survival and mean adult body weight (Milner et al. 2007). Despite the importance of genetic diversity for long-term population viability, the effects of recent human-induced population declines remain difficult to demonstrate, since genetic structures revealed with molecular markers can be driven by ancient demographic events that confound the signature of more recent population collapses (Harris et al. 2002). Evidence of lower breeder genetic diversity associated with population depletion through hunting is therefore scarce in mammals (but see Larson et al. 2002). Human-induced population collapses resulting in changes in population genetic structure has been revealed in Orangutans only (Goossens et al. 2006). In contrast, recent molecular surveys in large cebids (*Alouatta* and *Lagothrix*) may have exposed recent population bottlenecks, but these events were considered much older than the contemporary threats (Ruiz-Garcia 2005). Finally, population declines induced by hunting can be aggravated by the ravages of infectious diseases, such as Ebola hemorrhagic fever, as demonstrated in great apes in west Africa (Bermejo et al. 2006), but this has not been evidenced in Neotropical species.

Population declines can also result in marked effects on the dynamics of forest habitats as primates play a key role in many ecosystem processes. Their ecological traits, including dietary and habitat specialization, can help predict major extinction processes that would result from species extinctions. However, predicted effects of species loss in the neotropics—where species tend to share their main ecological functions with other mammals (Jernvall and Wright 1998)—would result in lesser impacts on ecosystems than in the paleotropics. Harvesting of key seed-dispersal agents, such as howlers, spider monkeys and woolly monkeys likely limit the quality and extent of seed deposition patterns (Nuñez-Iturri and Howe 2007; Stoner et al. 2007). Consequently the proportion of seeds dispersed is negatively affected, and may result in lower level of seedling recruitment, as tentatively demonstrated in several neotropical forest sites (Peres and van Roosmalen 2002; Serio-Silva and Rico-Gray 2002; Ratiarison and Forget 2005; Nuñez-Iturri and Howe 2007; Russo et al. 2006). This can also depress the gene flow and genetic diversity of plant populations facing a dramatic reduction in the aggregate pool of effective seed dispersal agents (Pacheco and Simonetti 2000).

15.3 Indigenous Groups and Hunting Practices in South America

Direct pressures on primate species depend on a complex set of interactions between historical, cultural, socioeconomic, and spatial factors related to the use of hunting catchment areas by neighboring communities, and/or other land uses within a given region (e.g., mining, logging) which may exacerbate local demand for wild game meat.

15.3.1 The Role of Primates in the Diet of Native Amazonians

The relative importance of wildlife meat and fish to Amazonian settlements is widely variable, ranging from 2 to 10% of the dietary protein intake of colonist groups along the Transamazon highway to 100% in the most isolated Amerindian populations (Redford and Robinson 1991). Primates are often the numerically dominant prey items harvested by indigenous groups throughout Amazonia, ranking higher than any other order of mammals (e.g., de Souza-Mazurek et al. 2000; Peres and Nascimento 2006). In French Guiana, primates represent the numerically dominant prey species (14–26% of prey items) harvested by Amerindian hunter-gatherer communities, below large ungulates (de Thoisy et al. 2005; Richard-Hansen et al. 2006). Conversely, as previously noted for Suriname (Mittermeier 1991), primates represent only around 10% of prey items captured by Bushnegro communities living along the main Maroni river of French Guiana, far below ungulates, rodents, birds and even xenarthrans. "Colonists" generally hunt fewer primates, and prefer large rodents and ungulates, mainly because these species resemble domestic livestock (Redford and Robinson 1987, but see Jerozolimski and Peres 2003). For instance, primates are not the principal source of game meat for extractive communities in the Jaú National Park, Brazilian Amazonia, who preferentially harvest ungulates and aquatic prey such, including fish and turtles (Barnett et al. 2002). In a small isolated non-tribal village of central French Guiana, ungulates, terrestrial frugivorous birds and rodents were the most frequently harvested prey (Richard-Hansen et al. 2004). More widely, primate biomass accounts for no more than 5% of total game biomass harvested by mixed communities in northern French Guiana, although, according to Creole tradition, some primate species are still regularly hunted for certain festive ceremonial occasions. This is particularly the case during the fruiting season, when food is abundant and animals become fat (Cormier 2006). On the other hand, primate meat was generally avoided by "white" people mainly because of their physical similarity to humans. This has been noted for other Amerindian groups elsewhere, but in other cases, primates were the favorite food *because* of this similarity (Mittermeier 1991; Cormier 2006).

Hunters largely target large-bodied species, such as *Ateles paniscus*, *Alouatta seniculus* and *Lagothrix lagotricha*, rather than small-bodied species (Peres 1990, Bodmer 1995, Shepard 2002; Franzen 2006). However, preferred midsized species, such as sakis and bearded sakis (*Pithecia* spp. and *Chiropotes* spp.) and capuchins (*Cebus* spp.), and even small-bodied species, such as squirrel monkeys (*Saimiri* spp.) and tamarins (*Saguinus* spp.) may be killed in far greater numbers than larger-bodied species regardless of the abundance of the latter. For example, 203 brown capuchins and 99 bearded saki monkeys were consumed in a single eastern Amazonian Indian village (Kayapó of A'Ukre) over 525 days of sampling, whereas only three howler monkeys were killed in the same period (Peres and Nascimento 2006). For the Waimiri Atroari Indians of central Amazonia, spider monkeys, tapirs and peccaries are also key target species (de Souza-Mazurek et al. 2000). Similarly, in Amerindian territories of southern French Guiana, howler monkeys, spider monkeys, and capuchin monkeys are also preferred target species (Renoux 1998; Richard-Hansen et al. 2006). In Manu National Park (Peru), Matsigenka tribal hunters show a clear preference for *Ateles* and *Lagothrix*, whereas smaller species are harvested either by young boys, and/or by adults returning from hunting trips empty-handed (Shepard 2002; Ohl et al. 2007). In the Yasuni National Park (Ecuador), Huaorani hunters also preferentially target howlers and spider monkeys (Franzen 2006).

15.3.2 What Regulates Harvests in Traditional Communities?

Considerations by ethnographers of Ameridian communities as "managers" of natural forest resources is an old debate fraught with difficulties (Balée 1989). Taboos and avoidance may involve all monkey species occurring in the catchment area of a community, but few communities avoid all primate species (Cormier 2006). Avoidance may be related to prey type and consumer status. For instance, communities may limit the consumption of some species (e.g., howler monkeys, uakaries) because of their similarity to humans (Cormier 2003; Kracke 1978). For others, avoidance may be related to age, gender, and reproductive and heath status of consumers. However, the cultural basis of prey avoidance may be entirely unrelated to conservation concerns over the status of prey populations. In some Amerindian communities, adult female *Ateles paniscus* dominate the harvest of this species (85–97%) and are strongly selected over males, since males of this species are not considered to be good enough to eat (e.g., Souza-Mazurek et al. 2000; Richard-Hansen et al. 2006). Conversely, brown capuchin kills may show a strong male-biased skew (25–35% of females only), which could be attributed to either the confrontational behavior of dominant males in the presence of hunters or an unbalanced sex-ratio in the population (Richard-Hansen et al. 2006).

Cultural limitation on the amount of forest and aquatic resources harvested (e.g., game, fish, non-timber plant products) are reported in Wayapi Indians and explained in terms of the fear of resource "wasting" and of "taking too much" (Renoux 1998). This may be interpreted as a conservation strategy, but this may be entirely unrelated to a proactive and conscious assessment of the risk of overharvesting. Similarly, Matsigenka Indians, who believe in the revenge of game spirits, report that the Saangarte spirits may hide animals when overhunting occurs (Shepard 2002). On the other hand, technological constraints may lead to overharvesting of females. For instance, Matsigenka bow hunters, who may prefer adult males, primarily kill female primates which often move slower and make easier targets (Shepard 2002). Other socioeconomic factors can limit the extent and intensity of hunting activities. Access to technology (outboard motors) fuel, and ammunition have a direct impact on hunting patterns, and consequently on harvests of monkey populations. Traditional livelihoods are often affected by infrastructure development and cash income. In some cases, smaller amounts of bushmeat in diets may be simply explained by a reduction of wildlife densities, changes in population structures, and/or cultural and economic changes influencing food preferences (Ayres et al. 1991). In Ecuador, for instance, roads provide wide accesses to large forest areas for Huaorani hunters, and allow persistent harvests of sensitive species such as spider monkeys (Franzen 2006). Improvement of socioeconomic conditions can also contribute to a decrease in subsistence hunting pressure (Jorgenson 2000), and depletion of primate populations by hunters may also result in shifts to alternative game species. Similar changes have been recorded in southern French Guiana: a comparison of harvests in the same area at two periods (1976–1977 and 1994–1995) showed decreased game harvests, related to either a decrease in the densities of target species or a shift to alternative resources (Renoux 1998).

15.4 Impact of Hunting on Primates: Two Cases Studies

Here we briefly illustrate the predominant impacts of subsistence or commercial hunting on neotropical primate populations. For this purpose, results of surveys conducted throughout lowland Amazonia (Peres 1990, 1999, 2000a,b; Peres and Palacios 2007) and in French Guiana (de Thoisy et al. 2005, 2006; C. Richard-Hansen, unpubl. data) are summarized in relation to local levels hunting pressure. These represent the most extensive studies anywhere in the tropics on the effects of hunting on forest primates using standardized line-transect censuses.

15.4.1 Lowland Amazonia

Humans have been hunting primates and other forest vertebrates in Amazonia since the arrival of the earliest paleoindians > 10,000 years BP, but consumption greatly increased following the first rubber boom in the late 19th century. Exploitation of primate meat by tribal and nontribal Amazonians has increased due to larger numbers of consumers, a greater spatial dispersion of these consumers, local scarcity of alternative sources of protein, changes in hunting technology, and because primates are often a preferred food. Peres (1999, 2000a,b) found that assemblage-wide primate biomass was strongly negatively correlated with hunting pressure, although, this effect size is also a function of forest habitat productivity and soil fertility (Peres 1997b, 1999, 2000a; Peres and Dolman 2000). At unhunted and lightly hunted forest sites, the densities of the three ateline genera, which are preferred targets of hunters, were consistently higher than those at moderately to heavily hunted sites. Peres and Palacios (2007) provide the first comprehensive large-scale meta-analysis of changes in vertebrate population densities in a large number of hunted and unhunted, but otherwise undisturbed, neotropical forest sites that takes into account differences in site productivity. Considering the variation in abundance among primate species at 101 Amazonian forest sites, population responses ranged from small-bodied species that on average more than doubled their abundance at higher levels of hunting pressure (cf. Peres and Dolman 2000), to midsized to large-bodied species that declined to less than half their abundance in intensively hunted sites. In the extreme, mean population densities of *Lagothrix* and *Ateles* in heavily hunted sites were only 1.8% and 8.7% of those in unhunted, but otherwise comparable, forest sites. Indeed, even moderate levels of hunting pressure can drive large atelines to local extinction, as documented in a number of forest sites surveyed throughout the Brazilian Amazon (Peres and Palacios 2007).

Peres (2000a) summarized new information on the average annual offtake of all game animals consumed by the rural, and usually unwaged, population of Brazilian Amazonia. Total game harvest throughout the region was estimated by multiplying species-specific per capita consumption rates by the size of the zero-income rural population across the entire region retaining forest cover. On the basis of estimates for primates, 3.8 million individuals are consumed annually in the Brazilian Amazon (range in estimates: 2.2–5.4 million), which represents a total biomass harvest of 16,092 tons. This is likely severely underestimated because it does not consider the fraction of lethally wounded animals that fail to be captured by hunters. Hunting rates are unsustainably high for several Amazonian primate species, often averaging over three times the maximum rate that could be sustained by a stable population (Peres 2000b). As a consequence, healthy population sizes of several large-bodied species can only be found in areas that are either effectively protected (e.g., strictly protected reserves and private forest set-asides) or extremely inaccessible to game hunters (Peres and Lake 2003).

This poses the difficult question as to what fraction of the original geographic range of different species still retains demographically viable populations. The extent of hunting-induced range contraction of several large-bodied taxa can be significant even in forest areas that remain relatively intact in terms of structural habitat disturbance detectable from satellite imagery (Peres et al. 2006). This is ecologically significant for all species regardless of geographic range size, but most serious for those range-restricted species, that are endemic to small parts of Amazonia. For example, the yellow-tailed woolly monkey, *Oreonax* (formerly *Lagothrix*) *flavicauda*, is endemic to the cloud forests of the Peruvian Andes at elevations of 1,700–2,700 m (Butchart et al. 1995). The remoteness of these areas had by default protected this species until the 1950s but since then agricultural colonization, road building projects, and logging have encroached relentlessly on its range (Leo Luna 1987). Yellow tails, like all woolly monkeys, are easy, attractive targets for hunters (Butchart et al. 1995). From the mid-1970s to the mid-1980s, at least 600 individuals had been killed by peasants and several populations had been driven to local extinction (Leo Luna 1987). Consequently, the estimated total population size of this monotypic genus is perhaps fewer than 250 animals placing it as critically endangered in the IUCN Red List (2006). Yellow-tailed woolly monkeys and other large-bodied primates endemic to small areas are obvious candidates for global extinction in the foreseeable future unless the largest remaining populations can be protected in reserves that are effectively protected from poachers.

In sum, the vast remaining forest cover that extends unbroken throughout the Amazon basin belies a scenario of partial to complete defaunation of large-bodied primates even in many relatively inaccessible areas that appear to remain structurally intact (Peres et al. 2006). These areas can no longer be considered as pristine primary forests because some key components of their large-vertebrate fauna have already been reduced to a pale shadow of their formerly intact condition (Peres and Lake 2003).

15.4.2 The Guianas

The Guianan shield is one of the largest pristine neotropical rainforest block and a floristically distinctive province compared to the Amazonian basin (Lindeman and Mori 1989). About 70 non-flying mammal species, including nine primates, are recorded in this region. This species richness is relatively low compared to western Amazonia, which may be explained by environmental unfavorableness, related to, for example, nutrient-poor soils (Emmons 1984). Eighty percent of the region is covered by uplands moist forests. The alluvial coastal plain covered by marsh forests, savannas, transition forests, herbaceous swamps and is rather narrow on this part of the Guianan shield (de Granville 1988). Compared to other neotropical countries, the forest conservation status of eastern Venezuela, Guyana, Suriname, French Guiana, and the Brazilian state of Amapá is still rather favorable, but recent increases of demographic pressures and a recent gold mining rush (Hammond et al. 2007) are serious threats on terrestrial biodiversity. French Guiana benefits from recent but extensive knowledge of primate communities in relation to habitats patterns. Eight primates species occur in the country: the red howler monkey *Alouatta seniculus macconnelli*, the black spider monkey *Ateles paniscus paniscus*, the tuffted capuchin *Cebus apella apella*, the wedge-capped capuchin *Cebus olivaceus castaneus*, the white-faced saki *Pithecia pithecia pithecia*, the bearded saki *Chiropotes satanas satanas*, the common squirrel monkey *Saimiri sciureus sciureus* and the golden handed tamarin *Saguinus midas midas*. All the species have a large distribution, except the bearded saki which is restricted to the south of the country. The spider monkey and the two sakis are protected by law, whereas other species can be hunted for subsistence use only. Primate hunting is widespread in French Guiana. In short, all human communities, whether or not indigenous, may hunt monkeys, either for subsistence or trade. Although prohibited, sales of monkey meat are common. Moreover, legal regulation of hunting is actually in course, resulting in widespread and often unsustainable game harvests, for instance in tapir (de Thoisy and Renoux 2004) and large primates (de Thoisy et al. 2005).

This case study provides a wide overview of the structure of primate communities in relation to hunting pressure and other features of sites. We examined patterns of primate richness and abundance at 41 forest sites (surveyed by BT and CRH) in relation to hunting pressure (classified as nil, medium or heavy), the type of game harvest (area hunted for traditional uses, or harvested by mixed communities facing other forms of disturbance such as logging), the other pressures on survey area (e.g., fragmentation), and forest vegetation types, including upland moist forest with or without pronounced relief, with low/high and continuous/discontinuous canopy (Fig, 15.1). In ten sites hunted by Amerindian and/or mixed communities, game harvests were monitored for 5 to 14 months. Among these sites, the number of hunters ranged from 13 to 105, and the size of the catchment area ranged from 225 to 1,250 km^2. Ethnographic and ecological data then provide an opportunity to relate the intensity of hunting pressure and local socioeconomic patterns to their effects on primate communities.

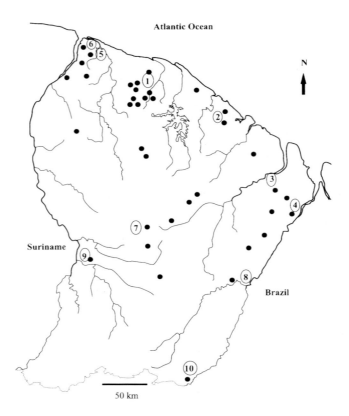

Fig. 15.1 Location of sites where primate surveys were implemented (*dots*) and where harvests were monitored (*numbers*)

The number of primate species per site ranged from as few as two in the most disturbed habitats, with tamarins and white-faced sakis dominating these communities, to as many as six species (howlers, spider monkeys, tufted and wedge-capped capuchins, white-faced sakis and tamarins) in pristine forests. Population abundance, expressed as the number of sightings of large primate species per km of transect walked was negatively correlated with levels of hunting pressure in the case of brown capuchins, howlers, and spider monkeys. A Canonical Correspondance Analysis (Fig. 15.2) provides an overview of the relationships between species richness and abundances and site descriptors. The axis 1 explained 72% of the variability in the multivariate pattern of species composition, with coordinates describing hunting and other habitat pressures being equivalent to 0.85 and 0.82. The abundance coordinates for large Cebids were −1.60, −0.69, −0.75 for spider monkeys, howler monkeys and capuchins, respectively, and the primate species richness coordinate amounting to −0.62. This symmetric distribution of primate community patterns on

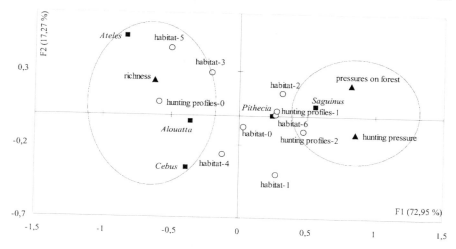

Fig. 15.2 Canonical Correspondance Analysis showing the relationships between primate species richness and abundance and the characteristics of 41 forest sites scattered across French Guiana. The negative relationship between hunting and habitat pressure and the richness and abundance of large-bodied species can be illustrated by their diametrically opposite multivariate responses

one hand, and threats on populations on the other hand, clearly underlines the relationships between levels of game harvests and population abundances of large frugivorous primates. Group sizes were also significantly affected by hunting pressure in the case of howler monkeys. For example, the mean ($\pm$ SD) group size recorded in unhunted, moderately, and heavily hunted areas was 4.6 $\pm$ 1.5, 3.8 $\pm$ 1.5, and 3.6 $\pm$ 2.0 individuals, respectively (Kruskall-Wallis test: $H_{obs} = 10.7$, p = 0.001). In contrast, abundances of small-bodied insectivorous/frugivorous species, such as the golden-handed tamarin, may be positively correlated with level of disturbance (Fig. 15.2).

The relative contribution of primate meat in the total harvest (primates vs. other game species) was highly variable among sites, ranging from less than 1% to more than 20% (Table 15.1). The proportion of primates in the overall harvest of prey species recorded depended not only on hunter preferences but also on local game abundances. For instance, in the northern French Guianan Amerindian settlements of Awala Yalimapo and Macouria, where threats on fauna were clearly important, harvest of primates were low in comparison to that observed at sites where Amerindian communities did not share their catchment areas. But even in the most remote southern part of the country, we found evidence of large primate depletion in all forest sites surveyed in the vicinity of large, isolated villages, and consequently even small-bodied species (golden-handed tamarins, squirrel monkeys) were hunted (Table 15.1). In contrast, the ethnic origin of hunters (i.e., indigenous vs. mixed communities) and the forest type had no detectable effect on the degree to which different primate populations were depleted. Primates were mainly hunted during "expeditions", rather than day hunts. On the other hand, we found that single-day

Table 15.1 Primate hunting at ten sites in French Guiana, including the ethnic group using the area, main use of game, estimates of game biomass harvested, and maximum sustainable harvest thresholds. Numbers refer to Fig. 15.1

Forest sites:	Local community	Use of game meat	Primate biomass/ total game biomass (%)	Primate biomass removed (kg/hunter /year/100 km^2)[1]	Number of primates taken/year	Sustainable threshold of species[2]
1. Counami	mixed	subsistence	1.9	0.5	Apa:1 Ase: 10 Cap:5	Apa: 12 Ase: 22 Cap: 24
2. Macouria	Amerindians	commerce, subsistence	6.4	1.5	Ase: 39 Cap: 52 Col: 7 Ppi: 4	Ase: 11 Cap: 56 Col: nd[3] Ppi: 20
3. Régina	mixed	commerce	0.5	0.2	Apa: 8 Cap: 14	Apa: 23 Cap: 58
4. St Georges	mixed	commerce	0.3	0.2	Apa: 1 Ase: 6 Cap: 22	Apa: 46 Ase: 83 Cap: 117
5. Mana	mixed	commerce, subsistence	19.6	5.8	Ase: 28 Cap: 40 Ppi: 24 Ssc: 28	Ase: 8 Cap: 9 Ppi: 7 Ssc: 9
6. Yalimapo	Amerindians	subsistence	1.2	0.7	Cap: 8 Col: 4	Cap: 12 Col: nd
7. Saül	mixed	subsistence	3.5	7.8	Apa: 3 Cap: 3	Apa: 5 Cap: 19
8. Camopi	Amerindians	subsistence	22.4	9.1	Apa: 132 Ase: 252 Cap: 220 Ppi: 16 Ssc: 16 Smi: 4	Apa: 40 Ase: 72 Cap: 27 Ppi: 36 Ssc: nd Smi: 370
9. Elahé	Amerindians	subsistence	12.6	29.9	Apa: 52 Ase: 56 Cap: 32 Col: 24 Ppi: 4 Ssc: 12	Apa: nd Ase: 5 Cap: 5 Col: 4 Ppi: 4 Ssc: nd
10. Trois Sauts	Amerindians	subsistence	9.9	6.4	Apa: 48 Ase: 65 Cap: 293 Ppi: 17 Ssc: 7 Smi: 29	Apa: nd Ase: 5 Cap: nd Ppi: nd Ssc: nd Smi: 132

[1] this variable allows comparisons of the hunting effort across communities with different number of hunters and different sizes of catchment areas.
[2] Apa = *Ateles paniscus*; Ase = *Alouatta seniculus*; Cap = *Cebus apella*; Col = *Cebus olivaceus*; Ppi = *Pithecia pithecia*; Ssc = *Saimiri sciureus*, Smi = *Saguinus midas*.
[3] nd = not determined, species not observed during surveys.

hunting trips were more profitable per unit of hunting effort than any given day of a multiple-day hunting expedition, regardless of the site, hunting method and the measure of yield considered (e.g., number or biomass of prey items captured per hunter-hour). It therefore appears that hunting effort allocated to multi-day expeditions in infrequently hunted areas primarily attempts to maximize yield of preferred (and locally depleted) prey species rather than the overall bag size (or biomass) of all potential prey species. The alternation of day hunting trips with expeditions farther afield lasting several days is cited in many hunting studies of native Amazonians (Smith 1976; Stearman 1990; Vickers 1991; Peres and Nascimento 2006). Beyond the social role of these expeditions, they represent a quest for preferred game species such as large primates which are already depleted in core hunting areas. From a wildlife management perspective, these expeditions are also extremely important because they disperse hunting activities into larger catchments, thereby diluting their impact on a per area basis. In French Guiana, Amerindians of French nationality have access to medical and social assistance and receive financial government aid, which guarantees them a regular cash income that can be used to purchase pirogues, motors, gasoline and firearms. Communities that have greater access to government handouts can easily enlarge their hunting areas, thus reducing the hunting impact even where human population density is high. Large primates are still hunted during long-range day trips, but these rely on motorboats which considerably expand the catchment area. However, these hunting zones are mainly located along major rivers, which still border very large source areas. In the most remote villages, access to money and fuel is limited, and hunters on foot can only reach a more restricted area surrounding the village in which sensitive wildlife populations have already collapsed. In these cases, hunting yields are typically very low and large primates are rarely killed.

To conclude, hunting pressure in French Guiana is a major factor explaining the variation in both the species richness and density of primate populations, as well as other vertebrate groups (de Thoisy et al. 2006). As shown in several sites of lowland Amazonia (Peres and Lake 2003), we found evidence of large primate depletion as soon as new areas become accessible to hunters. The potential roles of other factors related to habitat structure and quality (i.e., topography, geology, soil types, forest types) still remain difficult to demonstrate unambiguously, but the effects of harvests are much stronger and may obscure those of other environmental variables. The recent gold rush over the last decade (Hammond et al. 2007) resulted in a cryptic but exponentially growing harvest of sensitive species. With the newly decreed French Guiana National Park, the country presently contains a comprehensive and well- configured network of protected areas, contributing with a major role as wildlife refugia, which are likely to operate as source populations for large-bodied species. However, this National Park controversially remains legally open to hunting practices by tribal communities, under the policy rationale that both scientific monitoring and respect for aborigine livelihoods are part of the solution for nature resource conservation in inhabited Amazonian forests.

15.5 The Sustainability of Traditional Practices and the Future of Amazonian Primates

We now consider the question of whether subsistence hunting practices could coexist with viable prey populations of species characterized by slow life-histories such as primates. Sustainability of game hunting depends on the target species, game preferences, access to alternative sources of meat, either domestic or wild, and other uses of forest areas (e.g., logging) with potential direct or indirect impacts on wildlife populations. Predicted estimates of maximum sustainable yields suggest that harvest rates in most of Amazonia need not to be very high before they begin to drive primate populations to precipitous declines (Peres 2000a). In particular, *Ateles* and *Lagothrix* populations are almost always overhunted, whereas *Alouatta* and *Cebus* can sometimes be defined as sustainably exploited (Hill and Padwe 2000; Mena et al. 2000). However, midsized primates, including sakis, bearded sakis and capuchins, are typically harvested at rates exceeding their replacement capacity (Peres 1999; Bodmer and Robinson 2004). Based on data from French Guiana, we evaluated the degree to which traditional hunting practices could be considered to be sustainable using the Robinson and Redford offtake model. Briefly, unsustainable harvests are above a threshold expressed as the annual number of animals captured per unit area and calculated as the population density of any given species times the size of the catchment area x 0.03 (Robinson 2000). Surveys were conducted within hunting catchment areas, and population densities were calculated using the Leopold method (*see* de Thoisy 2000). This method is controversial because it is suspected to overestimate population densities (Brockelman and Ali 1987; Gonzales-Solis et al. 1996; Richard-Hansen and Niel 2005). Calculated sustainable offtake thresholds are therefore also consequently overestimated, resulting in a conservative diagnostic of when observed harvests exceed maximum sustainable harvests. Even in the sparsely populated southern part of the country, where only traditional Amerindian communities harvest wildlife, observed harvests were far above the predicted thresholds for the three largest primate species, capuchins, howlers and spider monkeys. In the north of the country, harvests by the Amerindian community of Macouria were also above thresholds, although harvests by other northern communities, either Amerindian or mixed, were below the critical thresholds (Table 15.1) (Fig. 15.3). This apparent underharvesting, however, could be interpreted as an example of inevitably small offtakes of previously depleted game populations, rather than a sustainable harvest per se. In Saül, a non-native isolated village in the central part of French Guiana, crude numbers of harvested mammals generally appear to be below their maximum sustainable harvest level. However, the percentage of production harvested as estimated from local abundance estimates based on line-transect censuses was far above the maximum sustainable level for *Ateles paniscus*, and the observed percentage of offtake (Robinson 2000) was also at the maximum predicted level. Levels of meat intake were not so high, but local wildlife densities were very low, and far below the 80% of carrying capacity required to meet a safe hunting strategy (Bodmer and Robinson 2004).

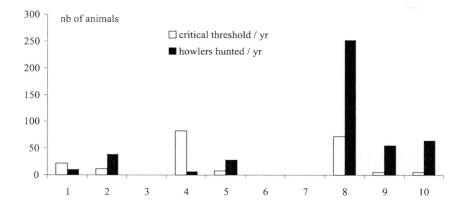

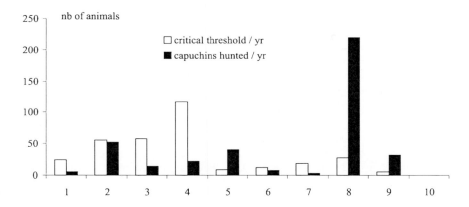

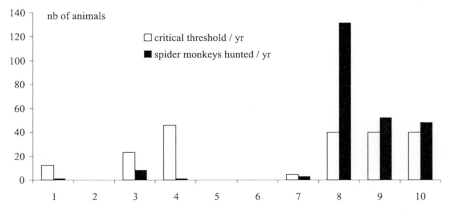

Fig. 15.3 Comparison of annual harvests and calculated maximum sustainable offtakes for three species in 10 French Guianan forest sites: the red howler monkey *Alouatta seniculus* (*above*), the brown capuchin *Cebus apella* (*middle*), and the black spider monkey *Ateles paniscus* (*below*). This is only a ratio of recorded harvests *vs.* theresholds for the three main species, on the ten sites. Numbers of sites refers to Table 15.1 and Fig. 15.1

In Peru, harvests of *Ateles paniscus* and *Cebus apella* by native subsistence hunters were close to, if not above, the predicted maximum sustainable offtakes, although the number of *Lagothrix lagotricha* and *Alouatta seniculus* kills were below the thresholds. Nevertheless surveys did not reveal any dramatic decrease in the abundance of target species (Alvard et al. 1997). Similarly, a study of a Siona-Secoya Indian community in Ecuador suggests no long-term large-scale depletion of primate populations driven by subsistence hunting (Vickers 1991). In contrast the use of long expeditions to hunt primates, as recorded in the Matsigenka communities in Peruvian Amazonia (Shepard 2002) and at some Indian villages of southern French Guiana (Richard-Hansen et al. 2006) suggests that these species are becoming scarce in the vicinities of villages, which was confirmed by line-transect censuses within core hunting areas (but *see* Ohl et al. 2007). Similar evidence of faunal depletion around indigenous settlements has been reported in Amazonia (de Souza-Mazurek et al. 2000, Peres and Nascimento 2006) and in Ecuador (Franzen 2006). However, the use of large hunting areas may also result in misinterpretation of hunting sustainability models, since faunal densities recorded in a small portion of the area may not reflect the entire area. This could be the case for some southern French Guianan villages, where surveys were conducted within the core of the hunting area near the villages, whereas most primate kills were obtained much farther from the settlement. This can partly explain the striking discrepancies between actual harvests and estimates of sustainable harvest thresholds (Fig. 15.2). Indeed, considering the mean primate population densities calculated on the basis of data from other areas subjected to moderate hunting pressure (e.g., Saül), harvests would appear to be sustainable for capuchins and howlers. Also, since these communities hunt across vast expanses of continuous habitats, there is a crucial need to include the non-hunted areas surrounding harvested areas to enable a better consideration of faunal dynamics according to different source-sink scenarios (Novaro et al. 2000).

The potential impact of traditional harvests on wild primate populations is concealed by a wide range of other concurrent pressures on forest habitat, which can confound assessments of the effects of hunting. Other stakeholders, including miners, loggers, and small farmers, may share an economic interest in harvest areas exploited by traditional communities, and often use the same target species. This may result in both interethnic conflicts and overharvesting, that cannot be attributed solely to the practices of traditional forest dwellers. Efficient conservation action plans targeting highly sensitive species will also require a wider ecological and socioeconomic research agenda, if they can claim to take into account the needs of traditional peoples. Field procedures to record population trends, population dynamics between source and sink areas, and harvest sustainability have to be improved. Ecological complexity models that can explicitly consider species distributions and abundances are a limiting factor for monitoring populations, with expected consequences on the reliability of sustainability indexes. Since habitat fragmentation is also a growing threat to forest species, the effectiveness of reserve corridors and importance of reserve design (Peres and Terborgh 1995; Ferrari et al. 1999; Azevado-Ramos et al. 2006) to maintain baseline patterns of faunal dispersal have to be better understood. Also, further work needs to be undertaken to better assess the ability of primate species to co-exist with other forest land uses implemented by local communities (e.g., agroforestry, ecotourism). Second, in many human-occupied protected areas, the protection of primate species cannot be effectively implemented without a "conservation community" approach (Hackel 1999). Long standing cultural and economic knowledge is necessary to propose realistic alternatives to primates hunting. For example, the Baboon Sanctuary project in Belize successfully promotes the "non-extractive" utilization of primates by local communities (Alexander 2000). Game harvest is an extractive activity that provides income not only for the hunters but also for the communities (Hill 2002); effective conservation plans will require economic returns to local communities. Scientific and charismatic value of the primates should help to promote conservation and fund-raising (Alexander 2000), some of which may help communities. Finally, no conservation plan can be implemented without the knowledge of the cost of long-term losses of depleting natural resources.

Combined with a significant role in the symbolism related to wild species, non-human primates are a widely used source of protein for Amerindian communities throughout Amazonia (Cormier 2006; Shepard 2002). Food choices vary among Amazonian communities, and the importance of primate meat in local diets is highly variable at a regional scale. Although most studies of tribal communities conclude that, with traditional use of space and respect for cultural beliefs, harvests are presumably sustainable, the status of primate populations in relation to traditional hunting pressure is inherently complicated by the fact that indigenous hunting interacts with other threats, such as road building, logging, and hunting by nontribal immigrants. Questions have been raised about the opportunity to maintain such traditions in forest landscapes increasingly facing other threats. In forest areas facing a high degree of hunting pressure, particularly where catchment areas are shared by several neighboring communities, we found that Amerindian communities are unable to harvest primates and other game species sustainably, even though very similar communities that are isolated in more remote areas can exercise sustainable harvests. Indeed, as densities of target species including ungulates and large gamebirds decrease, and hunting pressure on primates is expected to increase, there is a high risk of overharvesting several sensitive species (de Thoisy and Renoux 2004; de Thoisy et al. 2005). Our analysis of primate communities in French Guiana and concomitant studies on game hunting show that most primates are simply unable to coexist with poorly regulated hunting practices, even for the most benign subsistence purposes, as soon as human population densities increase. This is clearly at odds with the widespread belief that traditional aborigine communities share an intuitive and time-tested ability to ensure the sustainable use of natural resources. In most cases, apparently harmonious coexistence between indigenous groups and forest wildlife is more related to low densities of the indigenous population, and hence small offtakes exerted under conditions of negligible habitat changes, rather than an active body of adaptive knowledge guiding a successful resource management system. In addition, indigenous population growth is often inevitable, thus, placing greater pressures on natural resources, including sensitive game populations. Reconciling the subsistence needs of local peoples and the requirements of wild primate populations will therefore, always remain a difficult challenge. However, a renewed focus on the demarcation of indigenous territories, and subsequent enforcement of territorial rights, can provide adequate incentives for long-term resource management, particularly if successful partnerships can be implemented with conservation organizations (McSweeney 2005; Schwartzmann and Zimmerman 2005).

15.6 Summary

For millenia, coexistence between human and nonhuman primates has been mystical in all tropical forest regions. Many primate populations have, however, contributed to human diets, often resulting in drastic declines of several species. Although socio-cultural and religious taboos may still limit predation in contemporary times, harvesting of primate populations remains a frequent occurrence throughout the humid tropics. Population declines and local extinctions in relation to direct human exploitation are widely reported in South and Central America, with large-bodied species as the first target species being the most dramatically affected. We illustrate the relationships between offtakes by local communities and wild primate populations using two cases studies. Subsistence hunting has affected game populations throughout lowland Amazonia, with profound consequences to the size structure of primate assemblages, affecting even some of the most remote parts of the region. In French Guiana, the richness of primate communities and the abundance of large cebids were negatively correlated with levels of hunting pressure. Monitoring of harvests by both native and non-native communities revealed that the relative contribution of primate meat to the total harvest was highly variable, ranging from less than 1% to more than 20%. As shown in previous studies, predicted estimates of maximum sustainable yields suggest that harvests need not be very high before they begin to drive primate populations to precipitous declines. Although most studies of tribal communities conclude that, with traditional use of space and respect for cultural beliefs, harvests may be sustainable for some species, the status of primate populations facing subsistence hunting pressure by indigenous groups is profoundly complicated by the fact that harvests interact with other threats, such as road building, logging, and additive hunting by nontribal immigrants. All Amazonian studies show that most primates are simply unable to coexist with unregulated hunting, even for the most benign subsistence purposes, as soon as human population densities increase. We therefore question the widespread belief that traditional communities share an intuitive wisdom to ensure the sustainable use of natural resources. In most cases, harmonious coexistence between indigenous groups and forest wildlife is more related to low densities of the indigenous population and small offtakes exerted on habitats with negligible changes, rather than an active adaptive body of knowledge guiding a successful resource management system. Reconciling the subsistence needs of local peoples and the requirements of primate populations will therefore always remain a difficult challenge. However, demarcation of indigenous territories and subsequent enforcement of territorial rights can provide adequate incentives for long-term resource management. Efficient conservation action plans designed for sensitive species and respect for traditional cultures will also require further research and policy action. Ecological studies should include improved monitoring of population trends and the dynamics between source and sink areas. Finally, long-standing socioeconomic knowledge will be necessary to propose viable alternatives to primate hunting, and a "conservation community" approach should be promoted, with efficient economic returns to local communities.

References

Alexander, S. E. 2000. Resident attitudes toward conservation and Black howler monkeys in Belize: the Community Baboon Sanctuary. Environmental Conservation 27:341–350.

Alvard, M. S., Robinson, J. G., Redford, K. H., and Kaplan, H. 1997. The sustainability of subsistence hunting in the Neotropics. Conservation Biology 11:977–982.

Ayres, J. M., Lima, L. M., Martins, E. S., and Barreiros J. L. K. 1991. On the track of the road: changes in subsistence hunting in a Brazilian Amazonian village. In J. G. Robinson and K. H. Redford (eds), *Neotropical Wildife Use and Conservation* (pp 82–92). Chicago: University of Chicago Press.

Azevedo-Ramos, C., Domingues do Amaral, B., Nepstad, D. C., Filho, B. S., and Nasi, R. 2006. Integrating ecosystem management, protected areas, and mammal conservation in the Brazilian Amazon. Ecology and Society 11:art 17.

Balée, W. 1989. The culture of Amazonian forests. Advances in Economic Botany 7:1–21.

Barnett, A. A., Borges, S. H., de Castilho, C. V., Neri, F. M., and Shapley, R. L. 2002. Primates of the Jaú national park, Amazonas, Brazil. Neotropical Primates 10:65–70.

Bermejo, M., Rodriguez, J. D., Illera, G., Barroso, A., Vila, C., and Walsh, P. D. 2006. Ebola outbreak killed 5000 gorillas. Science 314:1564.

Bodmer, R. E. 1995. Managing Amazonian wildlife: biological correlates of game choice by detribalized hunters. Ecological Applications 5:872–877.

Bodmer, R., and Robinson, J. 2004. Evaluating the sustainability of hunting in the neotropics. In K. Silvius, R. Bodmer and J. Fragoso (eds.), *People in nature: wildlife conservation in South and Central America* (pp. 299–323). New York: Columbia University Press.

Bollen, A., and Donati, G. 2006. Conservation status of the littoral forest of south-eastern Madagascar: a review. Oryx 40:57–66.

Brockelman, W. Y., and Ali, R. 1987. Methods of surveying and sampling forest primate populations. In C. W. Marsh and R. A. Mittermeier (eds.), *Primate Conservation in the Tropical Rain Forest* (pp. 23–42).New York: Alan R. Liss, Inc.

Burney, D. A. 1999. Rates, patterns, and the processes of landscape transformation and extinction in Madagascar. In R. D. E. MacPhee (ed.), *Extinctions in Near Time: Causes, Contexts, and Consequences* (pp. 145–164). New York: Plenum Press.

Burney, D. A, Burney, L. P., Godfrey, L. R., Junters, W. J., Goodman, S. M., Wright, H. T., and Jull, A. J. 2004. A chronology for late prehistoric Madagascar. Journal of Human Evolution 47:25–53.

Butchart, S. H. M., Barnes, R., Davies, C. W. N., Fernandez, M., and Seddon, N. 1995. Observations of two threatened primates in the Peruvian Andes. Primate Conservation 1:15–19.

Cartelle, C., and Hartwig, W. C. 1996. A new extinct primate among the Pleistocene megafauna of Bahia, Brazil. Proceedings of the National Academy of Sciences 93:6405–6409.

Colell, M., Maté, C., and Fa, J. E. 1995. Hunting among Moka Bubis: dynamics of faunal exploitation at the village level. Biodiversity and Conservation 3:939–950.

Cormier, L. A. 2003. Kinship with monkeys: the Guajà foragers of Eastern Amazonia. New York: Colombia University Press.

Cormier, L. A. 2006. A preliminary review of neotropical primates in the subsistence and symbolism of indigenous lowland South American people. Ecological and Environmental Anthropology 2:14–32.

Costa, L. P., Leite, Y. L. R., Mendes S. L., and Ditchfield, A. D. 2005. Mammal conservation in Brazil. Conservation Biology 19:672–679.

Cowlishaw, G. 1999. Predicting the pattern of decline of African primate diversity: an extinction debt from historical deforestation. Conservation Biology 13:1183–1193.

de Granville, J. J. 1988. Phytogeographical characteristics of the Guianan forests. Taxon 37: 578–594.

de Thoisy, B. 2000. Line-transects: sampling application to a rainforest in French Guiana. Mammalia 64:101–112.

de Thoisy, B., and Renoux, F. 2004. Status of the lowland tapir in French Guiana: hunting pressure and threats on habitats. Second International Tapir Symposium, TSG/SSC/IUCN, Panama.

de Thoisy, B., Renoux, F., and Julliot, C. 2005. Hunting in northern French Guiana and its impacts on primates communities. Oryx 39:149–157.

de Thoisy, B., Brosse, S., Richard-Hansen, C., and Thierron, V. 2006. Rapid evaluation of relationships between impacts of forest anthropic activities and threats on biodiversity in French Guiana. VII Congresso Internacional sobre manejo de fauna silvestre na Amazônia e América latina. Ilhéus, Bahia, Brasil.

De Souza-Mazurek R. M., Pedrinho, T., Feliciano, X., Hilário, W., Gerôncio, S., and Marcelo, E. 2000. Subsistence hunting among the Waimiri Atroari Indians in central Amazonia, Brazil. Biodiversity and Conservation 9:579–596.

Emmons, L. H. 1984. Geographic variation in densities and diversities of non flying mammals in Amazonia. Biotropica 16:210–222.

Ferrari, S. F., Emidio-Silva, C., Aperecida Lopes, M., and Bobadilla, U. L. 1999. Bearded sakis in South-eastern Amazonia – back from the brink? Oryx 33:346–351.

Franzen, M. 2006. Evaluating the sustainability of hunting: a comparison of harvest profiles across three Huaorani communities. Environmental Conservation 33:36–45.

Freese, C. H., Heltne, P. G., Castro, N., and Whitesides, G. 1982. Patterns and determinants of monkey densities in Peru and Bolivia, with notes on distribution. International Journal of Primatology 3:53–90.

Fusari, A., and Carpaneto, G. M. 2000. Subsistence hunting and bushmeat exploitation in central-western Tanzania. Biodiversity and Conservation 9:1571–1585.

Fusari A., and Carpaneto, G. M. 2006. Subsistence hunting and conservation issues in the game reserve of Gile, Mozambique. Biodiversity and Conservation 15:2477–2495.

Garcia, G., and Goodman, S. M. 2003. Hunting of protected animals in the Parc National d'Ankarafantsika, north-western Madagascar. Oryx 37:115–118.

Goossens, B., Chikhi, L., Ancrenaz, M., Lackman-Ancrenaz, I., Andau, P., and Brudford M. W. 2006. Genetic signature of anthropogenic population collapse in Orang Utans. PLoS Biology 4:e25.

Gonzales-Solis, J., Mateos, E., Manosa, S., Ontanon, M., Gonzlez-Martin, M. and Guix, J. C. 1996. Abundance estimates of primates in an Atlantic rainforest area of southeastern Brazil. Mammalia 60:488–491.

Hackel, J. D. 1999. Community conservation and the future of Africa's wildife. Conservation Biology 13:726–734.

Hammond, D. S., Gond, V., de Thoisy, B., Forget P. M., and DeDijn, B. 2007. Causes and consequences of a tropical forest gold rush in the Guiana Shield, South America. Ambio 36:661–670.

Harris, R. B., Wall, W. A., and Allendorf, F. W. 2002. Genetic consequences of hunting: what do we know and what should we do? Wildlife Society Bulletin 30:634–643.

Haugaasen, T., and Peres, C. A. 2005. Primate assemblage structure in Amazonian flooded and unflooded forests. American Journal of Primatology 67:243–258.

Hill, C. M. 2002. Primate conservation and local communities – ethical issues and debates. American Anthropologist 104:1184–1194.

Hill, K., and Padwe, J. 2000. Sustainability of Aché hunting in the Mbaracayu reserve, Paraguay. In J. G. Robinson and E. L. Bennett (eds.), *Hunting for Sustainability in Tropical Forests* (pp. 79–105). New York: Columbia University Press.

Isaac, N. J. B., and Cowlishaw, G. 2004. How species respond to multiple extinction threats. Proceedings Royal Society of London B 271:1135–1141.

IUCN. 2006. 2006 IUCN Red List of Threatened Species. www.redlist.org. Accessed October 2006.

Jernvall, J., and Wright, P. C. 1998. Diversity components of impending primate extinctions. Proc. Natl Acad. Sci. 95:11279–11283.

Jerozolimski, A., and Peres, C. A. 2003. Bringing home the biggest bacon: a cross-site analysis of the structure of hunter-kill profiles in Neotropical forests. Biological Conservation 111: 415–425.

Jiang, X., Luo, Z., Zhao, S., Li, R, and Liu, C. 2006. Status and distribution pattern of black crested gibbon (*Nomascus concolor jingdongensis*) in Wuliang Mountains, Yunnan, China: implication for conservation. Primates 47:264–271.

Johns, A. D. 1991. Forest disturbance and Amazonian primates. In H. O. Box (ed.), *Primate Responses to Environmental Changes* (pp. 115–135). London: Chapman and Hall.

Johns A. D., and Skorupa, J. P. 1987. Responses of rain-forest primates to habitat disturbance: a review. American Journal of Primatology 8:157–192.

Johnson, K. 1996. Hunting in the Budongo Forest, Uganda. Swara Jan–Feb:24–27.

Jorgenson, J. P. 2000. Wildlife conservation and game harvest by Maya hunters in Quintina Roo, Mexico. In J. G. Robinson and E. L. Bennett (eds.), *Hunting for Sustainability in Tropical Forests*, (pp. 251–266). New York: Colombia University Press.

Kracke, W. H. 1978. Force and persuasion, leadership in an Amazonian society. Chicago: University of Chicago Press.

Kumara, H. N., and Singh, M. 2004. The influence of differing hunting practices on the relative abundance of mammals in two rainforest areas of the Western Ghats, India. Oryx 38:321–327.

Lahm, S. A. 1993. Utilization of forest resources and local variation of wildlife populations in Northeastern Gabon. In C. M. Hladik, A. Hladik, O. F. Linarea, H. Pagezy, A. Semple and M. Hadley (eds.), *Tropical Forest, People and Food* (pp. 213–226). Paris: Parthenon Publishing Group.

Larson, S., Jameson, R., Bodkin, J., Staedler, M., and Bentzen, P. 2002. Microsatellite DNA and mitochondrial DNA variation in remnant and translocated sea otter (*Enhydra lutris*) populations. Journal of Mammalogy 83:893–906.

Lehman, S. M. 2000. Primate community structure in Guyana: a biogeographic analysis. International Journal of Primatology 21:333–351.

Leo Luna, M. 1987. Primate conservation in Peru: a case study of the yellow-tailed woolly monkey. Primate Conservation 8:122–123.

Lindeman, J. C., and Mori, S. A. 1989. The Guianas. In D. G. Campbell and H. D Hammond (eds.), *Floristic Inventories of Tropical Countries: The Status of Plants Systematics, Collections and Vegatation, Plus Recommendations for the Future* (pp. 375–390). New York: New York Botanical Garden.

Lopes, M. A., and Ferrari, S. F. 2000. Effects of human colonization on the abundance and diversity of mammals in eastern Brazilian Amazonia. Conservation Biology 14:1658–1665.

Madhusudan, M. D., and Karanth, K. U. 2002. Local hunting and the conservation of large mammals in India. Ambio 31:49–54.

McSweeney, K. 2005. Indigenous population growth in the lowland Neotropics: social science insights for biodiversity. Conservation Biology 19:1375–1384.

Mena, V. P., Stallings, J. R., Regalado, B. J., and Cueva, L. R. 2000. The sustainability of current hunting practices by the Huaorani. In J. G. Robinson and E. L. Bennett (eds.), *Hunting for Sustainability in Tropical Forests* (pp. 57–78). New York: Columbia University Press.

Milner-Gulland, E. J., Bennett, E. L. and the SCB 2002. Annual Conference Wild Meat Group. 2003. Wild meat: the bigger picture. Trends on Ecology and Evolution 18:351–357.

Milner, J. M., Nilsen, E. B., and Andreassen H. P. 2007. Demographic side effects of selective hunting in ungulates and carnivores. Conservation Biology 21:36–47.

Mittermeier, R. A. 1991. Hunting and its effects on wild primate populations in Suriname. In J. G. Robinson and K. H. Redford (eds.), *Neotropical Wildlife Use and Conservation*, (pp. 93–106). Chicago, University of Chicago Press.

Mittermeier, R. A., Valladares-Pádua, C., Rylands, A. B, Eudey, A. A, Butynski, T. M., Jörg, U. Ganzhorn, J. U., Rebecca Kormos, R., Aguiar J. M., and Walker, S. 2005. The World's 25 Most Endangered Primates 2004–2006. IUCN/Primate Specialist Group and International Primatological Society.

Muchaal, P. K., and Ngandjui, G. 1999. Impact of village hunting on wildlife populations in the Western Dja Reserve, Cameroon. Conservation Biology 13:385–396.

Nijman, V. 2004. Effects of habitat disturbance and hunting on the density and the biomass of the endemic Hose's leaf monkey *Presbytis hosei* (Thomas, 1889) (Mammalia: Primates: Cercopithecidae) in east Borneo. Contributions to Zoology 73:art4.

Novaro, A. J., Redford, K. H. and Bodmer, R. E. 2000. Effect of hunting in source-sink systems in the Neotropics. Conservation Biology 14:713–721.

Nuñez-Iturri, G., and Howe, H. F. 2007. Bushmeat and the fate of trees with seeds dispersed by large primates in a lowland rainforest in western Amazonia. Biotropica 39:348–354.

Oates, J. F. 1996. Habitat alteration, hunting and the conservation of folivorous primates in African forests. Australian Journal of Ecology 21:1–9.

Ohl-Schacherer, J., Shepard, G. H., Kaplan, H., Peres, C. A., Levi, T; Yu, D.W. 2007. The sustainability of subsistence hunting by Matsigenka Native communities in Manu National Park, Peru. Conservation Biology 21:1174–1185

Olupot, W. 2000. Mass differences among male Mangabey monkeys inhabiting logged and unlogged forest compartments. Conservation Biology 14:833–843.

Pacheco, L. F, and Simonetti, J. A. 2000. Genetic structure of a Mimosoid tree deprived of its sed disperser, the spider monkey. Conservation Biology 14:1766–1775.

Peres, C. A. 1990. Effects of hunting on western Amazonian primate communities. Biological Conservation 54:47–59.

Peres, C. A. 1991. Humboldt's woolly monkeys decimated by hunting in Amazonia. Oryx 25: 89–95.

Peres, C. A. 1997a. Primate community structure at twenty western Amazonian flooded and unflooded forests. Journal of Tropical Ecology 13:381–405.

Peres, C. A. 1997b. Effects of habitat quality and hunting pressure on arboreal folivore densities in neotropical forests: a case study of howler monkeys (*Alouatta* spp.). Folia Primatologica 68:199–222.

Peres, C. A. 1999. Effects of hunting and habitat quality on Amazonian primate communities. In J. G. Fleagle, C. Janson and K.E. Reed (eds.), *Primate Communities* (pp. 268–283). Cambridge: Cambridge University Press.

Peres, C. A. 2000a. Effects of subsistence hunting on vertebrate community structure in Amazonian forests. Conservation Biology 14:240–253.

Peres, C. A. 2000b. Evaluating the impact and sustainability of subsistence hunting at multiple Amazonian forest sites. In J. G. Robinson and E. L. Bennett (eds.), *Hunting for Sustainability in Tropical Forests* (pp. 31–57). New York: Columbia University Press.

Peres, C. A., and Terborgh, J. W. 1995. Amazonian nature reserves: an analysis of the defensibility status of existing conservation units and design criteria for the future. Conservation Biology 9:34–46.

Peres, C. A., and Dolman, P. 2000. Density compensation in neotropical primate communities: evidence from 56 hunted and non-hunted Amazonian forests of varying productivity. Oecologia 122:175–189.

Peres, C. A., and van Roosmalen, M. 2002. Patterns of primate frugivory in Amazonia and the Guianan shield: implications to the demography of large-seeded plants in overhunted tropical forests. In D. Levey, W. Silva and M. Galetti (eds.) *Seed Dispersal and Frugivory: Ecology, Evolution and Conservation* (pp. 407–423). Oxford: CABI International.

Peres, C. A., and Lake, I. R. 2003. Extent of nontimber resource extraction in tropical forests: accessibility, to game vertebrates by hunters in the Amazon basin. Conservation Biology 17: 1–17.

Peres, C. A., and Nascimento, H. S. 2006. Impact of game hunting by the Kayapó of southeastern Amazonia: implications for wildlife conservation in Amazonian indigenous reserves. Biodiversity and Conservation 15:2627–2653.

Peres, C.A, Barlow, J., and Laurance, W. 2006. Detecting anthropogenic disturbance in tropical forests. Trends in Ecology and Evolution 21:227–229.

Peres, C. A., and Palacios, E. 2007. Basin-wide effects of game harvest on vertebrate population densities in Amazonian forests: implications for animal-mediated seed dispersal. Biotropica 39:304–315.

Perez, V., Godfrey, L. R., Nowak-Kemp, M., Burney, D. A., Ratsimbazafy, J., and Vasey, N. 2005. Evidence of early butchery of giant lemurs in Madagascar. Journal of Human Evolution 49:722–742.

Ratiarison, S., and Forget, P. M. 2005. Frugivores and seed removal at *Tetragastris altissima* (Burseraceae) in a fragmented forested landscape of French Guiana. Journal of Tropical Ecology 21:1–8

Redford, K. H., and Robinson, J. G. 1987. The game of choice: patterns of Indian and colonist hunting in the Neotropics. American Anthropologist 89:650–667.

Redford, K. H., and Robinson, J. G. 1991. Subsistence and commercial uses of wildlife in Latin America. In J. G. Robinson and K. H. Redford (eds.), *Neotropical Willdife Use and Conservation* (pp. 6–23). Chicago: University of Chicago Press.

Refish, J., and Koné, I. 2005. Impacts of commercial hunting on monkey populations in the Taï region, Côte d'Ivoire. Biotropica 37:136–144.

Renoux, F. 1998. Se nourrir à Trois Sauts: analyse diachronique de la prédation chez les Wayãpi du Haut-Oyapock. Journal d'Agriculture Traditionnelle et de Botanique Appliquée 40: 167–180.

Richard-Hansen, C., and Hansen. E. 2004. Hunting and wildlife management in French Guiana: current aspects and future prospects. In K. Silvius, R. Bodmer and J. Fragoso (eds.), *People in Nature: Wildlife Conservation in south and central America* (pp. 400–410). New York, Columbia University Press.

Richard-Hansen, C., Khazraie, K., Mauffrey, J.-F., and Gaucher, P. 2004. Pratiques de chasse dans un village isolé du centre de la Guyane: Evaluation de l'impact sur les populations animales. 6th International Wildlife Ranching Symposium, 6–9 juillet, Paris.

Richard-Hansen, C., and Niel, C. 2005. "Estimer les faibles densités d'espèces chassées en Guyane avec peu d'observations: Proposition de calcul d'une "largeur effective de comptage" (Effective Strip Width, ou ESW) spécifique." ONCFS, Rapport scientifique 2004:22–27.

Richard-Hansen, C., Gaucher, P., Maillard, J.-F., and Ulitzka, M. 2006. Análisis comparativa de la cacería en tres pueblos de comunidades indígenas en Guyana Francesa. VII Congreso Internacional sobre Manejo de Fauna Silvestre en la Amazonía y Latinoamérica, Ilhéus, Brazil.

Robinson, J. G. 2000. Calculating maximum sustainable harvests and percentage offtakes. In J. G. Robinson and E. L. Bennett (eds.), *Hunting for Sustainability in Tropical Forests* (pp. 521–524). New York: Colombia University Press.

Ruiz-Garcia, M. 2005. The use of several microsatellite loci applied to 13 neotropical primate revealed a strong recent bootleneck event in the woolly monkey (*Lagothrix lagothricha*) in Colombia. Primate Report 71:27–55.

Russo, S. E., Portnoy, S., and Augspurger C. K. 2006. Incorporating animal behavior into seed dispersal models: implications for seed shadows. Ecology 87:3160–3174

Schwartzmann, S., and Zimmerman, B. 2005. Conservation alliances with indigenous people of the Amazon. Conservation Biology 19:721–727.

Serio-Silva, J. C., and Rico-Gray, V. 2002. Interacting effects of forest fragmentation and howler monkey foraging on germination and dispersal of figs. Oryx 36:266–271.

Shepard, G. H. 2002. Primates in Matsigenka subsistence and World view. In A. Fuentes and D. Wolfe (eds.), Primates Face to Face: *The Conservation Implications of Human-Non Human Primate Interconnections* (pp. 101–136). Cambridge: Cambridge University Press.

Smith, N. J. H. 1976. Utilization of game along Brazil's Transamazon Highway. Acta Amazonica 6:455–466.

Stearman, A. M. 1990. The effects of settler incursion on fish and game resources of the Yuqui, a native Amazonian society of eastern Bolivia. Human Organization 49:373–385.

Stoner, K. E., Vulinec, K., Wright, S. J., and Peres, C. A. 2007. Hunting and plant community dynamics in tropical forests: a synthesis. Biotropica 39:385–392.

Sussman, R. W, and Phillips-Conroy, J. E. 1995. Survey of the distribution and density of primates of Guyana. International Journal of Primatology 16:761–791.

Urbani, B. 2006. A survey of primate populations in Northeastern Venezuelan Guayana. Primate Conservation 20:47–52.

Vaughan, C. 1993. Human population and wildlife: a Central American focus. Transactions of the 58th North American Wildlife and Natural Resources Conferences.

Vickers, W. T. 1991. Hunting yields and game composition over ten years in an Amazon Indian territory. In J. G. Robinson and K. H. Redford (eds.), *Neotropical Willdife Use and Conservation* (pp. 53–81). Chicago: University of Chicago Press.

Wadley, R. L., Carol, J., Pierce Colfer, C. J. P., and Hood, I. G. 1997. Hunting primates and managing forests: the case of Iban Forest Farmers in Indonesian Borneo. Human Ecology 25:243–271.

Walsh, P. D., Abernethy, K. A., Bermejo, M., Beyers, R., de Watcher, P., Akou, M. E., Huljbregts, B., Mambounga, D. I., Toham, A. K., Kilbourn, A. M., Lahm, S. A., Latour, S., Maisels, F., Mbina, C., Mihindou, Y., Oblang, S. N., Effa, E. N., Starkey, M. P., Telfer, P., Thibault, M., Tutin, C. E. G., White, L. J. T., and Wilkie, D. S. 2003. Catastrophic ape decline in western equatorial Africa. Nature 422:611–614.

Wilkie, D. S., Curran, B., Tshombe, R, and Morelli, G. A. 1998. Modeling the sustainability of subsistence farming and hunting in the Ituri Forest of Zaire. Conservation Biology 12:137–147.

Young, T. P., and Isbell, L. A 1994. Minimum group size and other conservation lessons exemplified by a declining primate population. Biological Conservation 68:129–134.

Chapter 16
Primate Densities in the Atlantic Forest of Southeast Brazil: The Role of Habitat Quality and Anthropogenic Disturbance

Naiara Pinto, Jesse Lasky, Rafael Bueno, Timothy H. Keitt, and Mauro Galetti

16.1 Introduction

16.1.1 Goals

Studies of variation in abundance within a species' geographic range provide the connection between the disciplines of ecology and biogeography. Empirical studies of various taxonomic groups show that density of a given species is unevenly distributed in space, with few "hotspots" and many "coldspots", where abundance is orders of magnitude lower (Brown et al. 1995). The typical explanation for this pattern is spatial variation in habitat suitability. In other words, variation in density is generated by how closely sites correspond to a species' niche (Brown et al. 1995). Like many ecological patterns, the correspondence between primate density and habitat suitability can be investigated at several spatial scales (Wiens 1989; Levin 1992). For example, coarse-scale studies comparing densities of howler monkeys (*Alouatta* spp.) across the Neotropics have shown that howler density is largely a function of primary productivity (Peres 1997). Fine-scale studies comparing neighboring forest fragments have also reported variation in howler density, but in this case the pattern is frequently attributed to anthropogenic pressure (Hirsh et al. 1994; Cullen et al. 2001; Chiarello 2003; Martins 2005).

In general, human impact on other primates can be direct via hunting, or indirect through habitat disturbance and fragmentation. However, some species thrive in disturbed habitats (Chiarello 1993, 2003; Rylands et al. 1993; Strier et al. 2000). This fact complicates the task of predicting changes in primate density across a gradient in land use. In the present chapter, we investigate the synergistic effects of environmental and anthropogenic factors on the density of five primate genera that inhabit the Atlantic forest of southeast Brazil. Our goal is not to produce distribution maps, but rather to: (i) synthesize available census information for the region; (ii) compare the genera's responses to anthropogenic impact; and (iii) map areas of high

N. Pinto (✉)
Jet Propulsion Laboratory, 4800 Oak grove ms 300–325 Pasadena, CA, 91105
e-mail: sardinra@jpl.masa.gov

P.A. Garber et al. (eds.), *South American Primates*, Developments in Primatology: Progress and Prospects, DOI 10.1007/978-0-387-78705-3_16,
© Springer Science+Business Media, LLC 2009

predicted densities based on available data. In this section, we introduce the reader to the Brazilian Atlantic forest, present the dataset used in the study, and describe the analytical tools used to study the determinants of primate density.

16.1.2 The Primates at the Brazilian Atlantic Forest

Studies in the Brazilian Atlantic forest provide an ideal opportunity to understand the interaction of anthropogenic factors and habitat quality on primate densities. This ecosystem is a biodiversity hotspot that occupies less than 8% of its original extent (Hirota 2003). Current studies estimate that 40% of the tree and shrub species in this ecosystem are endemic, as well as 22% of their bird and mammal species (Brooks et al. 2000) – and many new species are still being discovered in the region every year (Alves et al. 2006; Donha and Eliasaro 2006; Pontes et al. 2006). Due to its extensive elevational and latitudinal ranges, the Atlantic forest is recognized as a domain that includes several vegetation types (Oliveira-Filho and Fontes 2000). Exploitation of Atlantic forest species did not start recently, as it has been suggested that hunting and forest clearing were already widespread when the first Portuguese arrived in 1500 (Dean 1996). However, the anthropogenic pressure was intensified with the Portuguese colonization, expansion of the agricultural frontier (Dean 1996; Câmara 2003), and later establishment of Brazilian industrial centers in the area, which currently has a population of more than 130 million people (IBGE 2000). As a result of the intense land use in eastern Brazil, the distribution of forest remnants is very distinct from the fishbone pattern observed in the Brazilian Amazon, in which vast forest tracts are interrupted by a network of roads and pipelines. Rather, the Atlantic forest landscape is now an archipelago with small forest fragments embedded in a human-dominated matrix containing pastures, plantations, cities, and roads.

Twenty-three primate species are known to live in the Brazilian Atlantic forest, twenty of which are endemic to this ecosystem (Hirsh et al. 2006). According to the most recent IUCN Mammal Red List (IUCN 2006), three species are vulnerable, four are endangered, and nine are critically endangered (Table 16.1). While some primate populations in the Brazilian Amazon may be sustained via source-sink dynamics (Michalski and Peres 2005), these dynamics have never been documented for the Atlantic forest and are unlikely to be operating due to inter-fragment isolation and inhospitality of the matrix. Also, few fragments are large enough to sustain viable primate populations (Chiarello and Melo 2001; Bernardo and Galetti 2004; see also Marsden et al. 2005 for birds), and the extent to which existing conservation units are protecting primate populations against poaching remains unknown.

16.1.3 Census Data for Primate Species in the Brazilian Atlantic Forest

Data on primate abundance were compiled from a variety of sources including graduate theses, primary-literature publications, and grey-literature reports. In all cases,

Table 16.1 Primate species inhabiting the Brazilian Atlantic forest, their conservation (IUCN) status, number of sites that have been censused using the line-transect technique, and number of populations with 500 individuals or more. CE = Critically endangered, E = Endangered, V = Vulnerable

Species	IUCN Status	Census Sites	Viable (≥ 500) populations	Ref.
Alouatta guariba	CE	24	8	3, 4, 7, 8, 9, 10, 1113, 14, 15, 16
Alouatta belzebul	CE	0	–	–
Brachyteles arachnoides	E	9	3	4, 8, 10, 15
Brachyteles hypoxanthus	CE	2	0	6, 7
Callicebus barbarabrownae	CE	0	–	–
Callicebus coimbrai	CE	0	–	–
Callicebus melanochir	V	0	–	–
Callicebus nigrifons	–	3	1	1, 5, 12,
Callicebus personatus	V	8	5	2, 7, 8, 15, 16
Callithrix aurita	V	4	1	1, 8, 10
Callithrix flaviceps	E	2	1	7, 15
Callithrix penicillata	–	2	1	16, 17
Callithrix geoffroji	–	5	2	2
Callithrix jacchus	–	0	–	–
Callithrix kuhlil †	–	0	–	–
Cebus flavius	–	0	–	–
Cebus libidinosus	–	0	–	–
Cebus nigritus	–	25	10	1, 2, 3, 4, 7, 8, 10, 15, 17
Cebus xanthosternos	CE	0	–	–
Leontopithecus caissara	CE	0	–	–
Leontopithecus chrysomelas	E	0	–	–
Leontopithecus chrysopygus	CE	5	1	3
Leontopithecus rosalia	CE	0	–	–

(1) Sao Bernardo and Galetti 2004; (2) Chiarello 2000; (3) Cullen et al. 2001; (4) Martins 2005; (5) Romanini de Oliveira et al. 2003; (6) Strier et al. 2000; (7) Chiarello 2003; (8) Cosenza and Melo 1998; (9) Chiarello and Melo 2001; (10) Galetti et al. unpublished data; (11) Hirsh 1995; (12) Trevelin 2006; (13) Buss 2001; (14) Chiarello 1993; (15) Pinto et al. 1994; (16) Hirsh et al. 1994; (17) Bovendorp and Galetti 2007.

data were collected using the line-transect technique (Buckland et al. 2001). Values of population sizes were often calculated assuming no spatial variation in density within sites. Since this assumption is rarely met, we only show the number of viable populations (>500 individuals estimated) instead of attempting to calculate exact population sizes (Table 16.1).

There are a number of limitations inherent in the type of data used in this study. First, line-transect census data are available for only eleven of the twenty-three primate species that inhabit this ecosystem. Second, studies are mostly restricted to the states of São Paulo, Espírito Santo and Minas Gerais (Fig. 16.1). Intensive census studies are lacking for populations inhabiting states such as Paraná, Santa Catarina, Rio de Janerio, and northeast Brazil, where few forest fragments remain and some primate populations are believed to be on the brink of extinction, especially large-bodied species (Pontes et al. 2006).

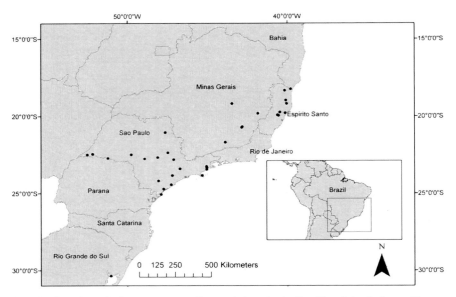

Fig. 16.1 Location of primate census studies carried out in the Brazilian Atlantic forest. The statistical analyses presented here focus on the southeast region, composed of the states of Minas Gerais, São Paulo, Rio de Janeiro, and Espírito Santo

16.1.4 Tools Used in the Present Study

Our task faces two challenges in addition to data scarcity and nonhomogeneous sampling across the ecosystem: first, dealing with nonlinear relationships and correlations between the independent variables, and second, the fact that the influence of a given environmental correlate can manifest itself at unknown spatial scales – for example, it is not possible to determine beforehand the area of influence of a city and therefore its potential impact on neighboring forest fragments. In the present chapter, we will apply tools that can help deal with the difficulties cited above: geographic information systems (GIS) and regression trees.

16.1.4.1 The Use of GIS in Conservation Studies

The use of remote sensing and GIS has recently increased among biologists, because these tools facilitate the analysis of large-scale associations between landscape patterns and biological outcomes. In the present work, three classes of maps are employed to model primate densities. First, maps of climate and elevation are used to differentiate between the evergreen coastal rainforest and the semideciduous forest (Oliveira-Filho and Fontes 2000). This distinction is extremely relevant for folivorous species (Peres 1997), because leaves from perennial trees are expected to be tougher (Coley 1983) and have lower nutritional content (Aerts 1996) than leaves from deciduous trees. Second, we used maps of human accessibility, land use, and social indicators, which can potentially serve as surrogates of anthropogenic disturbance and hunting pressure (Siren et al. 2006; Brashares et al. 2001; Laurance et al. 2005). Third, we used maps of fragment size. Note that climate and elevation maps reflect local habitat quality, whereas the other maps are based on information from the neighboring municipalities and road network that surround study sites.

16.1.4.2 Regression Trees

The statistical analysis of the relationship between environmental factors and population sizes is complicated by the existence of interactions (often nonlinear) among environmental predictors. For example, forest type is known to correlate with temperature, precipitation and elevation (Oliveira-Filho and Fontes 2000). Moreover, the exact shape of these relationships is unknown. Thus, we decided to use a data mining approach that enables us to look for environmental determinants of primate density while accommodating for nonlinear interactions between predictors and which does not require the specification of the relationship between the response and the predictors. Here, we will use Random Forest, a tree regression method (Breiman 2001; Liaw and Wiener 2002). This method recently started being applied in several areas of biology involving data mining, such as bioinformatics (Pang et al. 2006) and niche modeling (Garzon et al. 2006; Prasad et al. 2006). The algorithm works by iteratively splitting the group of data points. Each tree node represents a splitting rule (e.g., "elevation > 1500 m"), and nodes are followed by two branches representing the newly separated data points. More specifically, the splits are performed using the predictor variables to partition the response variable into two groups, so as to maximize the between-groups sum of squares. The output tree contains a series of branches representing the optimized sequence of splitting rules. Random Forest grows hundreds of trees, each one using a subset of the independent variables. The resulting trees are then averaged to obtain the final model, a procedure that reduces overfitting (Breiman 2001). As in other niche model and classification tools, data points are partitioned into a training set, used to construct the model, and a testing set, used to access model accuracy. For a very accessible review of regression tree methods, *see* Berk (2006).

16.2 Methods

16.2.1 Study Area

The study areas comprise four Brazilian states: São Paulo, Rio de Janeiro, Minas Gerais and Espírito Santo (Fig. 16.1). The region spans the two main Atlantic forest domains: the Atlantic rainforest and the Atlantic semideciduous forest. The former comprises areas up to 300 km inland that have high annual precipitation due to oceanic winds and mountain ranges, whereas the latter includes plateau areas with higher elevation and lower annual precipitation. For a detailed description of the forest types, *see* Oliveira-Filho and Fontes (2000).

16.2.2 Target Genera

We focus on five genera: (1) *Brachyteles* (muriqui), the largest species at 12 kg, a frugivore-folivore (Milton 1984; Strier 1991) that is distributed along the Brazilian southern states of São Paulo, Rio de Janeiro, Espírito Santo, and Minas Gerais, and the states of Paraná and Bahia; (2) *Alouatta* (howler monkey), a folivore (Glander 1978; Mendes 1989; Peres 1997) weighing 6.4 kg, distributed in the Brazilian south and all the way to the northeast along the coast; (3) *Cebus* (capuchin monkey), an insectivore-frugivore (Fragaszy et al. 2004) weighing 2.5 kg inhabiting the entire country except the extreme south; (4) *Callicebus* (titi monkey), a folivore-frugivore (Price and Piedade 2001) weighing 1.35 kg and inhabiting the Brazilian southeast, northeast and Amazon; and (5) *Callithrix* (marmoset), the smallest species at 0.30 kg. Neotropical marmosets feed on a large range of plant materials, including gums, fruits, and seeds, as well as animal preys (Correa et al. 2000). They are distributed along the Brazilian southeast, northeast and Amazon.

16.2.3 Compilation of Census Data

We compiled a list of census studies carried out between the years of 1993 and 2005 (Fig. 16.1; Table 16.1). In order to make the data comparable, we selected studies that used the line-transect technique (Buckland et al. 2001). This method basically consists of establishing transects distributed randomly or stratified according to habitat type and counting the number of individuals encountered. Information on straight-line distance to observed individuals is used to calculate the effective strip width (ESW) and estimate local density. Line-transect is considered one of the most precise census techniques and due to its simplicity and cost-effectiveness, it has been applied to census a broad range of animal and plant populations (Buckland et al. 2001). A total of 17 census studies using line-transect technique were found, and 16 were carried out within the Brazilian southeast. Out of those 16 studies, four were excluded: one study reported large within-site variation but did not provide separate density values for those sites (Hirsh et al. 1994); a second study was performed in a field site for which more recent information was available (Pinto et al. 1993); a third dataset (Chiarello 1993) reported extremely high density values for *Alouatta* in an urban park in São Paulo State. Preliminary models using this data point predict that all urban centers will have the highest howler densities. Although it is our intention to predict the impact of urbanization on primate densities, we believe that the conditions leading to the density value observed by Chiarello (1993) are probably tied to historical factors and latent variables that we are presently unable to measure. Last, we excluded data from Anchieta Island (Bovendorp and Galetti 2007) because this island has been a target of "repopulation" initiatives and several vertebrate species have been recently introduced in the area.

Table 16.2 List of GIS layers containing the independent variables used in the tree regression analysis

Variable Name	Units	Original Resolution (m)	Year(s) Data Collected	Ref.
Percent tree cover	%	500	2001	1
Mean annual temperature	Celsius * 10	800	1950–2000	2
Temperature seasonality	SD * 100	800	1950–2000	2
Total annual precipitation	mm	800	1950–2000	2
Precipitation seasonality	Coefficient of variation	800	1950–2000	2
Elevation	Meters	1000	various	3
Slope-based accessibility	Relative cost	1000	1996 (cities) and various (elevation)	3, 4
Road-based accessibility	Number of people/100,000	5000	1996 (census) and 2001 (roads)	4, 5
Industry	Number of units	Per city	1996	4
Crop area	Percent area devoted to permanent agriculture plots	Per city	1995	4
Median income	Median income for all people older than 10, in Reais	Per city	2000	4
Fragment size	Unitless (size classes from 1 to 6)	20	1999–2000	6

(1) Hansen et al. 2003; (2) Hijmans et al. 2005; (3) Danko 1992; (4) IBGE 1996; (5) DNIT 2007; (6) Eva et al. 2002.

16.2.4 GIS

For each primate genus, we obtained a grid map containing values of the dependent variable to be used in the tree regression, primate density (individuals/km^2). In order to locate study sites for which density information was available, we used a map of percent tree cover (Modis Vegetation Continuous Fields, Hansen et al. 2003), a forest inventory available for São Paulo state only (BIOTA 2006), and the figures available in the original publications. For large parks in São Paulo state, we used the location of transects buffered by a distance of 500 m. Data were pooled for small, contiguous fragments. Those fragments are (1) Sao Lourenço, Santa Lucia and Augusto Ruschi, and (2) M7 and Putiri, all of them in Espírito Santo state (*see* Chiarello 2003). In these cases, primate densities were averaged across fragments.

In addition, we obtained 12 grid maps representing the independent variables to be used in the tree regression (Table 16.2). Two grid maps are derived from least-cost path estimates used to model human movement across the landscape. The first one contains, for each cell, the number of people that can reach that cell when traveling by road for a maximum of 30 minutes. This was based on human census data for each municipality and a road network map. The model was built using the module Network Analyst within ArcGIS (ESRI, California). We assumed people departed city centroids and traveled along federal and state highways at a speed of 100 km/h. Since location of city streets and dirt roads was not available, it was assumed individuals leaving highways would travel to their final destinations along a straight line, at 50 km/h. A second grid map represents human accessibility, assuming people are moving by foot. The map contains the relative cost to reach each cell from the nearest city, assuming that cost is a function of distance and slope.

Although urban centers are obviously served by a large concentration of roads, some agricultural areas are also located near highways. In order to distinguish between these two land use types, we produced maps containing values of area devoted to agriculture, as well as degree of industrialization. In addition, a map of median income for each municipality (IBGE 1996) was produced in an attempt to obtain a surrogate for anthropogenic disturbance and/or hunting pressure. Last, forest fragments were mapped using a global land cover database (Eva et al. 2002). After excluding areas classified as "mosaic agriculture/degraded forest", the area for each fragment was calculated. We then assigned each cell with a value representing the size, in hectares, of the fragment where the cell is located. Six classes were used: (1) < 100; (2) > 100 and < 316; (3) >316 and < 1000; (4) >1000 and < 3162; (5) > 3162 and < 159,000; (6) > 159,000. All maps were rescaled to 500-m resolution. All GIS analyses were performed using ArcGIS 9.2 (ESRI, California). Map layers can be made available upon request to the first author.

16.2.5 Random Forest

The parameters used in the Random Forest run were: 3 independent variables (Table 16.2) could be used at each split; sampling was stratified, in such a way that all study areas were used to grow each tree; 500 trees were grown. After the model was run, we estimated the importance of all independent variables. Random Forest has two measures of variable importance: (i) mean percent increment in square error, calculated as the average increase in prediction error that results from shuffling the values of the predictor variable; (ii) percent increase in node impurity, the within-node variation (residual sum of squares) obtained after reshuffling values of the predictor variable (Breiman 2001; Prasad et al. 2006). Also, partial plots were constructed to study the relationship between the four most important environmental correlate and primate density. These plots are built by computing the relationship between the target predictor and the response averaged over the joint values of the other variables (Berk 2006). Last, the models were used with the entire range of values in the Brazilian southeast in order to predict density values for this region. All statistical analyses were performed in R (R Development Core Team 2007).

16.3 Results

16.3.1 General Aspects

When analyzing data for individual genera, we found no significant relationship between sampling effort (number of kilometers sampled) and density for *Alouatta* ($p = 0.76$), *Brachyteles* ($p = 0.47$), *Callicebus* ($p = 0.59$), *Cebus* ($p = 0.08$) or *Callithrix* ($p = 0.85$). For all genera, most sites were "coldspots" with lower densities and few sites were "hotspots." Within-genus variation in density reached three orders of magnitude for some genera: for *Alouatta*, density (individuals/km^2) ranges from 0.29 to 176.80 (mean $\pm$ SD: 23 $\pm$ 38, N $=$ 20). For *Brachyteles*, density ranged from 0.42 to 35.11 (9.63 $\pm$ 11.6, N $=$ 10). Density values for the genus *Cebus* ranged from 0.90 to 49.88 (16.63 $\pm$ 15.25, N $=$ 23). For *Callithrix*, density ranged from 1.83 to 110.3 (22.1 $\pm$ 29.4, N $=$ 10). Last, density for *Callicebus* ranged from 3.5 to 157 (24 $\pm$ 45.34, N $=$ 9). When comparing among all five genera, we did not observe any significant difference in mean density (Kruskal-Wallis rank-sum test, $p = 0.09$).

16.3.2 Determinants of Primate Density

A tree regression analysis using Random Forest was performed to study the effect of 12 variables (Table 16.2) on primate density. For all genera, the model was able to explain more than 90% of the variability in the training set (*see* Section 16.1.4.2). The output models produced by Random Forest were applied to the entire Brazilian southeast region (Fig. 16.2a–e). For all genera, the five most important predictors of primate density included precipitation and temperature, although genera responded differently to these climatic variables (Table 16.3). The five genera also displayed different responses to land use. For example, an increase in the area devoted to agriculture had a positive impact on the densities of *Callicebus* spp., but a negative impact on *Alouatta* spp.; also *Cebus* spp. displayed higher densities in the vicinity of industrialized cities (Table 16.3). In most cases, partial plots revealed a monotonic increase or decrease in primate density (shown as "+" or "−" on Table 16.3), but sometimes densities peaked at intermediate conditions (in this case, actual values are shown on Table 16.3). For example, density for *Callithrix* spp. was highest at intermediate values of median income and temperature (Table 16.3).

16.4 Discussion

16.4.1 Predicted Primate Density Hotspots

The analyses carried out in the present work enable us to tease apart the effects of anthropogenic impact and forest type on densities of primate species inhabiting a highly disturbed ecosystem. For all species, densities decreased with fragment size,

although this variable was not always an important predictor of primate density
(Table 16.3). Accessibility by road was not an important predictor of density for any
of the target genera (Table 16.3). Accessibility by foot was modeled as a function
of slope (*see* Methods) and had a positive impact on *Cebus* spp., *Callithrix* spp. and
Brachyteles spp. (Table 16.3), that is, areas considered accessible had higher primate densities. This variable is thus probably serving a substitute for slope. Overall,
results suggest that patterns of land use and social indicators from municipalities
where fragments are located provide better estimates of anthropogenic impact than
models of human movement.

For all genera, areas with non-zero predicted density extended beyond the distribution of the species used to train the model (Fig. 16.2a–e). This was expected

(a)   *Brachyteles*

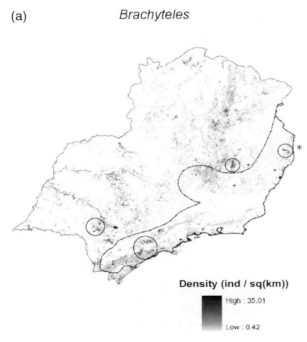

Density (ind / sq(km))

High : 35.01

Low : 0.42

Fig. 16.2 Map of predicted densities of primates in the Brazilian southeast. Species' ranges (from
Natureserve; www.natureserve.org) are delimited by an interrupted gray line, and locations of sampling points are shown by an arrow. (**a**) Muriquis (*Brachyteles* spp.). Circles show four predicated
hotspots of density (from west to east): semicididuous forest west of São Paulo, low rainforests
in São Paulo State, semideciduous forest in Minas Gerais, and low rainforests in Espírito Santo
muriguis are not found in the area indicated with a star (see discussion). (**b**) Howler monkeys
(*Alouatta* spp.). Ellipse indicates hotspots of density in semideciduous forest in Minas Gerais.
(**c**) Capuchin monkeys (*Cebus* spp.). Ellipses show four hotspots of density (from west to east):
semicididuous forest west of São Paulo state, low rainforests in São Paulo State, semideciduous
forest in Minas Gerais, and low rainforests in Espírito Santo. (**d**) Titi monkeys (*Callicebus* spp.).
Ellipses show predicted hotspots: Serra do Mar hill chain (*bottom*) and central Minas Gerais (*top*).
(**e**) Marmosets (*Callithrix* spp.). Circle indicates predicted hotspot in Espírito Santo forest

(b)   *Alouatta*

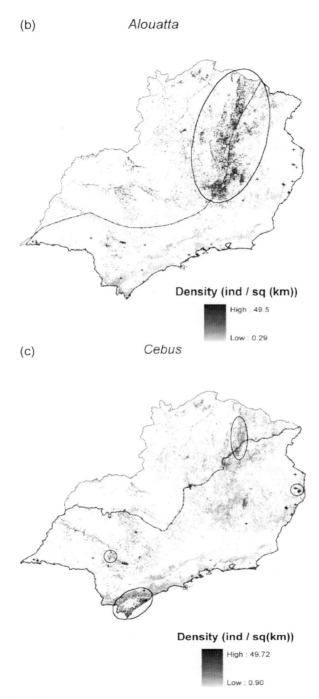

Fig. 16.2 (continued)

(d)   *Callicebus*

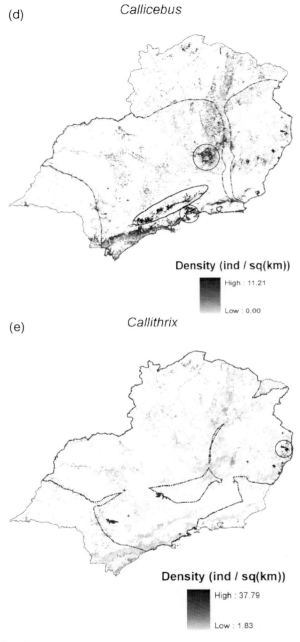

Fig. 16.2 (continued)

Table 16.3 List of the five most important determinants of primate density of five primate genera

Variable	Brachyteles	Alouatta	Callicebus	Cebus	Callithrix
Percent tree cover			−		
Slope-based accessibility	−				
Mean temperature		20°C	−		24°C
Variance in temperature	+			+	−
Variance in precipitation	+	+		+	
Precipitation		−	+	1400–1800 mm	1200 mm
Elevation	−			−	−
Median income	+		−		R$ 200–300
Industry				+	
Crop area		−	+		
Fragment size		3162–159000 ha			

given that models did not incorporate elements that can greatly influence range limits such as competition and historical factors. For *Callithrix* spp. and *Alouatta* spp., the Random Forest model most likely identified areas in the cerrado (the Brazilian savanna) with climate patterns similar to the Atlantic forest. The cerrado ecosystem is inhabited by primate species that have not been considered in our analyses but that nevertheless belong to the target genera studied here, such as *Callithrix penicillata* and *Alouatta caraya*. Accounts of species' ranges have changed over time, and the range maps shown here (www.natureserve.org) might not display the most current information. For example, *Callicebus* was considered present in the Paranapiacaba region (Rylands and Faria 1993; Hirsh et al. 2006), but is absent in this area (Mittermeier et al. 2008).

The largest genus, *Brachyteles*, did not display a clear preference for a particular forest type, as densities are predicted to be high in coastal zones as well as inland (Fig. 16.2a). Predicted hotspots are low, flat rainforest zones in São Paulo State as well as semideciduous forest in Minas Gerais and west of São Paulo (Fig. 16.2a). Fragments located in municipalities with low income displayed lower densities (Table 16.3). Income has been demonstrated to correlate with hunting pressure in other ecosystems (Shively 1997), although researchers differ in the procedure used to estimate income (Godoy et al. 2006) and many other factors such as employment stability might also play a large role in people's decision to consume wild meat (Siren et al. 2006). Rainforests in Espirito Santo (indicated by a star in Fig. 16.2a) are predicted hotspots. Still, these areas do not support murigui populations (Chiarello & Melo 2001). This suggests that historical factors or other unmeasured variables might be operating in Espirito Santo.

The most folivorous genus, *Alouatta*, showed a clear preference for areas with high precipitation seasonality, low annual precipitation, and high temperature seasonality (Table 16.3). Predicted hotspots are thus areas of semideciduous forest in Minas Gerais (Fig. 16.2b). This is in accordance with recent models developed for the Neotropics as a whole (Peres 1997), which showed that variation in density for *Alouatta* is largely governed by primary productivity. Fragments located in agricultural zones had lower *Alouatta* density, suggesting a negative effect of the landscape matrix that surrounds forest fragments, and/or that inhabitants of rural zones are more likely to engage in hunting activities.

The capuchin monkeys (*Cebus* spp.) showed a preference for areas with low elevation, high mean temperatures, high temperature seasonality, and high precipitation seasonality (Table 16.3). Industrialization had a positive impact on this genus, which is not surprising given its known diet flexibility and adaptability to urban habitats (Galetti and Pedroni 1994; Fragaszy et al. 2004). The hotspots for *Cebus* are low, flat areas in São Paulo and Espírito Santo, as well as semideciduous forests in São Paulo and Minas Gerais (Fig. 16.2c).

Densities for titi monkeys (*Callicebus* spp.) were higher in regions with relatively low mean temperatures, high precipitation, and in fragments embedded in agricultural zones (Table 16.3). The fact that titi monkey densities displayed a positive correlation with agriculture – as opposed to howlers – is interesting and exemplifies the importance of incorporating the landscape context on habitat suitability analyses. Although the mechanism driving these differences is not being examined here, it could be related to hunting pressure. Howlers are diurnal, extremely conspicuous species that forage in medium to large groups. On the other hand, titi monkeys are smaller canopy foragers that live in pairs, thus less likely to be spotted by poachers. The main predicted hotspots for *Callicebus* were the Serra do Mar hill chains in east São Paulo State, as well as central Minas Gerais (Fig. 16.2d). We also predicted high densities in the Paranapiacaba region, but more recent range maps for this genus indicate that it is absent in this area (Mittermeier et al. 2008).

Callithrix displays a preference for locations with intermediate values of climatic variables and income (Table 16.3). The highest estimated density values are associated with ranges of temperature and precipitation that compare favorably with studies done using presence-absence data for this genus (Grelle and Cerqueira 2006). As for the relationship between marmoset density and median income, it is possible that areas with low income have higher hunting pressure, whereas areas with high income also tend to be urbanized. In any case, social indicators proved to be better predictors of marmoset density than land use data. The predicted hotspots for *Callithrix* are the forests in Espírito Santo (Fig. 16.2e).

Overall, our analyses predict that semideciduous forests in Minas Gerais and São Paulo state have a large potential to support primate populations, despite the fact that most large forest tracts are located in the Serra do Mar and Serra da Paranapiacaba hill chains in coastal São Paulo.

16.4.2 Areas in Need of Future Research

Estimates of population sizes derived from the literature suggest that less than half of the study sites in the Brazilian southeast hold viable populations of the five genera studied here (Table 16.1). We assumed 500 individuals was the minimal viable population size (Franklin 1980), although some authors consider it an underestimate (Reed et al. 2003). In this scenario, more synthetic studies are needed to determine the drivers of primate abundance in the Brazilian Atlantic forest. We can identify three areas in need of future research. The first (and most obvious one) is the need to obtain more abundance data. Primate population studies are not yet available for states such as Santa Catarina, Rio de Janeiro, Paraná, and Bahia. These states still have large protected parks (e.g., Iguaçu, Itatiaia, Bocaina, Descobrimento, Una) that may hold large primate populations. The second issue arises in any comparative study: the need to evaluate whether some sites are more likely to violate the assumptions of the line-transect method (e.g., due to differences in forest type or topography).

Here, we encountered a main challenge when trying to scale up from local studies to landscape-scale predictions. Usually, census studies report one density estimate per forest fragment. This limits prediction in two ways: (i) extrapolation to neighboring fragments will be highly dependent on the choice of method to delineate fragment boundaries, and (ii) within-fragment environmental variation can be comparable to between-fragment variation. For example, Jacupiranga State Park covers an area of approximately $1552\,km^2$. In this park, slope ranges between 0 and 44%, whereas values for the entire study area range between 0 and 63%. The availability of density values associated with smaller, homogeneous areas (e.g., Buss 2001) should help bridge the gap between field studies and ecological modeling.

Finally, we found that variables such as land use and social indicators can serve as surrogates of anthropogenic impact. However, we are presently unable to tease apart the effects of hunting pressure and habitat disturbance. A wealth of socio-economic data is published by IBGE, the Brazilian Institute for Geography and Statistics (www.ibge.gov). If direct estimates of hunting pressure are made available, it would be possible to select the variables that more strongly correlate with hunting pressure.

16.5 Summary

In the present work, we focused on southeast Brazil's Atlantic forest and studied five primate genera: *Alouatta*, *Brachyteles*, *Callithrix*, *Callicebus*, and *Cebus*. After data were compiled from census studies that used the line-transect method, we applied regression trees in order to search for determinants of variation in primate density. Owing to its location in Brazil's most developed region, the Atlantic forest is not only highly fragmented, but also embedded in a landscape matrix encompassing a wide range of land use types and social contexts. Thus, the independent variables used in the regression analyses included not only surrogates of forest type (e.g., climate) and fragment size, but also data on social indicators and estimates of accessibility derived from human movement models. For all genera, we found that density was strongly influenced by forest type, and not influenced by our accessibility estimates. Interestingly, genera differed in their responses to land use and social indicators, a result that emphasizes the importance of incorporating information on the landscape matrix when performing habitat suitability analyses. The regression models produced here were used to construct maps of predicted primate density for the Brazilian southeast. Overall, the maps for all genera showed high predicted primate densities for the inland semideciduous forests, where primary productivity is expected to be higher. Finally, we suggest that more synthetic work is needed in our study area, and list a few topics in need of research.

Acknowledgments The program BIOTA FAPESP (2001/14463-5) funded the Laboratório de Biologia da Conservação (through a grant for MG) and the census in the protected areas in São Paulo State. MG receives a research fellowship from CNPq. The Teresa Lozano Long Institute of Latin American Studies (LILAS) at Austin, TX for promoting the exchange program between MG and THK. Dr. C. Joly and F. Kronka kindly provided the digital map of vegetation types for São Paulo State. S.K. Gobbo, R.M. Marques, C. Steffler, P. Rubim, and R. Nobre provided unpublished data from their study areas.

References

Aerts, R. 1996. Nutrient resorption from senescing leaves of perennials: are these general patterns? Journal of Ecology 84, 597–608.

Alves, A.C.R., Ribeiro L.F., Haddad C.F.B., and dos Reis, S.F. 2006. Two new species of *Brachycephalus* (Anura: Brachycephalidae) from the Atlantic forest in Parana state, southern Brazil. Herpetologica 62, 221–233.

Berk, R. A. 2006. An introduction to ensemble methods. Sociological Methods and Research 34, 263–295.

Bernardo C.S., and Galetti, M. 2004. Densidade e tamanho populacional de primatas em um fragmento florestal no sudeste do Brasil. Revista Brasileira de Zoologia 21, 827–832.

BIOTA 2006. Atlas Sinbiota: http://sinbiota.cria.org.br/index.

Bovendorp, R.S., and Galetti, M. 2007. Density and population size of mammals introduced on a land-bridge island in southeastern Brazil. Biological Invasions 9, 353–357.

Brashares, J.S., Arcese, P., and Sam, M.K. 2001. Human demography and reserve size predict wildlife extinction in West Africa. Proceedings of the Royal Society of London Series B-Biological Sciences 268, 2473–2478.

Breiman, L. 2001. Random Forests. Machine Learning 45, 5–32.

Brooks, T.M., Mittermeier, R.A., Mittermeier, C.G., da Fonseca, G.A.B, Rylands, A.B., Konstant, W.R., Flick, P., Pilgrim, J., Oldfield, S., Magin, G., and Hilton-Taylor, C. 2000. Habitat loss and extinction in the hotspots of biodiversity. Conservation Biology 16, 909–923.

Brown, J.H., Mehlman, D.W., and Stevens, G.C. 1995. Spatial variation in abundance. Ecology 76, 2028–2043.

Buckland, S.T., Anderson, D.R., Burnham, K.P., Laake, J.L., Borchers, D.L. and Thomas, L. 2001. *Introduction to Distance Sampling*, Oxford University Press, Oxford.

Buss, G. 2001. Estudo da densidade populacional do bugio-ruivo *Alouatta guariba clamitans* (Cabrera, 1940) (Platyrrhini, Atelidae) nas formacoes florestais do morro do campista, Parque Estadual de Itapua, Viamao, RS. Master Thesis. Universidade Federal do Rio Grande do Sul, Brazil.

Câmara, I.G. 2003. Brief history of conservation in the Atlantic forest. In: I. Gusmao-Camara and C. Galindo-Leal (Eds.), *The Atlantic Forest of South America. Biodiversity Status, Threats, and Outlook*. Island Press, Washington, pp. 31–42.

Chiarello, A.G. 1993. Home range of the brown howler monkey, *Alouatta fusca*, in a forest fragment in southeastern Brazil. Folia Primatologica 60, 173–175.

Chiarello, A.G. 2000. Density and population size of mammals in remnants of Brazilian Atlantic Forest. Conservation Biology 14, 1469–1657.

Chiarello, A.G. 2003. Primates of the Brazilian Atlantic forest: the influence of forest fragmentation on survival. In: L.K. Marsh (Ed.), *Primates in Fragments. Ecology and Conservation*. Kluwer Academic, New York, pp. 99–121.

Chiarello, A.G., and Melo, F.R. 2001. Primate population densities and sizes in Atlantic forest remnants of northern Espírito Santo, Brazil. International Journal of Primatology 22, 379–395.

Coley, P.D. 1983. Herbovory and defense characteristics of tree species in a lowland tropical forest. Ecological Monographs 53, 209–233.

Correa, H.K.M., Coutinho, P.E.G., and Ferrari, S.F. 2000. Between-year differences in the feeding ecology of highland marmosets (*Callithrix aurita* and *Callithrix flaviceps*) in south-eastern Brazil. Journal of Zoology 252, 421–427.

Cosenza, B.A.P., and Melo, F. 1998. Primates of the Serra do Brigadeiro State Park, Minas Gerais, Brazil. Neotropical Primates 6, 18–20.

Cullen Jr., L., Bodmer, R.E., and Valladares-Padua, C.V. 2001. Ecological consequences of hunting in Atlantic forest patches, São Paulo, Brazil. Oryx 35, 137–144.

Danko, D.M. 1992. Digital chart of the world. GeoInfo Systems 29–36.

Dean, W. 1996. *A Ferro e Fogo: A História e a Devastação da Mata Atlântica Brasileira*. Companhia das Letras, São Paulo.

DNIT 2007. Atlas das estradas do Brasil. www.dnit.br

Donha, C.G., and Eliasaro, S. 2006. Two new species of *Parmotrema* (Parmeliaceae, lichenized ascomycota) from Brazil. Mycotaxon 95, 241–245.

Eva, H.D., Miranda, E.E., Bella, C.M., Gond, V., Huber, O., Sgrenzaroli, M., Jones, S., Coutinho, A., Dorado, A., Guimaraes, M., Elvidge, C., Achard, F., Belward, A.S., Bartholome, E., Baraldi, A., Grandi,G., Vogt, P., Fritz, S., and Hartley, A. 2002. A vegetation map of South America. EUR 20159, European Commission, Joint Research Centre.

Fragaszy, D.M., Visalberghi, E., and Fedigan, L.M. 2004. Behavioral ecology: how to capuchins make a living? In: D.M. Fragaszy, E. Visalberghi, L.M Fedigan (Eds.), *The Complete Capuchin: the Biology of the Genus Cebus*. Cambridge University Press, Cambridge, UK, pp. 36–54.

Franklin, I.R. 1980. Evolutionary change in small populations. In: M.E. Soule and B.A. Wilcox (Eds.), *Conservation Biology: an Evolutionary-Ecological Perspective*. Sinauer, Sunderland, MA, pp. 135–150.

Galetti, M., and Pedroni, F. 1994. Diet of capuchin monkeys (*Cebus apella*) in a semideciduous forest in South-east Brazil. Journal of Tropical Ecology 10, 27–39.

Garzon, M.B., Blazek, R., Neteler, M., Sanchez de Dios, R., Ollero, H.S., and Furlanello, C. 2006. Predicting habitat suitability with machine learning models: the potential area of *Pinus silvestris* L. in the Iberian peninsula. Ecological Modeling 197, 383–393.

Glander, K.E. 1978. Howler monkey feeding behavior and plant secondary compounds: a study of strategies. In: G.C. Montgomery (Ed.), *The Ecology of Arboreal Folivores*. Smithsonian Press, Washington, pp. 561–574.

Godoy, R., Wilkie, D.S., Reyes-Garcia, V., Leonard, W.R., Huanca, T., McDade, T., Valdez, V., and Tanner, S. 2006. Human body-mass index (weight in kg/stature in m(2)) as a useful proxy to assess the relation between income and wildlife consumption in poor rural societies. Biodiversity and Conservation 15, 4495–4506.

Grelle, C.E., and Cerqueira, R. 2006. Determinantes da distribuicao geografica de *Callithrix flaviceps* (Thomas) (Primates, Callitrichidae). Revista Brasileira de Zoologia 23, 414–420.

Hansen, M.R., DeFries, R., Townshend, J. R., Carroll, M., Dimiceli, C., and Sohlberg, R. 2003. Vegetation Continuous Fields, MOD44B, 2001 percent tree cover, collection 3. University of Maryland, College Park, Maryland.

Hijmans, R.J., Cameron, S.E., Parra, J.L., Jones, P.G., and Jarvis, A. 2005. Very high resolution interpolated climate surfaces for global land areas. International Journal of Climatology 25, 1965–1978.

Hirota, M.M. 2003. Monitoring the Brazilian Atlantic forest cover. . In: I. Gusmao-Camara and C. Galindo-Leal (Eds.), *The Atlantic Forest of South America. Biodiversity Status, Threats, and Outlook*. Island Press, Washington, pp. 60–65.

Hirsh, A. 1995. Censo de *Alouatta fusca* Geoffroy, 1812 (Platyrrhini, Atelidae) e qualidade do habitat em dois remanescentes de Mata Atlantica em Minas Gerais. Master Thesis. Universidade Federal de Minas Gerais.

Hirsh, A., Dias, L.G., Martins, L.O., Resende, N.A.T. and Landau, E.C. 2006. Database of Georeferenced Occurrence Localities of Neotropical Primates. Department of Zoology, UFMG, Belo Horizonte. http://www.icb.ufmg.br/zoo/primatas/home_bdgeoprim.htm

Hirsh, A., Subira R., and Landau, E.C. 1994. Levantamento de primatas e zoneamento das matas da regiao do Parque Estadual do Ibitipoca, Minas Gerais, Brasil. Neotropical Primates 2, 4–6.

IBGE 1996, 2000. Instituto Brasileiro de Geografia e Estatistica. www.ibge.gov.br

IUCN 2006. Red List of Threatened Species. http://www.iucn.org/themes/ssc/redlist.htm

Laurance, W.F., Croes, B.M., Tchignoumba, L., Lahm, S.A., Alonso, A., Lee, M.E., Campbell, P., and Ondzeano, C. 2005. Impacts of roads and hunting on central African rainforest mammals. Conservation Biology 20, 1251–1261.

Levin, SA 1992. The problem of pattern and scale in ecology. Ecology 73, 1943–1967.

Liaw, A., and Wiener, M. 2002. Classification and regression by random Forest. R News 2, 18–22.

Marsden, S.J., Whiffin, M., Galetti, M. and Fielding, A.H. 2005. How well will Brazil's system of atlantic forest reserves maintain viable bird populations? Biodiversity and Conservation 14, 2835–2853.

Martins, M.M. 2005. Density of primates in four semideciduous forest fragments of São Paulo, Brazil. Biodiversity and Conservation 14, 2321–2329.

Mendes, S.L. 1989. Estudo ecologico de *Alouatta fusca* (Primates: Cebidae) na Estacao Ecologica Caratinga. Revista Nordestina de Biologia 6, 71–104.

Milton, K. 1984. Habitat, diet, and activity patterns of free-ranging woolly spider monkeys (*Brachyteles arachnoides* E. Geoffroy 1806). International Journal of Primatology 5, 491–514.

Michalski, F., and Peres, C.A., 2005. Anthropogenic determinants of primate and carnivore local extinctions in a fragmented forest landscape in southern Amazonia. Biological Conservation 124, 383–396.

Mittermeier R., Coimbra-Filho, A.F., Kierulff, M.C.M., Rylands, A.B., Pissinatti, A, and Almeida, L.M. 2008. Monkeys of the Atlantic Forest of Eastern Brazil. Pocket Identification Guide. Conservation International, Brazil.

Oliveira-Filho, A.T., and Fontes, M.A.L. 2000. Patterns of floristic differentiation among Atlantic forests in southeastern Brazil and the influence of climate. Biotropica 32, 793–810.

Pang, H., Lin, A., Holford, M., Enerson, B.E., Lu, B., Lawton, M.P., Floyd, E., and Zhao, H. 2006. Pathway analysis using random forests classification and regression. Bioinformatics 22, 2028–2036.

Peres, C.A. 1997. Effects of habitat quality and hunting pressure on arboreal folivore densities in Neotropical forests: a case study of howler monkeys (*Alouatta spp.*). Folia Primatologica 68, 199–222.

Pinto, L.P.S., Costa, C.M.R., Strier, K.B., and Fonseca, G.A.B. 1993. Habitat, density, and group size of primates in a Brazilian tropical forest. Folia Primatologica 61, 135–143.

Pontes, A.R.M., Malta, A., and Asfora, P.H. 2006. A new species of capuchin monkey, *Cebus* (Cebidae, Primates): found at the very brink of extinction in the Pernambuco endemis centre. Zootaxa 1200, 1–12.

Prasad, A.M., Iverson, L.R. and Liaw, A. 2006. Newer classification and regression tree techniques: bagging and random forests for ecological prediction. Ecosystems 9, 181–199.

Price E.C., and Piedade H.M. 2001. Diet of northern masked titi monkeys (*Callicebus personatus*). Folia Primatologica 72, 335–338.

R Development Core Team 2007. R: a language and environment for statistical computing. http://www.R-project.org

Reed, D.H., O'Grady, J.J., Brook, B.W., Ballou, J.D., and Frankham, R. 2003. Empirical estimates of minimum viable population sizes for vertebrates and factors influencing those estimates. *Biological Conservation* 113, 23–34.

Romanini de Oliveira, R.C., Coelho, A.S. and Melo, F.R. 2003. Estimativa de densidade e tamanho populacional de saua (*Callicebus nigrifons*) em um fragmento de mata em regeneracao, Vicosa, Minas Gerais. Neotropical Primates 11, 91–94.

Rylands, A.B., and Faria, D.S. 1993. Habitats, feeding ecology, and home range size in the genus *Callithrix*. In: A.B. Rylands (Ed.), *Marmosets and Tamarins: Systematics, Behavior, and Ecology*. Oxford University Press, Oxford, UK, pp. 262–272.

Rylands, A.B., Coimbra-Filho, A.F., and Mittermeier R.A. 1993. Systematics, geographic distribution, and some notes on the conservation status of the Callitrichidae. In: A.B. Rylands (Ed.), *Marmosets and Tamarins: Systematics, Behavior, and Ecology*. Oxford University Press, Oxford, UK, pp. 11–77.

Shively, G.E. 1997. Poverty, technology, and wildlife hunting in Palawan. Environmental Conservation 24, 57–63.

Siren, A.H., Cardenas, J.C., and Machoa, J.D. 2006. The relationship between income and hunting in tropical forests: an economic experiment in the field. Ecology and Society 11, 44.

Strier, K.B. 1991. Diet in a group of woolly spider monkeys, or muriquis (*Brachyteles arachnoides*) American Journal of Primatology 23, 113–126.

Strier, K.B. 2000. Population viabilities and conservation implications for muriquis (*Brachyteles arachnoides*) in Brazil's Atlantic forest. Biotropica 32, 903–913.

Trevelin, L.C. 2006. Aspectos da ecologia do saua (*Callicebus nigrifons*) no Parque Estadual da Cantareira, SP. Master Thesis. Universidade Estadual Paulista Julio de Mesquita Filho (UNESP).

Wiens, J. A. 1989. Spatial scaling in ecology. Functional Ecology 3, 385–397

Chapter 17
Ecological and Anthropogenic Influences on Patterns of Parasitism in Free-Ranging Primates: A Meta-analysis of the Genus *Alouatta*

Martin M. Kowalewski and Thomas R. Gillespie

17.1 Introduction

Parasites play a central role in tropical ecosystems, affecting the ecology and evolution of species interactions, host population growth and regulation, and community biodiversity (Esch and Fernandez 1993; Hudson, Dobson and Newborn 1998; Hochachka and Dhondt 2000; Hudson et al. 2002). Our understanding of how natural and anthropogenic factors affect host-parasite dynamics in free-ranging primate populations (Gillespie, Chapman and Greiner 2005a; Gillespie, Greiner and Chapman 2005b; Gillespie and Chapman 2006) and the relationship between wild primates and human health in rural or remote areas (McGrew et al. 1989; Stuart et al. 1990; Muller-Graf, Collins and Woolhouse 1997; Gillespie et al. 2005b; Pedersen et al. 2005) remain largely unexplored. The majority of emerging infectious diseases are zoonotic – easily transferred among humans, wildlife, and domesticated animals – (Nunn and Altizer 2006). For example, Taylor, Latham and Woolhouse (2001) found that 61% of human pathogens are shared with animal hosts. Identifying general principles governing parasite occurrence and prevalence is critical for planning animal conservation and protecting human health (Nunn et al. 2003). In this review, we examine how various ecological and anthropogenic factors affect patterns of parasitism in free-ranging howler monkeys (Genus *Alouatta*).

17.1.1 Evidence of the Relationships Between Howlers and Parasitic Diseases in South America

The genus *Alouatta* is the most geographically widespread non-human primate in South America, with 8 of 10 *Alouatta* species ranging from Northern Colombia

M.M. Kowalewski (✉)
Estacion Biologica Corrientes-MACN, Corrientes, Argentina; Department of Anthropology, University of Illinois, Urbana-Champaign, IL, USA
e-mail: mkowalew@illinois.edu

P.A. Garber et al. (eds.), *South American Primates,* Developments in Primatology: Progress and Prospects, DOI 10.1007/978-0-387-78705-3_17,
© Springer Science+Business Media, LLC 2009

to Argentina (Cortés-Ortiz et al. 2003). Howlers are classified as colonizers (Eisenberg 1972; Crockett 1998) due to their ability to adapt and survive in modified environments (Clarke et al. 2002; Bicca-Marques 2003; Zunino et al. 2007). In addition, in contrast to other New World primate taxa, data on patterns of parasitism in free-ranging howlers are available from variable environments throughout the geographic range of the Genus. This availability of published data on patterns of parasitism, coupled with their ability to survive in variable environments including those that bring them into increasing contact with human communities, make howlers an excellent model to study the dynamics of infectious disease transmission among wild primates, humans and domestic animals.

Howlers are known to be host to bacteria, protozoa, viruses, fungi, helminthes, and arthropods that also infect livestock and humans (Stuart et al. 1998). This shared susceptibility to infection has the capacity to lead to cross-species transmission in disturbed forest systems where howlers experience higher temporal and spatial overlap with livestock and humans.

Howler habitation of forest patches within plantations and cattle pastures or in close proximity to human settlements (Cabral et al. 2005; Estrada et al. 2006; Muñoz et al. 2006); frequent terrestrial travel (Young 1981; Kowalewski, Zunino and Bravo 1995; Delgado 2006; Pozo-Montuy and Serio-Silva 2007); and drinking from rivers and lagoons (Gilbert and Stouffer 1989; Bravo and Sallenave 2003), all increase opportunities for cross-transmission. In this chapter we integrate and compare data from studies on wild *Alouatta caraya, A. seniculus, A. guariba*, and *A. belzebul* inhabiting areas of undisturbed continuous forests and areas characterized by forest fragmentation and other anthropogenic pressure. We use these data to test a series of hypotheses concerning the relationship between habitat attributes and patterns of parasitism, and we consider how these processes may affect howler metapopulations.

17.1.2 Problems with Data Availability on Primates and Parasitic Diseases in South America

Information on parasites in New World primates is extremely fragmentary. Reviews have been published on the presence/absence of parasites in howlers, or with reference to howlers as parasite hosts (*see* Yamashita 1963; Diaz Ungria 1965; Thatcher and Porter 1968; Stuart et al. 1998; Stoner et al. 2005). However, several of these reviews include data on captive or semi-free ranging howlers. Such parasite lists are useful, but are unlikely to reflect the full range of host-parasite interactions in wild populations. Additional data on parasite prevalence in South American primates come from (a) examination of monkeys relocated or rescued from areas flooded by dam projects (*see* Fandeur et al. 2000; Volney et al. 2002; Duarte et al. 2006), (b) biomedical studies focusing on parasites that produce critical economic losses to human populations such as malaria, yellow fever and toxoplasmosis (Kumm and Laemmert 1950; Deane 1992; Lourenco de Oliveira and Deane 1995; Volney et al. 2002; Vasconcelos et al. 2003; Garcia et al. 2005), and (c) studies of a limited number of primate social groups to compare prevalence and presence of parasites in relation to habitat fragmentation (*see* Gilbert 1994; Santa-Cruz et al. 2000a,b; Godoy et al. 2004; Martins 2002; Delgado 2006; Kowalewski and Santa-Cruz, unpub. data).

17.1.3 *Effect of Deforestation on Parasite Infections*

Previous studies have shown that disturbance may alter the dynamics of parasite transmission (Gillespie et al. 2005a; Gillespie and Chapman 2006). A direct consequence of deforestation and increased fragmentation is the modification of forest structure and composition (Johns and Skorupa 1987; Plumptre and Reynolds 1994; Marsh 2003; Norconk and Grafton 2003; Rivera and Calme 2006). Selective logging is associated with the disappearance of species of economic value that may also be important in the diet of primates (Kowalewski and Zunino 1999, Gillespie et al. 2005a). Clear cutting of forests reduces the area of forest coverage drastically. Both selective logging and clear cutting allow the invasion of secondary forest, modifying forest composition and structure (Norconk and Grafton 2003; Zunino et al. 2007). These changes may result in dietary and nutritional stress on primate foragers, negatively affecting immune response and leaving individuals more susceptible to both parasitic infections and infectious diseases (Milton 1996; Solomons and Scott 1994; Chapman et al. in press). It has been shown that some species of primates living in logged forest have higher parasite prevalence and diversity (i.e., *Cercopithecus ascanius* [Gillespie et al. 2005a]; *Alouatta palliata* [Stoner 1996]; *Alouatta caraya* [Santa Cruz et al. 2000a]). Increasing contact between humans and primates and reduction of primate ranging areas result in increasing probabilities of infections for both humans and primates (Gillespie 2004, Stoner and Di Perro 2006)

17.1.4 *Goals of This Study*

A more detailed understanding of the relationship between patterns of parasitism and attributes of primate habitat will allow for more effective primate conservation and safeguarding of human and animal health. In this chapter, we take a meta-analysis approach to examine how various factors such as latitude, type of forest, altitude, annual precipitation, and degree of contact with human settlements affect parasite prevalence in populations of howler monkeys inhabiting different sites across South America. Some of the studies included in our analysis contain data on the prevalence of gastrointestinal, blood and ectoparasites of different parasite species studied at the same study site during different seasons or years (*see* Appendix 1). These comparisons are therefore not independent. However, we feel that to exclude non-independent comparisons may bias or limit our results more than their inclusion (Hedges and Olkin 1985; Gurevitch et al. 1992; Poulin 1994). Therefore we considered the studies in Appendix 1 in our analysis in order to answer the following questions: Do certain habitat features affect parasite prevalence and diversity in non-human primates across South America? Does the degree of contact between humans and nonhuman primates affect parasite prevalence in wild primates?

We also compare our howler-specific results with those of Nunn et al. (2005). In a comparison of 330 parasite species from 119 nonhuman primate species hosts, protozoan, but not helminth or virus, species richness was negatively correlated with distance from the equator. They argued that this effect may be caused by a greater abundance of arthropods serving as intermediate hosts in the tropics, as well as climate effects on both vectors and parasites. For example, warmer latitudes are associated with higher vector biting rates and more rapid parasite development (Liang, Linthicum and Gaydos 2002; Nunn et al. 2005).

17.2 Methodology

17.2.1 Meta-analysis: Parameters Used

We conducted a systematic review of published literature, dissertations, and personal communications with primate researchers in the field. Appendix 1 provides the following data: species of howler, study site, latitude of the study site, altitude and average annual rainfall, prevalence, type of parasite (blood, gastrointestinal or ectoparasites) and species (genera or group) of parasites found (we use species of parasites when possible, many studies provide only genera, family, or group of parasites present). Also included for each study site is forest type (fragmented or continuous); type of human contact; and if the parasite is present at any stage of its cycle in humans, domesticated dogs and cats, and/or cattle.

We categorized forests as either fragmented or continuous based on the definition provided in Marsh (2003). As such, fragments are defined as small (1–10 ha), medium (10–100 ha), and large (100–1,000 ha). Due to our limited database, we considered any forest with less than 1,000 ha fragmented and forests over 1,000 ha continuous. Types of human contact were classified as one of three types: (1) remote: site almost or totally isolated from humans (we included sites without nearby human settlement, but we cannot exclude the possibility that humans go into the forest for hunting or other subsistence activities), (2) rural: close to rural population or fishing camps or regularly visited by people (we consider rural sites when human settlements were located nearby assuming that interaction between humans and the forest was maintained on a regular basis, through, for example, selective logging, shade for cattle, trails, and hunting); and (3) urban: close to dense human populations, and intensely visited by tourists or local people – i.e., non-protected areas, certain natural preserves or national parks. The classification of types of human contact was based on the literature and information obtained from authors when necessary. We focused principally on studies in which more than 30 animals were sampled (however we have included three representative studies with less than 30 samples, considering them representative of the population under study). In Appendix 1 we provide the total number of samples collected in the study. Prevalence is defined as the number of individuals of the host species (howlers of different species in this case) with a particular parasite species divided by the number of hosts or samples analyzed, expressed as a percentage.

We divided the analysis into two sections. In the first section, we explore the effect of certain habitat attributes on the prevalence and diversity of parasite species (considering parasites collectively, only gastrointestinal parasites, and only blood parasites). In the second part of the analysis, we search for the most important habitat attributes that may influence parasite prevalence.

17.2.2 Meta-analysis: Statistical Analysis I

We used Kruskal-Wallis tests to study the effect of habitat type and degree of contact with humans on prevalence values (Zar 1999). We also explored associations between number of parasites species in each study site and latitude, altitude, and precipitation using non parametric Spearman correlation tests (Zar 1999). We ran analyses considering parasites collectively, and considering only blood parasites and only gastrointestinal parasites. Differences in type of collection (Appendix 1) were not further considered in the analyses as there were no statistical differences among prevalence values obtained through blood samples, fecal samples, and animals killed and examined (Kruskal Wallis Test: $N = 91$, $H = 1.09$, $P > 0.05$). We caution that our analyses integrate studies conducted at different scales. However, we believe that these results indicate the direction of important trends in the relationship between patterns of parasitism and certain environmental factors. We found only 4 studies that included ectoparasites -in *Alouatta caraya*- (Appendix 1), so we did not run separate tests for ectoparasites. To control for uneven sampling effort, we used the residuals from a linear regression of parasite species diversity or parasite prevalence (depending on the analysis) against sample size (Altizer et al. 2003). All tests are with alpha set at 0.05.

17.2.3 Meta-analysis: Statistical Analysis II

We ran a logistic regression (Johnson 1998) to identify the main variables influencing high and low parasite prevalence for howler populations considered. Prevalence was the categorical response variable, and latitude, altitude, type of contact, and precipitation were predictor variables. Dummy variables were generated for any categorical variable with more than one level (a separate dummy variable was generated to represent each of the categories for categorical variables, then the value of the dummy variable is 1 if the variable has that category, and the value is 0 if the variable has other category). We considered the value of 20% as an indicator of high or low prevalence values (response categorical variable) for different parasite species (in a histogram of frequencies, 20% was the value of the cut-off for the distribution of prevalence). Type of habitat was discarded in the analysis due to co-linearity. Backward elimination was used to determine which factors could be dropped from the multivariable model (Hosmer and Lemeshow 1989). The level of significance for a factor to remain in the final model was set at 5%.

17.2.4 Parasites Present in Howlers, Humans, and Domestic Animals

For each species or group of parasites found in a given howler species, we examined the literature (i.e., Acha and Szyfres 1986; Palmer et al. 1998; Pollard and Dobson 2000) to determine if the same species or group was present in humans, dogs or cats, and cattle. We compared the parasite diversity of each group to investigate the potential risk of cross- transmission of parasites among primates, humans and domestic animals.

17.3 Results

17.3.1 Factors Affecting Parasite Prevalence and Number of Parasite Species in Howler Monkeys

When parasites were analyzed collectively (gastrointestinal, blood, and ectoparasites), type of human contact affected the prevalence of different parasites (Kruskal-Wallis Test: $N = 91$, $H = 8.77$, $P < 0.05$; the same trend was found when comparing only gastrointestinal parasites Kruskal-Wallis Test: $N = 65$, $H = 9.34$, $P < 0.05$; no trend was found with blood parasites Kruskal- Wallis Test: $N = 20$, $H = 0.48$, $P > 0.05$). For example, Fig. 17.1 illustrates differences in average prevalence in groups of parasites in remote and rural areas. Bacteria and protozoa were higher in rural areas than in remote areas. In contrast average prevalence of helminths was slightly higher in remote areas than in rural areas.

We did not find any effect of forest type (fragmented vs. continuous) on prevalence of parasites considered collectively (Kruskal-Wallis Test: $N = 86$, $H = 0.18$, $P > 0.05$), on prevalence of gastrointestinal parasites (Kruskal Wallis Test: $N = 65$, $H = 0.14$, $P > 0.05$), or on prevalence of blood parasites (Kruskal Wallis Test: $N = 21$, $H = 0.02$, $P > 0.05$).

The relationship between gastrointestinal parasite diversity at a study site and average annual precipitation was positive and significant (Spearman correlation, $N = 10$, $r = 0.72$, $P = 0.018$, Fig. 17.2). However, no trend was found when comparing gastrointestinal parasite diversity with altitude or latitude ($P > 0.05$). No relationship was found when we consider blood parasite diversity ($P > 0.05$). This last result may be due to limited sample size ($N = 6$).

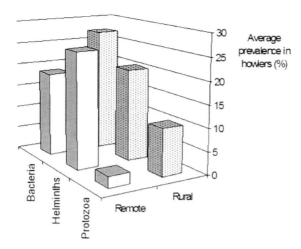

Fig. 17.1 Average prevalence of parasite groups (Bacteria, Helminths, Protozoa) in areas with different degree of contact with human settlements. A site was considered remote if the site was almost or totally isolated from humans and rural if the site was close to rural population or fishing camps, or regularly visited by people (*see* text for details)

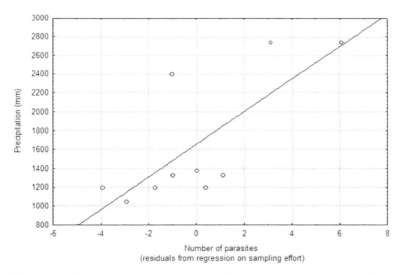

Fig. 17.2 Relationship between number of species of gastrointestinal parasites (residuals from regression on sampling effort) and precipitation measured in millimeters per year (mm) (from Appendix 1) (Spearman correlation test $N = 10$, $r = 0.72$, $p = 0.018$)

17.3.2 Looking for Predictors of Parasite Prevalence in Howler Monkeys

We used logistic regression analysis to investigate the relationship between environmental variables and the probability of transmission. Our model predicted 50% of

Table 17.1 Results from the logistic regression between high and low prevalence using the variables latitude and altitude

Variable	Estimate	p-value	Odds ratio	95% CI
Latitude	−382	0.0024	<0.001	<0.001–<0.001
Altitude	−0.0036	0.0072	0.996	0.994–0.999

the variance in parasite prevalence in the high prevalence category and 86.44% in the low prevalence category. These results indicate that latitude and altitude are the best predictors of level of prevalence. Based on odds ratio statistics, parasite prevalence is more likely to be high in sites at lower latitudes and altitudes (Table 17.1). Overall, the likelihood of having high parasite prevalence for howler populations used in this review was not related to average annual precipitation values (Wald Chi-square = 0.026, P > 0.05), as the inclusion of precipitation did not significantly improve the model. Finally we found that different kinds of contact with humans have equal risks of affecting prevalence (constituting a bad predictor of high or low prevalence). Similar results were found when we used only gastrointestinal parasites in the analysis.

Figure 17.3 illustrates a summary of the results from the meta-analysis. Factors such as high latitude, precipitation, humidity, degree of contact with human settlements, and low altitude seem to be associated to higher levels of parasite prevalence and number of species of parasites.

17.3.3 Parasites Shared by Howlers, Humans, and Domestic Animals

We examined how often observed howler parasites were zoonotic (producing infectious diseases in humans, domesticated dog and cats, and/or cattle; Fig. 17.4). In the case of gastrointestinal parasites species presented in this review (n = 29), we found that (1) parasites present in howlers could be potentially transmitted to humans in

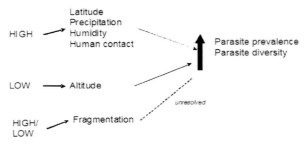

Fig. 17.3 Potential factors that could affect parasite prevalence and diversity in howler monkeys across South America. The figure was constructed based on the results obtained in this chapter. The predictions presented in the figure are limited because we considered several groups of parasites together but indicate trends across South America

Gastrointestinal parasites

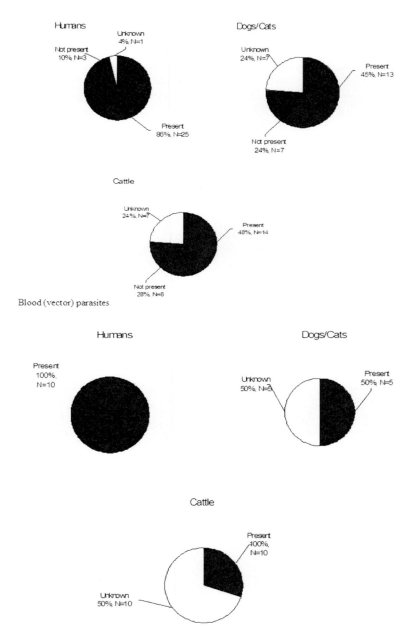

Fig. 17.4 Percentage of gastrointestinal parasite and blood (vector borne transmitted) parasite species found in howlers that also can be found in humans, domesticated dogs and cats and cattle

86% of the cases, they were never found in humans in 10%, and we could not find information on the parasites species and its relationship with other hosts in 4% of cases; (2) parasites present in howlers could be potentially transmitted to domesticated cats and dogs in 45% of cases, they were never found in domesticated cats and dogs in 31%, and we could not find data on transmission to cats and dogs in 24% of cases; and (3) parasites present in howlers could be potentially transmitted to cattle in 48% of cases, were never found in cattle in 28%, and we could not find data on transmission to cattle in 24% of cases. Some examples of potential zoonotic gastrointestinal parasites of howlers were *Ascaris* sp., *Fasciola* sp., *Giardia* sp., and *Trichuris* sp. (Appendix 1).

Considering blood-borne parasites (transmitted by insect vectors), we found (n = 10) that (1) parasites present in howlers could be potentially transmitted to humans in 100% of cases; (2) parasites present in howlers could be potentially transmitted to domesticated cats and dogs in 59% of cases, and we could not find data on transmission to cattle in 50% of cases; and (3) parasites present in howlers could be potentially transmitted to cattle in 30% of cases, and we could not find data on transmission to cattle in 70% of cases. Some of the zoonotic vector-borne parasites found in howlers were *Plasmodium* sp., *Trypanosoma* sp., and *Toxoplasma* sp. (Appendix 1).

The taxonomic distribution of parasites in howler species (Appendix 1), show that helminths (54%) comprised the taxomomic group most represented, followed by protozoa (36%) (Fig. 17.5). The pattern found was similar to that presented by Pedersen et al. (2005) in an extensive literature review on wild primates.

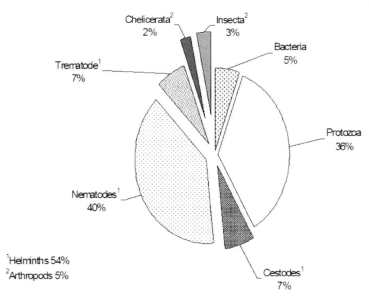

Fig. 17.5 Taxonomic distribution of the major groups of parasites represented in Appendix 1 (n = 91parasite species or groups)

17.4 Discussion

17.4.1 Patterns of Parasite Prevalence

Before discussing the results we want to point out the difficulty of obtaining complete data sets and the scarce amount of data on parasite diversity and transmission patterns in New World primates. We have concentrated our review on howler monkeys because they offer the most complete data set to look for patterns across South America. However, we confronted a series of limitations in our data set. For example, howler species such as *A. sara*, *A. macconelli* and *A. nigerrima* have not been sampled adequately for parasites, some studies are biased towards particular pathogens, identification techniques vary, and some studies provide information on only a small portion of a population (i.e., a social group) and others provide a "snapshot" from one-time sampling. We understand the limitations of our data set, and taking this into consideration, we outline a series of conclusions and directions for future research. The results presented here indicate that latitude, precipitation, altitude, and degree of human contact affect the prevalence of certain parasite species in howlers across South America. Prevalence of parasite species varied with the degree of contact with human settlements.

However parasite prevalence did not depend on the level of habitat fragmentation. We found a negative relationship between prevalence and latitude and altitude. Logistic regression models suggested latitude and altitude were mediators of the likelihood of having high or low parasitic prevalence (either higher or lower than 20%). We also found a positive relationship between parasite diversity at a study site and levels of precipitation. These results are in concordance with the idea of high levels of humidity or wetness (water vapor in the air) favoring certain species or groups of parasites (McGrew et al. 1989; Stuart et al. 1990, 1998; Stuart and Strier 1995; Stoner 1996; Semple et al. 2002; Nunn and Hyemann 2005; Eckert et al. 2006; Stoner and Gonzalez Di Perro 2006). In terms of latitude and altitude, prevalence is higher closer to the equator and in lower altitudes, where conditions are hotter, more humid, and less seasonal (especially vector borne diseases) than at higher latitudes and altitudes. These physical factors may be favoring persistence of certain species of parasites, but may also benefit the hosts or vectors involved in parasite cycles. We did not find a significant relationship between parasite diversity and altitude or latitude; however, in a meta-analysis of the Order Primates, Nunn et al. (2005) found a significant negative correlation between parasite diversity and latitude. This may reflect a Genus (*Alouatta*) or regional (South America) anomaly from the overall pattern for primates.

17.4.2 What Are the Important Factors to Take into Account when Looking for Patterns?

Patterns of parasite prevalence are complicated by characteristics of each study site, the type of collection, the number of animal samples, and the main goal of each independent research line. Our general analysis suggests that the prevalence of parasites did not vary across fragmented and continuous forests. This contrasts with findings from detailed field-based examination of this relationship. We stress that these comparisons are limited due to different scales of studies. Our database is composed of different studies across South America, and some of the studies presented below are comparisons of metapopulations in smaller areas. For example Gillespie and Chapman (2006) found, for metapopulations of red colobus (*Piliocolobus tephrosceles*) inhabiting nine fragments of different size, that forest fragmentation (especially the type of extraction or exploitation of the fragments, i.e., stump density) affected the prevalence of certain gastrointestinal parasites. Gilbert (1994) found that the rate of infection with gastrointestinal parasites in *Alouatta seniculus*, was higher in patches of small sizes than in larger patches or continuous forests, and also in habitats with higher density of primates and higher diversity of primate species. The same trend was found with howler monkeys (*A. palliata*) in Costa Rica (Stoner 1996). On the other hand, when two populations with different densities of muriquis (*Brachyteles arachnoides*) were compared, the lower density population had higher levels of infections. This effect appeared to be related to the relative humidity of the environment they inhabit (Stuart and Strier 1995). Although we did not include density as a variable in our analysis due to the lack of information across sites, we suggest that density should not be considered a good predictor of parasitic infections under all conditions.

Martins (2002) found, in a study of the relationship between habitat fragmentation and levels of gastrointestinal parasite infection in *A. belzebul*, that in general there was no strong effect of habitat fragmentation and density on the infection rates and diversity of parasites. On the other hand Santa Cruz et al. (2000a) found for *A. caraya* in northern Argentina, a positive relationship between infection indexes and area and degree of fragmentation of the habitat. Although at times equivocal, this data suggests that deforestation concentrates primates into small fragments, producing not only an increase in contact among individuals, but also greater overlap of travel routes (Freeland 1976; Gilbert 1994; Kowalewski and Zunino 1999, 2005; Santa Cruz et al. 2000a; Nunn and Dokey 2006). Studies of the golden monkey (*Leontopithecus rosalia*) demonstrate that groups that reside in logged areas are exposed to more diseases (Kleiman and Rylands 2002). Many parasite species or groups and their potential hosts have coexisted for long-term (Nunn and Altizer 2006). However, constant exposure to parasites may have lasting consequences on mortality and survival of wild primates (Stoner 1995; Gillespie and Chapman 2004; Gillespie 2006). Examples of parasites inducing high mortality in wild primates come from the description of epidemic outbreaks. It has been argued that, for example, yellow fever decimated populations of *Alouatta palliata* in Central America (Carpenter 1964; Galindo and Srihongse 1967; Stoner 1996), an unknown disease outbreak reduced *Alouatta seniculus* populations (almost 80%) in Venezuela (Pope 1998; Rudran and Fernandez-Duque 2003), and tick infestations resulted in 50% of infant deaths in *Papio ursinus* in Namibia (Brain and Bohrmann 1992). It has also been argued that the degree of virulence of pathogens remained unexplored and is complicated by factors such as competition among different strains of the same parasites species within a host, routes of parasite transmission, potential reduction of host fecundity, and unknown interaction with other parasites species within the host (Nunn and Altizer 2006).

Increased alteration of wild primate habitats is not only potentially affecting the conservation of primates, but also potentiates the likelihood of disease transmission between nonhuman primates and humans (Brack 1987; Lilly et al. 2002; Wolfe et al. 1998). From these results we conclude that the relationship between fragmentation and levels of parasite infection remains unresolved considering the range of variation found across species. Long-term research on specific populations would provide greater clarification. Other variables need to be addressed in the understanding of the dynamic of parasite transmission including for example life histories and transmission patterns of parasite species (Stuart et al. 1998). For example, we have shown that human activity or proximity is related to increasing prevalence possibly facilitating the interaction of different variables. Humans are potentially increasing the prevalence of certain parasitic infections in primates due to: (1) changes of the habitat through fragmentation, logging and other uses of the land, (2) direct contact with human feces, or contaminated water sources, (3) proximity and sharing of common vectors, and (4) increase of human population growth in tropical and subtropical countries in South America. Many of these factors may act synergistically to produce unexpected results in nonhuman primate populations, and possibly human populations.

Cross transmission of parasites between humans and primates have been shown in different parts of Africa and Asia (Wolfe et al. 1998; Bastone et al. 2003; Fuentes 2003; 2004; Ocaido et al. 2003; Gillespie et al. 2004; Vandamme 2004; Wallis and Lee 1999) and Central America (Vitazkova and Wade 2006). Despite the fact that many diseases are producing significant economic losses, little is known concerning wild primates' transmission dynamics and the relationship between increasing contact among wild primates, domesticated animals and humans. All species of howlers studied are known to be hosts for parasite stages that can infect human and domesticated animal health (Appendix 1, Fig. 17.4).

17.4.3 Parasites and Howlers in South America

Howlers are hosts to parasites that produce economic and health problems. Howlers are hosts of zoonotic gastrointestinal parasites such as *Giardia* sp. (Cabral et al. 2005), *Cryptosporidium* sp. (semicaptivity, Santa Cruz et al. 2003), and *Fasciola* sp. (Martins 2002, Cabral et al. 2005). Howler monkey groups are found living next to (and within) plantations, cattle fields, and small villages (Cabral et al. 2005; Estrada et al. 2006; Muñoz et al. 2006), travel on the ground, and drink water from rivers and lagoons (Young 1981; Gilbert and Stouffer 1989; Kowalewski et al. 1995; Bravo and Sallenave 2003; Pozo-Montuy and Serio-Silva 2007). All of these factors increase the likelihood of host contact with parasites and infections.

Howlers are also hosts to parasites involved in important infectious diseases. Several studies have explored *Plasmodium* sp. (*see* Davies et al. 1991, Lourenco de Oliveira and Deane 1995; Volney et al. 2002; Duarte et al. 2006), and *Toxoplasma* sp. (*see* for example Carme et al. 2002; Garcia et al. 2005). Other important human diseases are even more poorly explored as they relate to primates. For example, chagas disease, produced by protozoan *Trypanosoma cruzi,* is reported to be the most serious of the parasitic diseases in Latin America with important costs to public health, affecting millions of people with no vaccine or specific treatment for large-scale public health interventions (Dias et al. 2002). Trypanosomes were found in several species of howlers, especially populations living close to humans (Coppo, Moreira and Lombardero. 1979; Travi et al. 1986; Bar et al. 1999; Dereure et al. 2001). Moreover, some species of parasites may fail to appear in a study due to insufficient sampling effort or sampling limited to a specific pathogen (Gregory 1990, Walther et al. 1995; Nunn et al. 2003). Large studies (e.g., hundreds or thousands of individuals sampled) that measured prevalence of a single parasite may overrepresent sampling effort given to a host species (e.g., Hayami et al. 1984; Milton 1996). Since howlers are hosts to many infectious diseases, and also suffer epidemic outbreaks (i.e., yellow fever [Vasconcelos et al. 2003; Monath 2001]), the synergistic effects of habitat transformation, and increased contact with humans and vectors could increase the likelihood of new outbreaks in certain populations.

Howler monkeys are known to be generalists that demonstrate great behavioral plasticity in highly modified forests (Kowalewski and Zunino 1999, 2004; Estrada et al. 1999, 2002; Marsh 1999; Clarke et al. 2002; Bicca-Marques 2003). In this regard, howlers resemble black-and-white colobus (*Colobus guereza*) from Africa. Studies on the effects of selective logging and forest fragmentation on black-and-white-colobus have shown that this species persists in disturbed forest better (i.e., present in more forest fragments) than other species (i.e., red colobus [*Procolobus badius*] and red-tailed guenons [*Cercopitheus ascanius*]) (Chapman et al. 2000, 2006) and do not display altered patterns of parasitic infection as seen in other African primates (Gillespie et al. 2005a,b; Gillespie and Chapman 2006). These interspecific differences among African colobines open a set of questions and stress the necessity of future studies of specific differences between howler species and other Atelines. Spider and woolly monkeys have larger ranges than howlers and are assumed to be less behaviorally plastic (Di Fiore and Campbell 2007). In confronting anthropogenic alteration of habitat, these species may suffer different effects in terms of their parasitic infections (i.e., similar to red colobus). However detailed parasitic data on wild populations of these species is extremely scarce from South America (*see* for example Michaud et al. 2003 [*Lagothrix lagotricha*], Fandeur et al. 2000 [*Ateles paniscus*]).

Another line of research poorly explored in Latin America is nonhuman primate self- medication. Stoner and Gonzalez Di Perro (2006) found a negative relationship between individual intensity of infection and the time spent foraging *Ficus tecolutensis* in a study on three groups of *Alouatta pigra* in southern Mexico. These authors argued that the consumption of fig fruits may serve as an anti-parasitic medicine. Other food items related to medicinal use are bark, wood, and clay (Huffman 1997). For example, *Alouatta caraya* in northern Argentina was observed to eat bark and clay from the ground during the cold season usually associated with animals producing soft-liquid fecal material (Kowalewski, pers. obs.). *Cebus capucinus* was observed to rub their fur with certain species of plants in Costa Rica (Baker 1996). Fur rubbing in capuchins occurred more frequently during the rainy season. Baker (1996) suggested that this change in behavior was related to an increase in parasite infections due to an increase of humidity and temperature in the rainy season. Campbell (2000) reported fur rubbing in a group of *Ateles geoffroyi* in Costa Rica. The author argues that in this case, fur rubbing serves as a form of scent making. Several cases of self-medication have also been described in apes (*see* review in Huffman 1997). Although the relationship between self-medication and use of plants in New World primates remains speculative, it opens an interesting line of research in the relationships between parasites, habitat composition, and non-human primates. For example, it could be useful to explore primate parasite prevalence and intensity of infections in forests with different plant species composition caused by logging and clear cutting. Despite the unpredictability, low frequency, and methodological difficulties associated with collecting data on primate self-medication, we believe this will be an important area for future research (Huffman 1997; Nunn and Altizer 2006).

17.4.4 How Can We Improve the Study of Primate Parasitology in South America?

Consistent research on the relationship of parasite dynamics and infectious disease transmission among nonhuman primates in South America is currently lacking. Deforestation, habitat alteration, and degree of human contact with wild population are increasing in most parts of the world (Chapman et al. 2005). It would be helpful to intensify development of programs across countries, studying the effects of these changes on the relationships between human and nonhuman primates. We suggest (1) To focus studies on different population of primates that live under different degree of habitat alteration, (2) To record complete annual cycles of parasite profiles; after profiles most of the studies are "snap-shots" and hence limited in terms of providing sufficient information on parasite profiles and on host-parasite temporal dynamics and phenology (interannual comparisons), (3) To include in the analysis humans, domesticated animals and shared water and food sources, (4) To standardize data collection both in research design and methodology (*see* Gillespie 2006), (5) to involve local students, people, and local government in projects, (6) To identify parasites (molecular confirmation) to determine sources of infection and risk factors associated with transmission, and (7) to publish final results in journals or public access websites to allow comparisons (Table 17.2). Finally, considering the existence of generalist parasites common to humans, domesticated animals and howlers (Appendix 1), it is of vital importance for conservation and public-human health planning to develop long term research lines on the study of interactions and change of dynamics in systems where nonhuman primates come into contact with humans

Table 17.2 Suggestions to improve primate parasitological studies in South America

What to study?	What to include?	Who to involve?	How to do it?
Different populations of same species living under different degrees of habitat degradation and human contact.	Human samples Domestic animals samples Shared sources of food and water. Non-domesticated animals (when possible). GIS analysis.	Local students Local people and community (public schools) Administrative, health and educational personnel close to the study sites.	Standardize the collection of samples. Identify species Genetic identification of strains Identify source of infections (domesticated and non-domesticated animals).
Compare different species when sympatric and not with other primate species.			
Geographic distribution of disease risk.		Veterinary researchers to cover clinical aspect of health status in primates and to consider measures that need to be taken.	Publish results in public access websites to allow comparisons[1] Implement together with local people educational programs
Long-term studies to capture seasonal changes and interannual variation in parasites prevalence and diversity.		Molecular biologists to incorporate genetic identification of parasites and different strains	

[1]for example Gobal Mammal Database: http://www.mammalsparsites.org/ (Nunn and Altizer 2005)

(Table 17.2). We provide a short list of websites with information on public health status and research centers in Latin America (Appendix 2).

Our results provide a baseline for understanding causative factors for patterns of parasitic infections in wild primate populations and may alert us to iminent threats to primate conservation. Based on these results we present a series of recommendations for future parasitological studies in nonhuman primates (Table 17.2). Moreover, our results have implications for conservation of biodiversity, wildlife and human health.

17.5 Summary

Parasites play a central role in ecosystems, affecting the ecology and evolution of species interactions, host population growth and regulation, and community biodiversity. Howler monkeys are the most widespread nonhuman primates in South America, with 8 of 10 *Alouatta* species living in South America. We took a meta-analysis approach, integrating data from studies on wild *Alouatta caraya, A. seniculus, A. guariba*, and *A. belzebul,* to examine how various factors such as latitude, altitude, annual precipitation, continuous or fragmented forests, and degree of contact with human settlements affected parasite prevalence. We included in the analysis data on the prevalence of gastrointestinal parasites, blood parasites and ectoparasites, and number of parasite species. When all parasite types were analyzed together we found that type of human contact affected the prevalence of different parasites ($P < 0.05$). Our general analysis suggests that the prevalence of parasites did not vary across fragmented and continuous forests ($P > 0.05$). Logistic regression models suggested latitude and altitude were mediators of the likelihood of having high or low parasitic prevalence (either higher or lower than 20%) ($P < 0.05$). The relationship between gastrointestinal parasite diversity at a study site and average annual precipitation was positive and significant ($r = 0.72$, $P < 0.05$). Almost 86% of gastrointestinal parasites, and 100% of blood-borne parasites found in howlers are found in humans. Our results provide a baseline for understanding causative factors for patterns of parasitic infections in wild primate populations and may alert us to iminent threats to primate conservation.

Acknowledgments We are very grateful to Paul Garber, Alejandro Estrada, and two anonymous reviewers for their valuable input. We would like to thank the editors Alejandro Estrada, Paul Garber, Karen Strier, Julio C. Bicca-Marques and E. Heynmann for inviting us to contribute to this volume. We thank Juliane Hallal Cabral, Simone de Souza Martins, Ana Maria R. de C. Duarte and Michelle Viviane Sá dos Santos for contributing information on parasite prevalence and description of study sites. We also thank Julio C. Bicca-Marques for assistance with the bibliography. M.K. thanks A.M. Santa-Cruz and her Lab crew from the Universidad del Nordeste, Corrientes, Argentina for taking care of parasite identification in samples from Northern Argentina during 2003–2004. MK also thanks field assistants who helped in the collection of fecal samples: Vanina Fernandez, Nelson Novo, Romina Pave and Silvana Peker.

17.6 Appendix 17.1

In the table we present data on parasitic prevalence (%), parasite species, type of parasite, study site, location and characteristics of the study site, number of samples analyzed per study, type of sample collection. We considered the Type of forest: C continuous, F Fragmented; Alt. Altitude of study site in meters above sea level; Prec. Precipitation in millimeters per year, Type of parasites: G Gastrointestinal, B blood-borne, E ectoparasites, Humans, Dogs/cats, Cattle, Monkeys Yes indicates that the parasite species was found in these organisms; Collection 1 = blood, 2 = feces, 3 = blood and feces, 4 = blood, feces, and skin, 5 = killed and examined; Samples: number of samples analyzed in each study; for *Alouatta belzebul, Alouatta caraya, Alouatta guariba*, and *Alouatta seniculus.*

Appendix 17.1

Species	Site	Latitude	Type of forest	Alt. (m)	Prec. (mm/ year)	Prevalence (%)	Contact with people	Parasite Species (or taxa)	Type of parasite	Humans	Dogs/cats	Cattle	Monkey	Collection	Samples	Source
A. belzebul	Rio Tocantis, Tucurui, Brazil	3° 40′ S	C	75	2740	22.65	Rural	Amebidae	G	Yes	Yes	Yes	Yes	2	123	1
			C			5.55	Rural	Ancilostomidae	G	Yes	Yes	No	Yes	2	123	1
			C			9.00	Rural	Ascaris	G	Yes	No	Yes	Yes	2	123	1
			C			6.75	Rural	Endolimax	G	Yes			Yes	2	123	1
			C			1.15	Rural	Entamoeba	G	Yes	Yes	Yes	Yes	2	123	1
			C			1.15	Rural	Fasciola	G	Yes	No	Yes	Yes	2	123	1
			C			2.30	Rural	Giardia	G	Yes	Yes	Yes	Yes	2	123	1
			C			2.30	Rural	Hymenolepididae	G	Yes	?	?	Yes	2	123	1
			C			48.70	Rural	Iodamoeba	G	Yes	?	?	Yes	2	123	1
			C			24.15	Rural	Nematodeo (larvae)	G	Yes	Yes	Yes	Yes	2	123	1
			C			1.15	Rural	Trichuris	G	Yes	Yes	Yes	Yes	2	89	1
			C			23.80	Rural	Trypanoxyuris minutus	G	No	No	No	Yes	2	89	1
			F			2.50	Remote	Amebídeo*	G	Yes	Yes	Yes	Yes	2	89	1
			F			16.93	Remote	Ancilostomídeo	G	Yes	Yes	No	Yes	2	89	1
			F			13.70	Remote	Ascaris	G	Yes	No	Yes	Yes	2	89	1
			F			5.00	Remote	Endolimax	G	Yes			Yes	2	89	1
			F			4.77	Remote	Entamoeba	G	Yes	Yes	Yes	Yes	2	89	1
			F			46.03	Remote	Iodamoeba	G	Yes	Yes	Yes	Yes	2	89	1
			F			38.67	Remote	Nematodeo (larvae)	G	Yes	Yes	Yes	Yes	2	89	1
			F			0.83	Remote	Trichuris	G	Yes	Yes	Yes	Yes	2	89	1
			F			28.53	Remote	Trypanoxyuris minutus	G	No	No	No	Yes	2	89	1
A. caraya	Bella Vista, Corrientes Province, Argentina	28° 30′ S	F	60	1200	22.00	Remote	Trypanoxyuris minutus	G	No	No	No	Yes	5	88	2
			F			7.00	Remote	Bertiella macronata	G	Yes	No	No	Yes	5	84	2
			F			37.00	Remote	Pedicinus mjorbergi	E	No	No	No	Yes	5	302	2

Appendix 17.1 (continued)

Species	Site	Latitude	Type of forest	Alt. (m)	Prec. (mm year)	Prevalence (%)	Contact with people	Parasite Species (or taxa)	Type of parasite	Humans	Dogs/cats	Cattle	Monkey	Collection	Samples	Source
	Parana River, Chaco Province, Argentina	27°20' S	F	60	1200	7.00	Remote	Trypanosma sp	B	Yes	Yes	Yes	Yes	4	29	3
			F			3.45	Remote	Plasmodium sp	B	Yes	Yes	?	Yes	4	29	3
			F			21.00	Remote	Bertiella macronata	G	Yes	No	No	Yes	4	29	3
			F			7.00	Remote	Trypanoxuris sp	G	Yes	?	?	Yes	4	29	3
			F			9.80	Rural	Trypanoxyuris callithrix	G	No	No	No	Yes	2	51	4
			F			11.00	Rural	Cebalges gaudi	E	No	No	No	Yes	4	29	3
			F			24.00	Rural	Pedicinus mjorbergi	E	No	No	No	Yes	4	29	3
	Parana River, Chaco Province, Argentina	27°30' S	F	60	1200	52.00	Remote	Cestoda	G	Yes	Yes	Yes	Yes	2	384	5
			F			4.00	Remote	Trypanoxyuris minutus	G	No	No	No	Yes	2	384	5
			F			4.00	Remote	Giardia duodenalis	G	Yes	Yes	Yes	Yes	2	28	6
			F	55	1200	13.00	Rural	Trypanosoma cruzi	B	Yes	Yes	Yes	Yes	1	30	7
			F			5.00	Rural	Cestoda	G	Yes	Yes	Yes	Yes	2	256	8
			F			5.00	Rural	Trypanoxyuris minutus	G	No	No	No	Yes	2	256	8
			F			12.00	Rural	Bertiella macronata	G	Yes	No	No	Yes	5	110	9
			F			6.00	Rural	Trypanoxyuris minutus	G	No	No	No	Yes	5	110	9
			F			88.00	Rural	Pedicinus mjorbergi	E	No	No	No	Yes	5	110	9
	Parana River, Parana State, Brazil	22°46'20" S	F	252	1700	2.00	Rural	Amblyomma sp.	E	Yes	Yes	Yes	Yes	5	110	9
			F			17.60	Rural	Toxoplasma goldii	B	Yes	Yes	Yes	Yes	1	17	10

Appendix 17.1 (continued)

Species	Site	Latitude	Type of forest	Alt. (m)	Prec. (mm year)	Prevalence (%)	Contact with people	Parasite Species (or taxa)	Type of parasite	Humans	Dogs/cats	Cattle	Monkey	Collection	Samples	Source
	Porto-Primavera Hydroelectric, Sao Paulo-Mato Grosso do Sul, Brazil	21°15′ S	C	302	1500	8.10	Rural	*Plasmodium simiovale*	B	Yes	?	?	Yes	1	590	11
			F			8.30	Rural	*Plasmodium falciparum*	B	Yes	?	?	Yes	1	590	11
			F			7.60	Rural	*Plasmodium vivax*	B	Yes	?	?	Yes	1	590	11
			F			8.00	Rural	*Plasmodium vivax VK247* (Type II)	B	Yes	?	?	Yes	1	590	11
			F			7.60	Rural	*Plasmodium malariae/ brasiliensis*	B	Yes	Yes	?	Yes	1	590	11
	RPPN Nova Querencia, Mato Grosso do Sul, Brazil	20°43′34″ S	C	450	1379	16.60	Rural	*Trichuris* sp	G	Yes	Yes	Yes	Yes	2	59	12
			C			16.60	Rural	*Oesophagostomum* sp	G	Yes	No	No	Yes	2	59	12
			C			16.60	Rural	*Capillaria* sp	G	Yes	Yes	No	Yes	2	59	12
			C			16.60	Rural	*Enterobius vermicularis*	G	Yes	No	No	Yes	2	59	12
			C			16.60	Rural	*Trichostrongylus* sp	G	Yes	Yes	Yes	Yes	2	59	12
			C			33.30	Rural	*Eimeria* sp	G	No	No	Yes	Yes	2	59	12
	Tocantins	13°49′ S	F	460	1750	42.50	Rural	*Plasmodium simiovale*	B	Yes	?	?	Yes	1	42	11
			F			35.70	Rural	*Plasmodium falciparum*	B	Yes	?	?	Yes	1	42	11
	River, Goias, Brazil		F			26.50	Rural	*Plasmodium vivax*	B	Yes	?	?	Yes	1	42	11
			F			2.40	Rural	*Plasmodium vivax VK247* (Type II)	B	Yes	?	?	Yes	1	42	11
			F			14.30	Rural	*Plasmodium malariae/ brasiliensis*	B	Yes	Yes	?	Yes	1	42	11

Appendix 17.1 (continued)

Species	Site	Latitude	Type of forest	Alt. (m)	Prec. (mm year)	Prevalence (%)	Contact with people	Parasite Species (or taxa)	Type of parasite	Humans	Dogs/cats	Cattle	Monkey	Collection	Samples	Source
A. guariba	Morro Sao Pedro, Porto Alegre, RGS, Brazil	30°01'.59" S	F	230	1324	1.88	Rural	Ascaris spp	G	Yes	No	Yes	Yes	2	53	13
			F			5.66	Rural	Strongyloides sp.	G	Yes	Yes	Yes?	Yes?	2	53	13
			F			1.88	Rural	Enterobideos	G	Yes	No	No	Yes	2	53	13
			F			1.88	Rural	Trichuris sp	G	Yes	Yes	Yes	Yes	2	53	13
			F			11.32	Rural	Entamoeba sp	G	Yes	Yes	Yes	Yes	2	53	13
	Reserva Biologica Lami, Porto Alegre, RGS, Brazil	30°15'S	F	200	1324	8.77	Urban	Ascaris spp	G	Yes	No	Yes?	Yes	2	114	13
			F			30.70	Urban	Strongyloides sp.	G	Yes	Yes	No	Yes	2	114	13
			F			4.38	Urban	Enterobideos	G	Yes	No	?	Yes	2	114	13
			F			5.26	Urban	Entamoeba sp	G	Yes	?	?	Yes	2	114	13
			F			0.87	Urban	Fasciola sp	G	Yes	No	Yes	Yes	2	114	13
			F			0.87	Urban	Paragonimus sp	G	Yes	?	?	Yes	2	114	13
			F			0.87	Urban	Giardia sp	G	Yes	Yes	Yes	Yes	2	114	13
	Mata de Ribeirão Cachoeira, Campinas, SP, Brazil	22°50'13" S	F	650	1049	54.50	Remote	Nematodes	G	Yes	Yes	Yes	Yes	2	112	14
			F			69.20	Remote	Kathlaniidae	G	?	?	?	?	2	112	14
			F			84.60	Remote	Trypanoxyuris minutus	G	No	No	No	Yes	2	112	14
A. seniculus	Sinnamary River, Petit Saut Dam, French Guiana	5°04' N	C	45	3000	1.70	Remote	Plasmodium sp	B	Yes	?	Yes	Yes	1	117	15
			C			6.80	Remote	Plasmodium brasilianum	B	Yes	Yes	?	Yes	1	117	15
			C			97.50	Remote	Plasmodium malariae/ brasilianum	B	Yes	Yes	?	Yes	1	81	16
			C			71.60	Remote	Plasmodium falciparum	B	Yes	?	?	Yes	1	81	16
			C			37.04	Remote	Plasmodium vivax	B	Yes	?	?	Yes	1	81	16
			C			5.00	Remote	Toxoplasma goldii	B	Yes	Yes	Yes	Yes	1	50	17

Appendix 17.1 (continued)

Species	Site	Latitude	Type of forest	Alt. (m)	Prec. (mm year)	Prevalence (%)	Contact with people	Parasite Species (or taxa)	Type of parasite	Humans	Dogs/ cats	Cattle	Monkey	Collection	Samples	Source
	Balbina, Uatuma River, Amazonas State, BRA	1°55' S	C	34	2262	32.30	Rural	*Plasmodium brasilianum*	B	Yes	Yes	?	Yes	1	31	18
	Tambopata National Reserve, Madre de Dios department, PERU	13°08'10" S	C	250	2400	31.00	Rural	*Strongyloides* sp.	G	Yes	Yes	Yes?	Yes	2	16	19
			C			13.00	Rural	*Trichuris* sp.	G	Yes	Yes	Yes	Yes	2	16	19
			C			6.00	Rural	*Chilomastix* sp.	G	Yes	?	?	Yes	2	16	19
			C			6.00	Rural	*Blastocystis* sp.	G	Yes	Yes	Yes	Yes	2	16	19
			C			6.00	Rural	*Iodamoeba* sp.	G	Yes	?	?	Yes	2	16	19
	Biological Dynamics of Forest Fragments Project, Manaus, BRA	2°30' S	F	38	2606	29.30	Rural	*Spirurid species*	G	?	?	?	Yes	2	35	20
			F			47.33	Rural	*Trematodes (2 sp)*	G	?	?	?	Yes	2	35	20
			F			45.00	Remote	*Spirurid species*	G	?	?	?	Yes	2	35	20
			F			50.00	Remote	*Trematodes (2 sp)*	G	?	?	?	Yes	2	35	20
			C			34.00	Remote	*Spirurid species*	G	?	?	?	Yes	2	35	20
			C			46.00	Remote	*Trematodes (2 sp)*	G	?	?	?	Yes	2	35	20

[1] Martins 2002 [2] Pope 1966; [3] Santa Cruz et al. 2000a; [4] Prieto et al. 2002; [5] Kowalewski et al. in prep.; [6] Venturini et al. 2003; [7] Travi et al. 1986; [8] Delgado 2006; [9] Coppo et al. 1979; [10] Garcia et al. 2005; [11] Duarte et al. 2006; [12] Godoy et al. 2004 [13] Cabral et al. 2005; [14] Santos et al. 2005 [15] Fandeur et al. 2000; [16] Voley et al. 2002; [17] Carme et al. 2000; [18] Lourenco de Oliveira and Deane 1995; [19] Phillips et al. 2004; [20] Gilbert et al. 1994

17.7 Appendix 17.2

Websites with information about institutions dealing with human public health in Latin America (general information, databases, and centers of research).

Appendix 17.2

Name	Website	Goals
The Pan American Health Organization, USA	http://www.paho.org/	Provides data collected annually to characterize the health situation and trends in countries of Latin America (English and Spanish)
Centers for Disease Control and Prevention, USA	http://www.cdc.gov/	Provides information on parasites and associated diseases (English and Spanish)
American Society of Tropical Medicine and Hygiene, USA	http://www.astmh.org/	General information including grants opportunities (English)
Direccion General de Epidemiologia, Mexico	http://www.dgepi.salud.gob.mx/	General information -mainly Mexico (Spanish)
Instituto Nacional de Salud Publica, Mexico	http://www.insp.mx/	General information and links to health research centers in Mexico (Spanish)
Mais Saúde Brasil, Brazil	http://www.maissaudebrasil.com	General information (Portuguese)
Ministério da Saúde, Brazil	http://portal.saude.gov.br/saude/	General information (Portuguese)
Rede Nacional de Informações em Saúde, Brazil	http://www.datasus.gov.br/rnis/datasus.htm	General information, databases (Portuguese)
Administración Nacional de Laboratorios e Institutos de Salud "Dr. Carlos G. Malbrán", Argentina	http://www.anlis.gov.ar/anlis.htm	General information on health status and research in Argentina (Spanish)

References

Acha, P. N. and Szyfres, B. 1986. Zoonosis y Enfermedades Transmisibles Comunes al Hombre y a los Animales, OPAS 503. Oficina Sanitaria Panamericana, Oficina Regional de la Organización Mundial de la Salud, Washington.

Altizer, S., Nunn, C. L., Thrall, P., Gittleman, J. L., Antonovics, J., Cunningham, A., Dobson, A., Ezenwa, V., Pedersen, A. B., Poss, M. and Pulliam, J. R. C. 2003. Social organization and disease risk in mammals: Insights from comparative and theoretical studies. Annual Review of Ecology and Systematics 34:517–547.

Baker, M. 1996. Fur rubbing: use of medicinal plants by capuchin monkeys (*Cebus capucinus*). American Journal Primatology 38:263–270.

Bar, M. E., Alvarez, B. M., Oscherov, E. B., Damborsky, M. P. and Jörg, M. E. 1999. Contribución al conocimiento de los reservorios del *Trypanosoma cruzi* (Chagas, 1909) en la Provincia de Corrientes, Argentina. Revista da Sociedade Brasileira de Medicina Tropical 32(3):271–276.

Bastone, P., Truyen, U. and Loechelt, M. 2003. Potential of zoonotic transmission of non-primate foamy viruses to humans. Journal of Veterinary Medicine B50(9):417–423.

Bicca-Marques, J. C. 2003. How do howler monkeys cope with habitat fragmentation? In L. K. Marsh (ed.), *Primates in Fragments: Ecology and Conservation* (pp. 283–303). New York: Kluwer Academic/Plenum Publishers.

Brack, M. 1987. Agents Transmissible From Simians to Man. New York: Springer-Verlag. Brain. C. and Bohrmann, R. 1992. Tick infestation of baboons (*Papio ursinus*) in the Namib Desert. Journal of Wildlife Diseases 28(2):188–191.

Bravo, S. P. and Sallenave, A. 2003. Foraging behavior and activity patterns of *Alouatta caraya* in the northeastern Argentinean flooded forest. International Journal of Primatology 24(4):825–846.

Cabral, J. N. H., Rossato, R. S., de M. Gomes, M. J. T, Araújo, F. A. P, Oliveira, F. and Praetzel, K. 2005. Gastrointestinais de bugios-ruivos (*Alouatta guariba clamitans* Cabrera 1940) da região extremo-sul de Porto Alegre/RS - Brasil, diagnosticados através da coproscopia: implicações para a conservação da espécie e seus ha. Congresso Brasileiro de Parasitologia, Porto Alegre, RS, Brasil.

Campbell, C. J. 2000. Fur rubbing behavior in free-ranging black-handed spider monkeys (*Ateles geoffroyi*) in Panama. American Journal of Primatology 51:205–208

Carme, B., Aznar, C., Motard, A., Demar, M. and De Thoisy, B. 2002. Serologic Survey of *Toxoplasma gondii* in Noncarnivorous Free-Ranging Neotropical Mammals in French Guiana. Vector Borne and Zoonotic Diseases 2(1):11–17.

Carpenter, C. R. 1964. Naturalistic Behavior of Nonhuman Primates. Pennsylvania State University Press, University Park.

Chapman, C. A., Balcomb, S. R., Gillespie, T. R, Skorupa, J. P. and Struhsaker, T. T. 2000. Long-term effects of logging on primates in Kibale National Park, Uganda: A 28 year comparison. Conservation Biology 14:207–217.

Chapman, C. A., Gillespie, T. R. and Goldberg, T. L. 2005. Primates and the ecology of their infectious diseases: How will anthropogenic change affect host-pathogen interactions? Evolutionary Anthropology 14:134–144.

Chapman, C. A., Wasserman, M. D. and Gillespie, T. R., 2006. Behavioural patterns of colobus in logged and unlogged forests: the conservation value of harvested forests. In E. Newton-Fisher, H. Notman, V Reynolds and J.D. Patterson (eds.), *Primates of Western Uganda* (pp. 373–390) New York: Springer.

Chapman, C. A., Wasserman, M. D., Gillespie, T. R., Speirs, M. L. Lawes, M. J. and Ziegler, T. E. in press. Do nutrition, parasitism, and stress have synergistic effects on red colobus populations living in forest fragments? American Journal of Physical Anthropology.

Clarke, M. R., Crockett, C. M., Zucker, E. L. and Zaldivar, M. 2002. Mantled howler population of hacienda La Pacifica, Costa Rica, between 1991 and 1998: effects of deforestation. American Journal of Primatology 56:155–163.

Coppo, J. A., Moreira, R. A. and Lombardero, O. J. 1979. El parasitismo en los primates del CAPRIM. Acta Zoológica Lilloana 35:9–12.

Cortés-Ortiz, L., Birmingham, E., Rico, C., Rodríguez-Luna, E., Sampaio, I. and Ruiz-García, M. 2003. Molecular systematics and biogeography of the Neotropical monkey genus, *Alouatta*. Molecular Phylogenetics and Evolution 26:64–81.

Crockett, C. M. 1998. Conservation biology of the genus *Alouatta*. International Journal of Primatology 19(3):549–578.

Davies, C. R., Ayres, J. M. , Dye, C. and Deane L. M. 1991. Malaria infection rate of Amazonian primates increases with body weight and group size. Functional Ecology 5(5):655–662.

Deane, L. M. 1992. Simian malaria in Brazil. Memorias do Instituto Oswaldo Cruz 87(Suppl. 3):1–20.

Delgado, A. 2006. Estudio de patrones de uso de sitios de defecación y su posible relación con infestaciones parasitarias en dos grupos de monos aulladores negros y dorados (*Alouatta caraya*) en el nordeste argentino. Licenciatura Thesis, Universidad Nacional de Córdoba, Argentina (inedit).

Dereure, J., Barnabe, C., Vie, J. C., Madelenat, F. and Raccurt, C. 2001. Trypanosomatidae from wild mammals in the neotropical rainforest of French Guiana. Annals of Tropical Medicine and Parasitology 95(2):157–166.

Di Fiore, A. and Campbell, C. J. 2007. The Atelines: variations in ecology, behavior, and social organization..In C. J. Campbell, A. Fuentes, K. C. MacKinnon M. Panger and S. K. Bearder (eds.), *Primates in Perspective* (pp. 155–185). New York: Oxford Univ. Press.

Dias J. C. P., Silveira, A. C. and Schofield, C. J. 2002. The Impact of Chagas Disease Control in Latin America – A Review. Mem. Inst. Oswaldo Cruz, Rio de Janeiro 97(5):603–612.

Diaz-Ungria, C. 1965. Nematodes de primates venezolanos. Soc. Venezol. Ciencias Nat. 25: 393–398.

Duarte, A. M. R. C., Porto, M. A. L., Curado, I., Malafronte, R. S., Hoffmann, E. H. E., de Oliveira, S. G., da Silva, A. M. J., Kloetzel, J. K. and Gomes, A. C. 2006. Widespread occurrence of antibodies against circumsporozoite protein and against blood forms of *Plasmodium vivax, P. falciparum* and *P. malariae* in Brazilian wild monkeys. Journal of Medical Primatology 35(2):87–96.

Eckert, K. A., Hahn, N. E., Genz, A., Kitchen, D. M., Stuart, M. D., Averbeck, G. A., Stromberg, B. E. and Markowitz, H. 2006. Coprological surveys of *Alouatta pigra* at two sites in Belize. International Journal of Primatology 27(1):227–238.

Eisenberg, J. F., Muckenhirn, N. A. and Rudran, R. 1972. The relation between ecology and social structure in primates. Science 176:863–874.

Esch, G. and Fernandez, J. C. 1993. A Functional Biology of Parasitism: Ecological and Evolutionary Implications. London: Chapman and Hall.

Estrada, A., Juan-Solano, S., Ortiz Martinez, T. and Coates-Estrada, R. 1999. Feeding and general activity patterns of a howler monkey (*Alouatta palliata*) troop living in a forest fragment at Los Tuxtlas, Mexico. American Journal of Primatology 48(3):167–183.

Estrada, A., Mendoza, A., Castellanos, L., Pacheco, R., Van Belle, S., Garcia, Y., Muñoz, D. 2002. Population of the black howler monkey (*Alouatta pigra*) in a fragmented landscape in Palenque, Chiapas, México. American Journal of Primatology 58:45–55.

Estrada, A., Saenz, J., Harvey, C., Naranjo, E., Muñoz, D. and Rosales-Meda, M. 2006. Primates in agroecosystems: conservation value of some agricultural practices in Mesoamerican landscapes. In A. Estrada, P. A. Garber, M. S. M. Pavelka and L. Luecke (eds.), *New Perspectives in the Study of Mesoamerican Primates: Distribution, Ecology, Behavior, and Conservation* (pp. 437–470). New York: Springer.

Fandeur, T., Volney, B., Peneau, C. and De Thoisy, B. 2000. Monkeys of the rainforest in French Guiana are natural reservoirs for *P. brasilianum/P. malariae* malaria. Parasitology 120:11–21.

Freeland, W. J. 1976. Pathogens and the evolution of primate sociality. Biotropica 8:12–24.

Fuentes, A. 2003. The role of bio-cultural factors in assessing bi-directional pathogen transmission between human and nonhuman primates. American Journal of Physical Anthropology 36:97–98.

Fuentes, A. 2004. Human culture, macaque behaviour and global tourism: assessing the context and patterns of pathogen transmission risk in human-macaque interactions. Folia Primatologica 75(1):102–103.

Galindo, P. and Srihongse, S. 1967. Evidence of recent jungle yellow-fever activity in eastern Panama. Bulletin of the World Health Organization 36:151–161.

Garcia, J. L., Svoboda, W. K, Chryssafidis, A. L., de Souza Malanski, L., Shiozawa, M. M., de Moraes Aguiar, L., Teixeira, G. M, Ludwig, G., da Silva, L. R., Hilst, C. and Navarro, T. T. 2005. Sero-epidemiological survey for toxoplasmosis in wild New World monkeys (*Cebus* spp.; *Alouatta caraya*) at the Parana river basin, Parana State, Brazil. Veterinary Parasitology 133(4):307–11.

Gilbert, K. A. 1994. Endoparasitic infection in red howling monkeys (*Alouatta seniculus*) in the Central Amazonian basis: a cost of sociality? Ph. D. Thesis, The State University of New Jersey at New Brunswick Rutgers.

Gilbert, K. A. and Stouffer, P.C. 1989. Use of a ground water source by mantled howler monkeys (*Alouatta palliata*). Biotropica 21(4):380.

Gillespie, T. R., Greiner, E. C. and Chapman, C. A. 2004. Gastrointestinal parasites of the guenons of western Uganda. Journal of Parasitology 90:1356–1360.

Gillespie, T. R., Chapman, C. A. and Greiner, E. C. 2005a. Effects of logging on gastrointestinal parasite infections and infection risk in African primates. Journal of Applied Ecology 42: 699–707.

Gillespie, T.R., Greiner, E.C. and Chapman, C.A. 2005b. Gastrointestinal parasites of the colobine monkeys of Uganda. Journal of Parasitology 91:569–573.

Gillespie, T. R. 2006. Non-invasive assessment of gastro-intestinal parasite infections in free-ranging primates. International Journal of Primatology 27:1129–1143

Gillespie, T. R. and Chapman C. A. 2006. Prediction of parasite infection dynamics in primate metapopulations based on attributes of forest fragmentation. Conservation Biology 20: 441–448.

Godoy, K. C. I. , Odalia-Rimoli, A. and Rimoli, J. 2004. Infecção por endoparasitos em um grupo de Bugios-Pretos (*Alouatta caraya*), em um fragmento florestal no Estado de Mato Grosso do Sul. Neotropical primates 12(2):63–68

Gregory, R. D. 1990. Parasites and host geographic range as illustrated by waterfowl. Functional Ecology 4:645–654.

Gurevitch, J., Moorow, L. L., Wallace A. and Walsh, J. S. 1992. A meta-analysis of competition in field experiments. American Naturalist 140:539–572.

Hayami, M., Komuro, A., Nozawa, K., Shotake, T., Ishikawa, K., Yamamoto, K., Ishida, T., Honjo, S. and Hinuma, Y. 1984. Prevalence of antibody to adult T-cell leukemia virus- associated antigens (ATLA) in Japanese monkeys and other non-human primates. International Journal of Cancer 33:179–183.

Hedges, L. V. and Olkin, I. 1985. *Statistical Methods for Meta-Analysis.* Orlando, FL: Academic Press.

Hochachka, V. W. and Dhondt, A. A. 2000. Density-dependent decline of host abundance resulting from a new infectious disease. Proceedings of the National Academy of Science 97:5303–5306.

Hosmer, D. W. and Lemeshow, S. 1989. Applied logistic regression, New York: Wiley.

Hudson, P.J., Dobson, A.P. and Newborn, D. 1998. Prevention of population cycles by parasite removal. Science 282:2256–2258.

Hudson, P. J., Rizzoli, A., Grenfell, B. T., Heesterbeek, H. and Dobson, A. P. 2002. The Ecology of Wildlife Disease. Oxford: Oxford University Press.

Huffman. M. A. 1997. Current evidence for self-medication in primates: a multidisciplinary perspective. Yearbook Physical Anthropology 40:171–200.

Johns, A. D. and Skorupa, J. P. 1987. Responses of rain-forest primates to habitat disturbance: A review. International Journal of Primatology 8(2):157–191.

Johnson, D. E. 1998. Applied Multivariate Statistical Analysis. Belmont, CA: Duxbury Press.

Kleiman, D. G. and Rylands, A. B. 2002. Lion Tamarins: Biology and Conservation, Washington, DC: Smithsonian Institution Press.

Kowalewski, M. M., Bravo, S. P. and Zunino, G. E. 1995. Aggression between *Alouatta caraya* in forest patches in northern Argentina. Neotropical Primates 3(4):179–181.

Kowalewski, M. M. and Zunino, G. E. 1999. Impact of deforestation on a population of *Alouatta caraya* in northern Argentina. Folia Primatologica 70(3):163–166.

Kowalewski, M. M. and Zunino, G. E. 2004. Birth seasonality in *Alouatta caraya* in Northern Argentina. International Journal of Primatology 25(2):383–400.

Kowalewski, M. M. and Zunino, G. E. 2005. Testing the parasite avoidance behavior hypothesis with *Alouatta caraya*. Neotropical Primates 13(1):22–26.

Kumm, H. W. and Laemmert Jr., H. W. 1950. The geographical distribution of immunity to yellow fever among the primates of Brazil. American Journal of Tropical Medicine 30:733–748.

Liang, S. Y., Linthicum, K. J. and Gaydos, J. C. 2002. Climate change and the monitoring of vector-borne diseases. Journal of the American Medical Association 287:2286–2286.

Lilly, A. A., Mehlman P. T. and Doran D. 2002. Intestinal parasites in gorillas, chimpanzees, and humans at Mondika Research site, Dzanga-Ndoki National Park, Central African Republic. International Journal of Primatology 23(3):555–573.

Lourenco de Oliveira, R. and Deane, L. M. 1995. Simian malaria at two sites in the Brazilian Amazon. I. The infection rates of *Plasmodium brasilianum* in non-human primates. Memorias do Instituto Oswaldo Cruz 90(3):331–339.

Marsh, L. K. 1999. Ecological effect of the black howler monkey (*Alouatta pigra*) on fragmented forest in the Community Baboon Sanctuary. Ph. D. Thesis, Washington University, St. Louis.

Marsh, L. K. 2003. The nature of fragmentation. In L. K. Marsh (ed.), *Primates in Fragments: Ecology and Conservation.Marsh* (pp. 1–10). New York: Kluwer Academic/Plenum Publ.

Martins, S. de Souza 2000. Efeitos da fragmentação de habitat sobre a prevalência de parasitoses intestinais em *Alouatta belzebul* (Primates, Platyrrhini) na Amazônia Oriental. 2002. 86 f. Dissertação (Mestrado) – Curso de Mestrado em Zoologia, Universidade Federal do Pará, Museu Paraense Emílio Goeldi, Belém.

McGrew, W. C., Tutin, C. E. G., Collins, D. A. and File, S. K. 1989. Intestinal parasites of sympatric *Pan troglodytes* and *Papio* spp. at two sites: Gombe (Tanzania) and Mt. Assirik (Senegal). American Journal of Primatology 17(2):147–155.

Michaud, C., Tantalean, M., Ique, C., Montoya, E. and Gonzalo, A. 2003. A survey for helminth parasites in feral New World non-human primate populations and its comparison with parasitological data from man in the region. Journal of Medical Primatology 32(6):341–345.

Milton, K. 1996. Effects of bot fly (*Alouattamyia baeri*) parasitism on a free-ranging howler (*Alouatta palliata*) population in Panama. Journal of Zoology (Lond) 239:39–63.

Monath, T. P. 2001. Yellow fever. Lancet Infectious Diseases 1:11–20.

Muller-Graf, C. D. M., Collins, D. A. and Woolhouse, M. E. J. 1997. *Schistosoma mansoni* infection in a natural population of olive baboons (*Papio cynocphalus anubis*) in Gombe Stream National Park. Parasitology 115:621–627.

Muñoz, D., Estrada, A., Naranjo, E. and Ochoa, S. 2006. Foraging ecology of howler monkeys in a cacao (*Theobroma cacao*) plantation in Comalcalco, Mexico. American Journal of Primatology 68(2):127–142.

Norconk, M. A. and Grafton, B. W. 2003. Changes in forest composition and potential feeding tree availability on a small land-bridge island in Lago Guri, Venezuela. In L. K.Marsh (ed.), *Primates in Fragments: Ecology and Conservation* (pp. 211–227). New York: Kluwer Academic/Plenum Publ.

Nunn, C. L., Altizer, S., Jones, K. E. and Sechrest, W. 2003. Comparative Tests of parasite species richness in Primates. American Naturalist 162:597–614.

Nunn, C. L. and Altizer, S. M. 2005. The Global Mammal Parasite Database: An online resource for infectious disease records in wild primates. Evolutionary Anthropology 14:1–2.

Nunn, C. L. and Heymann, E. W. 2005. Malaria infection and host behavior: A comparative study of Neotropical primates. Behavioral Ecology and Sociobiology 59:30–37.

Nunn, C. L., Altizer, S. M., Sechrest, W. and Cunningham, A. A. 2005. Latitudinal gradients of parasite species richness in primates. Diversity and Distributions 11:249–256

Nunn, C. L. and Altizer, S. M. 2006. Infectious Diseases in Primates: Behavior, Ecology and Evolution. New York: Oxford University Press.

Nunn, C. L. and Dokey A. T.-W. 2006. Ranging patterns and parasitism in primates. Biology Letters 2:351–354.

Ocaido, M., Dranzoa, C. and Cheli, P. 2003. Gastrointestinal parasites of baboons (*Papio anubis*) interacting with humans in West Bugwe Forest Reserve, Uganda. African. Journal of Ecology 41(4):356–359.

Palmer, S. R., Soulsby, L, and Simpson, D. I. H. 1998. Zoonoses. Biology, clinical practice and public health control. Oxford: Oxford University Press.

Pedersen, A. B., Altizer, S. M., Poss, M., Cunningham, A. A. and Nunn C. L. 2005. Patterns of host specificity and transmission among parasites of wild primates. International Journal for Parasitology 35:647–657.

Phillips, K. A., Haas, M. E., Grafton, B. W. and Yrivarren, M. 2004. Survey of the gastrointestinal parasites of the primate community at Tambopata National Reserve, Peru. Journal of Zoology 264(2):149–151.

Plumptre, A. J. and Reynolds, V. 1994. The effect of selective logging on the primate populations in the Budongo Forest Reserve, Uganda. Journal of Applied Ecology 31(4):631–641

Pollard, A. J. and Dobson, A. R. 2000. Emerging infectious diseases in the 21st century. Current Opinion in Infectious Diseases 13:265–275.

Pope, B. L. 1966. Some parasites of the howler monkey of northern Argentina. Journal of Parasitology 52:166–168.

Pope, T. R. 1998. Effects of demographic change on group kin structure and gene dynamics of populations of red howling monkeys. Journal of Mammalogy 79(3):692–712.

Poulin, R. 1994. Meta-analysis of parasite induced behavioural changes. Animal Behaviour 48:137–146.

Pozo-Montuy, G. and Serio-Silva, J. C. 2007. Movement and resource use by a group of Alouatta pigra in a forest fragment in Balancán, México. Primates 48(2):102–107.

Prieto, O. H., Santa Cruz, A. M., Scheibler, N., Borda, J. T. and Gomez, L. G. 2002. Incidence and external morphology of the nematode Trypanoxyuris (Hapaloxyuris) callithricis, isolated from black-and-gold howler monkeys (Alouatta caraya) in Corrientes, Argentina. Laboratory Primate Newsletter 41(3):12–14.

Rivera, A. and Calme, S. (2006). Forest fragmentation and its effects on the feeding ecology of black howlers (Alouatta pigra) from the Calakmul area is Mexico. In A. Estrada, P. A. Garber, M. S. M. Pavelka and L. Luecke (eds.), New Perspectives in the Study of Mesoamerican Primates: Distribution, Ecology, Behavior, and Conservation (pp. 189–213). New York: Springer.

Rudran, R. and Fernandez-Duque, E. 2003. Demographic changes over thirty years in a red howler population in Venezuela. International Journal of Primatology 24(5):925–947.

Santa Cruz, A. C. M., Borda, J. T., Patiño, E. M., Gomez, L. and Zunino, G. E. 2000a. Habitat fragmentation and parasitism in howler monkeys (Alouatta caraya). Neotropical Primates 8(4):146–148.

Santa Cruz, A. C. M., Prieto, O. H., Roux, J. P., Patiño, E .M., Borda, J. T., Gomez, L. and Schiebler, N. 2000b. Endo y ectoparasitosis en el mono aullador (Alouatta caraya) (Humboldt, 1812), Mammalia, Cebidae, Informe Preliminar. Comunicaciones Científicasy Tecnológicas, Universidad Nacional del Nordeste.

Santa Cruz, A.C.M., Borda, J.T., Patiño, E.M., Prieto, O.H., Scheibler, N., González, A.O., Comolli, J.A., Zunino, G.E. and Gómez, L.G. 2003. Criptosporidiosis en mono aullador (Alouatta caraya Humboldt, 1812) en semicautiverio en Corrientes, Argentina. Comunicaciones Científicas y Tecnológicas, Universidad Nacional del Nordeste.

Santos, M. V. S., Ueta, M. T. and Setz, E. Z. F. 2005. Levantamento de helmintos intestinais em bugio- ruivo, Alouatta guariba (Primates, atelidae), na mata de ribeirão cachoeira no Distrito de Souzas/Campinas, SP. Congresso Brasileiro de Parasitologia, Porto Alegre, RS, Brasil.

Semple, S., Cowlishaw, G. and Bennett, P. M. 2002. Immune system evolution among anthropoid primates: parasites, injuries and predators. Proceedings of the Royal Society of London. Series B 269:1031–1037.

Solomons, N. W. and Scott, M. E. 1994. Nutrition status of host populations influences parasitic infections. In M. E. Scott and G. Smith (eds.), Parasitic and infectious diseases (pp. 101–114). New York: Academic Press.

Stoner, K. E. 1995. Dental pathology in Pongo satyrus borneensis. American Journal of Physical Anthropology 98(3):307–321.

Stoner, K. E. 1996. Prevalence and intensity of intestinal parasites in mantled howling monkeys (Alouatta palliata) in northeastern Costa Rica: Implications for conservation biology. Conservation Biology 10(2):539–546.

Stoner K. E., Gonzalez-Di Pierro, A. M. and Maldonado-Lopez, S. 2005. Infecciones de parásitos intestinales de primates: implicaciones para la conservación. Universidad y Ciencia (Sp. issue II):61–72.

Stoner, K. E. and Gonzalez -Di Perro, A. M. 2006. Intestinal parasite infections in *Alouatta pigra* in tropical rainforest in Lacandona, Chiapas, Mexico: implications for behavioral ecology and conservation. In A. Estrada, P. A. Garber, M. S. M. Pavelka and L. Luecke (eds.), *New Perspectives in the Study of Mesoamerican Primates: Distribution, Ecology, Behavior, and Conservation* (pp. 215–240). New York: Springer.

Stuart, M. D., Greenspan, L. L., Glander, K. E. and Clarke, M. 1990. A coprological survey of parasites of wild mantled howling monkeys, *Alouatta palliata palliata*. Journal of Wildlife Disease 26:547–549.

Stuart, M. D., Strier, K. B., and Pierberg, S. M. 1993. A coprological survey of wild muriquis, *Brachyteles arachnoides*, and brown howling monkeys, *Alouatta fusca*. Proceedings of the Helminthological Society of Washington 60(1):111–115.

Stuart, M. D. and Strier, K. B. 1995. Primates and parasites: A case for a multidisciplinary approach. International Journal of Primatology 16(4):577–593.

Stuart, M. D., Pendergast, V., Rumfelt, S., Pierberg, S., Greenspan, L., Glander, K. and Clarke M. 1998. Parasites of wild howlers (*Alouatta* spp.). International Journal of Primatology 19(3):493–512.

Taylor, L. H., Latham, S. M. and Woolhouse, E. J. 2001. Risk factors for human disease emergence. Philosophical Transactions of the Royal Society of London. Series B 356:983–989.

Thatcher, V. E. and Porter, J. A. Jr. 1968. Some helminth parasites of Panamanian primates. Transactions of the American Microscopical Society 87(2):186–196.

Travi, B. L., Colillas, O. J. and Segura, E. L. 1986. Natural trypanosome infection in Neotropical monkeys with special reference to *Saimiri sciureus*. In D. M. Taub and F. A. King (eds.), *Current Perspectives in Primate Biology* (pp. 296–306). New York: Van Nostrand Rehinold.

Vandamme, A. 2004. Frequent "natural" zoonotic transmission of simian foamy virus to humans. Aids reviews 6(2):118.

Vasconcelos, P. F. C., Sperb, A. F., Monteiro, H. A. O., Tortes, M. A. N., Sousa, M. R. S., Vasconcelos, H. B., Mardini, L. B. L. F. and Rodrigues, S. G. 2003. Isolations of yellow fever virus from *Haemagogus leucocelaenus* in Rio Grande do Sul State, Brazil. Transactions of the Royal Society of Tropical Medicine and Hygiene 97:60–62.

Venturini, L., Santa Cruz, A. M., González, J. A., Comolli, J. A., Toccalino, P. A. and Zunino, G. E. 2003. Presencia de *Giardia duodenalis* (Sarcomastigophora, Hexamitidae) en mono aullador (*Alouatta caraya*) de vida silvestre. Comunicaciones Científicas y Tecnológicas, Universidad Nacional del Nordeste.

Vitazkova, S. K. and Wade, S. E. 2006. Parasites of free-ranging black howler monkeys (*Alouatta pigra*) from Belize and Mexico. American Journal of Primatology 68(11):1089–1097.

Volney, B., Pouliquen, J. F., De Thoisy, B. and Fandeur, T. 2002. A sero-epidemiological study of malaria in human and monkey populations in French Guiana. Acta Tropica 82(1):11–23.

Wallis, J. and Lee, D. R. 1999. Primate conservation: the prevention of disease transmission. International Journal of Primatology 20:803–826.

Walther B. A., Cotgreave, P., Gregory, R. D., Price, R .D. and Clayton, D. H. 1995. Sampling effort and parasite species richness. Parasitology Today 11:306–310.

Wolfe, N. D., Escalante, A. A., Karesh, W. B., Kilbourn, A., Spielman, A. and Lal, A. A. 1998. Wild primate populations in emerging infectious disease research: the missing link? Emerging Infectious Diseases 4:149–158.

Yamashita J. 1963. Ecological relationships between parasites and primates. 1:Helminth parasites and primates. Primates 4:1–96.

Young, O. P. 1981. Chasing behavior between males within a howler monkey troop. Primates 22:424–426.

Zar, J. H. 1999. Biostatistical Analysis. 4th Edition. New Jersey: Prentice-Hall.

Zunino, G. E., Kowaleski, M., Oklander, L. and Gonzalez, V. 2007. Habitat fragmentation and population size of the black and gold howler monkey (*Alouatta caraya*) in a semideciduous forest in northern Argentina. American Journal of Primatology 69(9):966–975.

Chapter 18
Primate Conservation in South America: The Human and Ecological Dimensions of the Problem

Alejandro Estrada

18.1 Introduction

South America is the fourth largest of the Earth's seven continents (after Asia, Africa, and North America), occupying 17,820,900 sq km (6,880,700 sq mi), or 12 percent of the Earth's land surface. The region holds the world's interest as a result of having been the cradle of sophisticated civilizations of mankind such as the Valdivia (Ecuador), Chavin and: Chachapoyas-Aymaran kingdoms (Bolivia and southern Peru), Huari Empire (Central and northern Peru), Muisca (Colombia), and the Inca (Peru), among others (Encyclopedia Britannica 2006). South America is also considered to be one of the world's most important centers of the origin of genetic diversity and of domestication of important agricultural crops. Its indigenous peoples bred potatoes, maize, squash, various beans, manioc, and chili peppers, among others, from wild species endemic to the region (FAO 2001), and domesticated llamas and alpacas in the highlands of the Andes *ca* 3500 BC (The The Columbia Gazetteer of the World Online 2005; Microsoft Encarta Online Encyclopedia 2006).

A major geographical region contributing to the high biological diversity of South America is the Amazon basin. The basin encompasses an area about 7 million sq km in size. This vast equatorial ecosystem is home to one-fifth of the planet's plant and animal species (CBD/UNDP 2006). South America is characterized by high biological richness and high ethnic and cultural diversity. A unique feature of its fauna is a lack of affinity with the fauna of other continents, including North America north of the Mexican Plateau. Found throughout are families of mammals strictly confined to the region, including many species of primates. South America harbors a high diversity of primates represented by about 200 taxa, many of which are only found in specific regions (IUCN/SSC/PSG 2006), and new species are still being discovered (e.g., van Roosmalen, van Roosmalen and Mittermeier 2002). This fauna is an important component of the native vegetation in the region and

A. Estrada (✉)
Estación de Biología Tropical Los Tuxtlas, Instituto de Biología, Universidad Nacional Autónoma de México, México
e-mail: aestrada@primatesmx.com

P.A. Garber et al. (eds.), *South American Primates,* Developments in Primatology: Progress and Prospects, DOI 10.1007/978-0-387-78705-3_18,
© Springer Science+Business Media, LLC 2009

an integral part of the natural and cultural patrimony of South American countries. Rapid and extensive changes in the original distribution of primate habitats in the region as a result of human activity coupled with a general lack of knowledge of the basic biology, ecology and behavior makes conservation of South American primates a daunting task. Because of its high biological diversity the South American tropical vegetation is one of the world's greatest conservation challenges.

Pressures for land use have been pointed out as the major cause of tropical rain forest loss and fragmentation throughout the world (Donald 2004), and a major cause in increase in rates of species extinction in recent decades (Laurance et al. 2002). Fragmentation of habitat and stochastic forces, along with the increasing rarity of suitable habitat, also play an important role in further declines of animal populations and species at the local level (Henle et al., 2004a and references therein). Land-use patterns leading to the loss of primate habitats in South America are reported to be the result of human drivers such as cattle ranching, agricultural production, wood extraction, mining, colonization, hydroelectric projects, and oil exploration and development. In some cases these drivers may occur simultaneously, resulting in a synergistic destructive process. While the importance of these pressures may vary from region to region and country to country, three net results that impinge upon the persistence of primates in South America are diminished effective population size, isolation of populations, and local extinction of populations and species (Henle et al. 2004b).

The Millenium Ecosystem Assessment (MA 2005) definition of a driver is any natural or human induced factor that directly or indirectly causes a change in an ecosystem. A direct driver unequivocally influences ecosystem processes. An indirect driver operates more diffusely, by altering one or more direct drivers. The categories of indirect drivers of change are demographic, economic, sociopolitical, scientific and technological, and cultural and religious. Important direct drivers include climate change, plant nutrient use, land conversion leading to habitat change, invasive species, and diseases. For terrestrial ecosystems, the most important direct drivers of change in ecosystem health in the past 50 years have been land cover change (in particular, conversion to cropland and pasture), human population increase, and the application of new technologies which have facilitated the increased demand for food, timber, and fiber (MA 2005).

To understand and predict conservation pressures caused by human activity on primate habitats, populations, and species in a particular geographic region is not an easy task. We need to examine multiple factors that vary across spatial and temporal scales, and these are distinct from country to country depending on the particular historical, demographic, ethnic, political and economic conditions of each region and locality. However, this approach can provide an informed overview of the direct and/or indirect pressures exerted by anthropogenic drivers upon the persistence of primate habitats and primate populations and species. Equally important is to describe how regional conservation efforts may intervene in this process.

Given the complexity of conservation issues in tropical countries and regions (Laurance, 2007 and references therein), in this paper I briefly reviewed drivers of land-cover change including human population growth trends and density, levels of poverty, major land-use patterns as they relate to food production and to defor-estation rates, and regional conservation initiatives and their possible impact upon the persistence of primate populations and species and their habitats. I restricted projections of future trends to only a few of the anthropogenic drivers I examined because numerous assumptions about future human behavior would result in weak forecasts (*see* Laurance, 2007).

Appraisals were made for those countries with areas inside the Amazon basin (Brazil, Peru, Ecuador, Bolivia, Colombia, Venezuela, Guyana, French Guyana, and Suriname) and for those outside the Amazon basin (Argentina, Uruguay, and Paraguay). In our examination we also considered the small country island of Trinidad and Tobago, since its forests harbor populations of two primate species (*Alouatta seniculus insulanus* and *Cebus albifrons trinitatis*; Hirsch et al. 2002; IUCN/SSC/PSG 2006). However, because of its small size and insular nature, this country was excluded from certain analyses. In other cases, the number of countries under statistical scrutiny was reduced by 1–2 because of the lack of reliable statistics found in the sources listed in Appendix 1. These statistics were used as raw data or as transformed indices and variables to illustrate states, trends, and patterns for the region and for each country.

18.2 Biological Richness of South American Countries Harboring Primates

Tropical South America's biological diversity is noteworthy, especially if one con-siders that taxonomic inventories are not yet completed and that new species are discovered on a regular basis (CBD, 2006 http://www.biodiv.org/default.shtml). Although it is unusual to find new mammals, let alone primate species, since 1990, 24 new primate species have been recorded world-wide, 13 of which were found in Brazil. Among the most recent examples are two new species of titi monkeys, *Callicebus stephennashi* and *Callicebus bernhardi* (van Roosmalen et al. 2002). South American primates are important members of the mammal communities present in the habitats where they occur. An examination of the richness of higher plants, mammals, breeding birds, reptiles, and amphibians in tropical continental South America shows total species richness varying from 2,585 species in Uruguay to 58,623 species in Brazil (Appendix 2). Brazil and Colombia head the list with the highest number of species thus far recorded, followed by Ecuador, Peru. Countries such as French Guiana, Guyana, and Suriname have lower total species richness, but this is probably the result of insufficient inventories Bolivia (Appendix 2).

Information from the Database of Georeferenced Occurrence Localities of Neotropical primates (Hirsch et al. 2002) and the IUCN/SSCPSG (http://www.primate-sg.org/diversity.htm) shows that richness of primate taxa is not uniformly distributed across South American countries. Brazil has the largest number of doc-umented taxa (131), followed by Peru (51) and Colombia (50). These countries in turn are followed by Venezuela (26), Bolivia (23) and Ecuador (21); the rest of the countries have <10 taxa each (Appendix 3). While an association existed between total biological richness (measured by the number of higher plants, mammals, breeding birds, reptiles and amphibians in each country) and land area for the countries examined (r_s = 0.734 P = 0.003 n = 12), partial correlation analysis showed that number of primate taxa in each country was closely associated with forest cover (km^2 as of 2005) and not to the countries land mass (land mass constant, r forest cover = 0.740 P = 0.001, forest cover constant, r land mass = 0.26 P = 0.18 n = 13 in both cases). This suggests that primate-harboring countries with large amounts of remaining forest cover in proportion to their territories may have greater conservation responsibilities than those with less forest cover.

18.3 Human Dimension of the Conservation Problem

Land-use patterns mainly driven by production of food and by the extraction of other goods (e.g., wood, fiber) and services (e.g., hydroelectric projects, oil exploitation) for a large and fast growing human population have been identified as major drivers of tropical rain forest loss and fragmentation throughout the world (Donald 2004), and a major cause of increases in rates of species extinction in recent decades (Laurance et al. 2002; Henle et al. 2004a,b). Because social drivers are just as important as economic drivers in transformation of South American tropical forested landscapes, in the following paragraphs I provide information on human population size (including the indigenous component) and density, and on human population growth and projections as estimated by the U.N. Population Division (*see* Appendix 1). I further examined levels of human development and poverty as measured by the United Nations Development program (*see* Appendix 1).

18.4 Human Population

As of 2005 an estimated 357 million people inhabited the South American countries considered here. Brazil accounted for 50% of the population; Colombia and Argentina together contributed another 24%, and Peru, Venezuela and Ecuador accounted for 20%. The rest of the countries accounted for the remaining 6% (Appendix 4). Average growth rate of population for the period 1975–2004 was estimated at 1.7% (± 0.75%), but this varied from 0.6%/yr in Guiana to 2.8%/yr in Paraguay. An important social component of the human population in South America is its indigenous people. More than 350 indigenous groups inhabit South America (CIA 2006; LANIC 2006) and wild primates play an important role in the subsistence and symbolism of lowland South American indigenous people (Cormier 2006) and this impacts importantly on conservation (Peres 2000). The indigenous populations accounted in 2005 for *ca* 6.7% of the total population or about 23–24 million people. The Amerindians are not distributed uniformly across South American countries. Peru harbors the largest estimated numbers (53%) followed by Bolivia (20%) and Ecuador (13%). Brazil accounts for 7% of the Amerindian population, and Colombia and Venezuela for 2% each. Noteworthy, indigenous populations in Bolivia, Peru and Ecuador account for 55%, 45% and 25%, respectively, of their total population (Appendix 4).

The indigenous population represents an important cultural component of South America that is quickly vanishing as a result of rapid acculturation and displacement by nonindigenous groups. Poverty rates are higher than average amongst indigenous populations. Indigenous people have higher rates of illiteracy, chronic disease, and unemployment, and many live a marginal existence in remote forested areas with little attention paid to them by local governments (WRI/UNDP/UNEP/WB 2005; UNDP 2006a). Some of these groups still struggle for their physical survival, but many others have begun to demand ethnic recognition and assert their political visibility (Thorp et al. 2006). Much more research is needed on the ways indigenous populations manage their forests and primate wildlife, on their traditional ecological knowledge, and on ways to incorporate their interest in conservation plans (Altieri 2004; MA 2005; Cormier 2006).

18.5 Human Population Growth Trends, Projections and Population Density

Growing at an average annual rate of 1.7% for the period 1980–2000, the human population in the South American countries considered here has increased from 232 million in 1980 to ca 357 million in 2005, and projection estimates place the population at ca 500 million by 2030 (Appendix 4; Fig. 18.1). Projections show that Brazil will continue to be the demographically dominant region with a projected high population growth through 2050, followed by Colombia, Argentina, Peru, Venezuela and Ecuador. While populations will continue to be smaller in other countries, they nonetheless will undergo growth for the projected decades (Fig. 18.1). In short, all of the countries under consideration here will, in general, experience significant increases in the size of their populations in the next few decades, resulting in important demands for land, goods, and services coming from forested areas. Most of the population growth will be in urban areas, a trend that has been consistent for several decades and this will continue, at an accelerated rate, into the future (Fig. 18.1). Rapid growth of urban areas results in important demands on natural resources for water, food, construction materials, energy and the disposal of wastes (WRI/UNDP/UNEP/WB 2005). Urbanization transforms local landscapes as well as ecosystems both local and further afield, as "urban foot prints" extend well beyond their boundaries, especially in tropical nations, triggering landuse and land-cover changes in their zones of influence, sometimes over vast areas (MA 2005; UNFPA 2007).

As of 2005, the highest human population density was found in the country of Trinidad and Tobago (212 people/km^2) due to its very small size (5,139 km^2) and large population (1.4 million). In the continent, Ecuador (48.3 people/km^2), Colombia (43.9 people/km^2) and Venezuela (30.2 people/km^2) head the list. Intermediate densities were found in Brazil (21.6 people/km^2), Peru (21.9 people/km2)

(a)

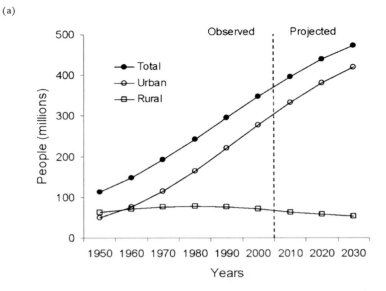

(b)

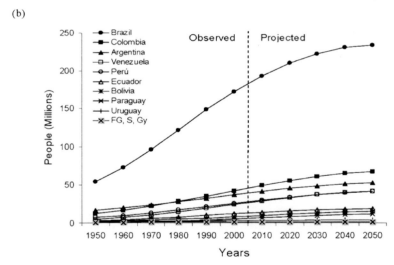

Fig. 18.1 (**a**) Human population growth trends and projections for rural and urban populations in South American countries harboring primates; (**b**) Population growth trends and projections for South American countries harboring primates (source of data *see* Appendix 1 and technical notes). FG French Guiana, S Suriname, Gy Guyana

and Uruguay (19.5 people/km^2). Lowest population densities were typical in countries such as French Guiana (2.1 people/km^2), Suriname (2.8 people/km^2) and Guyana (3.6 people/km^2). Population growth projections suggest important increases in population density in each of these countries (Fig. 18.1; Appendix 4).

18.6 Human Development and Poverty

Based on the rise and fall of national incomes (as measured by GDP or gross domestic product), we tend to equate human welfare with material wealth. However, the ultimate yardstick for measuring progress is people's quality of life (WRI/UNDP/UNEP/WB 2005; UNDP 2006a; WWF 2006). GDP continues to be measured in a way that does not take into account environmental degradation or the depletion of natural resources. Over the past decades there have been unprecedented increases in material wealth and prosperity across the world. At the same time these increases have been very uneven in the case of tropical countries, with vast numbers of people not participating in prosperity. It is ironic that amidst the enormous biological wealth of South American countries, mass poverty, deeply entrenched inequality, and lack of political empowerment are common, making the task of preserving primate habitats particularly difficult.

According to the United Nations' Human Development Index (HDI; UNDP 2006a) high poverty and low human development are typical of South American countries. The HDI looks beyond GDP to a broader definition of well being. The HDI provides a composite measure of three dimensions of human development: living a long and healthy life (measured by life expectancy), being educated (measured by adult literacy and enrolment at the primary, secondary and tertiary level) and having a decent standard of living (measured by purchasing power parity (PPP) income).

The HDI is not, in any sense, a comprehensive measure of human development. It does not, for example, include important indicators such as respect for human rights, democracy, and inequality. What it does provide is a combined measure of life expectancy, school enrolment, literacy and income to allow a broader view of a country's development than does income alone. The HDI varies from 0 to 1 (least to most human development). A comparison of the mean HDI for the top 25 developed countries in the world (HDI 0.94 ± 0.005) with those of the South American countries under consideration (HDI 0.76 ± 0.04) and those of the Mesoamerican region (HDI 0.73 ± 0.06), clearly shows the important gap that needs to be bridged to improve the quality of life of the human populations in countries harboring primates in the Neotropics. The HDI in South American varies from 0.68 (Bolivia) to 0.85 (Argentina), suggesting a perceivable gradient of human development across countries in the quality of life of their people (Appendix 5). Poverty is a pervasive condition in South American countries, where close to 25% of the population (about 80 million people) lives on less than US$2/day. The countries with the highest proportion of the population living on less that US$2/day are Bolivia and Ecuador (8.8 million people). Ranking lowest are Guyana and Uruguay (0.25 million people). Brazil alone has about 38 million people living on less than US$2/day (Appendix 5; Fig. 18.2).

Progress in human development is sometimes taken as evidence of convergence between the developed and the developing world. In broad terms, that picture is accurate: there has been a steady improvement in human development indicators for the developing world over several decades (UNDP 2006a; WWF 2006). But convergence is taking place at very different rates in different regions—and from

(a)

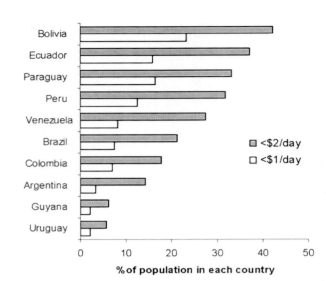

(b)

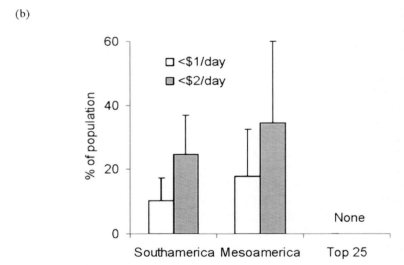

Fig. 18.2 (a) Proportion of the population living on less than $2US and $1US per day in each South American country harboring primates; (b) Proportion of the human population of South American and Mesoamerican countries harboring primates living on less than $2US and $1US per day. Same data shown for the top 25 developed countries in the planet; (source of data *see* Appendix 1)

different starting points. Inequalities in human development remain large, and for a large group of countries the disparity between rich and poor is widening. Inequalities in human development across countries and Neotropical regions are clearly seen when we compare the mean proportion of the population living on less than $1 and on less than US$2/day amongst countries in South America and when we profile this for South American and Mesoamerican countries (Fig. 18.2).

Challenges to sustainable economic growth are daunting. Despite immense resources, the distribution of income is highly skewed, with almost one-third of the region's people living in poverty. Most of the indigenous communities and rural populations in South America, the majority of whom are living below the poverty line, lack income opportunities and basic public services such as education, health, and housing. This lack of access to basic services has created pressure to overuse resources and encourage unsustainable practices that deplete environmentally fragile areas. Given that the Millennium Ecosystem Assessment (MA 2005) has predicted that rural poverty in developing regions will not improve (and may worsen) over the next 50 years, the pressure exerted by these people on their environment is unlikely to decrease in the near future.

18.7 Land-Use Patterns

The conversion of natural habitat to other land uses is the major driving force behind worldwide biodiversity loss (Sodhi et al. 2004; Laurance 2007). In recent years, as the world's population continues to grow and agricultural production must meet the rising demand for food. Agricultural expansion into forests and marginal lands, combined with overgrazing and urban and industrial growth, has substantially reduced levels of biological diversity over significant areas (Sodhi et al. 2004; Brook et al. 2006; Gardner et al. 2006; Laurance 2007). About 7,000 plant species have been cultivated and collected for food by humans since agriculture began about 12,000 years ago, but today only about 15 plant species and 8 animal species supply 90% of the food in the westernized world (USDA 2003a). A rapidly growing global human population and changing consumption patterns have stimulated the evolution of agriculture from traditional to modern, intensive systems. Nearly one-third of the world's land area is used for food production, making agriculture the largest single cause of habitat conversion on a global basis (CBD/UNDP 2006). In the following sections, I have mapped forest loss and agricultural land growth trends for the South American countries under consideration, but gave consideration to the fact that land use for food production coupled to other secondary drivers (e.g., logging) may result in the loss of enormous area of tropical forests, in fragmentation of native vegetation and in an alteration and degradation of the unique multilayered and closed tropical forest canopy (Whitmore, 1998).

18.8 Trends in Forest Loss

Extensive deforestation by human activity has been changing tropical landscapes in South America over several millennia (Wright and Muller-Landau 2006 and references therein). However, such changes have been significantly more rapid in the last 50–60 years, resulting in vast losses and, importantly, fragmentation of primate habitats and populations (Laurance et al. 2002; Marsh, 2003; Brook et al. 2006). For example, it is estimated that about 64% of the total land area (*ca* 16,836,950 km^2) of South American countries harboring primates was originally covered by forests. As of 2005, estimated forest cover was about 7.5 million km^2 or 45% of total land area (Appendix 6).

Estimated average original forest cover in South American countries (excluding Uruguay because of lack of reliable statistics) was 72%, and forest cover ranged from 30% in Paraguay to 99% in Guyana) (Appendix 6). Disappearance of original forest cover has not been uniform among the countries involved in this calculation, with Trinidad and Tobago, Brazil, Ecuador, Colombia, and Venezuela, losing more than others. French Guiana, Suriname and Bolivia still retain large proportions of their original forest cover (Fig. 18.3). Estimates of forest loss indicate that between 1960 and 2005 about 16% (1.3 billion ha) of existing forest was lost as a result of human activities (Fig. 18.3). Narrowing our examination to the period 1990–2005, we estimated that about 56,033,700 ha of forest vegetation were lost in this period. This is equivalent to about 3,735,580 ha of forest/year or about 10,234 ha/day (Appendix 6).

Average annual deforestation rate for the period 1990–2005 was estimated at –0.48%. Highest annual rates were found in Ecuador and Paraguay (–1.43% and –1.39%, respectively) and lowest in French Guiana and in Colombia (–0.02% and –0.08%, respectively) (Appendix. 6). All countries examined showed negative rates of change and these translate into forest vegetation being eradicated by human activities in each country. The country with the highest loss for the period 1990–2005 was Brazil (423,290,000 ha), followed by Bolivia (61,274,000 ha), Venezuela (43,130,000 ha) and Ecuador (29,640,000 ha). While these four countries as a group accounted for 92% of total estimated forest loss for the 1990–2005 period, Brazil alone accounted for 76% of this loss (Appendix 6).

18.9 Pastures and Arable and Permanent Crop Lands

Human induced pasturelands and arable and permanent crop lands dominate significant portions of the tropical South American landscapes. As of 2005, pasture and arable and permanent crop land covered an estimated 5,680 610 km^2 or about 34% of total land area (Appendix 7), with pasture accounting for about 80% of the land dedicated to these land uses. In 2005, Brazil accounted for 44% of pasture lands in existence in the South American realm involving primate harboring countries, with Colombia and Bolivia contributing 16%, and Paraguay, Venezuela, Peru, Uruguay and Ecuador another 17%. Brazil also dominated the contribution to arable and permanent crop land with 54%, followed by Argentina with 24% (Appendix. 7). Statistics from the FAO agency databases indicate that forest loss for the period 1960–2005 has been paralleled by significant increases in the extension of pasturelands and of arable and permanent cropland, with about 83,994,000 ha of pastures and 53,640,000 ha of arable and permanent crop land added in this period (Fig. 18.3).

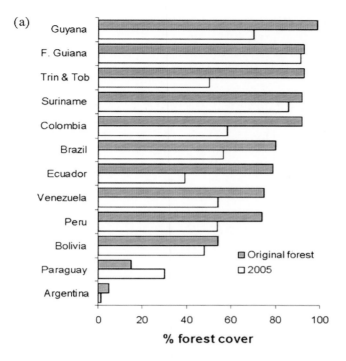

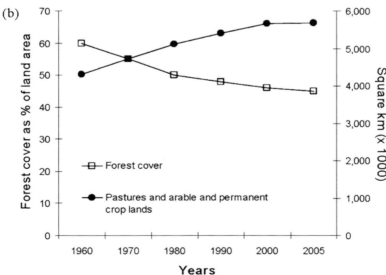

Fig. 18.3 (**a**) original forest cover 8000 years ago (assuming current climatic conditions) and forest cover as of 2005 in South American countries harboring primates; (**b**) trends in forest cover change and in pasture and arable and permanent crop lands for the period 1960–2005 (*see* Appendix 6 and Appendix 1 for technical notes)

18.10 Relationship Between Loss of Forest Cover and Other Variables

18.10.1 Population Density and Forest Cover

Recently, human population growth and associated high density, especially in rural areas, has been suggested to be a major driver of forest conversion to agriculture and to other forms of land use in tropical counties (Wright and Muller-Landau, 2006). The same authors proposed, in contrast to a recent array of studies providing an opposite view (Laurance, 2007 and references therein), that because there will be a strong trend toward urbanization in tropical America and Asia, pressures on forest cover will be reduced in rural areas, at least in this century (Wright and Muller-Landau, 2006). Responses to such arguments point out that (1) the relationship between rural and urban population densities and deforestation rate is likely to be extremely complex (Brook et al. 2006; Laurance 2007); (2) even if net tropical deforestation stops within the next few decades, most essential habitat for the majority of species will have already been eliminated or severely degraded (Gardner et al. 2006); (3) biological attributes (e.g., structure and composition) of forest fragments or regenerating forests are likely to be as or more critical for species persistence than total forest area (Brook et al. 2006); (4) even those people not residing in rural areas (and thereby not impacting forests directly) will nevertheless drive an increasing demand for basic necessities (food, timber for housing and fuel, among others) and the raw materials for economic development.

While it is evident that there is still little understanding about how human influence exactly scales with human population density, the number of people in a given area is frequently cited as a primary cause of declines in species and ecosystems, with higher human densities leading to higher levels of influence on nature (WRI/UNDP/UNEP/WB 2005; UNDP 2006a). Considering this and the caveats of the earlier paragraph, the data examined indicate that percent of forest cover remaining in 10 South American countries for which we have reliable data appear to be negatively related to total human population density ($r = -0.563$ $P = 0.044$ $n = 10$) (Fig. 18.4). Countries also can be ordered along a gradient of transformation, with French Guiana, Suriname and Guyana at one extreme with low population density and high forest cover to Ecuador with the highest population density and lowest forest cover in its territories, and countries such as Brazil, Peru, Colombia and Venezuela occupying intermediate positions (Fig. 18.4).

18.10.2 Forest Loss and Growth Trends in Pasture and in Arable and Permanent Crop Lands

Important threats to primate habitats in South American countries are brought about by industrial drivers of tropical deforestation such as large-scale cattle ranching, soy farming, oil-palm plantations, timber extraction, oil, gas and hydropower

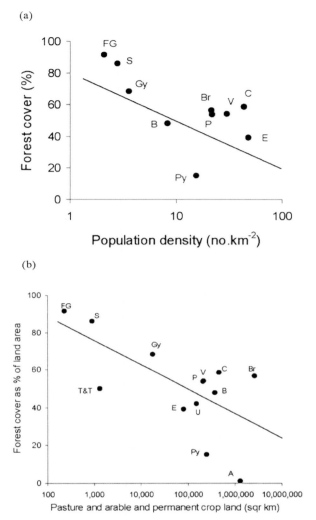

Fig. 18.4 (**a**) Social correlate of forest loss. The proportion of forest cover remaining in south American countries correlated with log population density in 2005; $r_s = -0.563$ P $= 0.044$. FG French Guiana, S Suriname, Gy Guyana, B Bolivia, Py Paraguay, Br Brazil, P Peru, V Venezuela, C Colombia, and E Ecuador; (**b**) socioeconomic correlates of forest loss. The proportion of forest cover remaining in south American countries correlated with existing area (log) of pastures and arable and permanent crop lands in 2005 (r $= -0.659$ P $= 0.007$). FG French Guiana, S Suriname, T & T Trinidad and Tobago, Gy Guyana, E Ecuador, P Peru, U Uruguay, V Venezuela, Py Paraguay, B Bolivia, C Colombia, A Argentina, Br Brazil. Log scale in "x" axis in both graphs

development, and major highways and infrastructure projects, among others. In addition, social drivers such as population growth and socio-political drives such as colonization, (Laurance et al. 2002; Sodhi et al. 2004) resulting in large demands for land for food production and related goods, together with global market demands, are severely impacting tropical forests (Laurance 2007). Consistent with this, we found that the proportion of forest cover remaining in 2005 in South American countries was inversely correlated with existing area of pastures and arable and permanent crop lands (r = −0.659 P = 0.007) (Fig. 18.4). Countries such as Brazil, Argentina, and Paraguay are at one extreme and have large areas of pasture and arable and permanent crop lands and reduced forest cover. At the other extreme are Guyana, Trinidad, Tobago, Suriname, and French Guiana, with relatively limited areas of pasture and arable and permanent crop lands and high forest cover (Fig. 18.4).

18.11 Forest Loss and Primate Taxa

Because forest loss caused by human activity will continue to exert significant pressures upon the persistence of primate populations and species in South American countries, it is important to project into the future the magnitude of such loss. Using an exponential decay model, I projected forest loss from 2005 at 50 and 100 years for each South American country harboring primates. In developing this model ($P = P_0(1 - r)^t$ – where P_0 is the initial quantity of y and t is time in units of the relative decay rate; Zar, 1998), I used average annual deforestation rates for 1990–2005 to project a "business as usual scenario". Such a scenario shows, for the countries as a group, a continued tendency for less forest cover (expressed as percent of land area), at 50 (42%) and at 100 years (36%). This tendency is even more evident when contrasted with the corresponding values for 2005 and with the estimated original forest cover for primate-harboring countries (Fig. 18.5). A country by country examination shows that, under this scenario, the greatest forest losses at 50 and 100 years are expected in Trinidad and Tobago, Suriname, Brazil, Ecuador, and Venezuela (Fig. 18.5 and Appendix 3). Contrasting scenarios can be projected reducing annual rate of forest loss by half or by assuming no net loss at 50 and at 100 years from 2005.

Between 1990 and 2005, an estimated 554,843 km^2 of forest disappeared in the territories of South American countries. Because such loss varied from country to country and because countries also vary in the number of primate taxa reported thus far in their territories, I estimated density of primate taxa (taxa/km^2 of forest cover as of 2005) for each country and further standardized the measure to number of taxa/1000 km^2 of forest cover (Appendix 3, 6). The working assumption is that, on average and assuming all conditions remain the same, countries with more species per unit of forest area would loose more species due to forest loss, than countries with lower species density, if both countries lost the same area of forest. Under this scenario primate taxa density and 2005 forest cover in each country were negatively correlated (r = −0.583 P = 0.01 n = 13), with Brazil, in spite of having the largest number of primate taxa (N = 131) and forest cover, being the lowest ranking country according to the density measure (Fig. 18.6). Countries such as Trinidad and Tobago, Paraguay, Ecuador, and Argentina ranked highest, either because they

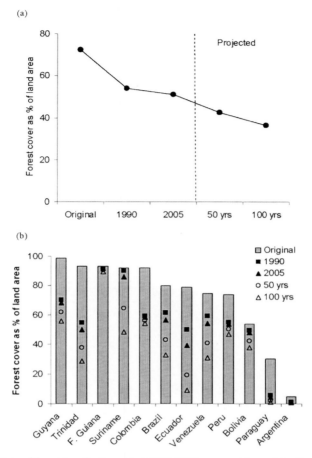

Fig. 18.5 (**a**) Rate of forest loss for the period 1990–2005 was used to model a "business as usual scenario", at 50 and at 100 years from 2005, using an exponential decay model; (**b**) same scenario for each South American country. Original forest cover in each case refers to forest cover 8,000 years ago, assuming current climatic conditions (*see* Appendix 1 for sources of data and technical notes)

contained small amounts of forest cover and a small number of primate taxa (i.e., Trinidad and Tobago, Uruguay, Argentina), or because they harbored a large number of primate taxa and a relatively low amount of forest cover (i.e., Ecuador) (Fig. 18.6; Appendix 3). In the case of Brazil, and assuming all habitats and primate taxa are uniformly distributed, we expect 1.0 primate taxa to exist in every 40,000 km^2 of forest cover. An estimated 423,290 km^2 of forest were lost in Brazil between 1990 and 2005 (Appendix 6) or 28,219 km^2/yr. Assuming that this yearly loss of forest continues for another 10 years ("business as usual scenario"), Brazil would lose an additional 200,000 km^2 of forest and about five primate taxa. In contrast, Colombia

(a)

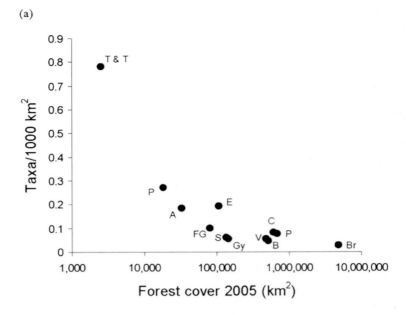

(b)

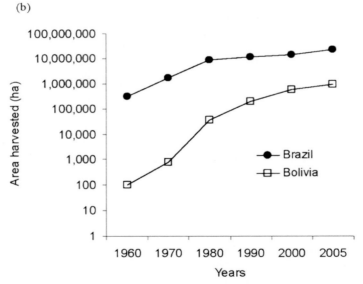

Fig. 18.6 (a) Relationship between log forest cover in 2005 and primate taxa (taxa/1000 km^2) in each south American country. Br Brazil, P Peru, C Colombia, E Ecuador, T & T Trinidad and Tobago, B Bolivia, V Venezuela, Gy Guyana, S Suriname, FG French Guiana, A. Argentina, Py Paraguay, and U Uruguay; (b) Expansion of soybean production (log area harvested) in Brazil and Bolivia for the period 1960–2005 (source of data in Appendix 1)

and Ecuador, would lose the same number of primate taxa with a loss of only about 60,000 km^2 and about 30,000 km ^{2} of forest, respectively (Appendix 8).

18.12 Forest Loss and Global Market Demands

While local demands for food and for other goods and services negatively impact forested lands, global market demands for consumable goods (e.g., crops) originating in South American countries also exert important pressures upon native habitats. Here I will focus on two crops, soybeans and palm oil that are increasingly important not only as a source of food products for humans and domestic animals, but also as a source of biofuel production.

18.12.1 Soybeans

Soybean (*Glycine max* L.) is a leguminous crop, which produces a number of important food and industrail products (Casson 2003). In Brazil, the soybean industry is considered to be of critical importance to the national economy because it is a dominant agricultural crop. Agriculture plays a significant role in Brazil and represents 14% of GDP and 33.5% of the value of exports. It also provides jobs for 13% of the labor force (Schnepf et al. 2001). The Brazilian government has taken a special interest in the expansion of soybean plantations since the 1960s because soybeans were identified as a viable export crop that could bring in valuable foreign exchange (Schnepf et al. 2001). Most of the demand for soybean products comes from the United States (24%), the European Union (24%) and China (6%), where soymeal is also used for livestock production (Oil World 2002; USDA 2003a). Soy oil (a byproduct of soymeal production), used by the food, detergents, cosmetics, and chemical industries, now ranks as the most important edible oil in the world with a global market share of 23% (Casson 2003).

Data from FAO (FAOSTATS 2006) for the period 1960–2005, shows that soybean production in Brazil and in Bolivia has undergone a steep increase in hectares harvested since the 1980s, gaining additional momentum between 2000 and 2005 (Fig. 18.6). This was the result of Brazilian crop researchers succeeding in breeding high-yield soybean varieties for every climate regime in the country, including tropical varieties for the equatorial lowlands (USDA 2003b). USDA has noted that tropical soybeans are already being cultivated near the major port of Santarem on the Amazon River in the state of Para. In the remote northern state of Roraima – where tropical rainforest is the dominant vegetation – three soybean crops can now be grown in a year (USDA 2003b). Until the 1980s, soybean plantations were primarily concentrated in the south-southeast region of Brazil (Paraná, Rio Grande do Sul & Santa Catarina). However, expansion has rapidly increased in the central-west states of Mato Grosso do Sul, Mato Grosso, and Goiás since the 1980s (Casson 2003), invading areas of Brazil harboring large expansions of tropical rain forest vegetation.

It has been argued that the latest increase in soybean production in Brazil has also resulted in more cattle ranchers and soy farmers, who require socio-environmental certification, and must comply with guidelines and regulations requiring forest reserves on private property. Although this could result in opportunities for conservation (Nepstad et al. 1999) , this kind of conservation does not consider preserving large tracts of forests or plans to interconnect coteries of isolated private reserves. Soy bean production has increased significantly in Brazil and in Bolivia, in terms of tons/year, however, this process has been at the cost of large expanses of forested land.

In Brazil, soybean expansion also has been linked to cattle ranch expansion and charcoal production. In the Cerrado region, for instance, soybean expansion has provided access to Cerrado trees, particularly those found nearby rivers in gallery forests (RBGE 2003). These trees are used by the Brazilian steel industry for charcoal production. It is estimated that 80% of the charcoal used in the Brazilian steel industry is derived from native Cerrado trees. The removal of Cerrado trees for the Brazilian steel industry is thought to result in the loss of 200,000 ha of gallery forests per year (ELC 2002). The removal of gallery forests has raised concern because these forests provide critical habitat for diverse fauna including a number of endemic species, and constitute corridors linking the Amazon and the coastal rainforests with the Cerrado on the central plateau (Tengnäs and Nilsson 2003). Soya cultivation is understood to be responsible for more Amazon destruction than any single other business in present times – including cattle ranching or logging (Biofuelwatch 2007; Boswell 2007).

18.12.2 Palm Oil

Among other commodities demanded globally, palm oil (*Elaeis guineensis Jacq.*) stands out in a similar fashion as soybeans. The oil palm is native to West Africa, and is now cultivated in large-scale plantations throughout the tropics. It is used in a number of commercial products including cooking oil, soap, cosmetics, and margarine. Globally, oil palm area increased by 43% from approximately 6 million ha in 1990 to 10.7 million ha in 2002 (Casson 2003). Large-scale oil palm and soybean plantations have a significant direct and indirect impact on biodiversity in bio-diverse countries such as Brazil because plantations are primarily large-scale, commercial monocultures. Development of monoculture plantations results predominantly in the total clearing of natural vegetation, and the use of pesticides and herbicides which largely eliminate remaining vestiges of indigenous biodiversity and significantly diminish the chances of habitat restoration.

18.12.3 Biofuels

A recent global interest is the production of biofuel from corn, sugar cane, soy beans, and palm oil. Brazil's bioethanol program goes back to the 1970s, and it is the only large-scale program which is now able to expand without government subsidies; 40% of Brazil's transport fuel comes from ethanol made from sugar-cane. Brazil produces one-third of global bioethanol, and is second only to the United States (corn ethanol) (RRU 2002; Boswell 2007). No country produces ethanol as cheaply and efficiently as Brazil. If global ethanol use is to soar, then much of this will come from Brazil. Brazil is also a regional leader in the production of biodiesel from soya beans and from palm oil (Biofuelwatch 2007; Boswell 2007). Because of rapidly growing markets in the EU, United States, China, Japan and elsewhere, Brazil is looking to double its bioethanol production in the next decade, and to vastly expand its biodiesel production for export, using soya, palm oil and castor oil (Biofuelwatch 2007; Boswell 2007). Such an expansion will require massive deforestation, additional water resources, heavy use of petroleum-based fertilizers and pest controllers, and fossil fuel energy to manufacture biofuels (Boswell 2007). This also will have a significant impact on local and global weather (RRU 2002; Righelato 2005).

18.13 Regional Conservation Initiatives

18.13.1 Natural Protected Areas and Primate Conservation

In spite of overpopulation, poverty and underdevelopment, and concerned with the conservation of their biodiversity, South American countries have taken important steps toward preserving their natural resources. All South American countries have ratified the International Convention on Biodiversity and have taken measures to protect natural ecosystems in their territories. Natural protected areas (NPAs) are internationally recognized as a major tool in conserving species and ecosystems. They also provide a range of goods and services essential to sustainable use of natural resources. South American countries have been building up their systems of NPAs, but the number of protected areas vary considerably from country to country, depending on national needs and priorities, and on differences in legislative, institutional, and financial support (Chape et al. 2003, 2005; Rodrigues et al. 2004; IUCN/WCPA/UNEP 2006).

The World Database on Protected Areas of the United Nations Environmental Program (Appendix 1) indicates that, as of December 2006, there were a total of 2,723 NPAs in the 13 South American countries under consideration here. These NPAs protect an estimated 3,032,897 km^2 or about 18% of the countries' cumulative territory (16,836,70 km^2) (Appendix 6). The same data set shows that the number of NPAs and the number of square kilometers protected has more than doubled since 1970, stressing the efforts by these countries to enhance conservation of their natural ecosystems (Fig. 18.7). A country-by-country examination shows that Brazil and Colombia contribute 66% of the total area protected by NPAs (3,032,897 km^2). However, Brazil is a major contributor with 54% of the area protected and 47% of the existing NPAs. In terms of area protected in proportion to their territories,

(a)

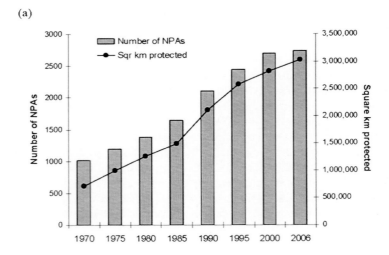

(b)

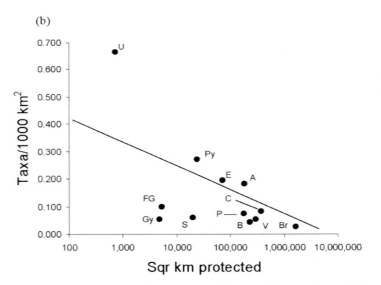

Fig. 18.7 (a) Increase in number of natural protected areas (NPAs) and square kilometers protected by South American countries harboring primates between 1970 and 2006; (b) Relationship between (log) square km protected and primate taxa density (taxa/1000 km^2) for South American countries (r = −0.732 P = 0.002 n =13). U Uruguay, Py Paraguay, E Ecuador, A Argentina, FG French Guiana, Gy Guyana, S Suriname, C Colombia, P Peru, B Bolivia, V Venezuela, Br. Brazil (*see* Appendix 1 for sources of data and technical notes)

Colombia and Venezuela rank highest with >30 and <40%, of their territories protected (Appendix 9).

An interesting question is whether establishment of NPAs increases the area (km^2) under protection. A positive association between number of NPAs and area protected ($r_s = 0.76$ P $= 0.001$ n $= 13$), suggests this to be the case. Partial correlation analysis, where the countries' land area was kept constant, showed that the number of NPAs was correlated with the number of square kilometers protected ($r = 0.674$ P $= 0.01$ df 10); when number of NPAs was kept constant, the correlation between land area and number of square kilometers protected was not significant ($r = 0.467$ P $= 0.12$ df 10). The number of NPAs and the number of km^2 under protection also were related to the countries biological richness, defined here as the total number of higher plants and vertebrates in each country (number of NPAs and biological richness $r_s = 0.881$ P $= ¡0.001$; km^2 protected and biological richness $r_s = 0.909$ P $=< 0.001$; n $= 12$ in each case), suggesting that as biological inventories are augmented, countries become interested in protecting their biological richness by increasing the number of NPAs and of km^2 under protection.

While natural protected areas are not the only method available to conservation planners, they are nonetheless appropriate in protecting primate biodiversity in South American countries. Consistent with this is the positive association found between number of square km protected and number of primate taxa in each country ($r_s = 0.694$ P < 0.01). However, when a density measure (taxa/1000 km^2 of forest cover) is used instead of number of primate taxa, a negative association ($r = -0.732$P $= 0.002$ n $= 13$) was found between these two variables (Fig. 18.7). This suggests that countries that display high taxa density values, tend, in general, to have a smaller number of square km protected, bringing up the need for more conservation efforts in their territories.

18.13.2 Caveats to Consider with Respect to NPAs

In spite of the efforts invested by South American countries in protecting their native vegetation and ecosystems, the South American system of NPAs suffers from several problems: few of the areas are actually protected, others remain as paper parks, the majority are less than 10,000 ha in size, half are not staffed, only a few have specific management plans, most are poorly delimited, research projects are only being carried out in a small number of them, and because deforestation rates in surrounding areas are particularly high, many protected areas already are or are becoming virtual islands of vegetation surrounded by altered landscapes (Defries et al. 2005; Rodrigues et al. 2004; UNDP 2006b).

Although protection of undisturbed habitat in natural protected areas (e.g., parks and reserves) in the Neotropics is crucial for primate conservation, these areas alone may not meet long-term conservation goals (Chape et al. 2005). To begin, the average landmass protected in South America is only about 18%, many of the natural protected areas may not be suitable for primate habitation (e.g., sand dune vegetation, high altitude forests, etc.), and in other cases primate species may not be found within park boundaries. The percentage of area currently protected in a given country may be a poor guide for additional conservation efforts. Instead regions with the greatest need for the growth of networks of protected areas are not necessarily those with a lower percentage of their area protected; rather, they typically are those with high levels of endemism (Rodrigues et al. 2004). Primate biodiversity, as is the case for the world's tropical biodiversity, is not evenly distributed across landscapes, regions, and countries. Detailed mapping of this biodiversity is still incomplete for the majority of the countries, regions and taxa (Gaston 2000).

18.14 Complimentary Conservation Scenarios

The inherent assumption that species richness is not supported in areas outside fully forested habitat is questionable (Ricketts 2001; Horner-Devine et al. 2003). The presence of surrounding land-use practices that are conducive to conserving species richness, despite loss of forest habitat, could enhance conservation capacity (Hughes et al. 2002). These areas may play an important role in long-term primate and biodiversity preservation and must be considered in landscape-level approaches to conservation (Schroth et al. 2004, Estrada et al. 2006). They also provide food and cash income for millions of rural households and comprise the basis of regional and national economies in many tropical countries (Ricketts 2001; Daily et al. 2003; Murphy and Lovett-Doust 2004). While primates that disperse into the anthropogenic landscapes as a result of habitat fragmentation and isolation may face important risks such as increased predation by humans and dogs, and increased exposure to disease (e.g., pathogens and parasites) from humans and domestic animals and contaminated sources of water, they also may find shelter, food and opportunities to disperse into new groups (Estrada et al. 2006). Such opportunities may allow primates to persist for many years in anthropogenic landscapes and thus active conservation efforts are required not only within but also outside of protected area boundaries, in the matrix of surrounding anthropogenic habitats (Naughton-Treves and Nick 2004; Estrada et al. 2006; Muñoz et al. 2006).

Studies of the ecological value of forest-dwelling primates have stressed their roles as potential pollinators, seed dispersers and as plant and seed predators. This reveals the importance of primate survival for continued forest dynamics and existence (Chapman and Onderdonk 1998; Bollen et al. 2004), and for local human population and economies, especially where humans use the fruits and seeds of such tree species (Lambert and Garber 1998). Primates may also provide important ecological services to the agroecosystems in which they live and/or visit. For example, the foraging activities of insect-eating monkeys such as squirrel monkeys (Saimiri sciureus) in palm plantations and of golden-headed lion tamarins (Leontopithecus chrysomelas) in cabruca cacao plantations may be important in ameliorating the impact of insect pests (Raboy et al. 2004). Primates such as howler monkeys living in cacao plantations may aid primary productivity via foliage and branch pruning and may also contribute important nutrients via their feces to the plantation's soil (Muñoz et al. 2006). Crop raiding may be less common in the Neotropics than in the Paleotropics, where primate species assemblages lack terrestrial forms and therefore, it is possible that relationships between humans and primates coexisting in forest-agricultural landscapes may be less antagonistic in this geographic region than in Africa or Asia (Estrada 2006).

18.15 Concluding Comments

This preliminary examination of the conservation problems faced by South American primates indicates a complex set of multidimensional factors that must be considered. While our approach has tried to look at regional and country-to-country patterns to identify major drivers of forest loss, historical, demographic and social issues intrinsic to each South American nation and region, also play an important and determining role. Diagnostic conservation approaches need to address two levels of inquiry: one regional and another more local. In this paper I have focused on the first approach in order to highlight major trends and processes.

A graphical summary of the interrelationships among the conservation issues examined here (Fig. 18.8), shows that while this graphic model addresses regional concerns, it also can be applied at the local level. The model also highlights overlapping and synergistic factors affecting conservation problems in primate habitats in South America. High human population growth (in both urban and rural populations) creates important demands for land use aimed at enhancing food production and the provisioning of other goods and services for the population. At the same time, social and economic pressures have led governments to political solutions,

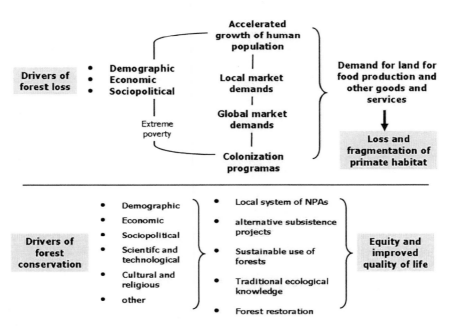

Fig. 18.8 Upper half, major social and economic drivers discussed in this chapter that threaten primate habitats in South American countries. Lower half, drivers of forest conservation. Much more research is needed to understand how these drivers operate and interlink with one another to produce tangible conservation initiatives

expanding colonization programs. In this scenario, global market demands (crops, timber, petroleum, minerals, etc.) also exert added pressures upon native habitats. These interactive processes have resulted in high and continuous rates of native habitat loss in the majority of South American countries harboring primates, which are also paralleled by high levels of poverty and low human development.

Notwithstanding the social and economic problems faced by South American countries, the undertaking of conservation initiatives also has become a national priority. These conservation efforts include (1) the establishment of a system of NPAs, along with more local initiatives such as community-based reserves, national parks, ecological reserves, and biological field stations; (2) programs promoting community-based sustainable use of the forest e.g., growing shade coffee andcacoa, spices, ornamental plants and ecotourism projects among others, (3) projects promoting restoration of native habitats in human-altered landscapes such as reforestation, establishment of biological corridors, and species reintroductions among others. Needless to stress, these initiatives must incorporate the interests and rights of indigenous groups and of rural people into conservation plans.

Effective conservation cannot succeed without within-country development of basic and applied research with primates. Such development depends also on having local professional primatologists. They in turn can also play a catalytic role in training and educating local people, university students, and government officials (*see* Strier et al. 2006 for an example). Such process will ensure continuity of nation-wide and local conservation efforts, enhancing the scientific knowledge of South American primates, and promoting development of diagnostic assessments that can lead to successful conservation and management of primate habitats, populations, and species.

To conclude, I believe that conservation primatologists need to give consideration to the direct and indirect impacts of demographic, economic, sociopolitical and cultural aspects of anthropogenic drivers that bring about tropical forest loss in South America. These drivers interact with one another in complex ways that differ from country to country, from region to region, and from locality to locality. Factored into this analysis, special attention must be given to the well being of the human population, including the culture and heritage of indigenous peoples. With this in mind, I list below a number of issues that merit attention in on-going and future primate conservation research in South America.

(1) the complexity of the interplay between human population growth and density, and transformation of forested landscapes (Laurance 2007), (2) the impact of industrial drivers and other potential uses of tropical land (e.g., biofuel production) upon the persistence of primate habitats (Boswell 2007); (3) surviving forests (i.e., primate habitats) are not merely shrinking in area, but they are also being extensively fragmented (Brook et al. 2006); (4) even if net tropical deforestation stops within the next few decades, most essential habitat for many primate species will have already been eliminated or severely degraded (Gardner et al. 2006); (5) biological attributes (e.g., structure and composition) of forest fragments or regenerating forests are likely to be as or more critical for primate species persistence than total forest area (Brooks et al. 2003; Brook et al. 2006); (6) the need for more regional

and/or local approaches in which consideration needs to focus on geographically restricted areas (e.g., the Atlantic forests and Cerrado forests of Brazil) with high species endemism, high habitat loss and fragmentation and/or high human population growth; (7) many endemic primate taxa also may be found outside of protected areas, and therefore may be at high risk of extinction (Dinerstein and Wikramanayake 1993; Rodrigues et al. 2004); (8) the extinction momentum implied by the species-area relationship, termed the "extinction debt" of past habitat loss (Tilman et al. 1994) coupled with small population sizes, is another critical threat to the persistence of primate populations and species (Brook et al. 2006); (9) many species, including primates, are becoming functionally, if not actually, extinct, and these losses are likely to have long-term impacts on ecosystem structure and functioning (Burnett and Flannery 2005); (10) the absence of a strong empirical foundation (data vacuum for most South American primate taxa) seriously hinders our ability to make accurate predictions about future threat and local extinctions as primate habitats face unparalleled threats from a range of factors, including land-use change, habitat fragmentation and isolation, habitat degradation, invasive species, wildfires, overhunting, increasing industrialization and globalization, together with ambiguous local conservation policies, and climate change (Laurance and Peres 2006; Gardner et al. 2006).

18.16 Summary

Competing pressure on land use has been identified as the major cause of tropical rain forest loss and fragmentation throughout the world, and a major cause of increases in rates of species extinction in recent decades. Because of its high primate diversity South America is one of the world's greatest conservation challenges. A large human population characterized by high levels of poverty, low human development, and high growth rates has created increased requirements for food and for other goods and services which together with global market demands have resulted in high levels of deforestation (rapid expansion of pasture and agricultural lands).in South America. Alarmingly, those countries with high losses of forests are those harboring high numbers of primate taxa. Regional conservation initiatives may mitigate these effects as they have resulted, as of 2006, in the establishment of 2,723 natural protected areas encompassing 18% of the territory of these countries. However, the skewed distribution of protected areas, suggests that these efforts may be insufficient to ensure the persistence of populations of all primate taxa, many of which are endemic to particular regions. Further diagnostics are urgently required at national and local levels to improve conservation approaches.

Acknowledgments Support for preparation of this chapter was kindly received from the Scott Neotropic Fund of the Cleveland Metro Park Zoo. I am also grateful to two anonymous reviewers and to Dr. Karen Strier and Dr. Paul Garber for helpful comments to improve this paper.

Appendix 1

Data drawn from data bases and sources listed below were compiled, tabulated and organized to calculate descriptive statistics and to depict trends over time. Also included in this appendix are technical notes concerning estimates of biodiversity, human population size and growth, poverty, forest area and forest loss and natural protected areas.

Biodiversity

Convention on Biological Diversity (CBD). UN Development Programme
http://www.biodiv.org/default.shtml
Earthtrends database www.earthtrends.org
World Resources Institute www.wri.org
GEO-3 Data Compendium. United Nations developmental Programme.
http://geocompendium.grid.unep.ch/data_sets/biodiversity/reg_biodiv_ds.htm
http://www.unep-wcmc.org/index.cfm
United Nations Environmental Program www.unep.rg
United Nations Development Programme www.undp.org
Mongabay www.mongabey.com
Conservation International www.conservation.org
World Widlife Fund www.panda.org; http://www.panda.org/about_wwf/where_we_
   work/latin_america_and_caribbean/about/index.cfm).
World Resources 2005: The Wealth of the Poor—Managing Ecosystems to Fight
   Poverty. 2005. World Resources Institute (WRI) in collaboration with United
   Nations Development Programme, United Nations Environment Programme, and
   World Bank. Washington, DC: WRI.
Human Development Report 2006. Published for the United Nations development
   Programme (UNDP). Beyond scarcity: Power, poverty and the global water crisis.
   Palgrave Macmillan, NY.
Living Planet Report 2006. World Wildlife Fund. Zoological Society of London,
   Global Footprint Network.
http://www.panda.org/news_facts/publications/living_planet_report/index.cfm
"South America." The Columbia Gazetteer of the World Online. New York:
   Columbia University Press, 2005. http://www.columbiagazetteer.org/
"South America," Microsoft® Encarta® Online Encyclopedia 2006
http://encarta.msn.com © 1997-2006 Microsoft Corporation. All Rights Reserved.
http://encarta.msn.com/encyclopedia_761574914/South_America.html

Technical notes biodiversity:

The Number of Known Species refers to the total number of known, described, and recorded species in a given country. Total numbers for all species groups include both endemic and non-endemic species (a species that is found in a particular region and nowhere else is said to endemic to that region). Numbers may also include introduced species. Figures are not necessarily comparable among countries because taxonomic concepts and the extent of knowledge about actual species numbers vary. Country totals of species are underestimates of actual species numbers. The number of Known Plants include vascular plant species (flowering plants, conifers, cycads and fern species), but do not include mosses. Known Mammals exclude marine mammals. Known Birds include only birds that breed in that country, not those that migrate or winter there. Data are collected by the United Nations Environment Programme World Conservation Monitoring Centre (UNEP-WCMC) from a variety of sources, including, but not limited to: national reports from the convention on biodiversity, other national documents, independent studies, and other texts. Data are updated on a continual basis as they become available; however, updates vary widely by country. While some countries (UNEP-WCMC estimates about 12) have data that were updated in the last 6 months, other species estimates have not changed since the data were first collected in 1992. The complete UNEP-WCMC dataset from which Known Species of Mammals, Birds, Plants, Reptiles, and Amphibians were extracted represents only about 2% of the total species of the world. As a result, the numbers reported here are vast underestimates of the actual species worldwide. Mammals and birds are better known and represented than other taxonomic groups. Invertebrates in the kingdom Animalia, the kingdom Protista, and the kingdom Monera are not included in these country profiles. Data on Known Species of Mammals, Birds, Plants, Reptiles, and Amphibians are based on a compilation of available data from a large variety of sources. They are not based on species checklists. Data have been collected over the last decade without a consistent approach to taxonomy. Additionally, while the number of species in each country does change, not all countries are updated systematically, and some data may not reflect recent trends.

Primate species richness

Hirsch, A.; Dias, L.G.; Martins, L. de O.; Campos, R.F.; Resende, N.A.T. and Landau, E.C. (2002). Database of Georreferenced Occurrence Localities of Neotropical Primates. Department of Zoology / UFMG, Belo Horizonte. http://www.icb. ufmg.br/zoo/primatas/home_bdgeoprim.htm and CD-Rom.
IUCN/SSC Primate specialist group http://www.primate-sg.org/diversity.htm

Human population

Earthtrends database www.earthtrends.org
World Resources Institute www.wri.org
United Nations Development Programme www.undp.org
http://www.unep-wcmc.org/index.cfm
Living Planet Report 2006. World Wildlife Fund. Zoological Society of London. Global Footprint Network.
http://www.panda.org/news_facts/publications/living_planet_report/index.cfm

Indigenous population

Latinamerican network information center U of Texas
http://www1.lanic.utexas.edu/la/region/indigenous/
http://www.state.gov/g/drl/rls/hrrpt/2001/wha/8297.htm
http://www.state.gov/g/drl/rls/hrrpt/2005/c17099.htm
The CIA World Factbook http://geography.about.com/library/cia/blcindex.htm
http://www.rethinkvenezuela.com/downloads/indigenous.htm
http://en.wikipedia.org/wiki/Main_Page
Human Development Report 2006. United Nations Development

Programme (UNDP). Beyond scarcity: Power, poverty and the global water crisis
Palgrave Macmillan, NY.
World Resources 2005: The Wealth of the Poor—Managing Ecosystems to Fight
Poverty. World Resources Institute (WRI) in collaboration with United Nations
Development Programme, United Nations Environment Programme, and World
Bank. Washington, DC: WRI.

Human population growth trends

FAOSTATS http://faostat.fao.org/site/418/default.aspx. Food and Agriculture Orga-
nization of the United Nations
GEO-3 Data Compendium. United Nations developmental Programme.
http://geocompendium.grid.unep.ch/data_sets/socio-economic/reg_soceco_ds.htm
The CIA World Factbook http://geography.about.com/library/cia/blcindex.htm
World Rosources Institute. Earthtrends database www.earthtrends.org
Human Development Report 2006. United Nations Development Programme.
UNDP http://hdr.undp.org/statistics/default.cfm
Living Planet Report 2006. World Wildlife Fund. Zoological Society of London.
Global Footprint Network.
http://www.panda.org/news_facts/publications/living_planet_report/index.cfm

Technical notes – human population Total Population refers to the de facto
midyear population of a country. The U.N. Population Division compiles and
evaluates census and survey results from all countries, adjusting data for the mis-
calculation of certain age and sex groups, misreporting of age and sex distributions,
and changes in definitions, when necessary. These adjustments incorporate data
from civil registrations, population surveys, earlier censuses, and population models based on information from socioeconomically similar countries. All projections
assume medium levels of fertility. View full technical notes at: http://earthtrends.wri.
org/searchable_db/variablenotes_static.cfm?varid=363&themeid=4

Population Density is calculated by WRI as the number of persons per square kilo-
meter of land area. Population data are from the United Nations Population division.
Total land area is from FAOSTAT.

Average Annual Population Growth Rate refers to the percentage growth in the
midyear population of each country. The values are estimated using demographic
models based on several kinds of demographic parameters: a country's population
size, age and sex distribution, levels of internal and international migration, fertility and mortality rates by age and sex groups, and growth rates of urban and
rural populations. Information collected through recent population censuses and surveys is used to calculate or estimate these parameters. View full technical notes at:
http://earthtrends.wri.org/searchable_db/variablenotes_static.cfm?varid=449&
themeid=4

Urban and **Rural** areas are defined by parameters that vary slightly from country to
country. Many countries define an urban area by the total number of inhabitants
in a population agglomeration. Typically the threshold for considering a region
urban is between 1,000 and 10,000 inhabitants. Any person not inhabiting an area

classified as urban is counted in the rural population. View full technical notes at: http://earthtrends.wri.org/searchable_db/variablenotes_static.cfm?varid=451& themeid=

For additional technical notes see: World Resources 2005: The Wealth of the Poor—Managing Ecosystems to Fight Poverty. 2005. World Resources Institute (WRI) in collaboration with United Nations Development Programme, United Nations Environment Programme, and World Bank. Washington, DC: WRI.

Poverty

http://www.unep.org

http://hdr.undp.org/reports/global/2003/indicator/index indicators.html

GEO-3 Data Compendium. United Nations developmental Programme.

http://geocompendium.grid.unep.ch/data_sets/socio-economic/reg_soceco_ds.htm

Earthtrends http://earthtrends.wri.org/datatables/population

UNDP http://hdr.undp.org/statistics/default.cfm

Human Development Report 2006. United Nations Development Programme. Beyond scarcity: Power, poverty and the global water crisis. Palgrave Macmillan, NY

Living Planet Report 2006. World Wildlife Fund. Zoological Society of London.

Global Footprint Network. http://www.panda.org/news_facts/publications/living_planet_report/index.cfm

World Resources 2005: The Wealth of the Poor—Managing Ecosystems to Fight Poverty. 2005. World Resources Institute (WRI) in collaboration with United Nations Development Programme, United Nations Environment Programme, and World Bank. Washington, DC: WRI.

Technical note – poverty

The **Human Development Index** is comprised of three sub-indices that measure health and lifespan, education and knowledge, and standard of living. It attempts to describe achievement of development goals related to quality of life using data that can be compared across countries and time. It is aggregated from 4 indicators: life expectancy, adult literacy, the gross school enrollment index, and GDP per capita. Life expectancy is the average number of years that a newborn baby is expected to live using current age-specific mortality rates. Adult literacy is defined as the percentage of the population aged 15 years and over which can both read and write, with understanding, a short, simple statement on their everyday life. The gross enrollment index measures school enrollment, regardless of age, as a percentage of the official school-age population. Gross Domestic Product (GDP) per capita measures the total annual output of a country's economy per person. These four indicators are classified in three separate categories—life expectancy, education, and GDP—which are indexed independently and then weighted equally to calculate the final index. More information is available at http://hdr.undp.org. See also: **World Resources 2005: The Wealth of the Poor—Managing Ecosystems to Fight Poverty.** 2005. World Resources Institute (WRI) in collaboration with United Nations Development Programme, United Nations Environment Programme, and World Bank. Washington, DC: WRI. **GDP, National Poverty Rates, International**

Poverty Rates, Income Inequality, and Unemployment Rates: World Bank. 2004. World Development Indicators Online. Washington, DC: The World Bank. Available at http://www.worldbank.org/ data/onlinedbs/onlinedbases.htm. **Human Development and Human Poverty Indices:** United Nations Development Programme. 2004. Human Development Report 2004. New York: United Nations. Available at http://hdr.undp.org/reports/global/2004/.

Forests

Forests, Pasture lands, Arable and permanent cropland

State of the World's Forests 2007. (2007). Food and Agriculture Organization of the United Nations. Rome, ISBN 978-92-5-105586-1

FAOSTAT http://faostat.fao.org/site/418/d.efault.aspx. Food and Agriculture Organization of the United Nations

Earthtrends database www.earthtrends.org

World Resources Institute www.wri.org

United Nations Development Programme www.undp.org

Convention on Biological Diversity (CBD_UNDP) http://www.biodiv.org/programmes/areas/agro/default.asp

GEO-3 Data Compendium. United Nations developmental Programme.

http://geocompendium.grid.unep.ch/data_sets/forests/reg_forest_ds.htm

http://geocompendium.grid.unep.ch/data_sets/land/reg_land_ds.htm

http://www.panda.org/news facts/publications/general/livingplanet/index.cf

http://www.panda.org/downloads/general/LPR 2002.pdf

UNDP http://geocompendium.grid.unep.ch/data sets/index nat dataset.htm

http://geocompendium.grid.unep.ch/geo3 report/index report.htm

http://www.worldbank.org/data/

Living Planet Report 2006. World Wildlife Fund. Zoological Society of London. Global Footprint Network. http://www.panda.org/news_facts/publications/living_planet_report/index.cfm

Deforestation trends

FAOSTATS http://faostat.fao.org/site/418/default.aspx. Food and Agriculture Organization of the United Nations

Earthtrends database www.earthtrends.org

World Recourses Institute http://earthtrends.wri.org/

http://www.panda.org/news facts/publications/general/livingplanet/

http://www.panda.org/downloads/general/LPR 2002.pdf

http://geocompendium.grid.unep.ch/data sets/index nat dataset.htm

World Resources Institute http://earthtrends.wri.org/

Global Enviornmental Facility http://www.gefweb.org/

UNDP http://geocompendium.grid.unep.ch/data sets/index nat dataset/htm; http://www.worldbank.org/data/

Forest loss and growth trends in pasture and agricultural lands

FAOSTATS http://faostat.fao.org/site/418/default.aspx. Food and Agriculture Organization of the United Nations

Earthtrends database www.earthtrends.org

GEO-3 Data Compendium. United Nations developmental Programme.
http://geocompendium.grid.unep.ch/data_sets/forests/reg_forest_ds.htm

World Resources 2005: The Wealth of the Poor—Managing Ecosystems to Fight Poverty. World Resources Institute (WRI) in collaboration with United Nations Development Programme, United Nations Environment Programme, and World Bank. Washington, DC: WRI.

Convention on Biological Diversity (CBD_UNDP) http://www.biodiv.org/programmes/areas/agro/default.asp

United States Department of Agriculture http://www.fas.usda.gov/

Technical notes – forests

Total Land Area is measured in thousand hectares and excludes the area under inland water bodies. Inland water bodies generally include major rivers and lakes. Data on land area were provided to the Food and Agriculture Organization (FAO) by the United Nations Statistical Division. **Forested Area** is calculated by WRI as a percentage of total land area using data from MODIS satellite imagery analyzed by the Global Land Cover Facility (GLCF) at the University of Maryland and from FAO's Global Forest Resources Assessment 2000. **MODIS Satellite Imagery** identifies the percent of tree crown cover for each 500- meter pixel image of land area based on one year of MODIS photography. Data were aggregated to country-level by the GLCF at the request of WRI. The values presented show the percentage of total land area with more than 10 percent or 50 percent of the ground covered by tree crowns. The Food and Agriculture Organization (FAO) estimates are drawn from FRA 2000. **Total forest area** includes both natural forests and plantations. Total Forest is defined as land with tree crown cover of more than 10 percent of the ground and area of more than 0.5 hectares. Tree height at maturity should exceed 5 meters. If no other land use (such as agro-forestry) predominates, any area larger than 0.5 hectares with tree crowns covering more than 10 percent of the ground is classified as a forest. Forest statistics are based primarily on forest inventory information provided by national governments; national gathering methodologies can be found at http://www.fao.org/forestry/fo/fra/index.jsp. FAO harmonized these national assessments with the 10-percent forest definition mentioned above. In tropical regions, national inventories are supplemented with high resolution Landsat satellite data from a number of sample sites covering a total of 10 percent of the tropical forest zone. Where only limited or outdated inventory data were available, FAO used linear projections and expert opinion to fill in data gaps. View full technical notes on-line at http://earthtrends.wri.org/searchable_db/variablenotes_static.cfm?varid=296& theme=9

Change in forest area is the total percent change in both natural forests and plantations between 1990 and 2005. Total forest is defined as land with tree crown cover of more than 10 percent of the ground and area of more than 0.5 hectares. Tree height at maturity should exceed 5 meters. In many cases, FAO projected forward or backward in time to estimate forest area in the two reference years and calculate change in area over the decade. View full technical notes on-line at

http://earthtrends.wri.org/searchable_db/variablenotes_static.cfm?varid=298&
theme = 9. See also State of the World's Forests 2007. 2007 Food and Agriculture
Organization of the United Nations. Rome, ISBN 978-92-5-105586-1 (download-
able from www.FAO.org).

Original forest as a percent of land area refers to the estimate of the percent of
land that would have been covered by closed forest about 8,000 years ago assuming
current climatic conditions, before large-scale disturbance by human society began.
Figures are based on a map of estimated forest cover developed by the World Conservation Monitoring Centre (WCMC). This map was developed by WCMC based
on numerous global and regional biogeographic maps. View full technical notes online at http://earthtrends.wri.org/searchable_db/variablenotes_static.cfm?varid =312
&theme=9

Forest area in 2005 as a percent of total land area is calculated by dividing total
forest area (see above) by total land area. See: State of the World's Forests 2007.
2007 Food and Agriculture Organization of the United Nations. Rome, ISBN 978-
92-5-105586-1 (dowloadable from www.FAO.org).

Pastures and arable and permanent crop lands

Arable and Permanent Cropland is calculated by WRI as a percent of total land
area. Arable land is land under temporary crops (double-cropped areas are counted
only once), temporary meadows for mowing or pasture, land under market and
kitchen gardens, and land temporarily fallow (less than five years). Abandoned land
resulting from shifting cultivation is not included in this category. Permanent cropland is land cultivated with crops that occupy the land for long periods and need
not be replanted after each harvest, such as cocoa, coffee, and rubber; this category
includes land under trees grown for wood or timber. Wherever possible, data on
agricultural land use are reported by country governments in questionnaires distributed by FAO. However, a significant portion of the data is based on both official
and unofficial estimates. **Permanent Pasture** is land used long-term (five years or
more) for herbaceous forage crops, either cultivated or growing wild. Shrublands
and savannas may be classified in some cases as both forested land and permanent
pasture.

Conservation initiatives/Natural Protected Areas and primate conservation

World Database on Protected Areas http://www.unep-wcmc.org/wdpa/index.htm
United Nations Environment Programme UNEP) and World Conservation Monitoring Centre (WCMC). The World Database on Protected Areas (WDPA) provides the most comprehensive dataset on protected areas worldwide and is managed
by UNEP-WCMC in partnership with the IUCN World Commission on Protected
Areas (WCPA) and the World Database on Protected Areas Consortium.

Technical notes – natural protected areas

A **Protected Area** is defined by the World Conservation Union (IUCN) as "an
area of land and/or sea especially dedicated to the protection and maintenance of
biological diversity, and of natural and associated cultural resources, and managed
through legal or other effective means." Since September 2002 the World Database

on Protected Areas (WDPA) consortium has been working to produce an improved and updated database, available to the public and maintained by the United Nations Environment Programme-World Conservation Monitoring Centre (UNEP-WCMC). The WDPA contains summary information for over 100,000 sites, including the legal designation, name, IUCN Management Category, size in hectares, location (latitude and longitude), and year of establishment. WRI calculated protected area data using the 2004 WDPA database. IUCN categorizes protected areas by management objective and has identified six distinct categories of protected areas.

Appendix 2

Estimated biodiversity of South American countries harboring primates as indicated by selected groups of organisms (sources of raw and/or compiled statistics and technical notes, see Appendix 1). Also shown is the estimated species richness for a selected group of northern hemisphere countries. Differences in mean values between South American and the other countries significant (t-test P<0.001).

	Land area (km^2)	Higher plants	Mammals	Breeding birds	Reptiles	Amphibians	Species richness
Brazil	8, 456, 510	56, 215	394	685	648	681	58, 623
Colombia	1, 038, 710	51, 220	359	708	517	623	53, 427
Venezuela	882, 060	21, 073	323	547	322	287	22, 552
Ecuador	276, 840	19, 362	302	640	415	434	21, 153
Peru	1, 280, 000	17, 144	460	695	347	352	18, 998
Bolivia	1, 084, 380	17, 367	316	504	257	162	18, 606
Argentina	2, 780, 690	9, 372	320	362	333	162	10, 549
Paraguay	397, 300	7, 851	305	233	136	75	8, 600
Guyana	214, 980	6, 409	193	242	133	103	7, 080
French Guiana	88, 150	5, 625	150	199	132	90	6, 196
Suriname	156, 000	5, 018	180	235	141	86	5, 660
Uruguay	176, 220	2, 278	81	115	69	42	2, 585
Trinidad and Tobago	5, 130	2, 259	100	131	93	34	2, 617
mean	1, 402, 653	18, 245	282	430	288	258	19, 502
± sd	2, 348, 126	17, 726	109	223	178	220	18, 347
Sweden	44, 996	1, 750	60	259	7	13	2, 089
Norway	33, 288	1, 715	54	241	7	5	2, 022
Canada	997, 011	3, 270	193	310	39	42	3, 854
Netherlands	4, 084	1, 221	55	192	13	17	1, 498
Belgium	3, 051	1, 550	58	191	12	17	1, 828
Japan	37, 780	5, 565	188	210	92	64	6, 119
Ireland	7, 028	950	25	143	6	4	1, 128
Germany	35, 698	2, 682	76	247	16	20	3, 041
Spain	50, 599	5, 050	82	281	67	32	5, 512
Italy	30, 127	5, 599	90	250	55	44	6, 038
Russia	1, 707, 540	11, 400	269	528	94	32	12, 323
Mean	268, 291	3, 704	104	259	37	26	4, 132
± sd	559, 224	3, 090	76	100	34	18	3, 282

Appendix 3

Richness of primate taxa in South American countries. Also shown is forest cover in 2005, density of primate taxa (number of taxa/km^2 forest cover and taxa/1000 km^2 of forest cover). See Appendix 1 and 6 for sources of data. T & T = Trinidad and Tobago.

Country	Primate taxa	Forest cover (km^2) 2005	Taxa/km^2	Taxa/1000 km^2
Brazil	131	4, 776, 980	0.000027	0.0274
Peru	51	687, 420	0.000074	0.0742
Colombia	50	607, 280	0.000082	0.0823
Venezuela	26	477, 130	0.000054	0.0545
Argentina	6	33, 021	0.000182	0.1817
Bolivia	23	520, 502	0.000044	0.0442
French Guiana	8	80, 630	0.000099	0.0992
Ecuador	21	108, 530	0.000193	0.1935
Suriname	8	134, 909	0.000059	0.0593
Guyana	8	146, 509	0.000055	0.0546
Paraguay	5	18, 475	0.000271	0.2706
Uruguay	1	1506	0.000664	0.6640
T & T	2	2, 565	0.000780	0.7797

Appendix 4

Human population size, density and growth projection (year 2030) for South American countries harboring primates (sources of raw and/or compiled statistics and technical notes, see Appendix 1). T & T = Trinidad and Tobago.

	Land area km2	Population (millions) 2005	People/km²	Avg annual (1980–2000) growth rate (%)	Population (millions)			Amerindians 2005	% of country's population
					1980	2005	2030		
Brazil	8,456,510	183	21.6	1.8	122	183	222	1.65	0.90
Colombia	1,038,710	46	43.9	2.0	28	46	61	0.46	1.00
Argentina	2,736,690	39	14.3	1.4	28	39	49	0.20	0.51
Peru	1,280,000	28	21.9	2.1	17	28	37	12.60	45.00
Venezuela	882,060	27	30.2	2.5	15	27	37	0.41	1.52
Ecuador	276,840	13	48.3	2.3	8	13	17	3.25	25.00
Bolivia	1,084,380	9	8.4	2.2	5	9	13	4.95	55.00
Paraguay	397,300	6	15.5	2.8	3	6	10	0.075	1.25
Uruguay	176,220	3	19.5	0.7	3	4	4	0.012	0.40
T & T	5,130	1	212	0.9	1	1	1	NA	NA
Guyana	214,980	0.8	3.6	0.6	1	1	1	0.056	7.00
Suriname	156,000	0.44	2.8	0.7	0.4	0.4	1	0.009	2.55
French Guiana	88,150	0.18	2.1	2.1	0.1	0.2	0.3	0.006	3.33

Appendix 5

Poverty levels of south American countries harboring primates as measured by the Human development Index (HDI*) and by the percent of the population in each country living on <$1 and <$2US dollar/day. T & T = Trinidad and Tobago. Also presented are the mean values of the Human Development Index (HDI) of 11 South American and 8 Mesoamerican countries (Guatemala, Belize, Honduras, Nicaragua, Costa Rica and Panama) harboring primates. and the 25 most developed countries in the planet (Norway, Iceland, Australia, Ireland, Sweden, Canada, Japan, United States, Switzerland, Netherlands, Finland, Luxembourg, Belgium, Austria, Denmark, France, Italy, United Kingdom, Spain, New Zealand, Germany, Hong Kong, China (SAR). Israel, Greece and Singapore). (sources of raw and/or compiled statistics and technical notes, see Appendix 1).

	Land area	Population (Millions) 2005	Population		HDI 2004	HDI rank (N = 177)	% of population	
			Urban (1000)	Rural (1000)			<$1/day	<$2/day
Brazil	8, 456, 510	183	154, 002	28, 796	0.78	72	7.5	21.2
Colombia	1, 038, 710	46	35, 293	10, 307	0.77	73	7.0	17.8
Argentina	2, 736, 690	39	35, 598	3, 713	0.85	34	3.3	14.3
Peru	1, 280, 000	28	20, 864	7, 105	0.75	89	12.5	31.8
Venezuela	882, 060	27	23, 474	3, 165	0.78	68	8.3	27.6
Ecuador	276, 840	13	8, 399	4, 980	0.74	100	15.8	37.2
Bolivia	1, 084, 380	9	5, 881	3, 258	0.68	114	23.2	42.2
Paraguay	397, 300	6	3, 603	2, 557	0.75	89	16.4	33.2
Uruguay	176, 220	3	3, 219	244	0.83	46	2.0	5.7
T & T	5, 130	1	999	312	NA	NA	NA	NA
Guyana	214, 980	0.8	296	472	0.72	104	2.0	6.1
Suriname	156, 000	0.44	342	101	0.78	67	NA	NA
French Guiana	88, 150	0.18	141	45	NA	NA	NA	NA
South América					0.76 ± 0.04			
Mesoamerica					0.73 ± 0.06			
Top 25 countries					0.94 ± 0.005			

Appendix 6

Forest cover in South American countries harboring primates (sources of raw and/or compiled statistics and technical notes regarding definitions of forest cover and estimates of original forest cover see Appendix 1). F. Guiana = French Guiana, T & T = Trinidad and Tobago. NA data not available.

	Land area km²	Original forest km²	Original forest as % of land area	1990 forests km²	1990 forest as % of land area	2005 forest km²	2005 forest as % of land area	% of total forests	Change 1990–2005	Difference 1990–2005 km²
Brazil	8,456,510	6,765,208	80	5,200,270	61.49	4,776,980	56.50	62.91	−0.54	423,290
Argentina	2,780,690	139,035	5	35,262	1.27	33,021	1.19	0.43	−0.42	2,241
Peru	1,280,000	947,200	74	701,560	54.81	687,420	53.70	9.05	−0.13	14,140
Bolivia	1,084,380	585,565	54	540,000	49.80	520,502	48.00	6.85	−0.24	19,498
Colombia	1,038,710	955,613	92	614,390	59.15	607,280	58.50	8.00	−0.08	7,110
Venezuela	882,060	661,545	75	520,260	58.98	477,130	54.10	6.28	−0.55	43,130
Paraguay	397,300	119,190	30	23,372	5.88	18,475	15.00	0.24	−1.40	4,897
Ecuador	276,840	218,704	79	138,170	49.91	108,530	39.20	1.43	−1.43	29,640
Guyana	214,980	212,830	99	151,040	70.26	146,509	68.20	1.93	−0.20	4,531
Suriname	156,000	143,520	92	147,760	94.72	134,909	86.00	1.78	−0.58	5,851
F. Guiana	88,150	81,980	93	80,910	91.79	80,630	91.50	1.06	−0.02	280
T & T	5,130	4,771	93	2,800	54.58	2,565	50.00	0.03	−0.56	235
Uruguay	176,200	3,624	2	NA	NA	1,506	0.85		NA	NA
Total	16,836,950	10,835,161	(64%)	8,155,794	(48%)	7,595,457	(45%)			554,843
Average %			72		48		51		−0.51	

68659299099365555257

Appendix 7

Pasture and arable and permanent crop lands in South American countries harboring primates. (sources of raw and/or compiled statistics and technical notes regarding definitions of forest cover and estimate of original forest cover, see Appendix 1). T & T = Trinidad and Tobago.

	Pastures (km^2) 2005	%	Arable and pernament crop land (km^2) 2005	%	Forest cover as % of land area 2005
Brasil	1, 970, 000	44	666, 000	56.17	56.5
Argentina	998, 470	22	289, 000	24.37	1.19
Colombia	420, 610	9	38, 500	3.25	58.5
Bolivia	338, 310	7	32, 560	2.75	48.0
Paraguay	217, 000	5	31, 360	2.64	15.0
Venezuela	182, 400	4	34, 000	2.87	54.1
Perú	169, 000	4	43, 100	3.64	53.7
Uruguay	135, 430	3	14, 120	1.19	42.0
Ecuador	50, 900	1	29, 850	2.52	39.2
Guyana	12, 450	0.276	5, 100	0.43	68.2
Suriname	210	0.005	680	0.06	86.0
T & T	110	0.002	1, 220	0.10	50.0
French Guiana	70	0.002	160	0.01	91.5
Estimated total	4, 494, 960		1, 185, 650		

Appendix 8

"Business as usual scenario" , estimated by using an exponential decay model, to project loss of forest at 50 and at 100 yrs from 2005 (see text). Paraguay excluded from this analysis due to uncertainty in available forest statistics for the period 1990–2005. F. Guiana = French Guiama, T & T = Trinidad and Tobago.

	Land area km^2	Original forest km^2	Anual change rate (%) 1990–2005	Forests as percent of total land area				
				Original	1990	2005	50 yrs	100 yrs
Brazil	8, 456, 510	6, 765, 208	−0.543	80.0	61.5	56.5	43.0	32.8
Argentina	2, 780, 690	139, 035	−0.424	5.00	1.3	1.2	1.0	0.8
Peru	1, 280, 000	947, 200	−0.134	74.0	54.8	53.7	50.2	46.9
Bolivia	1, 084, 380	585, 565	−0.241	54.0	49.8	48.0	42.6	37.7
Colombia	1, 038, 710	955, 613	−0.077	92.0	59.1	58.5	56.3	54.1
Venezuela	882, 060	661, 545	−0.553	75.0	59.0	54.1	41.0	31.1
Paraguay	397, 300	119, 190	−1.397	30.0	5.9	4.7	2.3	1.1
Ecuador	276, 840	218, 704	−1.430	79.0	49.9	39.2	19.1	9.3
Guyana	214, 980	212, 830	−0.200	99.0	70.3	68.2	61.7	55.8
Suriname	156, 000	143, 520	−0.277	92.0	90.2	86.0	64.7	48.3
F. Guiana	88, 150	81, 980	−0.023	93.0	91.8	91.5	90.4	89.4
T & T	5, 130	4, 771	−0.560	93.0	54.6	50.0	37.8	28.5
Average (%)			−0.488	70	54	51	43	37

Appendix 9

Natural protected areas (NPAs) in south American countries harboring primates (sources of raw and/or compiled statistics and technical notes see Appendix 1). T & T = Trinidad and Tobago.

	Land area km2	Number of NPAs	%	Square km protected	%	% of country's land area
Brazil	8, 456, 510	1, 287	47.06	1, 638, 547	54.03	19.40
Colombia	1, 038, 710	413	15.10	374, 600	12.35	36.10
Venezuela	882, 060	237	8.67	299, 900	9.89	34.00
Ecuador	276, 840	140	5.12	71, 978	2.37	26.00
Bolivia	1, 084, 380	54	1.97	231, 874	7.65	21.40
Peru	1, 280, 000	64	2.34	179, 257	5.91	14.00
Suriname	156, 000	18	0.66	19, 812	0.65	12.70
Argentina	2, 780, 690	328	11.99	182, 052	6.00	6.50
T & T	5, 130	86	3.14	322	0.01	6.30
Paraguay	397, 300	40	1.46	23, 664	0.78	6.00
French Guiana	88, 150	36	1.32	5, 306	0.17	6.00
Guyana	214, 980	3	0.11	4, 860	0.16	2.30
Uruguay	176, 220	29	1.06	725	0.02	0.40
Total		2, 735		3, 032, 897		

References

Altieri, M. A. 2004. Globally Important Ingenious Agricultural Heritage Systems (GIAHS): extent, significance, and implications for development. http://www.fao.org/ag/agl/agll/giahs/documents/backgroundpapers_altieri.doc

Biofuelwatch. 2007. Agrofuels – Toward a Reality check in nine key areas. Report submitted to the Twelfth Meeting of the Subsidiary Body on Scientific, Technical and Technological Advice (SBSTTA-12), 2–6 July 2007, Paris, France of the Convention on Biological Diversity . *Published by:* Biofuelwatch (http://www.biofuelwatch.org.uk/background.php)

Boswell, A. 2007. Biofuels for transport – a dangerous distraction. Scientists for Global Responsibility Newsletter (January No. 33). www.sgr.org.uk

Bollen, A., Van Elsacker, L. and Ganzhorn J. U. 2004.Tree dispersal strategies in the littoral forest of Sainte Luce (SE-Madagascar) Oecologia 139:604–616

Brook, B. W., Bradshaw, C. J. A., Koh, L. P. and Sodhi, N. S. 2006. Momentum drives de crash: mass extinction in the tropics. Biotropica 38:302–305.

Burnett, D. A. and Flannery, T. F. 2005. Fifty millennia of catastrophic extinctions after human contact. Trends Ecol. Evol. 20:395–401.

Casson A. 2003. Oil palm, Soybeans and Critical Habitat Loss: A Review. WWF Forest Conversion Initiative. Office Hohlstrasse 110 CH-8010 Switzerland

Chape, S., Fish, L., Fox, P. and Spalding, M. 2003. United Nations List of Protected Areas (IUCN/UNEP, Gland, Switzerland/Cambridge, UK.

Chape S., Harrison, J., Spalding, M. and Lysenko I. 2005. Measuring the extent and effectiveness of protected areas as an indicator for meeting global biodiversity targets. Phil. Trans. R. Soc. B 360:443–455.

Chapman, C. A. and Onderdonk D. A. 1998. Forests without primates: primate/plant codependency. Am. J. Primatol. 45:127–141.

CBD/UNDP. 2006. Convention on Biological Diversity-CBD/UNDP. United Nations.

CIA. 2006. World Factbook. http://geography.about.com/library/cia/blcindex.htm

Cormier, L. 2006. A Preliminary Review of Neotropical Primates in the Subsistence and Symbolism of Indigenous Lowland South American Peoples. Ecol. Envir. Anthrop. 2:14–32.

Daily, G. D., Ceballos G., Pacheco J., Suzan G. and Sanchez-Azofeifa, A. 2003. Country side biogeography of neotropical mammals: conservation opportunities in agricultural landscapes in Costa Rica. Cons. Biol. 17:1815–1826.

Defries, R., Hansen, A., Newton A.C. and Hansen M. C. 2005. Increasing isolation of protected areas in tropical forests over the past twenty years. Ecol. Appl. 15:19–26.

Dinerstein, E. and Wikramanayake, E. D.1993. Beyond hotspots: how to prioritize investments to conserve biodiversity in the Indo-Pacific region. Cons. Biol. 7:53–65.

Donald, P. F. 2004. Biodiversity impacts of some agricultural commodity production systems. Cons. Biol. 18:17–37.

Encyclopedia Britannica. 2006. www.britannica.com

ELC. 2002. Cerrado, Environmental Literacy Council, http://www.enviroliteracy.org/article.php/495.html.

Estrada, A. 2006. Human and nonhuman primate co-existence in the neotropics: a preliminary view of some agricultural practices as a complement for primate conservation. Ecol. Envir. Anthrop. 2:7–29.

Estrada, A, Sáenz, J., Harvey, C.A., Naranjo E., Muñoz D. and Rosales-Meda M. 2006. Primates in agroecosystems: conservation value of agricultural practices in Mesoamerican landscapes. In A. Estrada, P. A. Garber, M. S. M. Pavelka and L. G. Luecke (eds.), New Perspectives in the Study of Mesoamerican Primates: Distribution, Ecology, Behavior and Conservation (pp. 437–470). New York: Springer Press.

FAO. 2001. Food Security Document. URL, http:/www.faor.org/biodiversity

FAO. 2007. State of the World's Forests 2007. Rome: Food and Agriculture Organization of the United Nations.

FAOSTATS. 2006. http://faostat.fao.org/

Gardner, T. A., Barlow, J., Parry, L. W. and Peres, C. A. 2006. Predicting the uncertain future of tropical forest species in a data vacuum. Biotropica 39:25–30.

Gaston, K. J. 2000. Global patterns in biodiversity. Nature 405:220–227.

Henle, K., Lindemayer, D. B., Margules, C. R., Saunders, D. A. and Wissel. C. 2004a. Species survival in fragmented landscapes: where are we now? Biodiv. Cons. 13:1–8.

Henle, K., Davoes, K. F., Kleyer, M., Margules C. and Settele J. 2004b. Predictors of species sensitivity to fragmentation. Biodiv. Cons. 13:207–251.

Hirsch, A., Dias, L. G., Martins, L. de O., Campos, R. F., Resende, N. A. T. and Landau, E. C. 2002. Database of Georreferenced Occurrence Localities of Neotropical Primates. Department of Zoology / UFMG, Belo Horizonte. http://www.icb.ufmg.br/zoo/primatas/home_bdgeoprim.htm and CD-Rom.

Horner-Devine, C., Daily, G. C., Ehrlich, P. R. and C. L. Boggs. 2003. Countryside biogeography of tropical butterflies. Cons. Biol. 17:168–177.

Hughes, J. B., Daily, G. C. and Ehrlich, P. R. 2002. Agricultural policy can help preserve tropical forest birds in countryside habitats. Ecol. Lett. 5:21–129.

IUCN/SSC/PSG Primate Specialist Group. 2006. http://www.primate-sg.org/diversity.htm

IUCN/WCPA/UNEP. 2006. World Database on Protected Areas. WDPA (IUCN–WCPA and UNEP) http://www.unep-wcmc.org/wdpa/index.htm

Lambert, J. E. and Garber P. A. 1998. Evolutionary and ecological implications of primate seed dispersal. Am. J. Primatol. 45:9–28.

LANIC. 2006. Latinamerican Network Information Center-University of Texas. http://www1.lanic.utexas.edu/la/region/indigenous/

Laurance, W. F., Lovejoy, T. E., Vasconcelos, H. L. Bruna, E. M., Dirham, R. K. and Stoufer P. C. 2002. Ecosystem decay of Amazonian forest fragments: a 22 year investigation. Cons. Biol. 16:605–618.

Laurance, W. F. 2007. Have we overstated the tropical biodiversity crisis ? Trends Ecol. Evol. Online http://dx.doi.org/10.1016/j.tree.2006.09.014

Laurance, W. F. and Peres, C. A. Eds. 2006. Emerging Threats to Tropical Forests. Chicago: University of Chicago Press.

Marsh, L. (ed.). 2003. *Primates in Fragments*. New York: Kluwer-Plenum.

Microsoft Encarta Online Encyclopedia. 2006. http://encarta.msn.com © 1997–2006 Microsoft Corporation. http://encarta.msn.com/encyclopedia_761574914/South_America.html.

MA. 2005. Ecosystems and Human Well-being. Current State and Trends. The Millenium Ecosystem Assessment. New York: United Nations.

Muñoz, D., Estrada, A., Naranjo, E. and Ochoa, S. 2006. Foraging ecology of howler monkeys (*Alouatta palliata*) in a cacao (*Theobroma cacao*) plantation, Tabasco, Mexico. Am. J. Primatol. 68:127–142.

Murphy, H. T. and Lovett-Doust, J. 2004. Context and connectivity in plant populations and landscape mosaics: does the matrix matter? Oikos 105:3–14.

Naughton-Treves, L. and Nick, S. 2004. Wildlife conservation in agroforestry buffer zones: opportunities and conflict. In G. Schroth, G. Fonseca, C. Gascon, H. Vasconcelos, A. M. Izac and C. A. Harvey (eds.), *Agroforestry and Conservation of Biodiversity in Tropical Landscapes* (pp. 319–345). New York: Island Press Inc.

Nepstad, D., Verissimo A, Alencar A, Nobre C, Lima E., Lefebvre P., Schlesinger P., Potter C., Mountinho P., Mendoza E., Cochrane M., and Brooks V. 1999. Large-scale impoverishment of amazonian forests by logging and fire. Nature 398:505.

Oil World. 2002. The Revised Oil World 2002: Supply, Demand and Prices. Hamburg: ISTA Mielke GmbH.

Peres, C. 2000. Evaluating the impact of sustainability of subsistence hunting at multiple forest sites. In J. G. Robinson and E. L. Bennett (eds.), *Hunting for Sustainability in Tropical Forests* (pp. 31–57). New york: Columbia University Press.

Raboy B. E., Christman, M. C. and Dietz J. M. 2004. The use of degraded and shade cocoa forests by Endangered golden-headed lion tamarins, Leontopithecus chrysomelas. Oryx 38:75–83.

RBGE 2003. The Biodiversity of the Brazilian Cerrado, Royal Botanic Garden Edinburgh, http://www.rbge.org.uk/rbge/web/science/research/biodiversity/cerrado.jsp.

Ricketts, T. H. 2001. The matrix matters: effective isolation in fragmented landscapes. Am. Nat. 158:87–99.

Righelato, R. 2005. Just how green are biofuels. World Land Trust. http://www.worldlandtrust.org/

Rodrigues, A. S. L., Andelman, S. J., Bakarr, M. I., Boitani, L., Brooks, T. M., Cowling, R. M., Fishpool, L. D. C., da Fonseca, G. A. B., Gaston, K. J., Hoffmann, M., Long, J. S., Marquet, P. A., Pilgrim, J. D., Pressey, R. L.,, Schipper, J., Sechrest, W., Stuart, S. N., Underhill, L. G., Waller, R. W., Watts, M. E. J., and Yan, X. 2004. Effectiveness of the global protected area network in representing species diversity. Nature 428:640–643.

RRU. 2002. Evaluation of the Comparative Energy, Global Warming and Socio-Economic Costs and Benefits for Biodiesel, Resources Research Unit, School of Environment and Development, Sheffield Hallam University. Source: http://www.defra.gov.uk/farm/acu/research/reports/nf0422.pdf

Schroth, G., da Fonseca, G. A. B., Gascon, C., Vasconcelos, H., Izac, A. M. and Harvey, C. A. Eds. 2004. Agroforestry and Conservation of Biodiversity in Tropical Landscapes. New York: Island Press Inc.

Schnepf, R., Dohlman, E. and Bolling, C. 2001. Agriculture in Brazil and Argentina: Developments and Prospects for Major Field Crops. Market and Trade Economics Division, Economic Research Service, US Department of Agriculture, Agriculture and Trade Report, WRS-01-3.

Sodhi, L. P. Koh, B., Brook, W. and Ng, P. K. L. 2004. Southeast Asian biodiversity: an impending disaster. Trends Ecol. Evol. 19:654–659.

Strier, K. B., Boubli, J. P., Pontual, F. B. and Mendes, S. L. 2006. Human dimensions of northern muriqui conservation efforts. Ecol. Envir. Anthrop. 2:44–53.

Tengnäs, B. and Nilsson, B. 2003. Soybean: Where Is It From and What Are Its Uses? A report for WWF Sweden. Stockholm: WWF Sweden.

The Columbia Gazetteer of the World Online. 2005. South America. Columbia University Press New York. http://www.columbiagazetteer.org/

Tilman, D., May, R. M., Lehman, C. L.and Nowak M. A. 1994. Habitat destruction and the extinction debt. Nature 371:65–66.

Thorp, R., Caumartin C. and Gray-Molina, G. 2006. Inequality, Ethnicity, Political Mobilization and Political Violence in Latin America: The Cases of Bolivia, Guatemala and Peru. Bull. Lat. Am. Res. 25:453–480.

UNDP. 2006a. Beyond Scarcity: Power, Poverty and the Global Water Crisis. United Nations Development Programme (UNDP). New York: Palgrave Macmillan.

UNDP. 2006b. United Nations Environmental Program http://www.unep.org

UNFPA 2007. State of World Population 2007. Unleashing the Potential of Urban Growth. Rome:United Nations Population Fund.

USDA. 2003a. Agricultural Baseline Projections to 2012. Washington DC: United States Department of Agriculture,

USDA. 2003b. Brazil: Future Agricultural Expansion Potential Underrated, Production Estimates and Crop Assessment. Washington, DC: Division Foreign Agricultural Service. United States Department of Agriculture.

van Roosmalen, M. G. M., van Roosmalen, T. and Mittermeier, R. A. 2002. A taxonomic review of the titi monkeys, genus Callicebus Thomas, 1903, with the description of two new species, Callicebus bernhardi and Callicebus stephennashi, from Brazilian Amazonia. Neotrop. Primates 10 (Suppl):1–52.

Whitmore, T.C. 1998. Tropical Rain Forests. Oxford: Oxford University Press.

Wright, S. J. and Muller-Landau, H. C. 2006. The future of tropical forest species. Biotropica 38:267–301.

WRI/UNDP/UNEP/WB. 2005. The Wealth of the Poor – Managing Ecosystems to Fight Poverty. 2005. World Resources Institute (WRI) in collaboration with United Nations Development Programme, Washington, D. C.: United Nations Environment Programme and World Bank.

WWF. 2006. Living Planet Report 2006. World Wildlife Fund, Zoological Society of London and Global Footprint Network. http://www.panda.org/news_facts/publications/ living_planet_report/index.cfm

Zar, J. H. 1998. Biostatistical Analysis. 4th edition. Prentice Hall Inc, NJ.

Part V
Concluding Chapter

Chapter 19
Comparative Perspectives in the Study of South American Primates: Research Priorities and Conservation Imperatives

Alejandro Estrada and Paul A. Garber

19.1 Introduction

A major goal of this final chapter is to recognize the contributions of early biologists and mamalogists, as well as more recent primate specialists in advancing the development of Primatology in South America. We also provide a chronological overview of the richness of published scientific information on South American primates by country and major taxa in order to assess the current state of accumulated knowledge available to scholars and researchers. Finally, we outline what we consider important research priorities and conservation imperatives for the future.

19.1.1 Contribution by Early Primatologists to the Development of the Discipline in South America

An important component of the development of Primatology in South America has been the long-term efforts of an important group of biologists, mammalogists and primatologists. The significance of their efforts needs to be recognized here. Many of these scientists have served as catalysts in advancing critical research initiatives in taxonomy and systematics, behavior, ecology, molecular studies, and conservation biology. These scholars have devoted considerable time and effort in training and funding several generations of local students in the study of primates, and in editing classical volumes and contributing to the establishment of professional journals. In Brazil some of these pioneers are Milton Thiago de Mello and Adelmar Faria Coimbra-Filho, and Célio Murilo de Carvalho Valle, who, together with other Brazilian colleagues and foreign scholars such as Russell Mittermeier, Karen Strier, and Anthony Rylands, have been instrumental in the development of Brazilian primatology. Similarly, the seminal work of Filomeno Encarnación, Pekka

A. Estrada (✉)
Estación de Biología Tropical Los Tuxtlas, Instituto de Biología, Universidad Nacional Autónoma de México, México
e-mail: aestrada@primatesmx.com

P.A. Garber et al. (eds.), *South American Primates,* Developments in Primatology: Progress and Prospects, DOI 10.1007/978-0-387-78705-3_19,
© Springer Science+Business Media, LLC 2009

Soini, Rolando Aquino, and Warren Kinzey in conducting primate censuses and field studies throughout northeastern Peru, together with the community ecology and team research focus of John Terborgh at Manu National Park, have been critical in providing empirical data and conservation initiatives for the preservation of Peruvian primates and their habitats.

In Colombia, the long-term research commitments of Thomas Defler and Kosei Izawa have resulted in new and important information on the natural history, ecology and behavior of several primate species. Harboring the southernmost distribution of extant primates in the Neotropics, Argentina has had a long-history of primate research led by Gabriel Zunino and Marta Mudry. The work of these colleagues and their students has resulted in significant insights into the distribution, ecology, behavior, and population genetics of the primate taxa present in their country. We also note that the contributions of Roberta Bodini and her colleagues and students on the distribution and biogeography of primates in Venezuela have been critical in developing primatology in that country. In French Guiana, the long-term research program by Pierre Charles-Dominique has added importantly to the body of knowledge regarding primate ecology and rainforest regeneration, while the efforts by Mark van Roosmalen and others in Suriname, have advanced our knowledge of primate diversity ecology, behavior and conservation. In Ecuador, V.J. Albuja has contributed importantly to the knowledge of mammals, including the primate fauna, and in Bolivia the studies by Kosei Izawa , N. Masataka, and Paul Heltne provided the first primate studies in that country. Finally, the early work of Jody R. Stallings, resulted in a detailed survey of the primates of Paraguay.

In many South American countries, local scientists have generously collaborated with foreign colleagues to push forward research and conservation initiatives. In other countries, researchers from North America, Europe, and Japan initiated and were instrumental in promoting primate research. Overall, however, it has been the combined efforts of scholars indigenous to South America and scholars from other countries that have resulted in a significant growth of primate research in every South American country harboring primate populations. In three countries (Brazil, Colombia and Argentina), the presence of a critical mass of primate scientists and students has resulted in intellectually strong training programs and in the creation of local primate societies. The remaining countries lack the necessary infrastructure and critical mass of primate scientists needed to develop comprehensive research and training programs. Additional efforts by scientists in these countries, along with support from conservation agencies and collaborative agreements with foreign universities, represent important first steps in moving primate research and conservation forward in these countries.

19.1.2 Documentation: History, Countries and Taxa

The study of South American primates has a long history including knowledge shared by indigenous communities with early European naturalists. However, the study of *Alouatta caraya* by Hans Krieg published in 1928 is among the earliest field reports of a wild South American primate. We explored the recent history of South American primatology through published records found in the PrimateLit database (N = 4637; Nov 2007). These records begin between 1940 and 1945, with a series of studies by Brazilian naturalists reporting the results of expeditions to northwestern and northeastern Brazil (Travassos and Teixeira de Freitas 1941, 1943; *see* also data on museum specimens published by von Ihering 1914 and Vieira 1944, 1955), studies dealing with primate parasites (Gilmore 1943; Kreis 1945), and the first systematic naturalistic study of primates conducted by Alexander von Humboldt (1944). Significant contributions were made in 1949 by Philip Hershkovitz resulting from his survey of primates of northern Colombia (Hershkovitz 1949). Between 1950 and 1970 there are few published accounts of South American primates in the PrimateLit database, probably the result of expanding research activities with primates in Africa and Asia. During the early 1970s Warren Kinzey of the City University of New York along with his students, Peruvian colleagues, and Melvin Neville (then Director of the Peruvian Primate Project in Iquitos) initiated several primate field studies in Peru focusing on *Aotus* (Patricia Wright), *Cebuella* (Marleni Ramirez), *Saguinus* (Pekka Soini and Rogerio Castro) and *Callicebus* (Warren Kinzey). At this same time, Adelmar Coimbra-Filho was conducting research on *Leontopithecus* in the Atlantic Forest in Brazil. In 1976, Martin Moynihan published a volume titled *The New World Primates*, and in 1977 Philip Hershkovitz published his *Living New World Monkeys (Platyrrhini) Vol I*, and Devra Kleiman edited a volume titled The *Biology and Conservation of the Callitrichidae*. In 1978, Hartmut Rothe, Hans-Jürgen Wolters and John P. Hearn edited *Biology and Behaviour of Marmosets*. This renewed interest in the behavior and ecology of South American primates, along with information on fossil platyrrhines and comparative platyrrhine evolution and anatomy (spearheaded by Philip Hershkovitz, Alfred Rosenberger, Susan Ford, and Richard Kay) led to a second wave of field and laboratory research in the 1980s reaching a peak between 1991 and 2000 (Fig. 19.1). Studies during this latter period accounted for 65% of the PrimateLit citations, providing a wealth of new empirical data and theoretical perspectives on primate ecology, evolution, behavior, and mating systems. These contributions have continued into the first decade of the 21st century (Fig. 19.1).

A country by country examination of the PrimateLit (no PrimateLit records available for Trinidad and Tobago) for 1940–2006, indicates that Brazil accounted for 50% of published articles, with Peru and Colombia together contributing an additional 20%. The remaining South American countries (N = 9) combined, accounted for 30% (Fig. 19.2). The number of taxa present in each country and the proportion of PrimateLit citations ($r_s = 0.895$ P < 0.0001) were significantly correlated (the relationship held when removing Brazil from the calculation $r_s = 0.866$ P $= 0.003$), suggesting that, in general, those countries harboring a richer primate fauna are also those for which there are more published records available.

An examination of the PrimateLit citations at the genus level (keeping in mind that it is possible that a large segment of the published records on callitrichines, *Cebus*, *Saimiri* and *Aotus* are from captive studies given their importance in biomedical research) shows that almost 35% of the reports provide information on *Alouatta*,

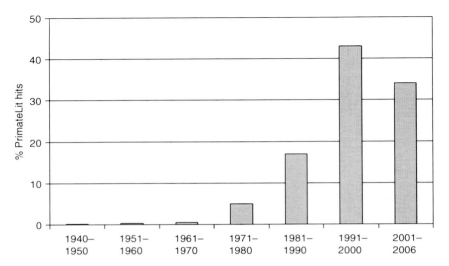

Fig. 19.1 Distribution of PrimateLit database hits on South American primates for the period 1940–2006

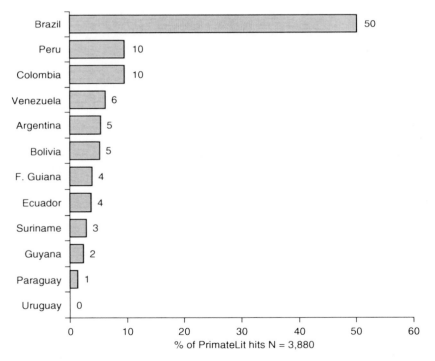

Fig. 19.2 PrimateLit database hits for the period 1940–2006 for each South American country

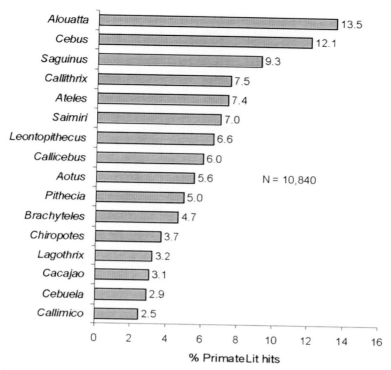

Fig. 19.3 PrimateLit database hits for the period 1940–2006 for each taxa for which there are reports

Cebus and *Saguinus*, and 22% on *Callithrix*, *Ateles* and *Saimiri*; the rest of the genera (n = 10) accounted for the remaining 43% of the PrimateLit citations (Fig. 19.3).

While the information from the PrimateLit database may not include important local scientific publications in South America such as Master and Ph.D theses, technical reports, and conference proceedings, and it may not accurately reflect the number of primate scientists and students present in each country, it can still be used as a general index of regional and country by country productivity in primate research. Examination of the database indicated that while there has been a significant accumulation of information on South American primates since the 1940s, there are still many countries, regions and taxa for which information is particularly scanty or virtually non-existent (*see* Figs. 19.2 and 19.3). The limited data for many South American primate taxa seriously hinders our ability to make accurate predictions concerning the current and future threat of local extinctions of primate populations, as primate habitats face unparalleled threats from factors such as land cover changes (in particular, conversion to cropland and pasture), habitat fragmentation and isolation, invasive species, wildfires, overhunting, increasing industrialization and globalization, and climate change (*see* below and Estrada, this volume; Laurance and Peres 2006; Gardner et al. 2006).

19.2 Conceptual Frameworks and Perspectives in Basic Science and Conservation Research

One key feature of tropical forest ecosystems is their high biological diversity and the primate species assemblages found within are a key component of such diversity (Rylands and Mittermeier, this volume; Estrada, this volume). Primate biodiversity, as is the case for the world's tropical biodiversity, is not evenly distributed across landscapes, regions, and countries, with many taxa endemic to specific regions and localities. Detailed mapping of primate biodiversity is still incomplete for the majority of South American countries, regions and taxa (Gaston 2000), an aspect of great concern considering the velocity with which native vegetation and landscapes are changing in the region as a result of human activity (Ryland and Mittermeier, this volume; Estrada, this volume).

Human population increase and the application of new technologies that have facilitated the increased demand for food, timber, and fiber, together with global market demands, make conservation of primate populations and habitats in the region a daunting task. This task is further complicated by the high primate diversity found in South America and the multifaceted nature of the human populations and cultures in the region. Hence, it is not surprising that we are faced with one of the world's greatest conservation challenges in which consideration needs to be given to a large human population characterized by high levels of poverty, low human development, and high growth rates.

It is evident that investigations are still needed on basic aspects of the biology and ecology of many primate populations and species. Regional mapping of the distribution of primate populations in protected and in continuous habitats, and the gathering of information on their demographic structure are necessary first steps to update distribution maps of populations and species (Fig. 19.4a). Such information also constitutes the baseline data needed to assess, with accuracy, the impact of human-induced habitat fragmentation and disturbance upon populations and their habitats. Similarly, there is a need for more long-term monitoring of selected populations to enhance our data banks on socio-demographic process affecting the stability, growth, and sustainability of primate populations. Community level research also is necessary, especially in those habitat areas of high primate and nonprimate species diversity.

Investigation on the basic ecology of primate populations in South America is fundamental for understanding their role in ecosystem dynamics and processes. Studies of primate–plant interactions are an elementary activity that will direct us to understand how primate populations sustain themselves in the habitats in which they live and to understand the role of primates in ecosystem processes such as primary productivity and habitat regeneration (Fig. 19.4b). Studies of primate–animal interactions at both micro and macro scales are required in order to more fully comprehend the dynamics of resource partitioning, commensalism, parasitism, and the effect of resource competition in primate communities (Fig. 19.4b). Research integrating primate social behavior, hormonal mechanisms associated with fertility and reproductive success, and factors affecting population productivity, and disease

(a)

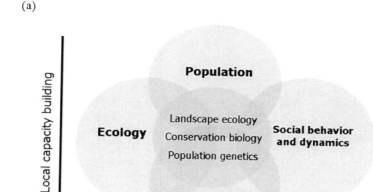

(b)

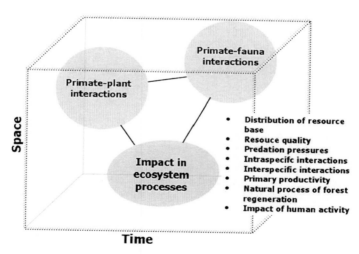

Fig. 19.4 (**a**) General areas of primate research. A multidisciplinary approach is essential in tackling investigations in each area. Integration of concepts and empirical information from disciplines such as landscape ecology, conservation biology and population genetics, among others, is also essential. Two added dimensions in the scheme are local capacity building and the input of social scientists when dealing with community oriented conservation approaches. (**b**) Basic ecological research focusing on primate–plant interactions and primate–fauna interactions and their relationship to ecosystem processes is needed to provide basic natural history information and ecological data. Such information is essential in assessments on the role of primates in ecosystem processes. This approach operates within a multidimensional framework in which various scales of time and space are particularly important as a context for examining the relevance of various sets of variables, some of which are listed in the diagram

transmission also are needed (*see* Ziegler and Strier, this volume). Of equal importance is the mapping, through long-term studies, of life-history traits as they relate to social systems, ontogeny of social behavior, reproductive strategies and the costs and benefits of kin and nonkin social bonds (*see*, Strier and Mendes, Blomquist et al. and Di Fiori this volume).

Conservation-oriented research needs to build on data obtained from population, ecological, and behavioral studies, and the results of conservation research need to be used by scientists with a less applied focus to inform their studies. Such a multidisciplinary approach will require incorporating theoretical, conceptual, and empirical information from the fields of landscape ecology, population genetics and conservation biology, among other disciplines (Fig. 19.4a). An added dimension of implementation phases of conservation research is the participation by human communities. Here the input of social scientists will be important in understanding local traditions, valuing local knowledge, and working with local communities to developing partnerships that are consistent with cultural beliefs and practices and offer collective benefits to individuals. This will require local capacity building at various levels – scientific, technical, and sociological. The most sustainable way to ensure that local people benefit from conservation is to educate them about the importance and necessity of conservation and to include them as an integral part of the decision-making process. Recognizing that deteriorating or poor socio-economic conditions are drivers of environmental degradation is critical as this information can empower local people by providing them with the necessary skills and know-how required to improve their conditions. Unless these socio-economic conditions are improved, conservation will never be a priority in the lives of local people (Young 2005).

Conservation studies also need to consider the knowledge and experience of indigenous people in sustainable use of native habitats and natural resources. Many indigenous groups continue to live within or at the edge of forested areas and possess an important reservoir of knowledge about forest biodiversity and sustainable management. Indigenous peoples have empirically experimented with medicinal plants and other forest resources for hundreds, and in some cases, thousands of years (e.g., Incan indigenous communities of South America). As such, these groups have, over many generations, identified sustainable use strategies (e.g., agroforestry practices) and harvesting techniques (traditional knowledge) that maximize land use and improve crop yield while simultaneously reducing the reliance of artificial herbicides and pesticides (Diemont et al. 2006a,b). Indigenous people practice forms of swidden agroforestry that both restore and conserve the rainforest. Their systems cycle through field and fallow stages that produce food, medicines, and raw materials, and regenerate tall secondary forest. In addition, these sophisticated sets of understandings, interpretations and meanings are part of a cultural complex that encompasses language, naming and classification systems, resource use practices, ritual, spirituality and worldview (MA 2005). Traditional knowledge is a valuable source of information about ecosystem condition, sustainable resource management, soil classification, land use and biodiversity (Altieri 2004; MA 2005; Cormier 2006). Much more research is needed on the ways indigenous and rural

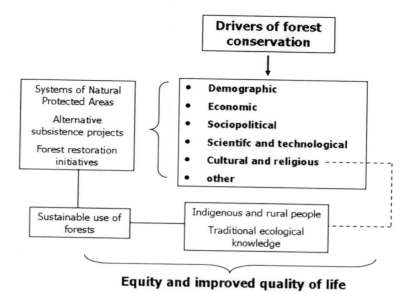

Fig. 19.5 Indirect drivers leading to land cover protection and restoration are basically the same drivers leading to forest loss, but working in the opposite direction. The synergistic effect of these drivers may result in various conservation scenarios, and may incorporate the contribution by indigenous peoples and culture

populations manage their forests and primate wildlife, on their traditional ecological knowledge, and on ways to incorporate their interest in primate conservation plans (Fig. 19.5).

19.2.1 Drivers of Forest Loss Versus Drivers of Forest Conservation

A considerable amount of effort has been invested in examining drivers of tropical ecosystem loss both at local and global scales (Laurance and Peres 2006; Estrada this volume). However, little attention is given to how these drivers affect tropical ecosystem conservation policy. We need to better understand ways of linking ecosystem loss with tangible conservation actions such as the establishment of systems of natural protected areas, community reserves, biological corridors, and areas of sustainable use. Drivers of tropical ecosystem conservation are among the same factors that drive land-cover changes leading to habitat loss (Fig. 19.5). Demographic drivers coupled with sound economic and sociopolitical decisions and adequate scientific and technological approaches can result in the design and implementation of ecologically, culturally, and educationally based initiatives leading to habitat conservation and recovery, while at the same time enhancing food production and human well-being (Fig. 19.4). Conservation research in Primatology needs to document, map, and examine the ways in which these drivers operate and interlink with basic human needs and culture, at the local, regional, continental, and global scales. Empirical and conceptual tools are necessary to develop conservation scenarios that can be effectively implemented in terms of the demographic, economic, sociopolitical, ecological, and cultural realities of each locality, without sacrificing equity and a reasonable quality of life for the human inhabitants.

19.3 Rethinking South American Primate Mating and Social Systems: Small Group Size and Female Reproductive Competition

Over a decade ago Karen Strier (1994b) published a seminal paper titled "The myth of the typical primate" in which she clearly demonstrated that many of our long-held notions of common or typical primate behavior, in fact, appear to be true for only a very small set of primate taxa. Building on that approach, a primary goal of this section is to explore a set of questions concerning the ancestral social and mating system of South American primates and female mating competition in order to encourage readers to rethink and re-examine a set of critical issues in primate socioecology.

New World primates represent a monophyletic group of some 199 taxa that are generally divided into 5 distinct subfamilies; Pitheciinae, Atelinae, Aotinae, Callitrichinae, and Cebinae (Singer et al. 2003). Although there has emerged a strong consensus concerning the phylogenetic position of most platyrrhine genera, the placement of *Aotus* remains problematic, with the night monkey regarded either as a sister taxa to the tamarin-marmoset-callimico clade (Singer et al. 2003; Schneider et al. 2001), a sister taxa to the *Cebus-Saimiri* clade (Schneider et al. 1996), or along with *Callicebus*, a basal member of the pitheciinae (Rosenberger 1992). Based on present evidence it appears that early in their evolution, each platyrrhine subfamily radiated into a distinct set of adaptive zones (Rosenberger 1992), which continue to distinguish these lineages today.

Morphological and molecular information suggest that certain lineages within each platyrrhine subfamily appear to more closely represent basal members of that clade. For example, *Alouatta* is regarded as the basal member of the Atelinae, *Callicebus* the basal member of the Pitheciinea, *Saguinus* or *Leontopithecus* the basal member of the callitrichinae, and *Aotus* the lone genus and therefore basal member of the Aotinae (or along with *Callicebus* a basal pitheciine) (Schneider and Rosenberger 1996; Schneider et al. 2001). Figure 19.6 outlines these evolutionary relationships and estimated divergence dates as summarized in Schneider et al. (2001). Information from these basal taxa can be used as a starting point to reconstruct the social and mating patterns of the earliest platyrrhines.

19.3.1 Reconstructing the Ancestral Platyrrhine Social and Mating System

Several authors have argued that a pair bonded group structure and a monogamous breeding pattern are the ancestral conditions for New World primates (Eisenberg 1981;

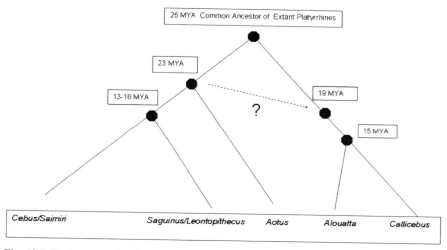

Fig. 19.6 Phylogeny and estimated divergence dates of basal genera of platyrrhine subfamilies. *Cebus* and *Saimiri* represent the Cebinae, *Saguinus* and *Leontopithecus* represent the callitrichinae, *Aotus* represents either the Aotinae or along with *Callicebus* a basal member of the pitheciinae, and *Alouatta* represents the atelinae

Kinzey 1987). This was based on a series of assumptions (1) that *Aotus* and *Callicebus* represent the most primitive of living platyrrhine taxa, (2) that the presence of limited sexual dimorphism in body mass is consistent with a monogamous breeding system, and (3) that male intrasexual tolerance or cooperation represents a derived condition in platyrrhini. However, more recent behavioral, reproductive, fossil, and molecular evidence fails to support these assumptions.Ancestral cebines (e.g., *Dolichocebus* and *Chilecebus*) are present in the fossil record since the early Miocene, dating to at least 20 MYA (Schneider and Rosenberger 1996; Fleagle and Tejedor 2002). This suggests that along with *Aotus*, *Cebus* and *Saimiri* represent long-lived platyrrhine lineages that diverged early from the common ancestor of extant platyrrhines. Although *Aotus* is characterized by a pair-bonded social structure, *Cebus* and *Saimiri* live in multimale–multifemale social groups.

Second, limited body mass dimorphism characterizes many taxa of South American primates, including most callitrichines, all species of *Ateles*, *Aotus*, *Brachyteles*, and *Callicebus*, several species of *Pithecia*, one species of *Chiropotes*, one species of *Cacajao*, and two species of *Saimiri* (Table 19.1; Ford and Davis 1992; Norconk 2007; Di Fiore and Campbell 2007; Fernandez-Duque 2007). These taxa exhibit a range of social and mating patterns including pair bonded groups, small harems, small multimale–multifemale groups, and large multimale–multifemale groups, and live in both highly cohesive groups and fission-fusion groups (Kinzey and Cunningham 1994). Based on a comprehensive review of South American primates, Ford (1994: 238) concluded that "a critical component of sexual selection and its effect on sexual dimorphism is not inter-male competition, per se, but the similarity or difference between levels of inter-male vs. inter-female competition."

Table 19.1 Sexual Dimporphism and Mating System in a Select Group of South American Primates

Species	Sexual Dimorphism in body Mass (AM:AF)	Social System
Callicebus torquatus	0.84–0.99	Pair bond
Ateles paniscus	0.84–1.08	Multimale–multifemale
Callicebus discolor	0.86	Pair bond
Leontopithecus rosalia	0.87	Polyandrous Multimale–multifemale
Ateles hybridus	0.90	Multimale–multifemale
Callicebus cupreus	0.90	Pair bond
Mico humeralifer	0.90	Polyandrous Multimale–multifemale
Callicebus personatus	0.91–1.03	Pair bond
Mico emiliae	0.94	Polyandrous Multimale–multifemale
Saguinus oedipus	0.95	Polyandrous Multimale–multifemale
Aotus azarai	0.95–1.0	Pair bond
Saguinus fuscicollis	0.96	Polyandrous Multimale–multifemale
Aotus zonalis	0.97	Pair bond
Saguinus labiatus	0.97	Polyandrous Multimale–multifemale
Saguinus nigricollis	0.98	Polyandrous Multimale–multifemale
Saguinus geoffroyi	1.0	Polyandrous Multimale–multifemale
Callicebus ornatus	0.99–1.01	Pair bond
Ateles belzebuth	0.99–1.05	Multimale–multifemale
Saguinus geoffroyi	1.0	Polyandrous Multimale–multifemale
Aotus nancymaae	1.01	Pair bond
Aotus vociferans	1.01	Pair bond
Ateles fusciceps	1.01	Multimale–multifemale
Saguinus bicolor	1.01	Polyandrous Multimale–multifemale
Cebuella pygmaea	1.03	Polyandrous Multimale–multifemale
Saguinus mystax	1.03	Polyandrous Multimale–multifemale
Pithecia irrorata	1.07	Pair bond
Aotus lemurinus	1.07	Pair bond
Ateles geoffroyi	1.07–1.10	Multimale–multifemale
Callithrix jacchus	1.08	Polyandrous Multimale–multifemale
Chiropotes satanas	1.08–1.19	Multimale–multifemale
Aotus griseimembra	1.09	Pair bond
Aotus triviragtus	1.10	Pair bond
Mico argentata	1.11	Polyandrous Multimale–multifemale
Brachyteles hypoxanthus	1.13	Multimale–multifemale
Saimiri sciureus	1.14	Multimale–multifemale
Pithecia pithecia	1.14	Harem
Callicebus moloch	1.15	Pair bond
Leontopithecus chrysomelas	1.15	Polyandrous Multimale–multifemale
Saimiri oerstedii	1.16	Multimale–multifemale
Cebus olivaceus	1.19	Multimale–multifemale
Cacajao calvus	1.19	Multimale–multifemale
Brachyteles arachnoides	1.20	Multimale–multifemale
Callithrix penicillata	1.23	Polyandrous Multimale–multifemale
Saguinus midas	1.35	Polyandrous Multimale–multifemale
Pithecia monachus	1.47	Pair bond
Callithrix geoffroyi	1.52	Polyandrous Multimale–multifemale

Data reported in this Table are from Ford and Davis 1992; Fernandez-Duque 2007; Norconk 2007; Jack 2007; Di Fiore and Campbell 2007

The fact that most species of South American primates are monomorphic or only slightly dimorphic (Ford 1994) may help to account for the fact that male sexual coercion, which is reported to be common in many species of Old World anthropoids, is rare or absent among platyrrhines (van Schaik et al. 2004). This important distinction between Old World catarrhines and New World platyrrhines may help to explain the prominent role that female mate choice, female promiscuity, and female intrasexual competition play in affecting the mating tactics of both male and female South American primates (Kowalewski and Garber, submitted; Strier 1994a, 1999, 2000). A further distinction between Old and New World monkeys is the apparently greater role, in platyrrhines, of olfactory signals in social and sexual communication (Heymann 2006).

Finally, although patrilocal spider monkeys and muriquis are noteworthy for high levels of male–male cooperation (Aureli et al. 2006) and tolerance (Strier 2004a), such behaviors also are commonly reported in callitrichines (Digby et al. 2007; Garber 1997; Huck et al. 2004), as well as certain species of howler monkeys (Kowalewski and Garber, submitted; Wang and Milton 2003, Pope 1990), capuchins (Jack and Fedigan 2004), bearded sakis (Veiga and Silva 2005), squirrel monkeys (Boinski 1994), and possibly uakaris (Barnett et al. 2005; Garber and Kowalewski in press). Male cooperation occurs in several contexts including when males remain in their natal groups to help care for younger siblings (moustached tamarins: Huck et al. 2005), when males disperse from their natal groups with partners that are known (golden lion tamarins: Bales, et al. 2006) or suspected to be brothers (Peruvian squirrel monkeys: Mitchell 1994) or father-son (moustached tamarins: Garber et al. 1993; Löttker and Heymann 2004), when unrelated males act co-operatively in mate defense (Kowalewski and Garber, submitted), and when unrelated males jointly participate in infant care (Garber et al. 1993). Thus, we are at a moment in the study of South American primates in which we need to re-evaluate previous ideas of platyrrhine behavior, ecology, and evolution by testing new theoretical perspectives using empirical data from long-term field studies and multiple studies of the same species in different habitats.

19.3.2 Exploring the Possibility that Ancestral South American Primates Lived in Small Multimale–Multifemale Groups

Although the exact form of the ancestral platyrrhine social and mating system remains uncertain, we explore the possibility that rather than being pair bonded, early South American primates lived in small, cohesive, social groups composed of 1– 4 adult males, 1–4 adult females, juveniles, and infants (Garber et al. 1993). In addition, we suggest that the ancestral platyrrhine mating system was characterized by a form of female reproductive competition that resulted in a limited number of female breeding opportunities within the group. We base this on the following information. First, a pair bonded mating system is rare among mammals including primates (Kappeler and van Schaik 2002). More importantly, given that most taxa of

South American primates are characterized by a considerable degree of mating and social flexibility, it is difficult to envision a set of factors in which intrasexual tolerance in both males and females would have evolved from a system characterized by extreme social inflexibility in which adult males are highly intolerant of other adult males and adult females are highly intolerant of other adult females (Kinzey 1987; Garber et al. 1993). Finally, several platyrrhine genera that are presumed to represent basal members of their subfamily are most commonly found to live in small multimale–multifemale social groups of 6–20 individuals with a relatively equal ratio of adult males to adult females (Table 19.2). For example, in all species of *Alouatta* (except some populations of *A. palliata*), the number of adult males in established groups is reported to range from 1 to 4 and the number of adult females from 1 to 6 (Di Fiore and Campbell 2007). In *Saguinus* these values are 1–4 for males and 1–4 for adult females (Digby et al. 2007; Garber et al. 1993; Löttker and Heymann 2004). In *Leontopithecus* groups are composed of 1–5 adult males

Table 19.2 Data on Group Size, Adult Composition, and Adult Sex Ratio in Select Species of South American Primates

Species	Group Size	No. Adult males	No. Adult females	F:M
Alouatta seniculus (8)	7.8	1.8	2.4	1.4:1
Alouatta caraya (5)	10.9	2.3	3.6	1.6:1
Alouatta guariba (3)	6.5	1.3	2.2	1.6:1
Alouatta pigra (8)	6.6	1.9	2.0	1.1:1
Cebus capucinus (28)	16.4	3.4	5.0	1.4:1
Cebus olivaceus (18)	21.0	3.5	6.9	1.9:1
Cebus apella (9)	17.1	5.6	5.3	0.9:1
Cebus albifrons (3)	18.2	6.5	5.9	0.9:1
Aotus azarai (175)	3.1	1	1	1.0:1
Aotus nancymaae (282)	3.9	1	1	1.0:1
Aotus nigriceps (9)	4.1	1	1	1.0:1
Aotus trivirgatus (11)	2.9	1	1	1.0:1
Aotus vociferans (115)	3.3	1	1	1.0:1
Aotus brumbacki (1)	3.0	1	1	1.0:1
Callicebus torquatus (10)	4.8	1	1	1.0:1
Callicebus ornatus (9)	3.2	1	1	1.0:1
Callicebus moloch (2)	4.1	1	1	1.0:1
Callicebus personatus (4)	2–6	1(2?)	1(2?)	1.0:1
Saguinus mystax (13)	7.0	2.2	2.0	0.9:1
Saguinus geoffroyi (5)	6.3	2.4	2.1	0.8:1
Saguinus oedipus (6)	5.8	2.7	1.7	0.6:1
Saguinus fuscicollis (47)	5.1	1.9	1.3	0.7:1
Leontopithecus rosalia (212)	5.4	1.8	1.5	0.8:1
Leontopithecus chrysomelas (4)	6.7	1.6	1(1.6)	0.6:1 – 1.0:1

Data for *Alouatta* are from Di Fiore and Campbell 2007 and Kowalewski and Garber, submitted; Data for *Cebus* are from Fragaszy et al. 2004; Data for *Callicebus* are from Norconk 2007; Data for *Aotus* are from Fernandez-Duque 2007; Data for *Saguinus* are from Garber et al. 1993; Dawson 1977; Neymann 1977, and Mendelson 1994; Data for *Leontopithecus* are from Dietz and Baker 1993 and Rylands 1982.

and 1–4 adult females (Digby et al. 2007). In *Cebus*, data presented by Fragaszy et al. (2004) indicate that across species mean group size ranges 18–21 and includes 2–7 adult males and 3–10 adult females (Table 19.2).

In contrast, *Callicebus* does not appear to fit this pattern. *Callicebus* represents a basal member of the pitheciine clade and is characterized by a pair bonded social system with a single adult male and a single breeding female. Assuming that the molecular evidence is correct and that *Callicebus* and *Aotus* are not closely related (Schneider et al. 2001; Singer et al. 2003) we offer the possibility that a pair bonded social system may have evolved independently twice from an ancestor living in small multimale–multifemale groups. Alternatively if the morphological evidence is correct, then *Aotus* and *Callicebus* represent each other's closest relatives (*see* Kinzey 1992; Rosenberger 1992, 2002), and monogamy is likely to have evolved once in the common ancestor of night monkeys and titi monkeys. In either case, the evolution of monogamy in *Callicebus* and *Aotus* represents a highly specialized reproductive pattern in which male investment (transport and provisioning) in a single offspring serves to increase the reproductive output of a single breeding female by shortening the interval between successive births and reducing maternal reproductive costs (Garber and Leigh 1997). Cooperative caregiving of twin infants, principally by adult male helpers in polygynously and polyandrously breeding tamarins and marmosets similarly is argued to play an important role in enhancing infant survivorship, reducing maternal reproductive investment, shortening the interval between successive births, and increasing the reproductive success of individual group members (Garber 1997; Digby et al. 2007). Garber et al. (1993) have argued that the highly specialized social, breeding, and infant rearing system of tamarins and marmosets, which includes reproductive suppression in subordinate adult females and high levels of cooperation among adult male group members, evolved from an ancestor living in small multimale–multifemale groups. We suggest that the monogamous mating system of *Callicebus* and *Aotus* may have evolved from such a small mulitmale–multifemale social system as well.

A final point is that despite considerable variation in body mass, ontogenetic development, social and mating systems, patterns of infant care, age at first reproduction, and potential reproductive output, these basal platyrrhine taxa all share in common the fact that each is characterized by a relatively small number of actively breeding females per group. We explore this issue more fully below.

19.3.3 Female Intrasexual Mating Competition

Compared with many species of Old World monkeys, basal South American primate taxa are characterized by social groups comprised of a smaller number of adult females, and an adult sex ratio more closely approaching 1:1 (Table 19.3). For example the ratio of adult females to males in *Cebus* spp. ranges from 0.9:1 to 1.4:1(Mean = 1.2:1, Fragaszy et al. 2004). In *Alouatta* (excluding *A. palliata*), these values are from 1.1:1 to 1.6:1 (Mean = 1.4:1). In *Saguinus, Leontopithecus,*

Table 19.3 Comparison of Adult Female to Adult Male Sex Ratio in Established Groups of Several Species of New World monkeys and Old World Monkeys

Genus	No. of Species	Ratio af:am
Alouatta	4	1.4:1
Cebus	4	1.2:1
Aotus	6	1.0:1
Callicebus	4	1.0:1
Leontopithecus	2	0.8:1
Saguinus	4	0.7:1
Erythrocebus	1	9.6:1
Cercopithecus	13	4.5:1
Macaca	14	3.5:1
Piliocolobus	5	3.4:1
Colobus	5	2.3:1
Procolobus	2	2.2:1

Data for *Alouatta* are from Di Fiore and Campbell 2007 and Kowalewski and Garber, submitted; Data for *Cebus* are from Fragaszy et al. 2004; Data for *Callicebus* are from Norconk 2007; Data for *Aotus* are from Fernandez-Duque 2007; Data for *Saguinus* are from Garber et al. 1993; Dawson 1977; Neymann 1977, and Mendelson 1994; Data for *Leontopithecus* are from Dietz and Baker 1993 and Rylands 1982; Data for *Procolobus*, *Colobus*, and *Piloocolobus* are from Fashing 2007; Data for *Macaca* are from Thierry 2007; and data from *Cercopithecus* and *Erythrocebus* are from Windfelder and Lwanga 2003; Isbell et al. 2002; and Cords 2000.

Aotus, and *Callicebus* these values are either male biased or closely approximate 1:1. This contrasts sharply with group structure in many species of Old World monkeys. Among African colobines, macaques, forest guenons, and patas monkeys, there is an average of from 2.2 to 9.6 adult females to adult males, in established social groups (Table 19.3). In many of these Old World monkeys, females are philopatric and form strong matrilineal relationships. In contrast, female philopatry is relatively rare in South American primates and found principally in the genus *Cebus* (Jack 2007). Although in some cases (e.g., *Callithrix jacchus*, *Saimiri boliviensis*, *Alouatta seniculus*) related females may co-reside and breed in the same group, many species of platyrrhines are characterized either by bisexual dispersal or male philopatry and female biased dispersal (Strier 1994a, b, 2000). Adult females, therefore, spend much of their reproductive life in social groups with unrelated females, and even among relatives, female relationships appear to be characterized by weaker and less tolerant intrasexual social bonds than are found in many Old World monkeys (Strier 1999; Di Fiore and Campbell 2007). Whether this reflects a difference in mechanisms of kin recognition, or in the value of kin bonds relative to other kinds of social bonds, is not clear, and studies aimed at identifying the conditions under which female playtrrhines behave nepotistically would be informative (Chapais 2001; Chapais et al. 2001).

In the absence of strong social bonds or alliances between adult females coresiding in the same group, and in response to incursions by migrating females attempting to enter established groups, adult female group members actively compete for access to a limited number of breeding positions. In platyrrhines, female breeding competition is most easily documented in tamarins, marmosets, night monkeys, and titi monkeys. Among most callitrichine species, regardless of the number of adult females present in the group, only a single female in each group breeds. Subordinate adult females are reproductively suppressed via physiological (failure to ovulate, naturally abort) and/or behavioral (dominance, infanticide) mechanisms (Garber 1997; Digby et al. 2007; Ziegler and Strier, this volume). Evidence of female-biased rates of scent marking in tamarins also are likely to be related to the intensity of reproductive competition among group females (Heymann 2003). In *Aotus* and *Callicebus*, intense female reproductive competition is associated with intrasexual intolerance and the enforced presence of only a single female per group. Fernandez-Duque (2007) reports that in 14 of 15 wild *Aotus* groups observed over a 3 year period, either the resident male or the resident female of the breeding pair was expelled and replaced by an invading same-sex conspecific. Thus, tamarin, lion tamarin, titi monkey, and night monkey females actively compete for male caregiving services and reproductive sovereignty.

In *Alouatta*, females aggressively compete to enter established social groups which normally contain only a small number of reproductively active females. In the case of red howler monkeys (*A. seniculus*), "dispersing females have often been ousted from their natal groups by aggression from older females and are seldom able to integrate themselves into established social groups (Di Fiore and Campbell 2007: 158). Moreover, red howler social groups rarely contain as many as four adult females. Although Crockett and Janson (2000) have argued that the risk of infanticide by invading adult males is the primary factor limiting the number of adult females per red howler social group, it also is likely that female intolerance plays a critical role in influencing group size and composition. This is supported by data from other howler species such as *A. pigra* and *A. caraya*. In populations of these species in which infanticide risk is reported to be absent or extremely low, females remain aggressive to immigrating females, and groups also contain only 1–4 adult females (Kowalewski 2007; Kowalewski and Garber submitted; Van Belle and Estrada 2006).

Capuchins appear to have diverged from this pattern to some degree. Female capuchins are philopatric and characterized by strong social bonds and coalitionary behavior (Jack 2007). Female capuchins actively solicit copulations from adult males by engaging in highly conspicuous proceptive behaviors (Linn et al. 1995; Fragaszy et al. 2004). In this regard, Izar et al. (this volume) argue that capuchin female reproductive strategies and female mate choice are designed to avoid male sexual coercion and reduce the risk of infanticide by mating during both fertile and nonfertile periods with particular adult males (Carnegie et al. 2006). In white-faced capuchins (*Cebus capucinus*), there is little evidence that resident males or females compete directly for access to reproductive partners (Jack and Fedigan 2004). In tufted capuchins (*Cebus apella*), however, females demonstrate a clear preference for soliciting the dominant male (Izar et al. this volume). In this regard, Linn et al. (1995) report that in captive tufted capuchins, higher ranking females had significantly greater rates of copulatory behavior with dominant males than did lower ranking females. In addition, although both dominant and lower ranking females were found to direct over 84% of solicitations to dominant males, 84% of dominant female copulations, but only 29% of low ranking females copulations, were with dominant males. Females were observed on 42 occasions to interfere with and disrupt other females during copulations, and such "interference signif-icantly altered the number of copulations"......that lower ranked females ...." had with dominant males " (Linn et al. 1995:51). Evidence that dominant females attempt to control or maintain preferential access to adult males, a possible form of female–female mating competition, also has been reported in *Cebus olivaceus* (O'Brien 1991). Based on grooming and proximity relationships between dominant females and dominant males, and attempts of higher ranking females to disrupt the copulatory behavior of lower ranking females when soliciting dominant males, Linn et al. (1995:53) conclude that "female–female reproductive competition is also a component of the breeding system of *C. apella*".

It appears that given limited or weak female–female social bonds, females in several primate species select mates exhibiting particular qualities, such as strong infant caregiving skills, the ability to co-operate with resident males in resource and mate defense, and the ability to tolerate resident females mating with other resi-dent males (Bicca-Marques 2003; Kowalewski and Garber, submitted). The preva-lence of female dispersal also permits females to adjust their group membership in response to the size and composition of males (Strier 2000, 2004b). We suggest that female–female mating competition was likely an important component of the mat-ing and social system of ancestral South American primates. We strongly encourage researchers to conduct field and captive research on South American primates to examine, test, and revise theories of sexual selection, female mate choice, and the benefits of male cooperation in primates.

19.4 Ending Comments

It is the hope of the editors that the chapters in this book provide a critical overview of current and future challenges in the study and conservation of South American primates. Some of these challenges have much to do with our growing but lim-ited data base for many South American taxa. Basic studies of systematics and taxonomy, combined with research on historical biogeography functional anatomy, ontogeny, and molecular evolution are needed to provide a framework for integrat-ing theoretical and empirical information on mating and social systems. Similarly, detailed field studies of the specific costs and benefits to individuals of kin and nonkin bonds, patterns of both short- and long-term intra and intersexual social relationships, and the benefits to infants, mothers, and practitioners of allomaternal infant caregiving are needed in order to evaluate more critically species life history traits and individual mating strategies. Documentation of the underlying hormonal basis of socio-sexual behavior, the phenology of hormonal profiles and the lability of hormonal systems in response to changing social and ecological contexts, need to be expanded.

Human induced fragmentation of primate habitats and populations together with the existence of large and expanding human populations brings humans and non-human primates into close spatial contact. Such "forced" proximity may alter host-parasite relationships, and parasite and disease transmission in still unknown ways that are of critical relevance to issues of public health and management of remnant primate populations. Moreover, the impact of climate change upon primate habitats has yet to be investigated. These are just a few of the challenges that South American Primatology faces in the coming decade and beyond. As we continue to improve our understanding of tropical ecosystems and the response of both human and nonhuman primates to habitat change, global warming, anthropogenic disturbance, and new vectors of infectious disease, we empower conservation-oriented research aimed at implementing models of land-use and conservation initiatives favorable for the persistence of the rich primate fauna of South America. We hope this volume represents an important step in achieving these goals.

Acknowledgments We are grateful to our co-editors Karen Strier, Eckhard Heymann, and Júlio César Bicca-Marques for constructive suggestions to improve this chapter. AE acknowledges the support from Universidad Nacional Autónoma de Mexico and from the Scott Neotropic Fund of the Cleveland Zoological Society. PAG thanks Chrissie, Sara, and Jenni for sharing their lives with me, and in teaching me about creatures large and small.

References

Altieri, M. A. 2004. Globally Important Ingenious Agricultural Heritage Systems (GIAHS): extent, significance, and implications for development. http://www.fao.org/ag/agl/agll/giahs/documents/backgroundpapers_altieri.doc

Aureli, F., Schaffner, C. M., Verpooten, J., Slater, K., and Ramos-Fernandez, G. 2006 Raiding parties of male spider monkeys: insights into human warfare? Am. J. Phys. Anthropol. 131: 486–497.

Bales, K. L., French, J. A., McWilliams, J., Lake, R. A., and Dietz, J. M. 2006 Effects of social status, age, and season on androgen and cortisol levels in wild male golden lion tamarins (*Leontopithecus rosalia*). Horm. Behav. 49:88–95.

Barnett, A. A., Volkmar de Castilho, C., Shapley, R. L., and Anicacio, A. 2005. Diet, habitat selection and natural history of Cacagao melanocephalus ouakary in Jau National Park, Brazil. Int. J. Primatol. 26:949–969.

Bicca-Marques, J. C. 2003. Sexual selection and the evolution of foraging behavior in male and female tamarins and marmosets. In C. Jones (ed.) *Sexual Selection and Reproductive Competition in Primates: New Perspectives and Directions* (pp. 455–475). Norman: American Society of Primatologists.

Boinski, S. 1994. Affiliation patterns among male Costa Rican squirrel monkeys Behaviour 130:191–209.

Carnegie, S. D., Fedigan, L. M., and Ziegler, T. E. 2006. Post-conceptive mating in white-faced capuchins, Cebus capucinus: hormonal and sociosexual patterns of cycling, noncycling, and pregnant females. In A. Estrada P. A. Garber, M. S. M. Pavelka and L. Luecke (eds.) *New Perspectives in the Study of Mesoamerican Primates: Distribution, Ecology, Behavior, and Conservation* (pp. 387–409). New York: Springer Press.

Chapais, B. 2001. Primate nepotism: what is the explanatory value of kin selection? Int. J. Primatol. 22:203–229.

Chapais, B., Savard, L., and Gauthier, C. 2001. Kin selection and the distribution of altruism in relation to degree of kinship in Japanese macaques (*Macaca fuscata*). Behav. Ecol. Sociobiol. 49:493–502.

Cormier, L. 2006. A Preliminary Review of Neotropical Primates in the Subsistence and Symbolism of Indigenous Lowland South American Peoples.. Ecol. Envir. Anthrop. 2:14–32.

Crockett, C. M. and Janson, C. H. 2000. Infanticide in red howlers: female group size, male membership, and possible link to folivory. In C. P. van Schaik and C. H. Janson (eds.), *Infanticide by Males and its Implications* (pp. 76–98). Cambridge: Cambridge University Press.

Di Fiore, A. and Campbell, C. J. 2007. The Atelines: variation in ecology, behavior, and social organization In C .J. Campbell, A. Fuentes, K. C. MacKinnon, M. Panger and S. K. Bearder (eds.), *Primates in Perspective* 155–185. New York: Oxford University Press.

Diemont S. A. W., Martin J. F., and Levy-Tacher S. I. 2006a. Energy evaluation of Lacandon Maya indigenous swidden agroforestry in Chiapas, Mexico. Agrof. Sys. 66:23–42

Diemont, S. A. W., Martina, J. F., Levy-Tacherb. S. L., Nighg R. B., Ramirez Lopez, P., and Golicherb D. J. 2006b. Lacandon Maya forest management: Restoration of soil fertility using native tree species. Ecol. Engin. 28:205–212

Digby, L. J., Ferrari, S. F., and Saltzman, W. 2007. Callitrichines: the role of competition in cooperatively breeding species. In C. J. Campbell, A. Fuentes, K. C. MacKinnon, M. Panger and S. K. Bearder (eds.), *Primates in Perspective* (pp. 85–106). New York: Oxford University Press.

Eisenberg, J. F. 1981. The Mammalian Radiations: An Analysis of Trends in Evolution, Adaptation, and Behavior. Chicago: University of Chicago Press.

Fernandez-Duque, E., 2007. Aotinae Social monogamy in the only nocturnal haplorhine. In C. J. Campbell, A. Fuentes, K. C. MacKinnon, M. Panger and S. K. Bearder (eds.), *Primates in Perspective* (pp. 139–154). New York: Oxford University Press.

Fleagle, J.G. and Tejedor, M. F. 2002. Early platyrrhines of southern South America. In W. C Hartwig (ed.), The Primate Fossil Record, (pp. 161–174). Canbridge: Cambridge University Press.

Ford, S. M. and Davis, L. M. 1992. Systematics and small body size: implications for feeding adaptations in New World monkeys. Amer. J. Phys. Anthropol. 88:415–468.

Ford, S. M. 1994. Evolution of sexual dimorphism in body weight in platyrrhines. Am. J. Primatol. 34:221–244.

Fragaszy D. M., Visalberghi, E., and Fedigan, L. M. 2004. The Complete Capuchin. Cambridge: Cambridge University Press.

Garber, P. A., Encarnacion, F., Moya, L., and Pruetz, J. D. 1993. Demographic and reproductive patterns in moustached tamarin monkesy (Saguinus mystax): implications for reconstructing platyrrhine mating systems. Amer. J. Primatol. 29:235–254.

Garber, P. A. 1997. One for all and breeding for one: cooperation and competition as a tamarin breeding strategy. Evol. Anthrop. 5:187–222.

Garber, P. A. and Kowalewski, M. M. (in press). Male cooperation in pitheciines: the reproductive costs and benefits to individuals of forming large mulitmale and multifmale groups. In L. Veiga, A. Barnett and M. A. Norconk (eds.), *Evolutionary Biology and Conservation of Titis, Sakis and Uakaris*, Cambridge: Cambridge University Press.

Garber, P. A. and Leigh, S. R. 1997. Ontogenetic variation in small-bodied New World primates: implications for patterns of reproduction and infant care. Folia Primat. 68:1–22.

Gardner, T. A., Barlow, J., Parry, L. W., and Peres, C. A. 2006. Predicting the uncertain future of tropical forest species in a data vacuum. Biotropica 39:25–30.

Gaston, K. J. 2000. Global patterns in biodiversity. Nature 405:220–227.

Gilmore, R. M. 1943. Mammalogy in an epidemiological study of jungle yellow fever in Brazil. J. Mammal. 24:144–162.

Hershkovitz, P. 1949. Mammals of northern Colombia. Preliminary report No. 4: Monkeys (primates), with taxonomic revisions of some forms. Proc. of the U.S. Nat. Mus. 98:323–427

Heymann, E. W. 2003. Scent marking, paternal care, and sexual selection in callitrichines. In C. B. Jones (eds.), *Sexual Selection and Reproductive Competition in Primates: New Perspectives and Directions* (pp. 305–325). Special Topics in Primatology 3. Norman: American Society of Primatologists.

Heymann, E. W. 2006. Scent marking strategies of New World primates. Am. J. Primatol. 69: 650–661.

Huck, M., Löttker, P., Heymann, E. W. 2004. Proximate mechanisms of reproductive monopolization in male moustached tamarins (*Saguinus mystax*). Am. J. Primatol. 64:39–56.

Huck, M., Löttker, P., Böhle, U. R., and Heymann, E. W. 2005. Paternity and kinship patterns in polyandrous moustached tamarins (*Saguinus mystax*). Am. J. Phys. Anthropol. 127: 449–464.

Humboldt, A. V. 1944. Memoirs on the monkeys of Amazonia. Memoir on the monkeys which live along the Orinoco, the Casiquiare and the Rio Negro. Rev. Acad. Col. Cienc. Exac.. Fis. y Nat. 5:506–527

Izar, P., Stone, A, Carneige, S, and Nakai, E. S. (2009) Sexual selection, female choice and mating systems. In P. A. Garber, A. Estrada, J. C. Bicca-Marques, E. Heymann and K. Strier (eds.), *South American Primates: Comparative Perspectives in the Study of Behavior, Ecology, and Conservation*. (pp. 157–1) New York: Springer Press.

Jack, K. M. 2007. The cebines: toward an explanation of variable social structure. In C. J. Campbell, A. Fuentes, K. C. MacKinnon, M. Panger and S. K. Bearder (eds.), *Primates in Perspective* (pp. 107–122). New York: Oxford University Press.

Jack, K. M. and Fedigan, L. M. 2004. Male dispersal patterns in white-faced capuchins, Cebus capucinus. Part 1: patterns and causes of natal emigration. Anim. Beh. 67:761–769.

Kappeler, P. M. and van Schaik, C. P. 2002. Evolution of primate social systems. Int. J. Primat. 23:707–741.

Kinzey, W. G. and Cunningham, E. P. 1994. Variability in platyrrhine social organization. Am. J. Primatol. 34:185–198.

Kinzey, W. G. 1992. Dietary and dental adaptations in the Pitheciinae. Am. J. Phys. Anthrop. 88:499–514.

Kinzey, W. G. 1987. A primate model for human mating systems. In W. G. Kinzey (ed.), *The Evolution of Human Behavior: Primate Models* (pp. 105–114). Albany: State University of New York Press.

Kowalewski, M. M. 2007. Patterns of affiliation and co-operation in howler monkeys: an alternative model to explain social organization in non-human primates. Ph. D. thesis, Dept. of Anthropology, University of Illinois at Urbana-Champaign

Kowalewski, M. M. and Garber, P. A. (in prep.). Mating promiscuity, energetics, and reproductive tactics in black and gold howler monkeys (*Alouatta caraya*).

Kreis H. A. 1945. Contributions to knowledge of parasitic nematodes. XXII. Parasitic nematodes from the tropics. Rev. Suis. Zool. 52:551–596.

Krieg, H. 1928. Scwarze brüllaffen (*Alouatta caraya* Humboldt). Zeitschrift für Säugetierkunde II:119–132.

Laurance, W. F. and Peres, C. A. (eds.). 2006. Emerging Threats to Tropical Forests. Chicago: University of Chicago Press.

Linn, G. S., Mase, D., LaFrancois, R. T., Okeefe, R. T., and Lipshitz, K. 1995. Social and menstrual cycle phase influences on the behavior of group-housed *Cebus apella*. Am. J. Primatol. 35:41–57.

Löttker P., Huck, M., Heymann, E. W. 2004. Demographic parameters and events in wild moustached tamarins (*Saguinus mystax*). Am. J. Primatol. 64:425–449.

MA. 2005. Ecosystems and Human Well-being. Current State and Trends. The Millenium Ecosystem Assessment. New York: United Nations.

Mitchell, C. L. 1994 Migration alliances and coalitions among adult male South American squirrel monkeys (*Saimiri sciureus*). Behaviour 130:169–190.

Norconk, M. 2007. Sakis, uakaris, and titi monkeys: behavioral diversity in a radiation of primate seed predators. In C. J. Campbell, A. Fuentes, K. C. MacKinnon, M. Panger and S. K. Bearder (eds.), *Primates in Perspective* (pp. 123–138). New York: Oxford Univesity Press.

O'Brien, T. G. 1991. Female–male social interactions in wedge-capped capuchin monkeys *(Cebus oliuaceus)*: Benefits and costs of group living. Anim. Beh. 41:555–567.

Pope, T. 1990. The reproductive consequences of male cooperation in the red howler monkey: Paternity exclusion in multi-male and single-male troops using genetic markers. Behav. Ecol. Sociobiol. 27:439–446.

Rosenberger, A. L. 1992. The evolution o ffeedin gnices in New World monkeys. Am. J. Phys. Anthrop. 88:525–562.

Rosenberger, A. L. 2002. Platyrrhine paleontology and systematics: the paradigm shifts. In W. C. Hartwig (ed.), *The Primate Fossil Record* (pp. 151–159). Cambridge: Cambridge University Press.

Schaik van, C. P., Pradhan, G. R., and van Noordwijk, M. A. 2004. Mating conflict in primates: infanticide, sexual harassment and female sexuality. In P. Kappeler and C. P. van Schaik (eds.), *Sexual Selection in Primates, New and comparative perspectives*, Cambridge: Cambridge University Press.

Schneider, H., Sampaio, I., Harada, M. L., Barroso, C. M. L., Schneider, M. P. C., Czelusniak, J., and Goodman, M. 1996. Molecular phylogeny of the new world monkeys (Platyrrhini, Primates) based on two unlinked nuclear genes: IRBP intron and 1 and e-globinsequences. Am. J. Phys. Anthropol. 100:153–179.

Schneider, H. and Rosenberger, A. L. 1996. Molecules, morphology, and platyrrhine systematics. In M. A. Norconk, A. L. Rosenberger and P. A. Garber (eds.), *Adaptive Radiations of Neotropical Primates* (pp. 3–19). New York: Plenum Press.

Schneider, H., Canavez, F. C., Sampaio, I., Moreira, M. A. M., Tagliaro, C. H., and Seuznez, H. N. 2001. Can molecular data place each neotropical monkey in its own branch? Chromosoma 109:515–523.

Singer, S. S., Schmitz, J., Schwiegk, C., and Zischler, H. 2003. Molecular cladistic markers in New World monkey phylogeny (Playtrrhini, Primates). Mol. Phyl. Evol. 26:490–501.

Strier, K. B. 1994a. Brotherhoods among atelins: kinship, affiliation, and competition. Behaviour 130:151–167.

Strier, K. B. 1994b. The myth of the typical primate. Yrbk. Phys. Anthrop. 37:233–271.

Strier, K. B. 1999. Why is female kin bonding so rare: comparative sociality of New World primates. In P. C. Lee (ed.) *Primate Socioecology* (pp. 300–319). Cambridge: Cambridge University Press.

Strier, K. B. 2000 From binding brotherhoods to short-term sovereignty: the dilemma of male Cebidae. In P. M. Kappeler (ed.), *Primate Males: Causes and Consequences of Variation in Group Composition* (pp. 72–83). Cambridge: Cambridge University Press.

Strier, K. B. 2004a. Patrilineal kinship and primate behavior. In B. Chapais and C. M. Berman (eds.), *Kinship and Behavior in Primates* (pp. 177–199). New York: Oxford University Press.

Strier, K. B. 2004b. Sociality among kin and nonkin in nonhuman primate groups. In R. W. Sussman and A. R. Chapman (eds.), *The Origins and Nature of Sociality* (pp. 191–214). New York: Aldine de Gruyter.

Travassos, L. and Teixeira de Freitas, J. F. 1941. Report of the fifth expedition of the Instituto Oswaldo Cruz, conducted to the area of Estrada de Ferro Norceste of Brazil in January, 1941. II. Parasitologic studies. Memorias do Instituto Oswaldo Cruz 36:272–295.

Travassos, L. and Teixeira de Freitas, J. F. 1943. Report of the seventh scientific expediation of the Instituto Oswaldo Cruz, conducted to the area of Estrada de Ferro in northeastern Brazil, in May, 1942. Memorias do Instituto Oswaldo Cruz 38:385–412.

Van Belle, S. and Estrada, A. 2006. Demographic features of *Alouatta pigra* populations in extensive and fragmented forests. In A. Estrada, P. A. Garber, M. S. M. Pavelka and L. Luecke (eds.), *New Perspectives in the Study of Mesoamerican Primates: Distribution, Ecology, Behavior, and Conservation* (pp. 121–142). New York: Springer Press.

Veiga, L .M. and Silva, S. S. B. 2005. Relatives or just good friends? Affiliative relationships among male southern bearded sakis *(Chiropotes satanas)*. In Livro de Resumos, XI Congresso Brasileiro de Primatologia, p. 174. Porto Alegre, Brasil.

Vieira, C. C. 1944. Os símios do Estado de São Paulo. Papéis Avulsos do Departamento de Zoologia IV:1–31.

Vieira, C. C. 1955. Lista remissiva dos mamíferos do Brasil. Arquivos de Zoologia VIII:341–374.

von Ihering, H. 1914. Os bugios do gênero *Alouatta*. Revista do Museu Paulista 9:231–256.

Wang, E. and Milton, K. 2003. Intragroup social relationships of male *Alouatta palliata* on Barro Colorado Island, Republic of Panana. Int. J. Primat. 24:1227–1244.

Young, C. 2005. A comprehensive and quantitative assessment of Belize Creole ethnobotany: Implications for forest conservation. Ph.D thesis University of Connecticut.

Author Index

Subject Index

Taxonomic Index

PRIMATES

A

Acrecebus fraileyi, 115
Adapiforms, 74, 104
Alaotran gentle lemur, 236
Alouatta
  A. arctoidea, 32
  A. belzebul, 32, 42, 58, 59, 113, 256, 267,
    270, 286, 287, 415, 434, 444, 449,
    465, 466
  A. caraya, 13, 32, 41, 42, 43, 97, 129, 130,
    131, 146, 201, 259, 348, 425, 434,
    435, 437, 444, 446, 449, 450, 511,
    522, 525
  A. discolor, 32, 42
  A. fusca, 97
  A. guariba
    A. g. clamitans, 32, 43, 201
    A. g. guariba, 32
  A. juara, 32, 42
  A. macconnelli, 32, 42
  A. nigerrima, 32, 42, 443
  A. palliata
    A. p. aequatorialis, 32
    A. p. coibensis, 32
    A. p. mexicana, 32
    A. p. palliata, 32
    A. p. trabeata, 33
  A. pigra, 33, 43, 259, 283, 446, 522, 525
  A. puruensis, 32
  A. sara, 32, 443
  A. seniculus
    A. s. amazonica, see Alouatta juara
    A. s. insulanus, see Alouatta
      macconnelli
    A. s. juara, see Alouatta juara
    A. s. puruensis, see Alouatta puruensis
    A. s. sara, see Alouatta sara

A. s. seniculus, 32, 41, 42, 97, 118, 197,
  198, 223, 236, 237, 253, 256, 258,
  263, 283, 286, 288, 293, 294, 295,
  305, 325, 326, 329, 391, 393, 400,
  403, 404, 434, 444, 449, 453, 465,
  522, 524, 525
  A. ululata, 32, 42
Alouattinae, 25
  See also Alouatta
Andean night monkey, *see Aotus miconax*
Andean saddle-back tamarin, *see Saguinus*
  fuscicollis leucogenys
Andean titi, *see Callicebus oenanthe*
Anthropoidea, anthropoids, 7, 10, 55, 62, 63,
  64, 70, 74, 82, 84, 94, 95, 191, 192,
  195, 197, 521
Antillothrix bernensis, 90, 101
Aotidae, 25, 30
  See also Aotus
Aotus
  A. azarae, 30, 39
    A. a. azarae, 30
    A. a. boliviensis, 30
    A. a. infulatus, 30
  A. brumbacki, 30, 522
  A. dindensis, 97, 99
  A. griseimembra, 30, 39, 520
  A. lemurinus, 30, 38, 39, 520
  A. miconax, 30
  A. nancymaae, 30, 520, 522
  A. nigriceps, 30, 369, 379, 522
  A. trivirgatus, 30, 38, 121, 198, 259,
    286, 522
  A. vociferans, 30, 38, 288, 290, 305,
    520, 522
  A. zonalis, 30, 38, 39, 520
Ape, 4, 129, 158, 455
  See also Gorilla, Pan, Pongo
Aripuanã marmoset, *see Mico intermedius*

553

O

Ochraceous bare-face tamarin, *see Saguinus martinsi ochraceus*
Olalla's titi, *see Callicebus olallae*
Old World monkeys, *see* Catarrhini
Orabassu titi, *see Callicebus moloch*
Orang-utan, *see Pongo pygmaeus*
Oreonax
  O. flavicauda, 33, 396
Ornate spider monkey, *see Ateles geoffroyi ornatus*
Ornate titi, *see Callicebus ornatus*
Owl monkey, *see Aotus trivirgatus*

P

Pachylemur insignis, 389
Pale weeper capuchin, *see Cebus olivaceus apiculatus*
Paleopropithecus ingens, 389
Pan, 44, 129, 455
Panamanian night monkey, *see Aotus zonalis*
Panamanian white-throated capuchin, *see Cebus capucinus imitator*
Papio
  P. hamadryas, 130
  P. ursinus, 444
Paraguyan yellow titi, *see Callicebus pallescens*
Paralouatta
  P. marianae, 78, 90, 101
  P. varonai, 77, 90, 97, 101
Parapithecids, 7
Patasola
  P. magdalenae, 100
Peruvian squirrel monkey, *see Saimiri boliviensis peruviensis*
Peruvian woolly monkey, *see Lagothrix cana tschudii*
Peruvian yellow-tailed woolly monkey, *see Oreonax flavicauda*
Pied bare-face tamarin, *see Saguinus bicolor*
Pig-tailed macaque, 389
Pithecia
  P. aequatorialis, 32, 40
  P. albicans, 31, 262
  P. irrorata, 31, 325, 326, 329, 520
    P. i. irrorata, 31
    P. i. vanzolinii, 31
  P. monachus, 31, 40, 298, 520
    P. m. milleri, 31
    P. m. monachus, 31
    P. m. napensis, 31, 40
  P. pithecia, 14, 23, 24, 25, 26, 46, 73, 86, 121, 198, 255, 262, 263, 283, 286,

288, 290, 293, 294, 295, 296, 297, 298, 300, 400, 520
    P. p. chrysocephala, 31
    P. p. pithecia, 31, 397
Pitheciidae, 25, 30
  See also Cacajao, Callicebus, Chiropotes, Pithecia
Pitheciinae, pitheciines, 3, 24, 25, 73, 92, 98, 101, 104, 105, 281, 286, 288, 291, 293, 296, 298, 303, 305, 309, 310, 518, 519
  See also Cacajao, Chiropotes, Pithecia
Platyrrhini, platyrrhines, 4, 6, 7, 8, 12, 23–45, 55, 56, 62, 63, 64, 69–89, 94, 101, 103, 105, 125, 131, 191, 193, 195, 197, 200, 204, 220, 223, 251–311, 511, 518, 519, 521, 524
Plesiadapiforms, 104
Poeppig's woolly monkey, *see Lagothrix poeppigii*
Pongo pygmaeus, 130
Prince Bernhard's titi, *see Callicebus bernhardi*
Procolobus badius, 446
Propithecus verreauxi, 203
Proteopithecus, 56
Proteropithecia, 78
Protopithecus
  P. brasiliensis, 97, 100, 452
Purús red howler monkey, *see Alouatta puruensis*
Pygmy marmoset, *see Cebuella pygmaea*

R

Red bald-headed uacari, *see Cacajao calvus rubicundus*
Red-bellied collared titi, *see Callicebus purinus*
Red-bellied tamarin, *see Saguinus labiatus*
Red-cap moustached tamarin, *see Saguinus mystax pileatus*
Red colobus, *see Procolobus badius*
Red-crowned titi, *see Callicebus discolor*
Red-faced black spider monkey, *see Ateles paniscus*
Red-fronted lemur, 220
Red-handed howler monkey, *see Alouatta belzebul*
Red howler monkey, *see Alouatta seniculus*
Red-mantle saddle-back tamarin, *see Saguinus fuscicollis lagonotus*
Red-tailed guenon, *see Cercopithecus ascanius*
Red titi, *see Callicebus cupreus*
Reed titi, *see Callicebus donacophilus*

OTHER ANIMALS